# Studies in Economic Design

**Series Editors**

Jean-François Laslier, CNRS - Paris School of Economics, Paris, France
Hervé Moulin, Adam Smith Business School, University of Glasgow, Glasgow, UK
M. Remzi Sanver, Université Paris-Dauphine, Paris, France
William S. Zwicker, Department of Mathematics, Union College, Schenectady, NY, USA

The book series *Studies in Economic Design* offers an outlet for scientific results from the study of economic, social and political institutions and mechanisms leading to the creative design of suitable legal-economic instruments.

Economic Design comprises the creative art and science of inventing, analyzing and testing economic as well as social and political institutions and mechanisms aimed at achieving individual objectives and social goals. The accumulated traditions and wealth of knowledge in normative and positive economics and the strategic analysis of Game Theory are applied with novel ideas in the creative tasks of designing and assembling diverse legal-economic instruments. These include constitutions and other assignments of rights, mechanisms for allocation or regulation, tax and incentive schemes, contract forms, voting and other choice aggregation procedures, markets, auctions, organizational forms such as partnerships and networks, together with supporting membership and other property rights, and information systems, including computational aspects.

The series was initially started in 2002 and with its relaunch in 2017 seeks to incorporate recent developments in the field and highlight topics for future research in Economic Design.

More information about this series at http://www.springer.com/series/4734

Jean-François Laslier · Hervé Moulin ·
M. Remzi Sanver · William S. Zwicker
Editors

# The Future of Economic Design

The Continuing Development of a Field
as Envisioned by Its Researchers

 Springer

*Editors*
Jean-François Laslier
CNRS - Paris School of Economics
Paris, France

Hervé Moulin
University of Glasgow
Glasgow, UK

M. Remzi Sanver
Université Paris-Dauphine
Université PSL, CNRS, LAMSADE
Paris, France

William S. Zwicker
Union College
Schenectady, NY, USA

ISSN 2510-3970          ISSN 2510-3989   (electronic)
Studies in Economic Design
ISBN 978-3-030-18052-2          ISBN 978-3-030-18050-8   (eBook)
https://doi.org/10.1007/978-3-030-18050-8

This Springer imprint is published by the registered company Springer Nature Switzerland AG
The registered company address is: Gewerbestrasse 11, 6330 Cham, Switzerland

*To Murat R. Sertel*

# Preface

It was in 1999 that Murat R. Sertel wrote *"Any modern account of the accomplishments of economics would easily bear witness to huge footprints of progress made in the field of design"* in a brilliant paper that gives an account of the developments in economic design, together with his profound understanding of the field.[1]

Twenty years later, we felt it would be useful to reconsider the field, regarding not only where we come from and where we are but also where we are going from here. Seventy-four papers collected in this volume are the result of a call that we extended in 2017 to colleagues from various generations:

> We are preparing a volume of short pieces tentatively titled The Future of Economic Design, and we thought of you as a possible contributor. Here is what we have in mind: Edited volumes normally collect completed research, and if a paper talks about future work at all, it is in relation to its particular contribution. But in informal "cocktail" conversations many of us engage in sweeping, high (or very high)-level assessments of our broad research field: Where do we come from? Where do we go from here? We would not risk submitting for review such deeply felt but hard to demonstrate opinions, and yet our colleagues and students are keen to hear them as they construct a high-level picture of their own research.
>
> This volume aims to collect a large number of these informal, short reflections on the future of economic design, or mechanism design broadly construed. Contributions will be of a few pages and not comprehensive. They could focus on historical, epistemological, or methodological aspects of any topic you see as related to economic design. Topics covered include, inter alia, theoretical foundations and practical applications relevant to market design, constitutional design, implementation theory, voting, auctions, fair division, matching, scheduling, network formation, law and economics, etc. An entry could be of a rather technical nature, for instance explaining a result that you feel has much more potential than has been recognized so far, or could be entirely nontechnical.

---

[1] Sertel, M. R. (1999), Introduction: Discoveries vs Inventions in Economics, in Contemporary Economic Issues Volume 4: Economic Behaviour and Design, edited by M. R. Sertel, Palgrave Macmillan.

Looking back at these paragraphs, we see several of our wishes being fulfilled: The reader will find in this volume a large number of contributions on a fireworks-like variety of topics and approaches, tackled by an impressive crowd of different authors.

Another idea was to have an original peer review process that took the form of a conversation using an online platform, where for three months all entries were visible to all contributors. The idea was to generate a flow of comments on the papers with the goal of improving rather than censoring them, because unsupported speculation and controversial views are the essence of this project. On that point, our wish only partially materialized: some contributions have sparked discussions and some have not, and it is not easy to detect any regularity or reason there.

Taken in its entirety, the collection clearly reflects some of the main trends of the field: the never-ending question of the role of theory, the emergence of bounded rationality considerations, the influence of computer science and computer scientists, and the spreading boundaries of the discipline to encompass neighboring fields such as law, political science, and philosophy. Of course, the collection also demonstrates that scholars of economic design, whose daily bread and butter consists of more narrowly technical investigations, are quite willing to widen their reflection, as our call invited.

It is hard, if it is even possible, to bring order to the splendid diversity you will find here. The interested reader might be best served by glancing through the entire volume. Nevertheless, we made an attempt to organize the contributions into eleven distinct themes: General Aspects, Aggregation and Voting, Algorithms and Complexity, Axiomatics, Behavioral Aspects, Fair Division, Implementation, Interpersonal Relations, Law, Matching and Markets, and New Technologies.

We wish to sincerely thank the more than 80 authors who accepted the challenge to show their view on the future of economic design in brief and concise contributions. We also express our gratitude to our editorial assistant Erhan Gümüş whose role was invaluable, and offer our enthusiastic thanks for his patience and attention to detail.

We were deeply sorry to hear about the passing of Dan Felsenthal, contributing author of this volume, on 20 February 2019. Dan was a staunch advocate of the majority principle, which he studied in great detail. His work on voting power with Moshé Machover was a major factor in the re-awakening of that field. Moshé, to whom we asked a few sentences following Dan's passing, writes that Dan *"was my close relative, friend and indispensable academic partner. He first led me into Social Choice and my contribution to it, for what it's worth, would have been unthinkable without his collaboration—contributing to our partnership his enormous creative energy, skill in calculation, forensic attention to detail and great insight."*

This volume appears in the book series "Studies in Economic Design". Four volumes had appeared under this series, before it paused after the untimely passing of its editor Murat R. Sertel in 2003. A volume with contributions in memory of

Leonid Hurwicz, edited by Walter Trockel, is forthcoming in the series. We hope that our edited volume will both contribute to a revival of the series and stand on its own merit as an original contribution to the field of economic design.

| | |
|---|---|
| Paris, France | Jean-François Laslier |
| Glasgow, UK | Hervé Moulin |
| Paris, France | M. Remzi Sanver |
| Schenectady, USA | William S. Zwicker |

# Contents

## Part II   Aggregation and Voting

## Part III   Algorithms and Complexity

# Part I
# General Aspects

# On the Future of Economic Design—The Personal Ruminations of a Matching Theorist

Szilvia Papai

**Abstract** This note is a brief personal rumination on the state of economic design, advocating the importance of theory and listing broadly desirable features of future theoretical research agendas.

Intrigued by the opportunity to participate in an initiative that is courageous enough to ask for personal and thought-provoking reflections, I foolishly accepted the invitation to write this short piece on the future of economic design, suppressing my instant thought of how inadequate my knowledge and understanding of the broad subject matter is. Far from being able to predict or recommend future developments in our field, this seemingly impossible assignment has triggered, for me, unanticipated series of reflections or, more realistically, ruminations. Since nowadays it's difficult to spend even a little bit of time on spontaneous contemplations, undertaking this task has turned out more enjoyable than I had imagined, despite my usual procrastination and the fear of exposing my (lack of) thoughts. Apart from generally applauding this refreshing initiative which will "generate a flow of comments," I have also noted that since "unsupported speculation and controversial views are the essence of this project," it captures Murat Sertel's inquisitive and playful spirit very well. Murat would have loved to see and argue about these personal reflections on economic design, and while doing so promoting the open mindedness and flexibility of thought that he cultivated in all of us, his colleagues, students, and friends.

***

I would like to first mention some of the positive recent changes and trends affecting economic design. As seen especially in the market design of matching institutions and in some important public auctions, it is an outstanding, and considering many other areas quite rare, development to have such close interaction between theory and practice, which has inspired some of the most fruitful efforts and outcomes in both realms. One need not elaborate on this, as it is well-known due to the famous

S. Papai (✉)
Department of Economics, Concordia University, 1455 de Maisonneuve Blvd.
West, Montreal, QC H3G 1M8, Canada
e-mail: szilviapapai@gmail.com; szilvia.papai@concordia.ca

© Springer Nature Switzerland AG 2019
J.-F. Laslier et al. (eds.), *The Future of Economic Design*, Studies in Economic Design, https://doi.org/10.1007/978-3-030-18050-8_1

FCC spectrum auctions and the celebrated matching applications of the National Resident Matching Program of hospital intern matching in the US, the school choice programs of New York City, Boston, and elsewhere, as well as due to the more recent important developments in kidney exchange, among others, that there has been an accelerating growth in auction and matching theory and practice, where practical design issues inspire theoretical work, and in turn theoretical findings help to advise the practical design of matching institutions in a most beneficial way. Related to this is an integration of different methodologies, the "engineering" approach as laid out by Roth (2002), the simultaneous and mutually reinforcing use of theoretical analysis, experiments and computations, complemented by empirical work, in order to be able to scientifically "engineer" solutions to complex real-life design problems that a purely analytical modeling approach alone cannot handle. Major successes resulting from this approach are well documented. Another welcome development is the increasing interaction among different disciplines: economics, mathematics, computer science, operations research, political science, etc., as evidenced by joint workshops, publications, and other collaborative efforts, albeit we are still far from having bridged the existing gaps stemming from different goals, approaches, and terminologies, and due to separate literatures.

In some fields of economic design the breadth of topics explored and the sheer array of new findings is quite astonishing: examples include the already mentioned areas of matching and auction design, and also the design of networks, e-commerce, and other online applications, among others. Other, less dynamic, areas have also produced many brilliant and worthwhile contributions. Research in economic design, as broadly understood, is alive and well, as demonstrated by related Nobel prizes, keynote speeches at general conferences, publications in general interest journals, and the significant impact on numerous real-life applications, and with a steady stream of bright new researchers joining each year the pursuit of inquiry into this research field.

***

Yet, despite all these positive developments and the overall well-being of our field, in some sense we have perhaps taken a step back. When I recently indicated during a talk that the model in our paper describes none of the potential applications exactly, someone in the audience audibly exclaimed: "at least you are honest." While the sympathy for my "admission" was apparent, possibly because the comment was made by a fellow theorist, I was nonetheless slightly taken aback. This was a misinterpretation of my remark, which intended to express, without elaborating, that the model is of theoretical interest and was deliberately kept simple in the hope of obtaining more far-reaching results than would have been possible had it been tailored closely to one particular application. Although it was my usual awkwardness in public speaking that had inadvertently allowed for the perception that we had failed to set up a model of "real" significance, my nagging feeling was that this kind of misunderstanding would have been less likely to happen a decade ago and that, possibly owing to the current erosion of the traditional academic environment, we are quietly moving away, or are at least expected to move away, from pursuing more abstract theory.

Even though we are not working in pure mathematics which often prides itself upon not having any applications (although, as it turns out, the future tends to turn up more practical uses than pure mathematicians would anticipate), do our results need to be instantly convertible to a policy recommendation, or to be immediately applicable to the design of a rule or institution in a specific context? Does our work need to have specific relevance for our own country, as has been at least implicitly suggested lately at my university, that research findings should be specifically useful for Canada and Canadians? Ought we to "descend" from our ivory-tower-inspired research inquiries and modify our agendas accordingly?

Is theory less popular in general than before? Or is it just a certain kind of theoretical inquiry, such as the axiomatic approach in economic design, that is less popular? Is this just a fleeting trend, a result of aspiring to do "design engineering" that is in vogue, or a more general product of our times that is here to stay? One general explanation is that theory is deep and time-consuming, and contemporary culture unfortunately does not prioritize deep thought. The time-crunch combined with an opportunistic attitude would also favor applications over theory. Even in academia, where deep thought should be at the heart of all activities and consistently practiced and promoted, the hurried pace of modern life no longer affords academics adequate time to do so, with available "free time" converging to zero for some of the most worthwhile undertakings, under administrative, teaching, professional and other constant daily pressures to perform.

To put this in a broader context, the so far subtle shift away from fundamental research may at least partially be due to the generally anti-academic political and social climate which promotes a managerial approach and consumerism and undervalues the academic enterprise on the whole with its implied expertise, resulting in a pervasively pragmatic view which is accepted and promoted by many from within academia as well. Pure theory seems to be less attainable when "measurable productivity" is becoming an imperative, as universities are more and more subject to a corporate business approach in which severe budget cuts to faculty and teaching resources are a daily reality, even as universities are becoming top-heavy with administrators and are able to find enough money for constructing new buildings and for opening new "multi-disciplinary" institutes and labs. In this anti-intellectual environment the ever-increasing importance of the applications of our work is expounded to us more than ever before.

What role does indeed "high-brow" theory play? Is it merely to express our exclusivity and superiority, as some would claim? Is it fair to say, as a supposedly young economist anonymously posted on the internet: "Theory sucks. No practical use with arbitrary assumptions."? I don't think so. While assumptions can be disputed and we all know that some are more realistic or salient than others, and while we realize that theoretical results are generally less accessible and immediately useful than the findings of applied or empirical work, the contribution may be much more substantial than a pragmatically minded researcher, a practitioner in some relevant area, or a journalist might be able to appreciate. In general, theory is fundamental to the deep and at the same time broad understanding of a subject matter, which provides an underlying fundamental structure on which to build more applied and specific analysis. Theoretical inquiry also serves as a catalyst for future significant work to be done at both the theoretical and applied level. While rather simplistically presented here, as one could elaborate upon more specific aspects of theoretical study in various areas in much more detail, these general statements apply to economic design quite well, in my opinion, where fundamental results are just as important as in other areas, such as in the various disciplines of natural science. Additionally, theory is becoming more important in aiding the evaluation of empirical findings, given increasing access to data and technological advances in computational power, as another contribution to this volume, Jackson (2019), points out.

*** 

Although huge advances have been made in the theory of economic design during the last decades, the broader objectives have not changed much since Leonid Hurwicz's lecture on resource allocation in the early 1970s[1]: we want to design new mechanisms, and we want to explore the constraints and trade-offs for the design of new mechanisms. We have found new and innovative approaches and methods to pursue such inquiries, but our fundamental goals have stayed the same, and many open questions remain. I trust that colleagues who are more knowledgeable than me will write about specific open questions as well as new problems to be explored in economic design in some detail, such as emerging new allocation problems based on online platforms, and also about potentially new areas of inquiry, for example, as pertaining to behavioral aspects resulting in more realistic depictions of the agents' behavior and in the use of new incentive axioms, or by inspecting the information structure and economic or social interactions more closely in existing areas of inquiry lacking such an approach, or by taking into account how the mechanisms themselves are understood and carried out, incorporating this explicitly into a theoretical perspective on economic design, among many other possibilities.

Instead, in the spirit of normative economics, I will briefly describe, without aiming to be exhaustive or specific, some of the broad type of developments and results that would be desirable for the future of economic design, illustrated from the narrow viewpoint of a matching theorist. These are obvious, on the one hand, but are becoming less realistic to achieve, on the other hand, in the current academic climate, and can be seen in some sense as the opposite of "literature-driven" (such as when axioms that are barely justifiable are used repeatedly based on previously published papers) or "publication-driven" (when little thought goes into an article or when it presents half-baked ideas) contributions which are unfortunately on the increase.

Broad context and foundations:

Apart from pursuing "standard" theory, broader foundations would also often be desirable, a thorough and thoughtful "intellectual approach" which would not only

---

[1] See Hurwicz (1973).

demonstrate an intellectual curiosity, but also justify the specific approach or assumptions, and place results in a broader context by analyzing the subject matter more widely, such as the social or political context, or by studying the philosophical or epistemological underpinnings of one's research, or whatever is appropriate for the particular topic. Authors of many of the classic papers, following in the tradition of Kenneth Arrow, William Vickrey, or Leonid Hurwicz, just to name a few, had the scope for drawing such an underlying fundamental picture, and their work has consequently gained more significance with a wider-ranging impact. While regular contributions to some specific area have their place and cumulatively lead to significant new knowledge, if we want to ensure that our more general contributions and their implications are well received and understood, we may want to take the time for such wider reflections when appropriate.

Unifying approach:
There is no need to explain the significance of papers with a unifying approach, or why they are relatively rare. The synthesis that these papers provide are often not just a technical accomplishment, but also come with a vision which allows them to make previously unknown connections and throw light on broader subject matter that enlighten us. We need more of these papers which lead to a deeper overall understanding, whether the unification connects different results (e.g., Lee and Sethuraman 2011), different mechanisms (e.g., Chen and Kesten 2017), or different models (e.g., Andersson and Svensson 2014; Alva and Manjunath 2019).

Deep insight:
Naturally, it is always disputed which papers or results contain deep insights, however there is no lack of consensus regarding the fact that papers with deep insights are rather desirable, but difficult to obtain. I will refer here to just a few recent papers in matching theory, perhaps in order to provoke some comments. Kesten (2010) has wide impact, since the EADAM rule it introduced has not only broken up the long-standing diarchy of the DA and TTC rules for solving matching problems, but also turned out to be a very important rule, as demonstrated by the more recent papers by Dur et al. (2015) and Ehlers and Morrill (2018), among others. As another example, following the genuine observations of Ergin (2002) and Kesten (2006) on reconciling stability and efficiency by restricting priorities in matching markets, Ehlers and Westkamp (2017) and Han (2018) are significant contributions which illuminate the possibility of finding constrained efficient and efficient strategyproof rules which are stable, given the possibility of weak priorities. Among many others these are contributions, I believe, that will stand the test of time.

***

I will conclude with a manifesto of sorts, betraying my Marxist origins by birth. The leisurely conversations of academics discussing both large and small issues of interest (while smoking pipes) are becoming relics of the past, with few notable exceptions such as during the one week spent on two boats at the Bosphorus Workshop on Economic Design. It is essential, as untimely it may seem, to preserve the spirit of free conversation and reflection, and resist the global corporatization of academia which fundamentally disrespects the academic pursuit, as well as, on a more mundane level, the frantic urges of our contemporary academic lives that make such ideas sound quaint and outdated. To this end, I would propose that we begin a collective conversation for which the current initiative is a perfect incubator. We ought to encourage, and participate in, such a dialogue by establishing an appropriate method of (virtual) communication, a platform where all ideas in the spirit of curiosity and discovery, no matter how "outrageous," are welcomed and debated by a friendly community of economic designers who are committed to and thrive on such interactions.

# References

Alva, S., & Manjunath, V. (2019). *Strategy-proof Pareto-improvement. Journal of Economic Theory, 181*, 121–142.

Andersson, T., & Svensson, L.-G. (2014). Non-manipulable house allocation with rent control. *Econometrica, 82*, 507–539.

Chen, Y., & Kesten, O. (2017). Chinese college admissions and school choice reforms: A theoretical analysis. *Journal of Political Economy, 125*, 99–139.

Dur, U., Gitmez, A., & Yilmaz, O. (2015). *School choice under partial fairness*. Working paper.

Ehlers, L., & Morrill, T. (2018). *(Il)legal assignments in school choice*. Working paper.

Ehlers, L., & Westkamp, A. (2017). *Strategy-proof tie-breaking in matching with priorities*. Working paper.

Ergin, H. (2002). Efficient resource allocation on the basis of priorities. *Econometrica, 70*, 2489–2497.

Han, X. (2018). Stable and efficient resource allocation under weak priorities. *Games and Economic Behavior, 107*, 1–20.

Hurwicz, L. (1973). The design of mechanisms for resource allocation. *American Economic Review, 63*, 1–30.

Jackson, M. O. (2019). The role of theory in an age of design and big data. *The Future of Economic Design*.

Kesten, O. (2006). On two competing mechanisms for priority-based allocation problems. *Journal of Economic Theory, 127*, 155–171.

Kesten, O. (2010). School choice with consent. *Quarterly Journal of Economics, 125*, 1297–1348.

Lee, T., & Sethuraman, J. (2011). *Equivalence results in the allocation of indivisible objects: A unified view*. Working paper.

Roth, A. E. (2002). The economist as engineer: Game theory, experimentation, and computation as tools for design economics. *Econometrica, 70*, 1341–1378.

# Principles of Economic Design

Nicolaus Tideman

**Abstract** There are three central principles of good economic design: (1) Freedom of choice leads to good outcomes. (2) Charging (crediting) people for the cost (benefits) of their choices yields efficient outcomes. (3) Goodwill is highly productive. These central principles lead to free markets, Pigouvian taxes and subsidies, lump-sum taxes on land, mechanisms that apply marginal-cost pricing to participation in decisions, and money creation by interest-free lending to citizens, in proportion to their expected demand for cash balances. However, every improvement in efficiency can be introduced in myriad ways, with different distributional consequences, leading to conflict that inhibits improvements. Therefore greater consensus about distributional equity is necessary to improve economic design.

I see three central principles of good economic design:

(1)   Freedom of choice leads to good outcomes.
(2)   Charging (crediting) people for the cost (benefits) of their choices yields efficient outcomes.
(3)   Goodwill is highly productive.

## 1   Freedom of Choice

Freedom of choice gives us the efficiency of competitive markets and the conviction that people should decide for themselves what to do with their lives, where to live, and whom to marry. Yet we know that there are limits to the merit of freedom of choice. We withdraw freedom of choice from those with temporary or permanent diminished mental capacity, and we offer it only gradually to children as they mature. Moreover, we know that most of us are likely to be subject to bouts of irrationality, such as hyperbolic discounting. Thus, an important issue in economic design is how

N. Tideman (✉)
Economics Department, Virginia Tech, Blacksburg, VA 24061, USA
e-mail: ntideman@vt.edu

© Springer Nature Switzerland AG 2019                                                     9
J.-F. Laslier et al. (eds.), *The Future of Economic Design*, Studies in
Economic Design, https://doi.org/10.1007/978-3-030-18050-8_2

to best accommodate the elements of irrationality that are found in humans, while achieving the greatest possible good from freedom of choice. The work on nudging (Thaler and Sunstein 2008) is an important contribution to this effort.

## 2 Ensuring that People Bear the Full Costs and Reap the Full Benefits of Their Actions

The Coase Theorem tells us that if people knew their interests perfectly and there were no bargaining costs, then bargaining would always lead to whatever Pareto-improving agreements were available. In such an ideal world, economic design would have only the task of distributing the initial entitlements. However, we do not live in a world of zero bargaining costs. Monopolists are rarely perfectly discriminating monopolists, so we have the inefficiency of monopoly pricing. Those exposed to negative as well as positive externalities frequently do not bargain with those who cause the externalities, so we have the inefficiency of uninternalized externalities. Taxes that are not intended to internalize externalities are rarely lump-sum taxes, so we have the excess burden of taxation.

Economic design applies principles of cost-based pricing to avoid these ineffi-ciencies. These principles lead to Pigouvian taxes and subsidies, and to lump-sum taxes on the sale value or the rental value of land. There is a similar principle for some monopolies. Consider a monopoly with the characteristics that (1) everyone wants the monopolist's product and (2) the benefit from the opportunity to buy the monop-olized product diminishes with distance and eventually goes to zero. The monopolist charges a price that exceeds the competitive price, reducing social surplus below its maximum. A variation on the Henry George Theorem (Arnott and Stiglitz 1979) implies that one can attain a Pareto improvement by paying the monopolist a subsidy to lower his price to marginal cost, financing the subsidy by collecting the increase in the rental value of the land that follows from the reduction in price and the subsequent increase in demand. Similarly, if the social benefits of a local public service such as a bus exceed its costs and everyone values the service, then one can fund this public service efficiently through a combination of marginal-cost pricing and the collection of the increase in the rental value of land that results from the public service. With intellectual property or other monopolies where distance is not a factor, it is more challenging to design an institution that can achieve efficiency.

The principle that marginal-cost pricing leads to efficient allocations has also been applied to participation in decisions, yielding a family of mechanisms (the pivotal mechanism, quadratic voting, and others) that motivate honest reports of preferences by charging individuals the (actual or expected) marginal social costs of their participation (Tideman and Plassmann 2017).

The principle of marginal-cost pricing breaks down when one attempts to apply it to activities that might have catastrophic consequences, such as driving a car in a way that might cause serious injury or death. There is no efficient price for driving at 60 miles an hour on residential streets, because there are no markets for voluntary purchase of what might be appropriated by such action. Calabresi (2008) has proposed a framework within which one might construct a socially acceptable system for managing the risks of accidents.

It is intriguing to entertain the possibility of extending marginal-cost pricing to an issue in macroeconomics. Consider a person who enters the economy with a positive demand for cash balances (or a person increases his or her demand for cash balances). This action generates a positive externality for the part of the economy that creates money (primarily the banking sector, through loans). A banking sector that operated as a monopoly would receive a corresponding monopoly profit. To the extent that the banking sector is competitive, the profit is dissipated when multiple banks compete for customers. To achieve efficiency, one must internalize the positive externality from increases in the demand for cash balances, by lending to every citizen the expected value of the quantity of cash that he or she would want to hold (based, say, on taxes paid). The citizen would not pay any interest, but would be required to buy an insurance policy that would guarantee repayment in the event of his or her death. (It costs essentially nothing to create cash, as long as people want to hold it rather than spend it.) Such a practice would not only internalize the externality, it would also provide the monetary authority with an arguably more direct and reliable way of managing the money supply.

One complication with respect to all applications of marginal-cost pricing is that they can be structured in myriad ways, with different distributional consequences. Because contention over the distributional outcome can lead to a failure to agree on any of the possibilities, it is important for economic design to attend to principles of distribution.

When dealing with allocative issues, economists often apply the Pareto principle, which prescribes that the outcome of an action is an improvement only if the action benefits at least some and causes no one harm. However, the limitations of the Pareto principle are often overlooked. The Pareto principle presumes that it is possible to see into the future, to learn what will happen if the currently contemplated action is not taken. This assumption is needed to give meaning to "the action causes no one harm." The Pareto principle also presumes implicitly that the distribution entailed in the scenario that will arise in the absence of the action is morally unobjectionable. This assumption is problematic, for example, when the action in question entails allocating pollution quotas to polluters without charge.

When it comes to pollution taxes, it is appealing to base relief to polluters on an alternative rationale. A competitive polluting industry does not profit from polluting, because it passes all benefits that arise in the absence of pollution taxes on to consumers, in terms of lower prices. Levying pollution taxes causes the price of output to increase and the demand to fall. An industry that uses durable equipment will then experience temporary excess capacity, and short-run equilibrium market prices will not cover fixed costs. In the long run, capacity will shrink, and expected profits will once again be zero. If the polluting industry could not reasonably have predicted that society will one day find its activities to be socially costly, then it would be reasonable for people to agree that the temporary loss from the period of excess capacity should be borne by taxpayers rather than by producers. In such a situation, it is reasonable to accompany the introduction of pollution taxes with lump-sum and transferable tax credits for existing producers, approximating their temporary losses from the shock of the unexpected introduction of pollution taxes. The transferability of the tax credits permits those producers who find it best to go out of business to benefit from the tax credits without needing to stay in business. This rationale for compensation does not apply if producers could have foreseen that their activity would be found to be socially costly or if they have been profiting from a privilege that kept prices above competitive levels. The general point with respect to economic design is that it is worth seeking consensus with respect to the question of whether distributional equity requires supplemental lump-sum transfers to accompany the adoption of efficiency-improving policies, because that consensus will make it easier to adopt such policies.

## 3   The Importance of Goodwill

The third principle of economic design, good use of the reservoir of goodwill, has been introduced into the thinking of economists only recently, and it is also the most subtle. It arises in experimental economics. One can observe this principle in experiments that invite individuals to choose between a benefit for themselves and a greater benefit for a group. Traditional economic analysis predicts that individuals will always act in their selfish interest. That prediction is not borne out. People often behave unselfishly.

If it were possible to create a society in which people always received the full value of their productive actions and always paid the full social costs of their costly actions, then the reservoir of goodwill would not be needed. Self-interest would lead to efficiency. But the difficulty of introducing pricing mechanisms with appropriate accompanying lump-sum transfers makes perfect pricing infeasible. For example, a heavy snowfall might be followed by appeals on the TV weather forecasts: "Please stay off the roads, so that the snowplows can clear them." People will sometimes do as they are asked, but often they will ignore the social costs of their decisions to drive on the snow-filled roads. One day, we might be able to introduce a congestion pricing system that can raise the cost of driving when there are snowstorms. But even a system that produced variable congestion prices that matched social costs perfectly at all times would not be perfect unless it also provided the lump-sum compensation that allowed those with medical emergencies to travel to hospitals without being unreasonably impoverished. A framework of goodwill permits efficiency and equity to emerge without the need for a system of taxes, subsidies, and lump-sum transfers.

The abysmal economic performance of countries with economic systems based on collectivism is a warning that there are limits to the reservoir of goodwill. Yet goodwill as an organizing principle works quite well in many families, monasteries, and groups like the Hutterites. For the future of economic design, it is important to understand what makes it possible to operate on a reservoir of goodwill, what causes such reservoirs to deteriorate, and how they can be rebuilt.

**Acknowledgements** I appreciate the comment by Gabriel Carroll. In response to it, I altered the text to make my idea clearer. Therefore the precise text he quotes is no longer present. Florenz Plassmann offered valuable suggestions.

# References

Arnott, R. J., & Stiglitz, J. E. (1979). Aggregate land rents, expenditure on public goods, and optimal city size. *The Quarterly Journal of Economics, 93*(4), 471–500.

Calabresi, G. (2008). *The cost of accidents: A legal and economic analysis.* New Haven: Yale University Press.

Thaler, R., & Sunstein, C. (2008). *Nudge.* New Haven: Yale University Press.

Tideman, N., & Plassmann, F. (2017). Efficient collective decision-making, marginal cost pricing, and quadratic voting. *Public Choice, 172*(1–2), 45–73.

# The Future of Decision-Making

Dorothea Baumeister

**Abstract** Decision-making is an important task in diverse fields and the used mechanisms often have a long tradition. This article describes situations from voting, scheduling a meeting, and online participation where collective decisions have to be taken. In all these fields digitization offers new possibilities that are not yet fully used. A task for the future is to use this potential in order to improve decision-making from various points of view.

## 1   Introduction

Collective decision-making is a very important issue that arises in many different situations. Examples range from citizens that collectively decide over the seats in a parliament b y going to the polls, over finding a date for a meeting, to the decision of what to do with the outcome of a debate on participatory budgeting. In all these fields different mechanisms can be used to determine the collective decision. Although the targets are very diverse, the mechanisms must all satisfy a certain number of criteria. In order to be accepted by the participants they have to respect their preferences, work efficiently, and be fair and transparent.

The former trust of participants in decision-making processes is in many situations no longer given. It often originated from the fact that most of the information was not accessible for all participants, and they did not have the means to evaluate them. Due to a proceeding digitization different aspects of decision-making processes receive more attention and the used methods are scrutinized by the participants. Furthermore, there is a higher demand that their whole preferences are respected and it is contested whether the actual form of preference elicitation is appropriate. In this chapter I will

Supported in part by the DFG-grant BA6270/1-1.

D. Baumeister (✉)
Institut für Informatik, Heinrich-Heine-Universität, Düsseldorf, Germany
e-mail: d.baumeister@hhu.de

© Springer Nature Switzerland AG 2019                                                    15
J.-F. Laslier et al. (eds.), *The Future of Economic Design*, Studies in
Economic Design, https://doi.org/10.1007/978-3-030-18050-8_3

describe three specific situations, namely elections, scheduling a meeting, and online participation, where decision-making is a very important issue. Additionally, I will mention directions for future research in all described fields.


## 2   Elections

In political elections it is very common to use some sort of plurality voting, where each voter only makes a cross for her most liked candidate. The way in which the seats in the parliament will then be distributed is often highly complicated. Considering the German Bundestag elections, the voters have two separate votes, and the seats will be distributed according to the proportions of the second vote, taking into account the direct mandates from the first vote, the five percent threshold and possible overhang mandates. All in all, it is a very complicated process that aims in fulfilling a certain number of fairness criteria. But as a result many Germans do not understand the process by which their votes determine the parliament's formation. When using some kind of plurality voting where only the most liked alternative counts, many people will give an untruthful vote in order to not waste the vote. Consider the small example in Table 1. We have the three parties $A$, $B$, and $C$, and 42 voters with the preferences given in the table. When using plurality voting the 5 voters who prefer $C$ might cast the vote $B > C > A$ instead of their sincere one. Their most preferred party $C$ has no chance of winning the election, so before wasting their vote they may prefer to support the second most-liked party $B$ in order to prevent the victory for their most despised party $A$. It is natural to ask whether it is enough to consider only the first choice of every participant if afterward a complicated process is used to form the parliament. There are of course systems that use more information from the voter than his first choice, for example a complete order over all candidates. But since the number of candidates may be fairly large, it would be too demanding to order all candidates. Even though in all these voting systems the participants do not express their preferences over possible outcomes, that is, coalitions that in the end will form the government.

   Hence one question for future research is to find a way to elicit the preferences of the voters in a way that allows us to take a decision that reflects the true preferences of the voters. In the case of political elections some very interesting experiments have been done recently during the French presidential elections in 2017. Instead of only voting for the most preferred candidate they tested various other voting systems, partly resulting in different outcomes, see Baujard et al. (2017) for a full summary (in French) of all voting experiments held during the French presidential

**Table 1**  The preference of 42 voters over the parties $A$, $B$, and $C$

| Number | Preference |
|--------|------------|
| 20 | $A > B > C$ |
| 17 | $B > A > C$ |
| 5 | $C > B > A$ |

elections in 2017. There is a tradition of organizing similar experiments in France (see Baujard et al. 2014 for a summary of the experiments from the French presidential elections in 2012), and also in other countries. But so far, this had no influence on the election system used in practice.

Going one step further and using the efforts of digitization it may be possible to create interactive systems where the user may express more diverse preferences, but only has to give the relevant parts of her preference. These relevant parts may be different for every voter and of course depend on what the others voted so far. One specific challenge is to design such systems in a transparent way to make it trustworthy. On the other hand, it has to be shown on the theoretical side, that the system has good properties. This implies that there is a need to classify newly developed procedures also from an axiomatic perspective, which means that a first step must be to identify the important axioms for a specific context. Consider for example neutrality in election systems. This property requires, that all candidates should be treated equally by the procedure. For political elections with a single winning candidate this seems to be a very fair requirement. But if a set of candidates has to be elected this property should not be transferred blindly. When forming a recruiting committee it may be important to elect members that are somehow diverse, i.e., they need to come from distinct disciplines, or a certain number of female/senior/young members must be included. Whereas classical single-winner elections is a well-studied field in computational social choice (see Part I in the Handbook of Computational Social Choice by Brandt et al. 2016), multi-winner elections received attention only recently; for an overview see the book chapter by Faliszewski et al. (2017). Future research should broaden the existing results in single-winner and multi-winner elections to newly developed voting mechanisms.

## 3  Scheduling a Meeting

Finding a date for a meeting in a larger group may be a complicated and inefficient task when performed by just asking everybody. Recently it became popular to make a Doodle poll (see http://doodle.com) in such cases. Here the participants may choose for every proposed date whether they can attend the meeting or not (additionally the option "If need be" can be chosen). A sample schedule is shown in Fig. 1.

It is often the case that not all time slots that are feasible for a person are equally good for her. For example, if there is a direct follow-up meeting afterward, the time slot will be less preferred. When only asking for a yes/no decision it may be the case that a date will be chosen that is a very bad option under the feasible ones for many attendees. Formally this means that an alternative that is Pareto-dominated by some other alternative may be chosen, which is a bad property for a decision-making procedure. The choice of "if need be" is a step in the direction of eliciting more diverse preferences that should at least resolve the above-described problem. A more advanced approach, but far less known than Doodle, is Whale[4] (see http://strokes.imag.fr/whale4/). Figure 2 shows a sample poll from Whale. Instead of

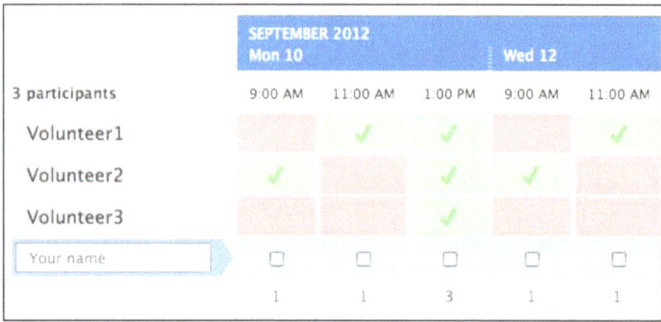

**Fig. 1** Sample Doodle poll (2017)

| | Japanese | French | Norwegian | American | Chinese | German | Senegalese | Iranian | Tunisian | Mexican |
|---|---|---|---|---|---|---|---|---|---|---|
| jeromelang | 3 | 8 | 4 | 7 | 6 | 5 | 9 | 1 | 10 | 2 |
| LeoPony1458 | 3 | 1 | 9 | 10 | 2 | 6 | 7 | 8 | 4 | 5 |
| BoubSter | 5 | 2 | 6 | 3 | 4 | 7 | 10 | 8 | 9 | 1 |
| natete | 9 | 8 | 5 | 4 | 6 | 3 | 10 | 2 | 7 | 1 |
| sleste34 | 6 | 1 | 9 | 10 | 7 | 8 | 2 | 5 | 3 | 4 |
| Mahdi Washha | 6 | 2 | 8 | 1 | 10 | 3 | 7 | 4 | 5 | 9 |

**Fig. 2** Sample Whale poll (2017)

scheduling a meeting, the participants here decide which country has the best food in the universe, by ordering them from the best (1) to the worst (10). In addition to approval/disapproval votes or complete linear orders several different preference expression modes are possible. The results will then be visualized using different decision-making methods. Some of them are more appropriate for experts on voting, but others are suitable for non-experts as well. One of the aims of this project is to provide election data for researchers. Hence this tool could be a first step in the direction of identifying how preferences should be elicited by a scheduling tool, and how they should be evaluated. The main challenges here are to find a way to elicit the preferences that is easy to perform for the users but expressive enough to make good decisions. Then, one has to identify a decision-making rule that has good properties, but is easy enough to follow also for non-expert users. The answers here should, of course, depend on the field of application. Digitization here already made the way to schedule a meeting more efficient, but as argued above there is still room for improvement.

## 4 Online Participation

The situation in online participation is a bit different. Such processes are often organized in a simple forum structure. Recently there are for example many online participation processes on participatory budgeting. Here the citizens have the possibility to discuss on which projects the city should spend money, or where money can be saved. In early projects the participation rate was fairly high because the participants were eager to be able to take part in processes in which they have not been involved before. Unfortunately, the enthusiasm did not hold due to several reasons. In some projects the discussions were not respected by the politicians when making the decisions, or the discussion ended with projects that were not realizable due to juridical reasons.

Another important point is, that the forum structure (possibly combined with some simple kind of voting) is not suitable for such processes. People that come in later may be confronted with very long discussions, and they will not go through it. Hence many arguments will be repeated and it is easy to lose track of important arguments. There are several attempts to make online participation more structured, for example by using pro/contra lists. But the problem is that either these systems are not able to represent an argumentation in its full complexity or they are not suitable for non-expert users. That is why most systems still rely on forums with the simple answer mechanism.

One attempt to tackle this problem from a different point of view is the dialog based online argumentation system described by Krauthoff et al. (2016). Instead of using a forum, this is an interactive system which asks you questions and then guides you through the present argumentation with the option to express your own opinion either by approving or disapproving current statements or by introducing new arguments or statements. This system is designed for non-experts, but still captures the full complexity of a real-world argumentation. The underlying structure is an argumentation framework, which can also be accessed by the users. So far there has been a field experiment with this system (see the work of Krauthoff et al. 2017) that shows that it is usable for non-experts in argumentation theory, but that there is still room for improvement. The tasks for future research in this area is to develop or to enhance systems that improve online argumentation processes in a way that they are accepted by the users. And related to decision-making, such processes should be able to compute the outcomes of such debates. This involves very interesting questions including: How can users express their preferences in such a system? And how can the preferences of the users be turned into a collective one?

## 5   Conclusion

All these examples show that the right form of elicitation of preferences and the process of finding a collective outcome is a very important task in diverse fields. These important research questions have to be answered by researchers from various disciplines. One has to consider it from a theoretical and from a practical point of view. The challenge is to find systems that are on the one hand easy to apply and understand for the people who use it. On the other hand, they must be accepted by the users and ideally also have good properties proven from a theoretical point of view.

A future challenge in all these fields is to develop and analyze these new processes such that they can become accepted by a wider public.

A very important issue in all described fields of decision-making is manipulation. Of course, there is also the possibility of manipulation from the technical side by different means (e.g.: introduce fake votes, delete votes, change existing votes...), but this will not be discussed here. Manipulation also occurs if voters try to legally manipulate the system for example by reporting an untruthful preference, as described in the previous section on elections. Especially when all information is available to everybody this is very important since a possible manipulator has all information to compute his manipulative vote. Manipulation in classical elections has been studied intensively by the computational social choice community; see the book chapter by Conitzer and Walsh (2016) for an overview. But when the decision-making method is very different, some of the known results do no longer hold. Manipulation for the case where the votes come in one by one has been studied by Hemaspaandra et al. (2014). Studying manipulation and other forms of legal interference in the above-described settingsfield.

# References

Baujard, A., Igersheim, H., Lebon, I., Gavrel, F., & Laslier, J.-F. (2014). Who's favored by evaluative voting? An experiment conducted during the 2012 French presidential election. *Electoral Studies*, *34*, 131–145.

Baujard, A., Blanch, R., Bouveret, S., Igersheim, H., Laruelle, A., Laslier, J.-F., et al. (2017). Compte-rendu de l'expérimentation "VOTER AUTREMENT" lors du premier tour de l'élection présidentielle française le 23 avril 2017 à Allevard-les-Bains, Crolles, Grenoble, Hérouville-Saint-Clair, Strasbourg et sur internet. Technical report. In French.

Brandt, F., Conitzer, V., Endriss, U., Lang, J., & Procaccia, A. (Eds.). (2016). *Handbook of computational social choice*. Cambridge: Cambridge University Press.

Conitzer, V., & Walsh, T. (2016). Barriers to manipulation in voting. In F. Brandt, V. Conitzer, U. Endriss, J. Lang, & A. Procaccia (Eds.), *Handbook of computational social choice* (pp. 127–145). Cambridge: Cambridge University Press.

Doodle. (2017). https://help.doodle.com/customer/portal/articles/761313-what-is-doodle-and-how-does-it-work-an-introduction. Retrieved 22 Sept 2017.

Faliszewski, P., Skowron, P., Slinko, A., & Talmon, N. (2017). Multiwinner voting: A new challenge for social choice theory. In U. Endriss (Ed.), *Trends in computational social choice*. AI Access Foundation. To appear.

Hemaspaandra, E., Hemaspaandra, L., & Rothe, J. (2014). The complexity of online manipulation of sequential elections. *Journal of Computer and System Sciences*, *80*(4), 697–710.

Krauthoff, T., Betz, G., Baurmann, M., & Mauve, M. (2016). Dialog-based online argumentation. In *Proceedings of the 6th International Conference on Computational Models of Argument* (pp. 33–40). IOS Press.

Krauthoff, T., Meter, C., & Mauve, M. (2017). Dialog-based online argumentation: Findings from a field experiment. In *CEUR Workshop Proceedings AI*IA Series*. To appear.

Whale. (2017). http://strokes.imag.fr/whale4/polls/viewPoll/00000000-0000-4000-8000-000000000000. Retrieved 22 Sept 2017.

# Theoretical Unification, Domain Restrictions and Further Applications: Some Comments on Future Research in Economic Design

Salvador Barberà

**Abstract** I suggest three promising avenues for mechanism design; (1) the further unification of the Arrowian approach with the one underlying Condorcet's Jury Theorem, (2) the search for useful domain restrictions embracing large classes of environments, and (3) the identification of further fields of application.

1. Let me start by mentioning a general line of research in the theory of voting that I find fascinating and at the same time very difficult to develop. It is the integration of the Arrowian point of view and the one that Arrow labeled in Social Choice and Individual Values as the idealist position, on which the Condorcet Jury Theorem is based. The Arrowian point of view (Arrow 1951, 1963) is that voters are endowed with given preferences, and that the purpose of voting is to satisfactorily aggregate the possibly diverging opinions they represent. The idealist position assumes, on the contrary, that the voters share an identical purpose and yet diverge in opinion regarding what action to take, due to differential information. Voting then becomes a method to aggregate information in search of an existing but unknown best decision (Young 1988).

Each of these points of view have given rise over the years to a large number of interesting contributions, adding different dimensions of discussion to the original statements of Arrow's impossibility result and that of the jury theorem. Incentives, communication and other aspects of the voting process have been amply explored, but following quite separate routes. Yet, I have no doubt that, in most voting situations, a combination of each of the two points of view would provide a better picture of the actual problems to solve by a mechanism designer. People's values and opinions permanently diverge on certain issues, but the same agents may also share common principles on other issues, and thus be ready to partially change attitudes on those, if presented with adequate evidence.

S. Barberà (✉)
MOVE, Universitat Autònoma de Barcelona
and Barcelona GSE, Edifici B, 08193 Bellaterra (Barcelona), Spain
e-mail: salvador.barbera@uab.es

© Springer Nature Switzerland AG 2019
J.-F. Laslier et al. (eds.), *The Future of Economic Design*, Studies in
Economic Design, https://doi.org/10.1007/978-3-030-18050-8_4

In the language of mechanism design, Arrow's model is one of private values: the type of each agent is given by her preferences, and these are fixed. The formulation of Condorcet's jury theorem and its extensions may in certain cases still be treated as a private values case, but often falls into the more complex context of interdependent types. For example, when agents communicate prior to voting, their expected payoffs may depend on the signals received by other agents. Moreover, in the latter theory, uncertainty enters the picture, due to the role of random signals (Austen-Smith and Feddersen 2006). So, even if both deal with voting, the two approaches are separated by a wide gap, and even the form of results diverge. Arrowian theory resorts to axiomatic characterizations, whereas in a Condorcet world the main objective is to identify the best estimators of the correct answer. And, to add to the list of differences, notice that preference aggregation is easy when only two alternatives are at stake, while the treatment of interdependent values is already complex even in that case, and quite inconclusive in the presence of three or more possible outcomes. Yet, recent efforts have helped to clarify the circumstances under which one rule or another may emerge as being the most adequate.

In spite of all these differences, I think it would be illuminating to find a unified treatment of the problems one must face when designing voting rules, one that would combine aggregation of preferences and of information in a world where agents' preferences are only partially sensitive to new evidence.

I am certainly not the first one to point at the tension between the aggregation and the epistemic purpose of voting. Work on the theory of Bayesian aggregation suggests that if treated jointly, combining these two objectives may lead to impossibilities (Hylland and Zeckhauser 1979; Mongin and Pivato 2015). Hence, the two functions of voting should probably be analyzed from different perspectives and subjected to separate sets of requirements, even if within a unifying perspective (Pivato 2013; Mongin and Pivato 2017). To mention another possible route, the flexible framework of judgement aggregation may be interpreted as serving the purposes of either of the two theories (List 2005; Dietrich and List 2007), though it does not seem to have undertaken yet the unification task that I propose.

I am uncertain about the analytical benefits from having a unified framework where axiomatic analysis and aggregation might be combined with optimization regarding those aspects on which all agents might agree. One of these benefits could be purely intellectual: the joy of unifying previously disconnected pieces of a great puzzle. The standing results in each of the two strands of literature should arise as special cases under additional conditions. But I hope there might be more. Much of Arrow's legacy has been to stimulate the search for rules that might satisfy part of the desiderata that he proved incompatible by his impossibility theorem. In particular, interesting, positive results may be obtained if dropping the universal domain assumption. I believe that, under appropriate restrictions suggested by specific design problems, it may be possible to unify these two now disjoint theories, and that this can help in raising new questions and results regarding mixed situations that are not covered by either one.

2. This leads me to emphasize a second general suggestion, which is related to a research program I have been developing with my co-authors Dolors Berga and Bernardo Moreno (see Barberà et al. (2010, 2016, 2018). We start from the obvious fact that essentially no mechanism can satisfy a long enough list of desiderata unless some restrictions are placed on its domain of definition. In the old literature on decentralization, as developed by masters like Hurwicz and Reiter, the extent of the class of economies that could be properly handled by means of a mechanism was referred to as its coverage. For example, roughly speaking, the coverage of the Walrasian mechanism would correspond to the set of convex economies. In social choice theory, we often concentrate on families of preferences satisfying certain properties, and study the possibility of defining satisfactory mechanisms that are at least able to operate when the agents' preferences lie within the family. The earlier and most celebrated example is given by the family of single peaked preferences.

In most applications of mechanism design, domain restrictions are also introduced (but much less explicitly) through the assumptions imposed, in each possible case, on the structure of the alternatives, on whether or not agents are selfish, and on other specific characteristics of their preferences. In joint work with Dolors and Bernardo, we have been trying to extract the common characteristics that underlie different sets of assumptions on domains that authors have used when modeling the restrictions suggested by specific models. Then, we exploit these common traits in order to explain the resemblance of results that arise in apparently unconnected worlds. In particular, we have dug into the reasons why incentive compatibility and efficiency can be achieved simultaneously in some cases and not in others, depending on the characteristics of the domains on which we require a mechanism to operate. But other connections between properties and results could be hopefully understood through a similar approach. Any specific problem of design suggests what domains are pertinent. We think that a broader perspective, leading to a better understanding of the role of domains and their characteristics would enhance our understanding of the trade-offs between attractive conditions that may or may not be possible to reconcile within a mechanism.

3. One of the reasons why mechanism design has been successful is because two very important classes of theoretical models that can be related to this approach have found useful applications. Auctions and matching have saved the day of a literature on implementation that at some point seemed to find comfort in the study of very hard to understand and unrealistic mechanisms. (Though, to be fair, authors never tried to oversell them, often presented them as examples that some possibility results were attainable and used them to illustrate different useful points regarding solution concepts and the need for further work).

The success of these applications is due to many factors, and one of them is probably the fact that, while seeking to accomplish very similar purposes of fairness or efficiency, different legislators have chosen different mechanisms to implement them. Take, for example, school choice problems. Districts allocate positions in schools through a variety of algorithms, and comparing the outcomes that may result from alternative methods, which may at first glance seem a matter of detail, has proven to be the source of rich theoretical developments and policy recommendations. The importance of details, in certain cases, or their irrelevance, in others, is also a crucial outcome of developments in auction theory.

Based on these positive experiences, I think it would pay off to identify additional fields of application, where detailed institutional prescriptions may emanate from theoretical results. Comparisons between the institutional arrangements used by different countries or firms to solve similar problems could open a long and fruitful road to travel. One possible way to guide the search for new applications is to enter well established fields and to introduce an additional emphasis on mechanism design. This is already happening. Issues in labor economics benefit from the insights provided by the literature on matching with contracts. Hervé Moulin and co-authors have consistently used the mechanism design approach to re-visit classical problems in operations research, with an added emphasis on fairness issues (see, for example, Bogomolnaia and Moulin 2010; Moulin 2009). Large parts of public economics could also be revised. But of course, all of this has to be done with respect and modesty: after all, the study of taxation is a centuries-old tradition which precedes mechanism design by name, but actually is part of the activities of economists as social engineers.

Let me finish with a personal experience that supports my call for re-visiting traditional subjects, and doing it with great respect. Years ago, having studied the axiomatic literature on bankruptcy problems, Clara Ponsatí and I organized a series of seminars with legal scholars who specialized in the study of legislation regarding the treatment of firms that were in trouble. Our initial goal was to share with them the discoveries of our fellow economists regarding how to solve distribution problems when the claims of agents exceed the size of the pie (Thomson 2003). To our surprise, they were quite unmoved by these developments. One of the reasons was that in Spanish legislation almost all claims are lined up in an order of priority, and the distribution of the residual assets, if any, is of secondary importance. Establishing priorities, and avoiding their manipulation by insiders who knew about the trouble before others did, was a much larger concern for them. Even further, their view of the process of deterioration of a firm's position was much broader than ours, since their overall concern, when analyzing these cases, was dynamic and centered on how to avoid bankruptcy, rather than managing it once it had fatally arrived. Of course, other broad views regarding the subject have been developed by economists (Hart 1995). But ever since then, I believe that the field of law and economics is a fertile territory to explore, as it contains many of the ingredients that economic design needs for success in applications: consequential institutions, and a variety of case studies, by country, culture and history.

**Acknowledgements** I am thankful to the editors, and to Dolors Berga, Gabriel Carroll, Marcus Pivato, Matthew Jackson, Antonio Miralles, Bernardo Moreno, and Flip Klijn for their critical remarks and constructive comments.

# References

Arrow, K. (1951, 1963). *Social choice and individual values* (2nd ed.). New York: Wiley.

Austen-Smith, D., & Feddersen, T. J. (2006). Deliberation, preference uncertainty, and voting rules. *American Political Science Review, 100,* 209–217.

Barberà, S., Berga, D., & Moreno, B. (2010). Individual versus Group strategy-proofness: When do they coincide? *Journal of Economic Theory*, 145–164.

Barberà, S., Berga, D., & Moreno, B. (2016). Group strategy-proofness in private good economies. *American Economic Review*, 1073–1099.

Barberà, S., Berga, D., & Moreno, B. (2018). Restricted environments and incentive compatibility in interdependent values models. Barcelona GSE WP 1024.

Bogomolnaia, A., & Moulin, H. (2010). Sharing a minimal Cost spanning tree: Beyond the Folk Solution. *Games and Economic Behavior, 69*(2), 238–248.

Dietrich, F., & List, C. (2007). Arrow's theorem in judgment aggregation. *Social Choice and Welfare, 29*(1), 19–33.

Hart, O. (1995): *Firms, contracts and financial structure*. Clarendon Lectures in Economics, Oxford University Press.

Hylland, A., & Zeckhauser, R. (1979). The impossibility of Bayesian group decision making with separate aggregation of beliefs and values. *Econometrica, 47,* 1321–1336.

List, C. (2005). Group knowledge and group rationality: A judgment aggregation perspective. *Episteme, 2*(1), 25–38.

Mongin, P. (1995). Consistent Bayesian aggregation. *Journal of Economic Theory, 66*(2), 313–351.

Mongin, P., & Pivato, M. (2015). Ranking multidimensional alternatives and uncertain prospects. *Journal of Economic Theory, 157,* 146–171.

Mongin, P., Pivato, M. (2017). Social Preference under Twofold Uncertainty. (preprint).

Moulin, H. (2009). Pricing traffic in a spanning network. In *10th ACM Conference on Electronic Commerce* (pp. 21–30). California, USA.

Pivato, M. (2013). Voting rules as statistical estimators. *Social Choice and Welfare, 40*(2), 581–630.

Pivato, M. (2017). Epistemic democracy with correlated voters. *Journal of Mathematical Economics, 72,* 51–69.

Thomson, W. (2003). Axiomatic and game-theoretic analysis of bankruptcy and taxation problems: A survey. *Mathematical Social Sciences, 45*(3), 249–297.

Young, H. P. (1988). Condorcet's theory of voting. *American Political Science Review, 82*(2), 1231–1244.

# Design for Weakly Structured Environments

**Gabriel Carroll**

**Abstract** This essay argues that, in order for economic design to speak to a broader range of applications, it should move toward models with less parametric structure. This will in turn require a cultural shift away from the current emphasis on being able to solve exactly for optimal mechanisms. I offer some ideas about what other kinds of valuable theoretical contributions would look like, drawing on existing recent work for examples.

Economic theory abounds in simple stylized models. How often we see consumers' preferences parameterized by one-dimensional types; uncertainty represented by two possible states of nature, with agents receiving conditionally independent signals (or perhaps, normally distributed states, observed with normal noise); workers choosing a one-dimensional level o f effort.

Economic design, as a field, has aspirations to inform big applications: not only auctions and matching markets, where theory has already been quite influential, but also design of protocols for legislative bargaining, judicial decision-making, or international negotiations; tax codes, patent systems, financial markets. In many cases, practical design in these areas does draw on high-level ideas from economic theory. But if theory is to offer anything like the kind of detail-level contributions it has made in (say) school choice, it will have to confront an obvious gap: Real-world preferences and information are far more complex than in traditional models. Just for an example, in committees or juries that deliberate to make a decision, members do not have a single signal about the likelihood of two states of nature; they may have knowledge ("signals") from various sources militating in favor of one decision or another, but also knowledge about which sources are credible, how to interpret the signals, and so forth. Negotiators may fail to reach instant agreement due not only

This essay benefited from discussions and comments from Rohan Pitchford, Kemal Yıldız, and especially Nicole Immorlica. Much of the writing was done while the author was visiting the Research School of Economics at the Australian National University, and their hospitality is gratefully acknowledged. The author is supported by a Sloan Foundation Fellowship.

G. Carroll (✉)
Stanford University, Stanford, CA, USA
e-mail: gdc@stanford.edu

© Springer Nature Switzerland AG 2019
J.-F. Laslier et al. (eds.), *The Future of Economic Design*, Studies in
Economic Design, https://doi.org/10.1007/978-3-030-18050-8_5

to private information about the issues they bargain over (and there are typically many simultaneous interacting issues), but also about the negotiating process itself and their strength within it. The point is not that we need to add more parameters to existing models; it is that *no* parameterized structural model can hope to capture all such complexities.

It is undoubtedly useful for economists to study stylized models, just as geneticists study fruit flies: as a simple starting point. Structural assumptions make it possible to give closed-form solutions and understand a model exhaustively. But eventually we have to come to the real world, and either understand when and why the simple model provides a good first-order approximation, or identify specific ways in which it is seriously deficient. Either way, we need modeling tools to think seriously about the richness that exists in real design problems, relatively free of structural assumptions.

My expectation—and my hope—is that progress in mechanism design over the coming decades will come from developing more useful general models of preferences, information, and actions in complex environments, relatively free of structural assumptions; and developing conceptual tools to argue for why certain kinds of mechanisms will work well in such environments. An exemplar of such a weakly-structured framework is the Arrow-Debreu model for general equilibrium: the goods can be arbitrary; preferences can take arbitrary form (aside from being monotone, and usually convex). The framework is broad enough to express that some conclusions are quite general (such as the First Welfare Theorem), while others demand much more restrictive assumptions (such as convergence of tatônnement dynamics). Decision theory is another existing area of economics where modeling is usually quite free of structural assumptions; but the typical end point in that field is to give alternative representations or interpretations of a given primitive (choice behavior), and it is not clear whether this methodology can be ported to mechanism design, where the goal is to recommend new mechanisms.

Making this progress toward more free-form models will require a shift in the criteria by which research in economic theory is evaluated. There is a strong cultural norm in economics that favors models that can be solved exactly. This is true both in the axiomatic branch of the design literature—where the ideal is a list of criteria that pick out a unique mechanism, or a small class of mechanisms, for some application—and in the Bayesian branch—where one writes down a numerical objective, typically expected revenue or expected welfare, and solves for the exact optimum. If a theorist proposes a non-optimal mechanism, the instinctive reaction is "plenty of things are non-optimal; what's so special about yours?"

But in any situation that remotely approaches the complexity of most real-world design problems, there is no hope of describing the exact optimum. Instead, design is and should be pluralistic. Economic design is not so different from design of physical products: a consumer may not find a perfect car to drive, or even a perfect sandwich for lunch, but having a range of options on the market makes it more likely she will find a choice she can be pretty happy with; and the same can be said for voting rules or negotiation protocols.

Accordingly, the criterion to evaluate a theoretical contribution should be: does it offer a new idea that will plausibly be helpful in eventually informing design (or in understanding mechanisms that are in use)? This is of course a subjective question. Perhaps the field focuses on exact optimization simply because it is unambiguous whether a particular work meets this criterion. A challenge for the mechanism design theory community, then, will be to reach some amount of agreement on what other kinds of ideas to recognize as valuable contributions, so that future researchers are less compelled to focus on the confining class of problems that have exact solutions.

What might such contributions look like? I will suggest a few pointers for inspiration based on existing work (of course, citations are selective due to space constraints). This is not meant to be an exhaustive list, and I hope that future work will discover new directions.

I find it helpful to classify work in economic design into two main kinds, "principles" and "engineering." "Principles" research uses stylized models to offer general insights. For example, the Myerson-Satterthwaite theorem (Myerson and Satterthwaite 1983) showed us that there is generally no way to ensure efficient outcomes when parties have private information about preferences; subsequent work such as Cramton et al. (1987) and Segal and Whinston (2011) showed how this conclusion can be overturned if the outside option is sufficiently symmetric. "Engineering" works out technical details to tailor mechanisms to specific contexts (which can mean either designing mechanisms for specific real-world uses in practice, or applied-theory models inspired by an application). Thus, for example, the two-sided matching literature, which has produced a stream of research adapting versions of the Gale-Shapley algorithm to the idiosyncrasies of different applications (examples include Ashlagi et al. 2014; Hafalir et al. 2013; Sönmez and Switzer 2013).

Both of these branches can contribute to designing for weakly structured environments. On the principles side:

- One kind of contribution will consist of offering new tools for the designer's toolkit. An example is Chassang's (2013) work on calibrated contracts. He considers a very general dynamic moral hazard model, with almost no structural assumptions. The principal would like to incentivize an agent (a wealth manager) to make good investment decisions, using a linear contract, so that the agent shares in the gains and losses he produces. But limited liability prevents her from making the agent share the losses. Chassang adapts tools from no-regret learning to show how this infeasible linear contract can nonetheless be closely approximated by adjusting the share fraction up or down over time depending on past performance.
  Thus, the emphasis is not on finding an optimal contract, but on providing a tool that can plausibly be useful for getting around a specific obstacle in incentive design.
- There are a number of approximation results that show how simple mechanisms can perform reasonably well across a broad range of environments. For just one of many examples, consider selling two goods to a buyer with unknown values. Finding the revenue-maximizing mechanism is a difficult and ill-understood problem in general; unlike with just one good, the optimum for two goods can be quite complicated, e.g. offering infinitely many different probabilistic bundles at various prices (Daskalakis et al. 2013). But Hart and Nisan (2017) showed that if the

values for the goods are independently distributed, then just selling the goods separately gives at least $1/2$ the maximum possible revenue. Subsequent works have extended this in many directions (for example Babaioff et al. 2014; Rubinstein and Weinberg 2015).

Related to this is the price-of-anarchy literature, studying welfare in equilibria of simple mechanisms with multiple agents. For example, Christodoulou et al. (2016) showed that when $m$ buyers buy $n$ goods via simultaneous second-price auctions held separately for each good, welfare in Bayes-Nash equilibrium is a $1/2$-approximation of first-best welfare as long as buyers' preferences over goods are submodular. This, too, has spawned many follow-ups and generalizations (Roughgarden et al. 2017 gives a survey).

Such approximation results are bread and butter in computer science. In economics they are often greeted with the skeptical reaction that getting $1/2$ of the optimum is not very satisfying. I might point out in reply that when exact optimization is hard and the starting point is zero, getting to $1/2$ is a pretty decent first step. In any case, the usual criterion for evaluating economic theory is not whether the results map literally into practice—they rarely do—but whether the analysis gives new (and perhaps even surprising) ideas that help us better understand the situation at hand. By this criterion, approximation results surely have a role to play. As Hartline (2012) argues, by studying whether a particular class of mechanisms can or cannot give an approximation guarantee, we have a principled way to formalize ideas about which properties of mechanisms are of first-order importance and which ones are not. In addition, the technical analysis underlying approximation guarantees sometimes leads to new conceptual frameworks for understanding classes of mechanisms, such as the "smoothness" framework for price-of-anarchy results.

• Some headway can also be made by considering models that retain strong structure on some dimensions while dropping it on others. An example of this is the work of Bergemann et al. (2017). They study a traditional first-price auction for a single object, but make no structural assumptions about the players' information (either about their own value for the object, others' values, or others' information). They show how one can nonetheless obtain lower bounds for equilibrium revenue in such a mechanism. Intuitively, the revenue cannot be very low, since then any bidder could simply deviate to a moderately low bid, and win at a profit with high probability. It turns out that this idea can be pushed to its logical conclusion to give a tight lower bound for revenue as a function of the value distribution, and to describe the worst-case information structure. The fact that a tight bound can be obtained seems specific to the application, but the idea of allowing information to be unstructured is in line with the theme here. (Carroll 2016 performs an exercise in a similar spirit with a binary decision mechanism.)

On the engineering side, it is hard to envision details about specific applications at a distance. Yet, there may be some categories of methodological advances that are likely to have widespread impact across many applications. I will make a few guesses:

- For mechanisms that are used repeatedly, it is natural to try to learn about distributions over agents' preferences, information, and so forth based on previous runs of the mechanism, and dynamically adjust parameters of the mechanism accordingly. This creates a need for tools to efficiently learn these parameters from past data; and indeed, there is demand for such tools in e-commerce. Recent theoretical work on data-driven design, such as Cole and Roughgarden (2014), Morgenstern and Roughgarden (2015), explores this topic in the canonical auction model.

  One could readily envision expanding these ideas to other domains of incentive design where there are extensive past interactions to learn from, such as government procurement (Laffont and Tirole 1986) or regulation of utilities (Baron and Myerson 1982). In these domains, more so than in e-commerce auctions, interactions are complex, and the prevailing theoretical tools are stylized models that are hard to interpret literally. Hence, many new questions arise about what a designer can learn from quantitative data about the structure of the environment, as well as how best to learn it.

- In some problems, just finding a good outcome is already an engineering challenge. This challenge by itself falls within the traditional territory of operations researchers, not economists. Yet, there are situations where incentives interact with the optimization problem in substantive ways, creating new theoretical challenges, for which economics has something to contribute. This has been a theme of theoretical research in algorithmic game theory for several years, but is just starting to be a serious consideration in practical applications, with the recent FCC incentive auction to reallocate radio spectrum in the United States as a headline example (Leyton-Brown et al. 2017). As both the theory of mechanism design and the adoption of algorithms to automate decision-making move into more complex domains, the interaction between optimization and incentives is likely to become more widespread.

- The design of the interface by which participants interact with a mechanism is also important. Currently, much of mechanism design starts by applying the revelation principle and assumes that agents' full preferences can be elicited. As design moves toward applications in which preferences and information are more complex and unstructured, this assumption will be less and less realistic, and it will be increasingly useful to have principled ways of thinking about how to efficiently elicit the most important information.

  Also important, and currently more neglected, is communication in the other direction: especially if one hopes to engineer automated decision-making mechanisms for settings where participants can refuse to accept the mechanism's output (or can stop using it for future interactions), one challenge can be explaining clearly to participants why they can't have everything they want, and persuading them to be satisfied with the outcome of the mechanism. As technology develops to bring increasingly detailed and quantitative input into large-scale collective decisions (such as participatory budgeting, Cabannes 2004; Goel et al. 2016), this latter task may well also become technical enough to have a role for formal theory.

Finally, given that I have started from an ambitious vision of the scope of possible applications for mechanism design, I feel it is important to temper ambition with humility: Some of the areas of application I have described are relatively removed from the traditional domain of economics—the allocation of material resources—and are more the province of experts in political science, law, or social psychology. While I have predicted that economics will, by developing more general and flexible models, increase its ability to inform actual design in these areas, economics will not and should not supplant domain-specific expertise. Economics has some specific comparative advantages to offer: the ability to think quantitatively about cost-benefit tradeoffs; the habit of thinking about strategic incentives, and especially about how changing the rules of the game will make the players play differently. These strengths should complement the perspectives and tools that come from other disciplines.

# References

Ashlagi, I., Braverman, M., & Hassidim, A. (2014). Stability in large matching markets with complementarities. *Operations Research*, *62*(4), 713–732.

Babaioff, M., Immorlica, N., Lucier, B., & Weinberg, S. M. (2014). A simple and approximately optimal mechanism for an additive buyer. In *FOCS'14* (pp. 21–30).

Baron, D. P., & Myerson, R. B. (1982). Regulating a monopolist with unknown costs. *Econometrica*, *50*(4), 911–930.

Bergemann, D., Brooks, B., & Morris, S. (2017). First-price auctions with general information structures: Implications for bidding and revenue. *Econometrica*, *85*(1), 107–143.

Cabannes, Y. (2004). Participatory budgeting: A significant contribution to participatory democracy. *Environment and Urbanization*, *16*(1), 27–46.

Carroll, G. (2016). Informationally robust trade and limits to contagion. *Journal of Economic Theory*, *166*, 334–361.

Chassang, S. (2013). Calibrated incentive contracts. *Econometrica*, *81*(5), 1935–1971.

Christodoulou, G., Kovács, A., & Schapira, M. (2016). Bayesian combinatorial auctions. *Journal of the ACM*, *63*(2), 11.

Cole, R., & Roughgarden, T. (2014). The sample complexity of revenue maximization. In *STOC'14* (pp. 243–252).

Cramton, P., Gibbons, R., & Klemperer, P. (1987). Dissolving a partnership efficiently. *Econometrica*, *55*(3), 615–632.

Daskalakis, C., Deckelbaum, A., & Tzamos, C. (2013). Mechanism design via optimal transport. In *EC'13* (pp. 269–286).

Goel, A., Krishnaswamy, A. K., Sakshuwong, S., & Aitamurto, T. (2016). Knapsack voting: voting mechanisms for participatory budgeting, unpublished paper, Stanford University.

Hafalir, I. E., Yenmez, M. B., & Yildirim, M. A. (2013). Effective affirmative action in school choice. *Theoretical Economics*, *8*(2), 325–363.

Hart, S., & Nisan, N. (2017). Approximate revenue maximization with multiple items. *Journal of Economic Theory*, *172*, 313–347.

Hartline, J. (2012). Approximation in mechanism design. *American Economic Review*, *102*(3), 330–336.

Laffont, J.-J., & Tirole, J. (1986). Using cost observation to regulate firms. *Journal of Political Economy*, *94*(3), 614–641.

Leyton-Brown, K., Milgrom, P., & Segal, I. (2017). Economics and computer science of a radio spectrum reallocation. *Proceedings of the National Academy of Sciences*, *114*(28), 7202–7209.

Morgenstern, J. H., & Roughgarden, T. (2015). On the pseudo-dimension of nearly optimal auctions. In *NIPS'15* (pp. 136–144).

Myerson, R. B., & Satterthwaite, M. A. (1983). Efficient mechanisms for bilateral trading. *Journal of Economic Theory*, *29*(2), 265–281.

Roughgarden, T., Syrgkanis, V., & Tardos, É. (2017). The price of anarchy in auctions. *Journal of Artificial Intelligence Research*, *59*, 59–101.

Rubinstein, A., & Weinberg, S. M. (2015). Simple mechanisms for a subadditive buyer and applications to revenue monotonicity. In *EC'15* (pp. 377–394).

Segal, I., & Whinston, M. D. (2011). A simple status quo that ensures participation (with application to efficient bargaining). *Theoretical Economics*, *6*(1), 109–125.

Sönmez, T., & Switzer, T. B. (2013). Matching with (branch-of-choice) contracts at the United States Military Academy. *Econometrica*, *81*(2), 451–488.

# Game of Trons: Normatrons and Positrons Contribute to Economic Design

Roberto Serrano

**Abstract** Is Economic Design a Normative or a Positive Agenda? This paper provides some thoughts about this question. Readers should feel free to disagree with all of them.

Is Economic Design a Normative or a Positive Agenda? There are several camps of scholars that are currently contributing to this important field. Although this is not meant to be a universal statement, there might be some degree of misunderstanding among some of them regarding the purpose of the intellectual enterprise in which they are embarked. My attempt in these few paragraphs, beyond acknowledging this fact, is to state my strong appreciation for the different interpretations and approaches to research in economic design. As such, this short piece is just a modest collection of unstructured thoughts, and it is hence far from providing a comprehensive treatment to all relevant aspects in the opening question.

Let us amiably refer to scholars favoring the normative and positive approach as *n*ormatrons and *p*ositrons, respectively. This is not necessarily a partition of the researchers in the field, as some may have difficulties alining themselves with only one of these options.

The positrons have lately been interested, for example, in better understanding the properties of variants of the Vickrey auction or the deferred-acceptance algorithm for matching because of their real-world empirical relevance. This is no doubt a valuable exercise. However, the normatrons have sometimes criticized the positrons for being too narrow-minded, too driven by what they think is empirically plausible. One can see a hypothetical normatron writing a referee report on a paper by a positron pushing the latter to study the efficiency or strategy-proofness of perturbations of

There are no connections between these pages and chemistry, as far as the author knows. Thanks to Hervé Moulin for his kind invitation to join this interesting project, and to coauthors Antonio Cabrales, Geoffroy de Clippel, Kfir Eliaz, Takashi Kunimoto, Rene Saran, and Rajiv Vohra for many helpful discussions over the years. Bobby Pakzad-Hurson and Jesse Shapiro provided comments and encouragement.

R. Serrano (✉)
Department of Economics, Brown University, Providence, Rhode Island, USA
e-mail: roberto_serrano@brown.edu

© Springer Nature Switzerland AG 2019
J.-F. Laslier et al. (eds.), *The Future of Economic Design*, Studies in
Economic Design, https://doi.org/10.1007/978-3-030-18050-8_6

those mechanisms, for the sake of robustness. And one can hypothetically see the positron dismissing that request, because those perturbed mechanisms have not been used in practice. In my view, both are right. The positron has the right to include in her/his paper whatever s/he thinks is relevant, and the normatron should perhaps write a nice paper about such perturbations. Indeed, I do not think that a necessary condition for a great paper is to perform every robustness check, nor do I think that normatrons should never write papers with a positive spin about specific mechanisms.

Many normatrons have been interested in depicting the landscape of what is possible in terms of decentralizing economic decisions on the basis of some theory of behavior at the individual level. This has led to the identification of conditions (with a varying degree of normative content) that are necessary and/or sufficient for implementability in a host of game-theoretic solution concepts and in a variety of allocation problems. Establishing such possibility or impossibility results is important, as it formalizes some of the key constraints facing the economic authority.

For the sufficiency part of the argument, the literature has often relied on canonical mechanisms, whose role has sometimes been misunderstood or underappreciated. One may have received, again hypothetically, thoughtful referee reports on some of one's own work, probably written by positrons, dismissing any result that is proved by using such unrealistic mechanisms. What this criticism misses is that such general mechanisms are simply a unifying method of proof over many environments, and hence they are incredibly powerful tools in the task of surveying the landscape of possibilities. Having said that, it is indeed important to listen to the criticism and insist on better properties of the implementing mechanisms, perhaps excluding integer games or other devices that some may think unnatural. Adding such constraints to the design will likely enhance the realism of its applicability.

Inspired by some recent work that combines mechanism design and bounded rationality, I have to question the use of canonical mechanisms for a different reason. Any canonical mechanism that "works" to prove an implementability theorem tries to make truth-telling "focal," or in general, to hint to the agents that they should use the messages under which the right outcome arises. But then the question is whether that "focal" feature of those message profiles is actually consistent with the theory of boundedly-rational behavior assumed. The question extends to classic mechanisms as well, say Maskin's for Nash implementation. In fact, experimental/empirical evidence must be brought to bear (and there is already some nice work about it). That is, we may be facing a new fixed-point problem to resolve this "chicken-and-egg" dilemma: given a theory of behavior assumed, can we construct an implementing mechanism? and, given that implementing mechanism, is the theory of behavior assumed plausible? It seems that the resolution of this fixed-point problem may necessitate a nice blend of theoretical and experimental work, where normatrons and positrons could enjoy working together.

By the way, against the opinion of an editor in one of the so-called top journals, expressed in a hypothetical rejection letter, let me humbly assert that the research agenda that combines mechanism design and bounded rationality should be extremely important for a general audience. One only ought to take a look at recent political results in different countries to realize that designing institutions that are not vulnerable to manipulation or that are to be used by less than rational individuals is an endeavor of great relevance.

I will conclude with a word on the term "robustness," which, when applied to mechanism design, I think has been ill-used, by confining it to describe the testing of mechanisms that can survive more demanding requirements on informational assumptions, especially concerning higher-order beliefs. In particular, hypothetical editors and referees should stay away from that narrow identification and accept the obvious fact that there are many other dimensions along which results may be asked to be "robust." To be fair, this Wilson doctrine sort of approach has been successful, both applied to specific mechanisms by positrons and in the drawing of the landscape of possibilities by normatrons. This is so, even though the latter has, for the most part, simply served to confirm our previous understanding, i.e., negative results have been turned into even more negative ones, once the robustness to high-order beliefs has been demanded, while some more permissive results have still stood their ground.

Some normatrons and some positrons may disagree in their interpretations of the field, or even in the meaning of the overall research agenda, but they all will surely agree on striving to produce high-quality work and apply those high standards with an open mind to evaluate the work of others. That is what in the end matters. Since I tend to think that all coherent views are valid, I hope the reader will ponder the ones expressed here as well. They are meant to be suggestive in the continuation of our collective collaborative project, with the aim of achieving a better understanding of human institutions. This is a good desideratum, it seems to me, for the future of economic design. And to be sure, all hypothetical events that have been described in previous paragraphs are not necessarily hypothetical.

Disclaimer: The most interesting scenes and passages related to the title have been censored by the National Science Foundation (this statement is to be taken also as probably hypothetical).

# Reverting to Simplicity in Social Choice

Nisarg Shah

**Abstract** Social choice theory studies the design of rules for eliciting individual preferences and making collective decisions in various domains. The domain of interest—be it election of a candidate based on citizens' votes, allocation of goods to agents, or matching of agents to other agents—dictates the space of possible decisions as well as the criteria that make certain decisions socially more desirable. This often leads to the design of domain-specific solution concepts. In contrast, certain conceptually simple solutions from microeconomics (e.g., maximization of a social welfare function) seem to naturally apply to a broad range of domains, after suitable modeling of agent preferences, and achieve compelling properties. In this article, I point to recent advances in the understanding of such solution concepts in different domains and advocate for reverting to simplicity in social choice.

The past few decades have seen an accelerating shift from analysis of elegant theoretical models to treatment of important real-world problems, which often bear complexity in the forms of constraints, priorities, or endowments. This has paved the way for the design of complex algorithmic solutions. In microeconomics, complex solutions also stem from normative economics, where the goal is to find *some* solution that satisfies a combination of axioms, and no emphasis is placed on the simplicity or the intuitive appeal of the solution itself. On the one hand, having too few axioms could permit a solution to make arbitrary choices from a wide range of possibilities, and on the other hand, having nonessential axioms could unnecessarily complicate the resulting solution. For instance, with the rise of computational economics, polynomial time computability became a popular desideratum. But for many problems, it has become dispensable due to the availability of fast integer programming solvers.

In this article, I examine the appeal of reverting to conceptually simple solutions, and its practical implications on the future of economic design. I draw on my experience of research on social choice theory, which studies societal decision making based on individual preferences. I begin with the positive case of *fair division*, where the power of conceptually simple solutions is relatively well understood, and then discuss *voting*, where the appeal is less clear.

N. Shah (✉)
University of Toronto, Toronto, ON, Canada
e-mail: nisarg@cs.toronto.edu

© Springer Nature Switzerland AG 2019                                                    39
J.-F. Laslier et al. (eds.), *The Future of Economic Design*, Studies in
Economic Design, https://doi.org/10.1007/978-3-030-18050-8_7

First, what do I mean by "conceptually simple solutions"? While there is no clear definition, such solutions are typically easy to describe, intuitively appealing, and importantly, defined for a wide range of domains. In theoretical computer science, greedy algorithms and dynamic programming are recognized to be conceptually simple. In this article, I focus on an approach that stems from welfare economics: maximizing a collective utility function (or simply, welfare maximization). In this approach, one first defines, for each participant, a utility function that maps each possible outcome to a real number, then defines a collective utility function (CUF) that aggregates individual utilities, and finally chooses the outcome maximizing the collective utility. Three popular choices for the CUF are utilitarian (sum of utilities), Nash (product of utilities), and egalitarian (minimum utility).[1]

Let me discuss their efficacy in fair division applications, where the goal is to fairly divide a common pool of resources among participants. The study of fair division begins by answering what *fair* means. One of the most compelling notions of fairness is envy-freeness (Foley 1967), which requires that no participant prefer what another participant receives to what she receives. Take the classic setting of rent division, where $n$ rooms and a total rent are to be divided among $n$ roommates. Svensson (1983) showed that envy-free outcomes are guaranteed to exist when individuals have quasi-linear utilities, i.e., if their utility is their value for the assigned room minus the rent paid. When participants can be charged payments for receiving goods, utilitarian CUF is typically preferred. Indeed, for rent division the First Welfare Theorem implies that in any envy-free outcome, the allocation of rooms must maximize utilitarian CUF. There still exist multiple divisions of rent that guarantee envy-freeness; Gal et al. (2016) showed that choosing the rent division that maximizes egalitarian CUF (subject to envy-freeness) provides additional guarantees. It is worthwhile remarking that this solution concept has the added benefit of being polynomial time computable.

Let us now turn our attention to goods allocation *without* money. The lack of money makes interpersonal comparison of utilities less meaningful, which in turn makes Nash CUF compelling because, under mild conditions, it is uniquely independent of individual utility scale (Moulin 2003).[2] Indeed, consider the cake-cutting setting (Steinhaus 1948), where a divisible heterogeneous good ("cake") is to be allocated. It is commonly assumed that participants have additive utilities; in fact, cake-cutting is the quintessential fair division setting with additive utilities. It has been shown that cutting the cake by maximizing Nash CUF satisfies most desiderata considered in the literature:

- It is equivalent to a market equilibrium approach (strong competitive equilibrium from equal incomes, or s-CEEI) (Sziklai and Segal-Halevi 2015).
- It satisfies group envy-freeness (Berliant et al. 1992), which generalizes envy-freeness and Pareto optimality (Weller 1985).

---

[1]The leximin method is a refinement, in which, after maximizing the minimum utility, ties are broken in favor of higher second minimum utility, remaining ties are broken in favor of higher third minimum utility, and so on.

[2]That is, scaling the utilities of an individual does not alter the outcome maximizing Nash CUF.

- It produces an outcome in the core,[3] which generalizes a different fairness notion called proportionality, and Pareto optimality.
- It satisfies intuitive properties such as resource monotonicity (dividing more cake cannot be worse for anyone) and population monotonicity (dividing between more participants cannot be better for anyone) (Sziklai and Segal-Halevi 2015).

The intractability of computing this outcome, except in special cases Aziz and Ye (2014), has led researchers to explore finite, bounded, and polynomial time computable solutions.[4] Such solutions are often conceptually more intricate than simply maximizing Nash CUF; take, for example, the polynomial-time Even-Paz protocol (1984) for proportional cake-cutting or the (surprisingly complex) bounded-time Aziz-Mackenzie protocol (2016) for envy-free cake-cutting. It is unclear if one can instead compute the outcome maximizing Nash CUF up to an accuracy sufficient for real-world problems.

For allocating *indivisible* goods without money, under additive utilities, there is no outstanding method due to the impossibility of achieving strong fairness notions like envy-freeness. Recently, Caragiannis et al. (2016a) showed that maximizing Nash CUF achieves envy-freeness up to one good: no participant would envy another participant if the former got to remove at most one good from the latter's bundle. It remains to be seen whether maximizing Nash CUF leads to (relaxations of) stronger guarantees for this setting as it does for cake-cutting. Nonetheless, we have deployed this approach to our fair division website, http://www.spliddit.org, due to the simplicity of the solution concept and the usefulness of envy-freeness up to one good in explaining fairness of the chosen outcome to the participants.

In the literature on fair division with non-additive utilities, one prominent method is the leximin method, which maximizes egalitarian CUF. Kurokawa et al. (2015) studied the leximin method, motivated by the real-world problem of allocating unused classroom space to charter schools. Under mild conditions, they showed that when utility functions satisfy an "optimal utilization" requirement,[5] the leximin method satisfies proportionality, envy-freeness, and Pareto optimality, along with a strong game-theoretic desideratum called group strategyproofness.

Welfare maximization, while prevalent in fair division applications, is largely overshadowed by the axiomatic approach in voting applications, although it has had its advocates (Harsanyi 1955; Hillinger 2005). One potential reason is that while fair division deals with preferences over an exponential outcome space, and thus imposes restricted utilities for tractability, voting applications typically deal with a small outcome space, and thus ask voters to report rankings over possible outcomes. Unfortunately, in this ordinal model, no conceptually simple, or intricate, solution dominates due to celebrated impossibility results like Arrow's (1951) impossibility.

The situation improves a little if one adheres to cardinal utility theory, which posits that voter preferences have intensities, which can be represented by cardi-

---

[3]This is an easy derivation given the equivalence to s-CEEI.

[4]Geometric requirements such as contiguity of allocated pieces of cake have also inspired a significant body of research.

[5]Leontief and dichotomous utilities are examples that satisfy this requirement.

nal utilities. Nonetheless, in most voting applications, such as political elections, it is cognitively difficult for voters to specify a real-valued utility for an alternative. Procaccia and Rosenschein (2006) reconciled this conflict by introducing the *implicit utilitarian voting* framework, where voters still report ranked preferences, but these are treated as proxies for underlying cardinal utilities. To address the lack of full information and derive a unique solution concept, they combined welfare maximization with an elegant solution concept from theoretical computer science: optimization of worst-case approximation, where the worst case is over all possible full information (cardinal utilities) consistent with the reported ordinal information. Boutilier et al. (2015) showed the promise of this framework for selecting a single alternative. Caragiannis et al. (2016b) extended the approach to selecting a subset of alternatives and replacing worst-case approximation by another simple concept from learning theory, *minimax regret*. While they showed that implicit utilitarian rules perform well on real data, the framework currently lacks the strong axiomatic justification that welfare maximization has in fair division. One direction for the future is to study complex voting problems with exponential outcome space, such as participatory budgeting (Cabannes 2004; Benade et al. 2017), where restricted utility forms may again be imposed. Another direction is to bridge the gap between voting and fair division by treating the outcomes of a voting process as public goods. Recent work has shown that some of the fairness notions considered in the fair division literature are well-defined for voting problems, provide non-trivial guarantees, and can be achieved (Conitzer et al. 2017; Fain et al. 2018; Aziz et al. 2017).

Let me now cast a wider net. In this article, I examined the appeal of using conceptually simple solutions in social choice, surveyed their success in fair division applications, and contrasted with the relative lack thereof in voting applications. There are also practical implications of gravitating towards such solution concepts.

Perhaps the most obvious implication is that we need to invest more effort to understand the limits of their capabilities—the domains for which different solution concepts are suitable, and the practical considerations they can incorporate. For instance, it would be interesting to study if the attractiveness of the leximin method or Nash CUF extends beyond the "optimal utilization" domain (Kurokawa et al. 2015) and the additive utility domain (Weller 1985; Varian 1974; Berliant et al. 1992; Sziklai and Segal-Halevi 2015; Caragiannis et al. 2016a), respectively. Kurokawa et al. (2015) show that the leximin method can incorporate arbitrary external constraints on feasible outcomes as long as the outcome space remains convex. Would a similar result hold for Nash or utilitarian CUF? Also, most CUFs can easily incorporate priorities for participants in the form of real-valued weights.[6] Are there other forms of priorities they can incorporate? Finally, implicit utilitarian voting uses worst-case approximation and minimax regret to deal with partial information. Would these be useful to deal with partial information in fair division?

---

[6]Utilitarian, Nash, and egalitarian CUFs admit weighted variants that use sum of utilities multiplied by weights, product of utilities to the power weights, and minimum of utilities divided by weights, respectively.

Going one step further, there is also a subtle methodological implication for social choice researchers. Instead of starting from a set of axioms and designing *some* solution concept that satisfies them, one may want to examine fundamental solutions that emerge (often uniquely) from simple concepts such as welfare maximization, worst-case approximation, or minimax regret. Budish (2012) contrasts the prevalence of a similar optimization-based approach in the mechanism design literature to the prevalence of the axiomatic approach in the applied matching literature. Note that axioms can still be useful for justifying the choice of one solution concept over another, and for explaining appropriateness of the chosen outcome to the participants (cf. the article in this volume by Ariel Procaccia). Also, the axioms could be setting-dependent even if the solution concept is more generally defined.[7] It is also worthwhile remarking that even if a solution concept is simple, the algorithm for computing its outcome may be complicated; see, e.g., the algorithm for maximizing Nash CUF (Aziz and Ye 2014; Caragiannis et al. 2016a) or the algorithms for implicit utilitarian voting (Boutilier et al. 2015; Caragiannis et al. 2016b).

Conceptually simple solutions are easy to convey to participants and intuitively appealing, and therefore have a practical advantage over complex solutions. They have thus been advocated in other areas of computational economics as well. For instance, in algorithmic mechanism design literature, simple auctions have been shown to be approximately optimal for many complex settings (Hartline and Roughgarden 2009; Daskalakis and Pierrakos 2011; Greenwald et al. 2017), and have been advocated for their robustness (Hartline 2013). I believe such solution concepts have potential for significant real-world impact in social choice applications.

# References

Arrow, K. (1951). *Social choice and individual values*. New York: Wiley.

Aziz, H., & Mackenzie, S. (2016). A discrete and bounded envy-free cake cutting protocol for any number of agents. In *Proceedings of the 57th Symposium on Foundations of Computer Science (FOCS)* (pp. 416–427).

Aziz, H., & Ye, C. (2014). Cake cutting algorithms for piecewise constant and piecewise uniform valuations. In *Proceedings of the 10th Conference on Web and Internet Economics (WINE)* (pp. 1–14).

Aziz, H., Bogomolnaia, A., & Moulin, H. (2017). Fair mixing: the case of dichotomous preferences. arXiv:1712.02542.

Benade, G., Nath, S., Procaccia, A. D., & Shah, N. (2017). Preference elicitation for participatory budgeting. In *Proceedings of the 31st AAAI Conference on Artificial Intelligence (AAAI)* (pp. 376–382).

Berliant, M., Thomson, W., & Dunz, K. (1992). On the fair division of a heterogeneous commodity. *Journal of Mathematical Economics, 21*(3), 201–216.

---

[7]Conversely, sometimes an axiom is broadly defined, but is achieved in different settings by different solution concepts. For instance, the core is achieved by maximizing the Nash CUF in allocation of public goods Fain et al. (2018); Aziz et al. (2017), by the top trading cycles mechanism in housing markets (Shapley and Scarf 1974), and through stable matching algorithms in two-sided matching markets (Gale and Shapley 1962).

Boutilier, C., Caragiannis, I., Haber, S., Lu, T., Procaccia, A. D., & Sheffet, O. (2015). Optimal social choice functions: A utilitarian view. *Artificial Intelligence*, *227*, 190–213.

Budish, E. (2012). Matching versus mechanism design. *ACM SIGecom Exchanges*, *11*(2), 4–15.

Cabannes, Y. (2004). Participatory budgeting: a significant contribution to participatory democracy. *Environment and Urbanization*, *16*(1), 27–46.

Caragiannis, I., Kurokawa, D., Moulin, H., Procaccia, A. D., Shah, N., & Wang, J. (2016a). The unreasonable fairness of maximum Nash welfare. In *Proceedings of the 17th ACM Conference on Economics and Computation (EC)* (pp. 305–322).

Caragiannis, I., Nath, S., Procaccia, A. D., & Shah, N. (2016b) Subset selection via implicit utilitarian voting. In *Proceedings of the 25th International Joint Conference on Artificial Intelligence (IJCAI)* (pp. 151–157).

Conitzer, V., Freeman, R., & Shah, N. (2017). Fair public decision making. In *Proceedings of the 18th ACM Conference on Economics and Computation (EC)* (pp. 629–646).

Daskalakis, C., & Pierrakos, G. (2011). Simple, optimal and efficient auctions. In *Proceedings of the 7th Conference on Web and Internet Economics (WINE)* (pp. 109–121).

Even, S., & Paz, A. (1984). A note on cake-cutting. *Discrete Applied Mathematics*, *7*, 285–296.

Fain, B., Munagala, K., & Shah, N. (2018). Fair allocation of indivisible public goods. In *Proceedings of the 19th ACM Conference on Economics and Computation (EC)*. Forthcoming.

Foley, D. (1967). Resource allocation and the public sector. *Yale Economics Essays*, *7*, 45–98.

Gal, Y., Mash, M., Procaccia, A. D., & Zick, Y. (2016). Which is the fairest (rent division) of them all? Manuscript.

Gale, D., & Shapley, L. S. (1962). College admissions and the stability of marriage. *Americal Mathematical Monthly*, *69*(1), 9–15.

Greenwald, A., Oyakawa, T., & Syrgkanis, V. (2017). Simple vs optimal mechanisms in auctions with convex payments. arXiv:1702.06062.

Harsanyi, J. C. (1955). Cardinal welfare, individualistic ethics, and interpersonal comparisons of utility. *Journal of Political Economy*, *63*(4), 309–321.

Hartline, J. D. (2013). *Mechanism Design and Approximation*. Book Draft. http://jasonhartline.com/MDnA.

Hartline, J. D., & Roughgarden, T. (2009) Simple versus optimal mechanisms. In *Proceedings of the 10th ACM Conference on Economics and Computation (EC)* (pp. 225–234).

Hillinger, C. (2005). The case for utilitarian voting. Discussion papers in economics, University of Munich, Department of Economics.

Kurokawa, D., Procaccia, A. D., & Shah, N. (2015). Leximin allocations in the real world. In *Proceedings of the 16th ACM Conference on Economics and Computation (EC)* (pp. 345–362).

Moulin, H. (2003). *Fair division and collective welfare*. Cambridge: MIT Press.

Procaccia, A. D., & Rosenschein, J. S. (2006). The distortion of cardinal preferences in voting. In *Proceedings of the 10th International Workshop on Cooperative Information Agents (CIA)* (pp. 317–331).

Shapley, L., & Scarf, H. (1974). On cores and indivisibility. *Journal of mathematical economics*, *1*(1), 23–37.

Steinhaus, H. (1948). The problem of fair division. *Econometrica*, *16*, 101–104.

Svensson, L.-G. (1983). Large indivisibles: An analysis with respect to price equilibrium and fairness. *Econometrica*, *51*(4), 939–954.

Sziklai, B., & Segal-Halevi, E. (2015). Resource-monotonicity and population-monotonicity in cake-cutting. arXiv:1510.05229.

Varian, H. (1974). Equity, envy and efficiency. *Journal of Economic Theory*, *9*, 63–91.

Weller, D. (1985). Fair division of a measurable space. *Journal of Mathematical Economics*, *14*(1), 5–17.

# A Probabilistic Approach to Voting, Allocation, Matching, and Coalition Formation

Haris Aziz

**Abstract** Randomisation and time-sharing are some of the oldest methods to achieve fairness. I make a case that applying these approaches to social choice settings constitutes a powerful paradigm that deserves an extensive and thorough examination. I discuss challenges and opportunities in applying these approaches to settings including voting, allocation, matching, and coalition formation.

## 1 Introduction

Suppose two agents have opposite preferences over the two possible social outcomes. What should be a fair resolution for this problem?

If the outcome is required to be deterministic, then it is patently unfair to one of the agents. However, one can regain fairness by at least three different approaches: (1) resort to randomisation so that each of the social outcomes has equal probability, (2) treat the outcomes as divisible and resort to time sharing where each social outcome has half of the time share or, (3) use a uniform frequency distribution if there will be multiple occurrences of the discrete outcomes. Mathematically, all three resolutions towards fairness are equivalent because the outcomes have equal probability, time-share, or frequency. In the rest of the article, when I will describe a probabilistic approach to social choice, I will use it abstractly so as to model approaches (1), (2), and (3).

I argue that although a probabilistic approach has been applied in several social choice settings in both theory and practice (see e.g. Stone 2011), there is potential to revisit fundamental social choice settings such as voting, allocation, matching, and coalition formation with this powerful paradigm. Considering the natural aversion of many people to important decisions being based on the toss of the coin, such an approach may be especially useful for time and budget sharing scenarios. A related chapter in this book is by Brandt (2019).

H. Aziz (✉)
UNSW Sydney and Data61 CSIRO, Sydney, NSW, Australia
e-mail: haris.aziz@data61.csiro.au

© Springer Nature Switzerland AG 2019  45
J.-F. Laslier et al. (eds.), *The Future of Economic Design*, Studies in Economic Design, https://doi.org/10.1007/978-3-030-18050-8_8

## 2   A Case for Probabilistic Social Choice

I first list some of the compelling reasons why probabilistic social choice is a powerful
and useful approach.

(i) **Modeling time-sharing** Some of the settings and their corresponding results
    in the literature ignore the possibility of time-sharing. For example, several
    results in voting suppose that only one alternative is selected. However, the
    voting could be about deciding the fraction of time different genres of music
    are played on radio. Similarly most of the results in matching and coalition
    formation suppose that agents form exclusively one coalition or set of part-
    nerships (Manlove 2013). Many of these results need to be re-examined when
    we allow the flexibility of time-sharing.

(ii) **Participatory Budgeting** Voting can also be used to decide on which projects
     get how much budget (see e.g. Fain et al. 2016; Aziz et al. 2017). The approach
     is getting traction as grassroots participatory budgeting movements grow in
     stature (Cabannes 2004). In this context, a probabilistic social choice view is
     useful because the probability of an alternative can represent the fraction of
     the budget allocated to it.

(iii) **Achieving fairness** As explained in the example above, a probabilistic or
      time-sharing approach to social choice is geared towards achieving fairness.
      In the example above, no deterministic mechanism can simultaneously be
      anonymous (treat agents symmetrically) and neutral (treat social outcomes
      symmetrically). On the other hand, a probabilistic approach easily overcomes
      this impossibility. Suppose that in the example, the two social outcomes are
      allocating one item each to the agents where one item is valuable to both and
      the second item is useless to both. Then the only way to avoid envy is to use
      a probabilistic approach in a broad sense.

(iv) **Incentivizing participation** Another reason to take a probabilistic approach is
     to provide participation incentives (Brandl et al. 2015; Aziz et al. 2017, 2018).
     For the example described above, at least one of the agents appears to have no
     strict incentive to participate if the decision is made deterministically. On the
     other hand, probabilistic rules can be designed that give each voter the ability
     to at least make an epsilon difference (in expectation) to the outcome.

(v) **Achieving strategyproofness without resorting to dictatorship** Some of
    the most striking results in social choice give the message that if one wants
    agents to have incentives to report truthfully, then one has to resort to dicta-
    torship. In our running example it would mean selecting the preferred social
    outcome of one pre-specified agent. However, with a probabilistic o r ime-
    sharing approach, one can still achieve strategyproofness and also circumvent
    the prospect of a single agent with over-riding power (Aziz et al. 2014; Chen
    et al. 2013; Procaccia 2010; Procaccia and Tennenholtz 2013).

(vi) **Achieving stability** In much of the social choice and cooperative game theory
     literature, a theme of results involves the lack of outcome which satisfies an
     appropriate notion of stability. In voting, the most prominent result within

this theme is Condorcet's theorem says that it can be possible that for any given social outcome, a majority of people prefer another outcome. However, Condorcet's cycles vanish when the probabilistic 'maximal lottery' rule is used (Aziz et al. 2013; Brandl et al. 2016).[1] Similarly, core stable outcomes may not exist for general settings such as hedonic coalition formation in which agents have preferences over coalitions they are members of. However, if we allow for probabilistic outcomes or for time-sharing arrangements, there exist stable outcomes (Aharoni and Fleiner 2003).

(vii) **Circumventing impossibility results** Social choice is at times notorious for some of the bleak impossibility results in its literature. Several results showing that no apportionment method simultaneously satisfies basic monotonicity axioms (Young 1994). However, these problems disappear if a bit of randomisation is used. Similarly, there are results pointing out that no deterministic voting rule satisfies some basic consistency properties. However, this is not anymore the case if one uses the *maximal lotteries* randomised voting rule (Brandl et al. 2016).[2]

(viii) **Better welfare guarantees** When considering cardinal preferences over outcomes, a probabilistic approach may achieve better approximation welfare guarantees while simultaneously achieving other axiomatic properties (Anshelevich et al. 2015; Anshelevich and Postl 2016; Procaccia 2010). In some cases, randomization may allow for better welfare or ex ante Pareto improvement while satisfying stability constraints (Manjunath 2016).

# 3  Research Challenges and Opportunities

I outlined several advantages of considering a probabilistic approach. At an abstract level, resorting to randomisation means that one can consider the full continuous space of outcomes. This can both be a challenge as well as an opportunity for new and exciting research.

(i) **Formalizing and exploring a range of solution concepts and axioms** When considering probabilistic or time sharing approches, there are several ways in which a solution concept for deterministic settings can be extended to probabilistic settings. Take for example pairwise stability for the classic stable matching problem in which we want to pair men and women in a way so that no man and woman not paired with each other want to elope. When considering probabilistic outcomes, there is a hierarchy of stability concepts that are all generalisations of deterministic stability (Aziz and Klaus 2019; Dogan and Yildiz 2016; Kesten and Unver 2015; Manjunath 2016). Understanding the nature and structure of these properties is already a significant research direction. More

---

[1] The argument for the existence of such a lottery invokes von Neumann's minimax theorem.

[2] For further discussion on probabilistic approaches to circumvent impossibility results in voting, we refer to the survey by Brandt (2017).

importantly, the potential to generalize important axioms based on stability, Pareto optimality, and fairness in several different ways gives useful creative leeway for institution designers for exploring the tradeoffs and compatibility between different levels of properties.

Similar to the potential of defining a several levels of axiomatic properties, one can explore different levels of properties of mechanisms. A case in point is strategyproofness and participation incentives.

(ii) **Eliciting, representing, and reasoning about preferences** In most voting or matching settings, agents express preferences over deterministic outcomes. As we move from deterministic to probabilistic settings, there is an interesting challenge to elicit and represent agents' risk attitudes towards different lotteries. One possible approach that involves compact preferences is to extend the agents' preferences over discrete outcomes to preferences over probabilistic outcomes by *lottery extension* methods such as first order stochastic dominance (Brandl 2013; Brandt 2017; Bogomolnaia and Moulin 2001; Aziz et al. 2013; Cho 2016).

(iii) **Designing time-sharing mechanisms** Since modeling time-sharing is one of the most important motivations of probabilistic social choice, it is important to come up with compelling time-sharing mechanisms. Although voting has been studied for decades, a probabilistic perspective leads to interesting and meaningful new voting rules (see e.g., Aziz and Stursberg 2014). When allowing for fractional outcomes, several well-known mechanisms such as Gale-Shapley Deferred Acceptance or Gale's Top Trading Cycles need to be generalized (Kesten and Unver 2015; Athanassoglou and Sethuraman 2011; Bogomolnaia and Moulin 2004).

(vi) **Efficiency issues** When trying to achieve fairness via randomization, a straightforward approach is to uniformly randomize over reasonable deterministic outcomes or reasonable deterministic mechanisms. However, such a naive approach such as randomizing over Pareto optimal alternatives can lead to loss of ex ante efficiency. This phenomenon is starkly highlighted by the fact that random serial dictatorship that involves uniform randomization over the class of serial dictatorships can lead to unambiguous loss of welfare (Bogomolnaia and Moulin 2001; Aziz et al. 2013). This issue motivates the need to design interesting new mechanisms that are not victim to such a phenomenon.

(v) **Computational complexity** Generally speaking, optimising in continuous environments is computationally more tractable than in discrete environments. However, when considering time-sharing outcomes that are implicitly convex combinations of a potentially exponential number of discrete outcomes, computing the time shares can be a computationally arduous task (Aziz et al. 2014). Therefore, when formulating time-sharing mechanisms for different social choice settings, computational complexity rears its head as a potential challenge as well an opportunity for innovative algorithmic research.

(vi) **Instantiating a lottery** As mentioned earlier, uniformly randomizing over desirable outcomes can result in loss of efficiency or computational intractability. Therefore, mechanisms may first use an alternative way to find an *expected* 'fractional' outcome—say a weighted matching signifying probabilities for partnerships. If approach (1) is being used, then the expected outcome needs to be instantiated via a concrete lottery. Finding a suitable lottery is trivial in single-winner voting and easy for simple assignment settings[3] but can become a challenge for richer settings with more complex constraints (Budish et al. 2013). When instantiating a lottery, an interesting challenge that arises is to instantiate over deterministic outcomes that *also* satisfy some weaker notions of stability, fairness, or other properties (see e.g. Teo et al. 2001; Akbarpour and Nikzad 2017).

To conclude, a probabilistic approach to social choice in particular voting, allocation, matching, and coalition formation leads to several interesting research questions and directions.

**Acknowledgements** The author is supported by a Julius Career Award and a UNSW Scientia Fellowship. He thanks Gabrielle Demange, Jörg Rothe, Nisarg Shah, Paul Stursberg, Etienne Billette De Villemeur, and Bill Zwicker for useful feedback. He thanks all of his collaborators on this topic in particular Florian Brandl, Felix Brandt, and Bettina Klaus for several insightful discussions. Finally he thanks Hervé Moulin and Bill Zwicker for encouraging him to write the chapter.

# References

Aharoni, R., & Fleiner, T. (2003). On a lemma of scarf. *Journal of Combinatorial Theory Series B, 87*, 72–80.

Akbarpour, M., & Nikzad, A. (2017). Approximate random allocation mechanisms. Technical report 422777, SSRN.

Anshelevich, E., & Postl, J. (2016). Randomized social choice functions under metric preferences. In *Proceedings of the 25th International Joint Conference on Artificial Intelligence (IJCAI)* (pp. 46–59). AAAI Press.

Anshelevich, E., Bhardwaj, O., & Postl, J. (2015). Approximating optimal social choice under metric preferences. In *Proceedings of the 29th AAAI Conference on Artificial Intelligence (AAAI)* (pp. 777–783).

Athanassoglou, S., & Sethuraman, J. (2011). House allocation with fractional endowments. *International Journal of Game Theory, 40*(3), 481–513.

Aziz, H., & Klaus, B. (2019). Random matching under priorities: stability and no envy concepts. Social Choice and Welfare.

Aziz, H., & Stursberg, P. (2014). A generalization of probabilistic serial to randomized social choice. In *Proceedings of the 28th AAAI Conference on Artificial Intelligence (AAAI)* (pp. 559–565). AAAI Press.

Aziz, H., Brandt, F., & Brill, M. (2013). On the tradeoff between economic efficiency and strategy proofness in randomized social choice. In *Proceedings of the 12th International Conference on Autonomous Agents and Multi-agent Systems (AAMAS)* (pp. 455–462). IFAAMAS.

Aziz, H., Brandt, F., Brill, M., & Mestre, J. (2014). Computational aspects of random serial dictatorship. *ACM SIGecom Exchanges, 13*(2), 26–30.

---

[3] Any fractional bipartite matching can be represented as a convex combination of discrete matchings via Birkhoff's algorithm.

Aziz, H., Bogomolnaia, A., & Moulin, H. (2017). Fair mixing: the case of dichotomous preferences. Working paper.

Aziz, H., Luo, P., & Rizkallah, C. (2018). Rank maximal equal contribution: A probabilistic social choice function. In *Proceedings of the 32nd AAAI Conference on Artificial Intelligence (AAAI)* (pp. 910–916).

Bogomolnaia, A., & Moulin, H. (2001). A new solution to the random assignment problem. *Journal of Economic Theory, 100*(2), 295–328.

Bogomolnaia, A., & Moulin, H. (2004). Random matching under dichotomous preferences. *Econometrica, 72*(1), 257–279.

Brandl, F. (2013). Efficiency and incentives in randomized social choice. Master's thesis, Technische Universität München.

Brandl, F., Brandt, F., & Hofbauer, J. (2015). Incentives for participation and abstention in probabilistic social choice. In *Proceedings of the 14th International Conference on Autonomous Agents and Multi-agent Systems (AAMAS)* (pp. 1411–1419). IFAAMAS.

Brandl, F., Brandt, F., & Seedig, H. G. (2016). Consistent probabilistic social choice. *Econometrica, 84*(5), 1839–1880.

Brandt, F. (2017). Rolling the dice: Recent results in probabilistic social choice. In U. Endriss (Ed.), *Trends in computational social choice* (chap. 1, pp. 3–26). AI Access.

Brandt, F. (2019). Collective choice lotteries: Dealing with randomization in economic design. In J.-F. Laslier, H. Moulin, R. Sanver, & W. S. Zwicker (Eds.), *The future of economic design*. Springer-Verlag. Forthcoming.

Budish, E., Che, Y.-K., Kojima, F., & Milgrom, P. (2013). Designing random allocation mechanisms: Theory and applications. *American Economic Review, 103*(2), 585–623.

Cabannes, Y. (2004). Participatory budgeting: A significant contribution to participatory democracy. *Environment and Urbanization, 16*(1), 27–46.

Chen, Y., Lai, J. K., Parkes, D. C., & Procaccia, A. D. (2013). Truth, justice, and cake cutting. *Games and Economic Behavior, 77*(1), 284–297.

Cho, W. J. (2016). Incentive properties for ordinal mechanisms. *Games and Economic Behavior, 95*, 168–177.

Dogan, B., & Yildiz, K. (2016). Efficiency and stability of probabilistic assignments in marriage problems. *Games and Economic Behavior, 95*, 47–58.

Fain, B., Goel, A., & Munagala, K. (2016). The core of the participatory budgeting problem. In *Web and Internet Economics - 12th International Conference, WINE 2016, Montreal, Canada, Proceedings* (pp. 384–399). 11–14 Dec 2016.

Kesten, O., & Unver, U. (2015). A theory of school choice lotteries. *Theoretical Economics*, 543–595.

Manjunath, V. (2016). Fractional matching markets. *Games and Economic Behavior, 100*, 321–336.

Manlove, D. F. (2013). *Algorithmics of matching under preferences*. Hackensack: World Scientific Publishing Company.

Procaccia, A. D. (2010). Can approximation circumvent Gibbard-Satterthwaite? In *Proceedings of the 24th AAAI Conference on Artificial Intelligence (AAAI)* (pp 836–841). AAAI Press.

Procaccia, A. D., & Tennenholtz, M. (2013). Approximate mechanism design without money. *ACM Transactions on Economics and Computation, 1*(4), 1–26.

Stone, P. (2011). *The luck of the draw: The role of lotteries in decision making*. Oxford: Oxford University Press.

Teo, C.-P., Sethuraman, J., & Tan, W.-P. (2001). Gale-shapley stable marriage problem revisited: Strategic issues and applications. *Management Science, 49*(9), 1252–1267.

Young, H. P. (1994). *Equity: In theory and practice*. Princeton: Princeton University Press.

# Collective Choice Lotteries

## Dealing with Randomization in Economic Design

**Felix Brandt**

**Abstract** Randomization is playing an ever increasing role in economic design with examples ranging from fair allocation to matching markets to voting. I propose and briefly discuss three interdisciplinary and interrelated research questions that deserve further attention: (i) when are collective choice lotteries acceptable, (ii) how do agents compare lotteries, and (iii) how can randomized rules be implemented.

## 1 Introduction

Economic design has seen the emergence of a number of attractive randomized rules for allocation, matching, and even voting. In this essay, I would like to raise three fundamental questions in the context of randomization in economic design, which I think have not been sufficiently addressed in the literature.

I will take the perspective of social choice theory and refer to randomized collective choice rules, which map a collection of individual preference relations to a socially most-preferred, representative, or otherwise adequate lottery over the alternatives. However, since alternatives may represent allocations (with or without payments), matchings, coalition structures, or any other type of economic outcome, the following observations should be equally relevant for mechanism design, market design, auction theory, matching markets, random assignment, fair division, and so forth.

When aggregating the preferences of multiple agents into a single collective choice, it is easily seen that certain cases call for randomization or other means of tie-breaking. For example, if there are two alternatives, $a$ and $b$, and two agents such that one prefers $a$ and the other one $b$, there is no deterministic way of selecting a single alternative without violating basic fairness conditions (referred to as *anonymity* and *neutrality*). However, *ex ante* fairness can easily be restored by returning an even chance lottery over $a$ and $b$. When allowing for more randomization than is necessary to break ties, classic impossibilities such as *Arrow's theorem*,

F. Brandt (✉)
Institut für Informatik, Technische Universität München, Munich, Germany
e-mail: brandtf@in.tum.de

© Springer Nature Switzerland AG 2019

J.-F. Laslier et al. (eds.), *The Future of Economic Design*, Studies in
Economic Design, https://doi.org/10.1007/978-3-030-18050-8_9

the *Gibbard-Satterthwaite theorem*, the *no-show paradox*, or the *incompatibility of Condorcet-consistency with population-consistency* can be circumvented under suitable assumptions about the agents' preferences over lotteries (e.g., Gibbard 1977; Brandl et al. 2016, 2019; Brandl and Brandt 2019). This is reminiscent of game theory, where the availability of randomized strategies is a crucial requirement of fundamental results like the minimax theorem and the Nash equilibrium existence theorem. In the context of voting, two classes of rules that turned out to be particularly desirable from an axiomatic perspective are *random (serial) dictatorships* and rules that return *maximal lotteries* (whose existence follows from the minimax theorem) (see, e.g., Fishburn 1984a; Brandl et al. 2016; Aziz et al. 2018). These rules have also been considered in matching and allocation subdomains of the general voting domain where random serial dictatorship is known as *random priority* and maximal lotteries as *mixed popular matchings* or *popular random assignments* (see, e.g., Bogomolnaia and Moulin 2001; Brandt et al. 2017). Further interesting possibilities emerge in the voting domain when assuming that the agents' preferences adhere to certain structural restrictions (such as single-peakedness or dichotomousness) (e.g., Ehlers et al. 2002; Bogomolnaia et al. 2005).

Curiously, the use of lotteries for the selection of officials goes back to the world's first democracy in Athens, where it was widely regarded as a principal characteristic of democracy (Headlam 1933), and has recently gained increasing attention in political science (see, e.g., Goodwin 2005; Dowlen 2009; Stone 2011; Guerrero 2014).[1] Randomization has also turned out to be a valuable tool to achieve fairness in matching markets, in particular when individual preferences may contain ties. Bogomolnaia and Moulin (2004) have identified attractive randomized matching rules for the important case of dichotomous preferences. Randomization is perhaps most common when assigning objects to agents. When objects are *indivisible*, it is impossible to deterministically assign objects such that agents with identical preferences receive the same objects (*equal treatment of equals*). This problem is usually avoided by randomization, i.e., by assigning lotteries over objects to the agents. Randomization is particularly natural in the unit-demand (aka house allocation) case, where each agents receive at most one object, because it is not possible to compensate agents via bundles of objects. Besides random priority, the *probabilistic serial rule* by Bogomolnaia and Moulin (2001) has gathered a lot of interest. Even when objects are *divisible* (such as in *cake cutting* problems), randomization has been exploited to achieve *ex ante* fairness and strategyproofness for piecewise valuation functions (e.g., Chen et al. 2013). In settings that involve the approximation of some global measure, such as social welfare in combinatorial auctions, it is well-known that randomized rules can outperform deterministic ones (see, e.g., Nisan and Ronen 2001; Dobzinski and Dughmi 2013; Fischer and Klimm 2015).[2]

---

[1]For further philosophical discussions of using lotteries to achieve fairness, see, for example, Broome (1991); Nissan-Rozen (2012).

[2]For a related discussion of the benefits and challenges of randomization, see Haris Aziz's piece in this volume (Aziz 2019).

In summary, randomized collective choice rules have emerged in various areas of economic design. At the same time, a number of pressing interdisciplinary research questions remain unresolved.

(i)  When are collective choice lotteries acceptable?
(ii)  How do agents compare lotteries?
(iii)  How can randomized rules be implemented?

In the remainder of this essay, I will comment on these questions.

**When are collective choice lotteries acceptable?** Whether randomization is inadmissible, acceptable, or even desirable obviously depends on the application. While electing a political leader via lottery would be controversial and perhaps considered by some a failure of deliberative democracy, randomly selecting an employee of the day, a restaurant to go to, or background music for a party seems quite natural. Important factors in this context are how frequently collective choice rules are executed and how much randomization is entailed by the rule. The interplay between these properties can be explained by risk aversion on behalf of the agents. For example, most people would probably be more content with randomization for daily collective decisions than for annual ones.

Interestingly, humans appear to have less reservations against randomization in matching and allocation than in voting. In fact, randomized voting rules are rarely used in the real world while randomized matching and allocation rules are widely applied. This is partly due to the difference of the *public good* nature of voting versus the *private good* nature of matching and allocation. In private good settings, extensive randomization is often accepted in order to satisfy fairness conditions such as *envy-freeness* (which have no equivalent in the public good setting). There also seem to be cultural and psychological factors influencing an agent's stance towards lotteries, which are largely unexplored.

To the best of my knowledge, there is no formal analysis of the *degree of randomization* of collective choice rules. While randomization is likely more acceptable if lotteries are only invoked in rare cases to break otherwise unresolvable ties, the degree of randomization can also be considered for a single lottery. Intuitively, an even chance lottery over two alternatives entails more uncertainty than a lottery with almost all probability on the first alternative and negligible probability on the second alternative. Suitable metrics for the degree of randomization may include Shannon entropy, the distance to the nearest degenerate lottery, the variance of the lottery, or just the size of the support.

**How do agents compare lotteries?** Central concepts such as Pareto efficiency (no agent can be made better off without making another one worse off) or strategyproofness (no agent can obtain a more preferred outcome by misrepresenting his preferences) can only be meaningfully defined by making reference to the agents' preferences over lotteries. These preferences are typically defined by assuming the existence of a *von Neumann-Morgenstern (vNM) utility function* which assigns cardinal utility values to alternatives and asserting that a lottery is preferred to another lottery if the former yields more expected utility than the latter.

There are at least two problems with this approach. First, various experimental studies have shown that vNM utility theory is systematically violated by human decision makers. An alternative model that, in my view, is much under-appreciated is Fishburn's *skew-symmetric bilinear utility (SSB)* theory, a significant generalization of vNM utility theory which assigns a utility value to each *pair* of alternatives and dispenses with the controversial independence and transitivity axioms (see, e.g., Fishburn 1984b, 1988; Aziz et al. 2015; Brandl and Brandt 2019). The second problem is that asking agents to submit their vNM utility functions—or, equivalently, their complete preferences over lotteries—is usually impractical. I would even argue that most agents are not even aware of these preferences in the first place. (Even if agents *think* they can competently assign vNM utilities to alternatives, these assignments are prone to be based on arbitrary choices because of missing information and the inability to fully grasp the consequences of these choices.) One approach to bypass this problem is to systematically extend the agents' preferences over alternatives to (possibly incomplete) preferences over lotteries via so-called *lottery extensions*.[3] The most influential preference extension is *first-order stochastic dominance*, which is obtained by quantifying over all consistent vNM utility functions. However, there are also other sensible ways to extend preferences to lotteries. For example, by quantifying over all consistent SSB utility function, one obtains the *bilinear dominance* extension, which is coarser than stochastic dominance and therefore leads to weaker notions of efficiency and strategyproofness (Aziz et al. 2015, 2018). Another very intuitive, but little studied, lottery extension is given by postulating that lottery $p$ is preferred to lottery $q$ if and only if $p$ is more likely to return a better alternative than $q$ (see, e.g., Brandl and Brandt 2019). This extension corresponds to the canonical SSB utility representation consistent with the ordinal preference over alternatives and has been supported by recent experimental evidence (Butler et al. 2018).

**How can randomized rules be implemented?** An often neglected problem in collective choice, especially with more sophisticated rules, is that agents need to be convinced that their preferences were taken into account properly and that the outcome was computed correctly. These concerns are of particular importance for *randomized* rules because the randomization itself needs to be implemented in a verifiable way. This issue can for example be tackled using distributed cryptographic protocols that allow agents to jointly generate random numbers and to verify the correctness of the rule's outcome (including the randomization) (see, e.g., Brandt and Sandholm 2005).

As an alternative to the use of intricate protocols from cryptography, physical, publicly verifiable randomization procedures remain of great interest. This can be based on the simple observation that, even in 2019, most randomization procedures of public interest such as drawing lottery numbers on live television or deciding kickoffs and penalty kick orders in soccer matches are still realized via simple physical devices such as urns, dice, or coins. In a similar vein, random priority is a popular allocation rule because it is associated with a natural allocation procedure: each of the

---

[3]For other approaches to come up with coherent collective decisions when individual decision-makers violate traditional decision-theoretic assumptions, see Regenwetter et al. (2009) and Danan et al. (2016).

agents is asked for his most preferred of the remaining objects in random order. Apart from its simplicity, the random priority procedure has the advantage of only asking agents to explore those parts of their preferences that are required to compute the outcome. Curiously, this procedure runs in polynomial time even though computing the probabilities resulting from random priority is #P-complete (Aziz et al. 2013; Saban and Sethuraman 2015). An interesting question is whether similar procedures (for example, adaptive urn processes) exist for other randomized rules such as the probabilistic serial rule.

**Acknowledgements** This material is based on work supported by the Deutsche Forschungsgemeinschaft under grant BR 2312/12-1. I am grateful to Florian Brandl, Ashley Piggins, Marcus Pivato, and Bill Zwicker for helpful comments.

# References

Aziz, H. (2019). A probabilistic approach to voting, allocation, matching, and coalition formation. In J.-F. Laslier, H. Moulin, R. Sanver, & W. S. Zwicker (Eds.), *The Future of Economic Design*. Berlin: Springer.

Aziz, H., Brandt, F., & Brill, M. (2013). The computational complexity of random serial dictatorship. *Economics Letters, 121*(3), 341–345.

Aziz, H., Brandl, F., & Brandt, F. (2015). Universal Pareto dominance and welfare for plausible utility functions. *Journal of Mathematical Economics, 60*, 123–133.

Aziz, H., Brandl, F., Brandt, F., & Brill, M. (2018). On the tradeoff between efficiency and strategyproofness. *Games and Economic Behavior, 110*, 1–18.

Bogomolnaia, A., & Moulin, H. (2001). A new solution to the random assignment problem. *Journal of Economic Theory, 100*(2), 295–328.

Bogomolnaia, A., & Moulin, H. (2004). Random matching under dichotomous preferences. *Econometrica, 72*(1), 257–279.

Bogomolnaia, A., Moulin, H., & Stong, R. (2005). Collective choice under dichotomous preferences. *Journal of Economic Theory, 122*(2), 165–184.

Brandl, F., & Brandt, F. (2019). Arrovian aggregation of convex preferences. Working paper.

Brandl, F., Brandt, F., & Hofbauer, J. (2019). Welfare maximization entices participation. *Games and Economic Behavior. 14*, 308–314.

Brandl, F., Brandt, F., & Seedig, H. G. (2016). Consistent probabilistic social choice. *Econometrica, 84*(5), 1839–1880.

Brandt, F., & Sandholm, T. (2005). On correctness and privacy in distributed mechanisms. *Revised selected papers from the 7th AAMAS Workshop on Agent-Mediated Electronic Commerce (AMEC)* (Vol. 3937, pp. 212–225)., Lecture Notes in Artificial Intelligence (LNAI).

Brandt, F., Hofbauer, J., & Suderland, M. (2017). Majority graphs of assignment problems and properties of popular random assignments. *Proceedings of the 16th International Conference on Autonomous Agents and Multiagent Systems (AAMAS)* (pp. 335–343). IFAAMAS.

Broome, J. (1991). Fairness. *Proceedings of the Aristotelian Society, New Series, 91*, 81–101.

Butler, D., & Pogrebna, G. (2018). Predictably intransitive preferences, *Judgment and Decision Making. 13*(3), 217–236.

Chen, Y., Lai, J. K., Parkes, D. C., & Procaccia, A. D. (2013). Truth, justice, and cake cutting. *Games and Economic Behavior, 77*(1), 284–297.

Danan, E., Gajdos, T., Hill, B., & Tallon, J.-M. (2016). Robust social decisions. *American Economic Review, 106*(9), 2407–2425.

Dobzinski, S., & Dughmi, S. (2013). On the power of randomization in algorithmic mechanism design. *SIAM Journal on Computing*, *42*(6), 2287–2304.

Dowlen, O. (2009). Sorting out sortition: A perspective on the random selection of political officers. *Political Studies*, *57*(2), 298–315.

Ehlers, L., Peters, H., & Storcken, T. (2002). Strategy-proof probabilistic decision schemes for one-dimensional single-peaked preferences. *Journal of Economic Theory*, *105*(2), 408–434.

Fischer, F., & Klimm, M. (2015). Optimal impartial selection. *SIAM Journal on Computing*, *44*(5), 1263–1285.

Fishburn, P. C. (1988). *Nonlinear preference and utility theory*. The Johns Hopkins University Press.

Fishburn, P. C. (1984a). Probabilistic social choice based on simple voting comparisons. *Review of Economic Studies*, *51*(4), 683–692.

Fishburn, P. C. (1984b). SSB utility theory: An economic perspective. *Mathematical Social Sciences*, *8*(1), 63–94.

Gibbard, A. (1977). Manipulation of schemes that mix voting with chance. *Econometrica*, *45*(3), 665–681.

Goodwin, B. (2005). *Justice by lottery* (1st ed. 1992). Chicago: University of Chicago Press.

Guerrero, A. (2014). Against elections: The lottocratic alternative. *Philosophy and Public Affairs*, *42*(2), 135–178.

Headlam, J. W. (1933). *Election by Lot at Athens*. Cambridge: Cambridge University Press.

Nisan, N., & Ronen, A. (2001). Algorithmic mechanism design. *Games and Economic Behavior*, *35*(1), 166–196.

Nissan-Rozen, I. (2012). Doing the best one can: A new justification for the use of lotteries. *Erasmus Journal for Philosophy and Economics*, *5*(1), 45–72.

Regenwetter, M., Grofman, B., Popova, A., Messner, W., Davis-Stober, C. P., & Cavagnaro, D. R. (2009). Behavioural social choice: a status report. *Philosophical Transactions of the Royal Society*, *364*(1518), 833–843.

Saban, D., & Sethuraman, J. (2015). The complexity of computing the random priority allocation matrix. *Mathematics of Operations Research*, *40*(4), 1005–1014.

Stone, P. (2011). *The luck of the draw: The role of lotteries in decision making*. Oxford: Oxford University Press.

# Part II
# Aggregation and Voting

# Interactive Democracy: New Challenges for Social Choice Theory

**Markus Brill**

**Abstract** *Interactive Democracy* (aka *e-democracy* or *digital democracy*) is an umbrella term that encompasses a variety of approaches to make collective decision making processes more engaging and responsive. A common goal of these approaches is to utilize modern information technology—in particular, the Internet—in order to enable more interactive decision making processes. An integral part of many interactive democracy proposals are online decision platforms that provide much more flexibility and interaction possibilities than traditional democratic systems. This is achieved by embracing the novel paradigm of delegative voting, often referred to as *liquid democracy*, which aims to reconcile the idealistic appeal of direct democracy with the practicality of representative democracy. The successful design of interactive democracy systems presents a multidisciplinary research challenge; one important aspect concerns the elicitation and aggregation of *preferences*. In this article, I argue that the emergence of online decision platforms and other interactive democracy systems leads to new challenges for social choice theory.

## 1 Introduction

In her 2014 TED talk *How to upgrade democracy for the Internet era*, Pia Mancini poignantly states that "we are 21st-century citizens, doing our very, very best to interact with 19th century-designed institutions that are based on an information technology of the 15th century" (Mancini 2014). Mancini goes on to observe that the way democratic societies make collective decisions is highly outdated. This leads to the question: "If Internet is the new printing press, then what is democracy for the Internet era?" Mancini and her collaborators approach this question by developing an app, *DemocracyOS* (Mancini 2015), that allows users to propose, debate, and

---

An extended version of this article has appeared in the proceedings of the 17th International Conference on Autonomous Agents and Multiagent Systems (Brill 2018).

M. Brill (✉)
TU Berlin, Berlin, Germany
e-mail: brill@tu-berlin.de

© Springer Nature Switzerland AG 2019
J.-F. Laslier et al. (eds.), *The Future of Economic Design*, Studies in
Economic Design, https://doi.org/10.1007/978-3-030-18050-8_10

vote on issues. *DemocracyOS* is only one example of a quickly growing number of approaches that aim to reconcile established democratic processes with the desire of citizens to participate in political decision making.[1] Another example is the software *LiquidFeedback* (Behrens et al. 2014), which is developed by the *Association for Interactive Democracy*.[2] Currently, these tools are mainly used for decision making within progressive political parties (Blum and Zuber 2016, p. 162) or in the context of community engagement platforms such as *WeGovNow* (Boella et al. 2018). A common goal of these approaches, often summarized under the umbrella term *Interactive Democracy*[3] (henceforth ID), is to utilize modern information technology—in particular, the Internet—in order to enable more interactive decision making processes.

When designing a platform for interactive collective decision making, there are lots of design decisions to be made, regarding, for example, *issue selection* (which issues are voted on?), *option generation* (which options are on the ballot?), *interaction opportunities* (how is deliberation and delegation organized?), *ballot structure* (in which format can participants express their preferences?), and *aggregation methods* (which method is used to tally the votes?). There is no shortage of concrete suggestions of how ID platforms could be implemented (see Ford 2014 for an overview). Most of these suggestions, however, are rather ad hoc in nature and little attention is devoted to a principled comparison and evaluation of methods. This increases the risk of employing methods with unintended flaws.

In this article, I argue that insights and tools from the theory of *preference aggregation* (aka *social choice theory*) can be employed to aid the design of online decision platforms and other ID tools. *Computational social choice* (COMSOC), an interdisciplinary subfield at the intersection of economics and computer science, seems to be particularly relevant in this endeavor. Even though research in COMSOC has made tremendous progress in recent years (Brandt et al. 2016; Endriss 2017), the practical impact of the field has remained rather limited.[4] This is partly due to the fact that many of the rather sophisticated preference handling and preference aggregation mechanisms that are routinely studied in the COMSOC literature, though superior in theory, are rarely used in practice. The novel application area of interactive democracy, as described in this article, has the potential to change that.

---

[1] *DemocracyOS* has since been superseded by *Sovereign*, developed by the *Democracy Earth Foundation* (http://www.democracy.earth).

[2] http://www.interaktive-demokratie.org/index.en.html.

[3] The field is lacking a unified terminology. For example, Interactive Democracy is sometimes referred to as *iDemocracy* (Carswell 2012) or *participatory democracy* (Aragonès and Sánchez-Pagés 2009). The terms *liquid democracy* and *delegative democracy* usually refer to the paradigm of delegative voting (see Sect. 2.1). And terms like *e-democracy* (Shapiro 2018), *digital democracy* (Hague and Loader 1999), and *Internet democracy* (Margolis and Moreno-Riaño 2013) emphasize the role of information technology.

[4] Notable exceptions are websites like *Spliddit* (Goldman and Procaccia 2014) and *RoboVote* (http://robovote.org).

## 2   Challenges

In this section, I give three examples of how the emergence of ID systems gives rise to novel challenges for (computational) social choice.

### 2.1   Delegative Voting

Participants of online decision platforms (henceforth called *voters*) can often choose whether they want to vote directly on a particular issue or whether they want to *delegate* their vote to somebody they trust. Delegations can be specified either on an issue-by-issue basis, for whole topic areas, or even globally. Crucially, delegations are transitive and decisions whether to vote directly, to delegate, or to abstain can be changed at any time.[5] This paradigm of delegative voting, which is often referred to as *liquid democracy*, aims to reconcile the idealistic appeal of *direct democracy* (where every voter votes directly on every issue) with the practicality of *representative democracy* (where voters vote for delegates, who then vote on the voters' behalf on all issues) by giving voters the opportunity to have their say on all issues, but not requiring them to get informed on each issue.[6]

Delegative voting gives rise to several novel questions in voting theory. For instance, in order to successfully implement a delegative voting infrastructure, one has to think about potential problems such as *delegation cycles* (what if voter 1 delegates her vote to voter 2, who delegates to voter 3, who delegates back to voter 1?) and *abstentions* (what if a voter delegates to somebody who abstains from voting?). Moreover, it has been observed by Blum and Zuber that the flexibility of the delegative voting paradigm, which allows voters to delegate their vote to different representatives depending on the area of the issue, can lead to outcomes that are not consistent on a global level (Blum and Zuber 2016, pp. 178–179). Christoff and Grossi (2017), who have studied this problem in the formal context of binary aggregation, suggest to employ techniques from opinion diffusion (DeGroot 1974; Grandi 2017) in order to resolve such inconsistencies. (They also propose a way to resolve delegation cycles and abstentions, by requiring voters to specify "default" votes that are used to overrule delegation decisions should the latter lead to cycles or abstaining voters.) And Brill and Talmon (2018), who have studied the special case in which each pairwise comparison between two alternatives can be delegated to a different voter, suggest to employ the framework of distance rationalization (Elkind and Slinko 2016).

---

[5]For details, see the articles by Ford (2002), Green-Armytage (2015), and Blum and Zuber (2016). Some of the ideas behind delegative voting can be traced back to the works of Dodgson (1884), Tullock (1967), and Miller (1969). For an historical overview of ideas, see the surveys by Ford (2014) and Behrens (2017).

[6]The question whether the delegative voting paradigm actually leads to "superior" voting outcomes (as compared to direct and representative democracy) has been addressed from a variety of perspectives (Alger 2006; Green-Armytage 2015; Cohensius et al. 2017; Kahng et al. 2018; Gölz et al. 2018; Kling et al. 2015).

In general, there appears to be a tradeoff between flexible and fine-grained delegation possibilities on the one side and increased potential of inconsistent (or underspecified) outcomes on the other side. In this context, it will also be interesting to explore generalizations of delegative voting such as *statement voting* (Zhang and Zhou 2017).

## 2.2  Proportional Representation

A defining feature of ID systems is that all participants are allowed—and encouraged—to contribute to the decision making process, either directly by participating in discussions and voting on issues, or indirectly by delegating their decision power. In particular, in a context where one (or more) out of several competing options needs to be selected, each participant can propose their own option if they are not satisfied with the existing ones. This may lead to situations where a very large number of alternative options needs to be considered. Since it cannot be expected that every participant looks at all available options before making a decision, the *order* in which competing options are presented (e.g., on a website that facilitates the discussion and voting process) plays a crucial role (see Behrens et al. 2014, Chapter 4.10). A natural approach is to order competing options by their "support", i.e., by the number of participants that have expressed their approval of the option in question.

This gives rise to what Behrens et al. (2014) describe as the "noisy minorities" problem: relatively small groups of very active participants can "flood" the system with their contributions, creating the impression that their opinion is much more popular than it actually is. This is problematic insofar as alternative options (that are potentially much more popular) run the risk of being "buried" and not getting sufficient exposure. In order to prevent this problem, the mechanism ordering competing options needs to ensure that the order adequately reflects the opinions of the participants. The search for orderings that are "representative" in this sense leads to challenging algorithmic problems not unlike those underlying the problem of choosing representative committees (see, e.g., Chamberlin and Courant 1983; Monroe 1995; Aziz et al. 2017). In a recent paper, Skowron et al. (2017) approach the problem by formalizing the notion of *proportional rankings* and evaluating the representativeness of common ranking methods. Roughly speaking, a proportional ranking is one in which the number of top positions allocated to options supported by a particular group is proportional to the size of that group.

## 2.3  New Forms of Aggregation

In traditional voting theory, it is usually assumed that there is a finite set of alternatives and the preferences of voters are given as rank-orderings over this set. This framework, while very general in theory, is not always practical. Often, the space of alternatives has some combinatorial structure, and exploiting this structure is necessary for both eliciting and aggregating preferences in a meaningful way (Lang and Xia 2016). Two example scenarios that arise in ID applications are the aggregation of *societal tradeoffs* (Conitzer et al. 2015), where the goal is to aggregate numerical tradeoffs between different kinds of socially undesirable activities and the collection of aggregate tradeoffs is required to be consistent, and *participatory budgeting* (Cabannes 2004), where the goal is to allocate budgetary spending of a local government based on citizens' preferences, and a combinatorial structure of the solution space is imposed by a budget constraint.

For those and related scenarios, the structure of the solution space makes it impractical to elicit preferences directly. Rather, there are various ways in which users can specify their preferences. Each preference format presents a tradeoff between *expressive power* (does the format allow to express fine-grained preferences?), *succinctness* (can preferences be represented compactly?), and *aggregatability* (does the preference format allow for a computationally efficient and axiomatically desirable aggregation mechanism?), among other things.[7] Techniques developed in COMSOC facilitate a much-needed principled comparison of preference formats and aggregation methods. First steps in that direction have recently been made by Goel et al. (2016) and Benade et al. (2017).[8]

# 3   Conclusion

The emergence of citizen participation systems in general—and of online voting platforms in particular—appears to be an irreversible development. The question is not if, but rather when, these systems become standard components of the democratic process. A multidisciplinary research program is necessary for making these systems secure, equitable, inclusive, user-friendly, and computationally reliable. In this article, I have argued that insights and tools from (computational) social choice are relevant for this important endeavor.

It is well known that even a well-intended design often results in voting systems that exhibit unexpected flaws (e.g., see Fishburn 1974). Moreover, once a voting system is established, it is very hard to change it due to its incumbent position. The emergence of interactive democracy thus presents a unique opportunity to influence the future of our democracies for the better. COMSOC is but one piece of the puzzle; albeit an important one. Let us ensure that its voice is added to the conversation.

---

[7]Identifying reasonable formats in which voters can express their preferences has similarities to the search for bidding languages in combinatorial auctions (Nisan 2006).
[8]Other approaches towards participatory budgeting have been proposed by Fain et al. (2016), Shapiro and Talmon (2018), and Aziz et al. (2018), among others.

**Acknowledgements** This material is based upon work supported by a Feodor Lynen research fellowship of the Alexander von Humboldt Foundation and by the Deutsche Forschungsgemeinschaft under grant BR 4744/2-1. The author would like to thank Dorothea Baumeister, Jan Behrens, Steven Brams, Vincent Conitzer, Paul Gölz, Umberto Grandi, Davide Grossi, Michel Le Breton, Rolf Niedermeier, Dominik Peters, Marcus Pivato, Ariel Procaccia, Ehud Shapiro, Nimrod Talmon, and Bill Zwicker for helpful comments.

# References

Alger, D. (2006). Voting by proxy. *Public Choice, 126*(1–2), 1–26.

Aragonès, E., & Sánchez-Pagés, S. (2009). A theory of participatory democracy based on the real case of Porto Alegre. *European Economic Review, 53*(1), 56–72.

Aziz, H., Lee, B. E., & Talmon, N. (2018). Proportionally representative participatory budgeting: Axioms and algorithms. *Proceedings of the 17th International Conference on Autonomous Agents and Multiagent Systems (AAMAS)* (pp. 23–31). IFAAMAS.

Aziz, H., Brill, M., Conitzer, V., Elkind, E., Freeman, R., & Walsh, T. (2017). Justified representation in approval-based committee voting. *Social Choice and Welfare, 48*(2), 461–485.

Behrens, J. (2017). The origins of liquid democracy. *The Liquid Democracy Journal, 5*, 7–17.

Behrens, J., Kistner, A., Nitsche, A., & Swierczek, B. (2014). *The Principles of LiquidFeedback*.

Benade, G., Nath, S., Procaccia, A. D., & Shah, N. (2017). Preference elicitation for participatory budgeting. *Proceedings of the 31st AAAI Conference on Artificial Intelligence (AAAI)* (pp. 376–382). AAAI Press.

Blum, C., & Zuber, C. I. (2016). Liquid democracy: Potentials, problems, and perspectives. *Journal of Political Philosophy, 24*(2), 162–182.

Boella, G., Francis, L., Grassi, E., Kistner, A., Nitsche, A., Noskov, A., et al. (2018). WeGovNow: A map based platform to engage the local civic society. *WWW 2018: Companion Proceedings of the The Web Conference 2018* (pp. 1215–1219). International World Wide Web Conferences Steering Committee.

Brandt, F., Conitzer, V., Endriss, U., Lang, J., & Procaccia, A. (Eds.). (2016). *Handbook of computational social choice*. Cambridge: Cambridge University Press.

Brill, M. (2018). Interactive democracy. *Proceedings of the 17th International Conference on Autonomous Agents and Multiagent Systems (AAMAS) Blue Sky Ideas track* (pp. 1183–1187). IFAAMAS.

Brill, M., & Talmon, N. (2018). Pairwise liquid democracy. *Proceedings of the 27th International Joint Conference on Artificial Intelligence (IJCAI)* (pp. 137–143). IJCAI.

Cabannes, Y. (2004). Participatory budgeting: A significant contribution to participatory democracy. *Environment and Urbanization, 16*(1), 27–46.

Carswell, D. (2012). *The end of politics and the birth of iDemocracy*. Biteback Publishing.

Chamberlin, J. R., & Courant, P. N. (1983). Representative deliberations and representative decisions: Proportional representation and the Borda rule. *The American Political Science Review, 77*(3), 718–733.

Christoff, Z., & Grossi, D. (2017). Binary voting with delegable proxy: An analysis of liquid democracy. *Proceedings of the 16th Conference on Theoretical Aspects of Rationality and Knowledge (TARK)* (pp. 134–150).

Cohensius, G., Mannor, S., Meir, R., Meirom, E., & Orda, A. (2017). Proxy voting for better outcomes. *Proceedings of the 16th International Conference on Autonomous Agents and Multiagent Systems (AAMAS)* (pp. 858–866). IFAAMAS.

Conitzer, V., Brill, M., & Freeman, R. (2015). Crowdsourcing societal tradeoffs. *Proceedings of the 14th International Conference on Autonomous Agents and Multiagent Systems (AAMAS) Blue Sky Ideas Track* (pp. 1213–1217). IFAAMAS.

DeGroot, M. H. (1974). Reaching a consensus. *Journal of the American Statistical Association*, *69*(345), 118–121.

Dodgson, C. L. (1884). *The principles of parliamentary representation*. Harrison and Sons.

Elkind, E., & Slinko, A. (2016). Rationalizations of voting rules. In F. Brandt, V. Conitzer, U. Endriss, J. Lang, & A. D. Procaccia (Eds.), *Handbook of Computational Social Choice* (pp. 169–196). Cambridge: Cambridge University Press. Chap. 8.

Endriss, U. (Ed.). (2017). *Trends in computational social choice*. AI Access.

Fain, B., Goel, A., & Munagala, K. (2016). The core of the participatory budgeting problem. *Proceedings of the 12th International Workshop on Internet and Network Economics (WINE)* (pp. 384–399)., Lecture Notes in Computer Science (LNCS).

Fishburn, P. C. (1974). Paradoxes of voting. *The American Political Science Review*, *68*(2), 537–546.

Ford, B. (2002). Delegative democracy. Unpublished manuscript. http://www.brynosaurus.com/deleg/deleg.pdf.

Ford, B. (2014). Delegative democracy revisited. Blog post. http://bford.github.io/2014/11/16/deleg.html.

Goel, A., Krishnaswamy, A. K., Sakshuwong, S., & Aitamurto, T. (2016). Knapsack voting: Voting mechanisms for participatory budgeting. Unpublished manuscript.

Goldman, J., & Procaccia, A. D. (2014). Spliddit: Unleashing fair division algorithms. *SIGecom Exchanges*, *13*(2), 41–46.

Gölz, P., Kahng, A., Mackenzie, S., & Procaccia, A. D. (2018). The fluid mechanics of liquid democracy. In *Proceedings of the 14th International Workshop on Internet and Network Economics (WINE)* (Vol. 11316, pp. 188–202), Lecture Notes in Computer Science (LNCS). Springer.

Grandi, U. (2017). Social choice and social networks. In U. Endriss (Ed.), *Trends in Computational Social Choice* (pp. 169–184). AI Access. Chap. 9.

Green-Armytage, J. (2015). Direct voting and proxy voting. *Constitutional Political Economy*, *26*(2), 190–220.

Hague, B. N., & Loader, B. D. (Eds.). (1999). *Digital democracy: discourse and decision making in the information age*. Routledge.

Kahng, A., Mackenzie, S., & Procaccia, A. D. (2018). Liquid democracy: An algorithmic perspective. *Proceedings of the 32nd AAAI Conference on Artificial Intelligence (AAAI)* (pp. 1095–1102). AAAI Press.

Kling, C. C., Kunegis, J., Hartmann, H., Strohmaier, M., & Staab, S. (2015). Voting behaviour and power in online democracy: A study of LiquidFeedback in Germany's Pirate Party. *Proceedings of the 9th International AAAI Conference on Web and Social Media (ICWSM)* (pp. 208–217). AAAI Press.

Lang, J., & Xia, L. (2016). Voting in combinatorial domains. In F. Brandt, V. Conitzer, U. Endriss, J. Lang, & A. D. Procaccia (Eds.), *Handbook of Computational Social Choice*. Cambridge: Cambridge University Press. Chap. 9.

Mancini, P. (2014). How to upgrade democracy for the internet era [video file]. Retrieved from http://www.ted.com/talks/pia_mancini_how_to_upgrade_democracy_for_the_internet_era.

Mancini, P. (2015). Why it is time to redesign our political system. *European View*, *14*(1), 69–75.

Margolis, M., & Moreno-Riaño, G. (2013). *The prospect of internet democracy*. Routledge.

Miller, J. C. (1969). A program for direct and proxy voting in the legislative process. *Public Choice*, *7*(1), 107–113.

Monroe, B. L. (1995). Fully proportional representation. *The American Political Science Review*, *89*(4), 925–940.

Nisan, N. (2006). Bidding languages for combinatorial auctions. In P. Cramton, Y. Shoham, & R. Steinberg (Eds.), *Combinatorial Auctions*. MIT Press.

Shapiro, E., & Talmon, N. (2018). A participatory democratic budgeting algorithm. Technical report. http://arxiv.org/abs/1709.05839v6.

Shapiro, E. (2018). Point: Foundations of e-democracy. *Communications of the ACM*, *61*(8), 31–34.

Skowron, P., Lackner, M., Brill, M., Peters, D., & Elkind, E. (2017). Proportional rankings. *Proceedings of the 26th International Joint Conference on Artificial Intelligence (IJCAI)* (pp. 409–415). IJCAI.

Tullock, G. (1967). *Towards a mathematics of politics*. University of Michigan Press.

Zhang, B., & Zhou, H. (2017). Brief announcement: Statement voting and liquid democracy. *Proceedings of the 36th ACM Symposium on Principles of Distributed Computing (PODC)* (pp. 359–361). ACM.

# New Electoral Systems and Old Referendums

Gabrielle Demange

**Abstract** I discuss the future of social choice theory in the design of electoral systems. Two routes are promising. First, thanks to computing facilities, new voting procedures can be designed. I present two positive recent experiments, and hope some are to come. Second, the well-known referendum, which is being increasingly popular on a variety of situations, needs to be investigated more thoroughly; I discuss some issues and directions for improvement.

## 1 Introduction

With the development of computing facilities, the social choice approach is being used to design and implement rules in a large and growing number of areas such as market design, assignment mechanisms, recommendation systems and rating procedures on Internet, or electoral systems. I focus here on the latter.

Changing an electoral system is presumably more difficult than setting new venues of exchanges or allocation mechanisms. Though, promising experiments allowing for a broader and fairer voters' expression than the current rules have been successful. I describe two of them in Sect. 2. At the same time, I have been struck—as many others-by recent voting events such as the EU referendum in UK, 2016 (the so-called Brexit). In Sect. 3, I first discuss the UK process in light of social choice theory, the surveys and statistical analysis that have been conducted after the vote, and then raise some issues that could be investigated to improve the use of referendums.

## 2 Changing the Electoral System: Two Experiments

A set of experiments has been conducted in France to assess the acceptance of new systems for electing the French President. The current system is a two-round vote. According to polls, several times, a Condorcet winner has been eliminated in the first

G. Demange (✉)
Paris School of Economics-EHESS, 48 boulevard Jourdan, 75014 Paris, France
e-mail: demange@pse.ens.fr

© Springer Nature Switzerland AG 2019                                                      67
J.-F. Laslier et al. (eds.), *The Future of Economic Design*, Studies in
Economic Design, https://doi.org/10.1007/978-3-030-18050-8_11

round. There is a call for changing the system to remedy this drawback as well as to allow for a more nuanced and broader voters' expression. In particular, experiments have been conducted to test how voters understand and accept new voting rules. The first experiments tested approval voting in 2002 'in situ', i.e., in some voting posts (see the account in Laslier and Van der Straeten 2008). In the selected voting posts, after receiving some explanation on the method, voters could vote under approval voting if they wanted to. These votes had no influence on the electoral outcome as voters also cast their vote to the 'real' election. The test was successful in the sense that the participation was large. Other rules—Borda scores and majority judgment—have been tested recently, taking advantage of Internet.[1] These tests are encouraging, as they show an interest from the population and the media.

The second experiment, conducted in Switzerland, is a real one. During a trial period starting in 2004, a new method—called the New Apportionment Procedure (NAP)—has been used in the Zurich canton to allocate seats to parties and districts. The method is now definitely adopted in the Zurich canton as well as in some others. The main motive for changing the system was to offer a solution to the problem of 'lost ballots', which was highlighted in 2002 by the Swiss Federal Court. In small districts, the low number of seats did not give any chance to small parties to get a seat. As a result, some supporters of small parties complained their votes were lost and that they were not treated on an equal footing with supporters of the same parties but in a larger district in which the number of seats at stake enabled their representation.[2]

The problem of lost ballots arises in a large number of elections for representatives of distinct areas in an assembly. Such elections entail two dimensions, a geographical one linked to the areas (districts in Switzerland or countries in EU) and a political one linked to parties; a representation is called a bi-apportionment to refer to these two dimensions. In many instances, the allocation of seats to areas is pre-determined. When the sizes of the areas widely differ, as is the case for districts in Switzerland or countries in EU, small areas get fewer seats than the large ones, even though the allocation of seats to areas is distorted in their favor, meaning that the ratios of the number of assigned seats to population size decrease with the size. As a result, not only there are lost ballots but also the parties' representation is distorted: as soon as the votes across areas are not proportional between each other, the distortion in the geographical allocation of seats induces a distortion in the political representation. In particular, a proportional representation of the parties is far from being feasible when the votes are counted per area.[3] In some countries, electoral rules try to correct the distortion by introducing cumbersome features such as a variable number of seats, resulting in 'bugs' as in Italy and Germany.

---

[1]For experiments conducted in the 2017 election, see 'Voter autrement' https://vote.imag.fr/ and http://www.jugementmajoritaire2017.com/.

[2]Of course, some ballots are necessarily lost in an election, those in favor of a looser. It was the 'unfair' treatment of supporters of the same party that triggered the reform.

[3]For a recent account of bi-apportionment problems see Demange (2012); for a review of electoral systems in various countries see Simeone and Pukelsheim (2006).

Such flaws can be solved by a 'bi-divisor' method.[4] The New Apportionment Procedure in Switzerland is based on bi-divisor methods, introduced in Balinski and Demange (1989a, b). At that time, I doubted that the methods would be used for real political elections, mainly because they are not computable by hand. Pukelsheim (2006) made a tremendous job by implementing the procedure and obtaining the support of the politicians and the population, who accepted the NAP through referendum. Roughly speaking, the method works as follows: the votes are adjusted by 'divisors', one for each party and one for each district, and then rounded up to obtain the number of seats of each party in each district. The divisors are necessary because of the distortions in the allocation of seats to districts and (to a lower extent) the indivisibility in the seats. They introduce a link between the votes in different areas, thereby avoiding losing ballots, while keeping the prescribed allocation of seats to districts. Although the rule is not computable by hand, voters can easily check the outcome once they know the divisors. This might explain why the NAP was adopted by referendum in Zurich.

These experiments are promising: voters are open to changes in the voting systems and new rules can be defined with the help of computing facilities. Designing new voting procedures is definitely a promising line of research for social choice theory.

I now turn to one of the oldest rules, the referendum.

# 3  Referendums

The number of referendums has increased in Western Europe (for an analysis of this trend and a classification of referendums, see e.g. Setala 1999). One factor explaining this trend is mechanically induced by proposed changes in EU (enlargement, new constitution), which trigger referendums in many countries. Another factor put forward by political scientists would be related to citizens' preferences, specifically to the 'unfreezing' of alignments, according to which voters no longer 'belong' to a party. Misalignments could explain why voters may not want to delegate all decisions to their representatives and increasingly demand referendums on specific topics (as revealed in surveys, see Donovan and Karp 2006; Bowler et al. 2007).[5]

From the point of view of social choice theory, the referendum setting, in which people choose between two alternatives, works well for aggregating either preferences or dispersed information. When individuals differ only in their preferences, the majority rule is well defined and strategy-proof. When they share the same preferences but are uncertain as to which alternative is preferred, the majority rule is more likely than each single individual to select the correct alternative if votes are sincere (the Condorcet jury theorem). Various works have scrutinized these positive results. The Condorcet jury theorem has been challenged because sincere voting might not

---

[4]The bi-divisor methods could be applied in other contexts than the political one.

[5]These works refer to referendums at a national level. There is also an increasing demand for participatory democracy at the local level.

constitute an equilibrium behavior under private information, resulting in a biased outcome (Austen-Smith and Banks 1996). The argument however relies on rather sophisticated voters; furthermore, it has not much bite when the electorate is large and no player is pivotal.

The reality is naturally often more complex than these two polar cases, mixing heterogeneity in preferences and private information on the alternatives.[6] So what about the EU referendum in UK, 2016?

## 3.1   The EU Referendum in UK, 2016

Many think that something went wrong. Assuming information to have been determinant in the votes and applying Austen-Smith and Banks argument, the result would have been biased if, knowing that Leave had a chance to win, a voter thought that the proponents of Leave had an information of better quality than those of Remain. But detailed analyses show[7] that votes were mainly dictated by education, age and ethnicity, and recent changes in immigration in their area, all of which determine preferences.

Obviously, information, definitely incomplete, turned out to be a crucial issue. The referendum was incredibly ill prepared with an electorate under fundamental uncertainty. The overall consequences are still to be 'interpreted' as stated in 'The six flavours of Brexit', The Economist, 22 July 2017. As some recent analyses indicate,[8] a large majority would prefer a 'soft' Brexit, suggesting that the question was not well formulated. Though, overall, the EU referendum seems to have worked well in the sense that it has correctly aggregated voters' preferences. This is supported by the surveys, which indicate that Leave would have still won a few months after the referendum.[9] Furthermore, most of those who voted Remain accept the result and do not want a second vote.[10]

Has social choice theory something to add, apart from the obvious facts that a referendum should be prepared and the electorate should vote under a clear picture of the consequences? Given the trend towards more referendums (or, more generally, towards participative democracy) there are still open issues worth investigating.

---

[6]Also, it has been known for a long time that voters may not answer to the question but rather signal their mistrust in the current authorities. Though this force may have played a role in the UK referendum, it does not seem to explain the votes, as shown in the references in Footnote 8.

[7]http://www.bbc.com/news/uk-politics-38762034#share-tools, https://www.economist.com/news/britain/21720576-migration-good-economy-so-why-are-places-biggest-influxes-doing-so      and Becker et al. (2017).

[8]http://uk.businessinsider.com/yougov-british-people-have-turned-against-a-hard-brexit-2017-6.

[9]https://yougov.co.uk/news/2017/03/29/attitudes-brexit-everything-we-know-so-far/.

[10]http://whatukthinks.org/eu/questions/if-a-second-eu-referendum-were-held-today-how-would-you-vote/.

## 3.2    How to Revisit Referendums?

How can the use and design of a referendum be improved? By its very definition, a referendum is simple: it applies to a set of voters who face a binary decision, typically the status quo against a reform. Despite its simplicity, a variety of options pertaining to its organization and design have not been much investigated in social choice theory. Here are a few ones.[11]

### 3.2.1    Who Is Allowed to Vote?

Most of the studies start with 'Let N be the set of voters'. The problem of a changing electorate is accounted for in studies in public goods by the 'Voting with their feet' made famous by Tiebout, which reflects a rather extreme form of voting. The design of an electorate has not been much analyzed from a normative point of view (as far as I know). Due to the increased mobility of people, this issue becomes more and more important especially in the decisions involving the splitting or merging of populations.[12] The Scottish independence referendum, 2014, and the vote for independence in Catalyuna, 2014, adopted very different participation rules than the EU referendum in UK, 2016. What rules should determine the electorate of such referendums?

### 3.2.2    Which Information on the Reform Is Available?

To help voters assess properly the issue at stake and to give incentive to the proponents of a reform to be precise, the period before the referendum is crucial. Politicians, standard media or social networks provide information. Information can also been provided by introducing sequential or iterative procedures (keeping in mind that incentive properties are altered). One possibility suggested to me by K. Nehring is to organize a first round bearing on whether a referendum should be organized on a given issue. In the case of UK, if the first round vote on contemplating exiting EU had been positive, this would have forced more discussion and public deliberation on the Brexit program, and presumably a more concrete proposal. Gersbach et al. (2017) propose two-rounds votes, in which a (randomly drawn) sample of voters votes in the first round and the remaining population in the second. They argue that

---

[11]In the case of a reform versus the status quo, a qualified majority can be used instead of a simple one, raising the issue of how to choose the threshold that determines the acceptance of a reform. This issue has been investigated theoretically from different perspectives and is typically tackled in a constitution, contrary to the other issues raised in the text.

[12]A similar issue arises in other contexts than the political ones. On Internet, users rate products, and online surveys compete with controlled polls. What rules should govern these expressions to make them more valuable?

such a procedure may help solving low turnout due to costly participation. It might help raising voters' concerns as well.

### 3.2.3  What Type of Issues Can Be Solved by a Referendum and Who Can Initiate a Referendum?

Are there better systems than referendums to decide on the splitting or merging of areas? Who should decide on initiating such a referendum? The huge discrepancy in the rules governing the use of a referendum across Western countries explains the important differences in their numbers, with Switzerland and Italy standing apart due to the possibility of 'active' referendums initiated by citizens (Setälä 1999). Peter (2016) empirically studies the interaction between the type of referendum and the turnout in representative elections. She shows that the turnout in representative elections tends to increase when citizens can initiate referendums but not when only politicians can initiate them. This suggests investigating thoroughly the rules governing referendums in conjunction with the other elections taking place in a country.

More generally a promising route for social choice theory would be to analyze the choice of the set of voting rules governing a jurisdiction together with their architecture. This would fill an important gap since social choice theory has so far focused on the choice of a single voting rule.

**Acknowledgements**  I thank H. Aziz, H. Gersbach, K. Nehring, H. Nurmi and M. Pivato for their comments. I am most grateful to Julie Godefroy for helpful comments and lively discussions on the Brexit.

## References

Austen-Smith, D., & Banks, J. S. (1996). Information aggregation, rationality, and the Condorcet jury theorem. *American Political Science Review, 90*(1), 34–45.
Balinski, M. L., & Demange, G. (1989a). Algorithms for proportional matrices in reals and integers. *Mathematical Programming, 45*(1–3), 193–210.
Balinski, M. L., & Demange, G. (1989b). An axiomatic approach to proportionality between matrices. *Mathematics of Operations Research, 14*(4), 700–719.
Becker, S. O., Fetzer, T., & Novy, D. (2017). Who voted for Brexit? A comprehensive district-level analysis. *Economic Policy, 32*(92), 601–650.
Bowler, S., Donovan, T., & Karp, J. A. (2007). Enraged or engaged? Preferences for direct citizen participation in affluent democracies. *Political Research Quarterly, 60*(3), 351–362.
Demange, G. (2012). On party-proportional representation under district distortions. *Mathematical Social Sciences, 63*(2), 181–191.
Donovan, T., & Karp, J. A. (2006). Popular support for direct democracy. *Party Politics, 12*(5), 671–688.
Gersbach, H., Mamageishvili, A., & Tejada, O. (2017). Assessment Voting in Large Electorates. WP CER-ETH 17/284.

Laslier, J. F., & Van der Straeten, K. (2008). A live experiment on approval voting. *Experimental Economics, 11*(1), 97–105.

Peters, Y. (2016). Zero-sum democracy? The effects of direct democracy on representative participation. *Political Studies, 64*(3), 593–613.

Pukelsheim, F. (2006). Current issues of apportionment methods. In Simeone, F. Pukelsheim (Eds.), op. cited.

Setälä, M. (1999). Referendums in Western Europe–A wave of direct democracy? *Scandinavian Political Studies, 22*(4), 327–340.

Simeone, & Pukelsheim, F. (2006). (Eds.). Mathematics and democracy: Recent advances in voting systems and collective choice. Springer.

# Social Choice 2.0 and Customized Multiwinner Voting

**Piotr Faliszewski**

**Abstract** We describe the need for custom-made multiwinner voting for individual applications, and argue that the problem of evaluating such rules requires new techniques, beyond classic axiomatic analysis.

## 1 A New Perspective

In the early days of the World Wide Web, web pages were either prepared by motivated enthusiasts or by professional authors, such as journalists whose main outlets were in the physical world, but whose articles also appeared on-line. The Web 2.0 revolution came from encouraging all the Internet users to provide web content by posting comments, using social media, writing blogs, sharing their photographs, or engaging in many other activities (O'Reilly 2017). Some of the main ideas of Web 2.0 are that everyone is an author and that everyone receives customized Internet experience. We argue that social choice theory is facing a similar kind of revolution, where elections become a common tool, used by a broad audience for many different tasks.[1] The multiplicity of these tasks and the fact that elections may be run by non-specialists mean that we need easily customizable voting rules, and we need to provide intuitive means of arguing about them.

Multiwinner elections seem to have a particularly important role to play in this upcoming revolution. On the one hand, such elections have an impressive set of applications, ranging from shortlisting, through various business decision problems, to electing parliaments (or, more plausibly, to electing local decision-making bodies such as electoral colleges at universities). On the other hand, the landscape of multiwinner rules is much more complicated than that of single-winner ones and finding an appropriate rule for a given task is not easy. Put differently, designing customized

---

[1] In fact, Doodle already is an extremely popular site where people use Approval voting to schedule meetings (Doodle 2019). LiquidFeedback (http://liquidfeedback.org) is another example of a very related project.

P. Faliszewski (✉)
AGH University, Kraków, Poland
e-mail: faliszew@agh.edu.pl

© Springer Nature Switzerland AG 2019
J.-F. Laslier et al. (eds.), *The Future of Economic Design*, Studies in
Economic Design, https://doi.org/10.1007/978-3-030-18050-8_12

multiwinner rules for individual applications is possible, but difficult, and requires specialized knowledge. Making it easy is a necessary step to make the Social Choice 2.0 revolution possible.

## 2 Single-Winner and Multiwinner Voting

Below we compare the single-winner and multiwinner settings from the perspective of the types of goals that one may seek in each of them; this comparison is largely inspired by the one included in our recent overview (Faliszewski et al. 2017b).

In a typical election scenario we have a set of candidates $C = \{c_1, \ldots, c_m\}$ and a collection of voters $V = (v_1, \ldots, v_n)$, each with his or her preferences expressed in some way (e.g., as a ranking of the candidates from the best to the worst, or as an approval set, containing those candidates that the voter finds acceptable). In the single-winner elections the goal is to choose the single best candidate, that is, the one that is judged as highly as possible by as many voters as possible. However, in general the requirements of "being judged highly" and of "having broad support" are mutually exclusive and different voting rules find compromises between them in different ways. For example, the Plurality rule puts all the weight on the former criterion, whereas Approval voting is largely geared toward the latter. Indeed, this is one of the reasons why Plurality is often used for presidential elections (where the president should have a strong mandate from the majority and it is acceptable for large minorities to remain dissatisfied), whereas Approval is used for scheduling meetings (Doodle 2019) (where it is more important to ensure that everyone can attend, than to ensure that the meeting is at a time particularly convenient for a certain group of people). Thus, on the intuitive level—and with many simplifications!—the choice between single-winner rules reduces to picking a position on the spectrum between the two extremes: We choose to what extent we care about the winner receiving broad support and to what extent we care how highly this winner is judged by its supporters (e.g., by the majority).

The situation with multiwinner voting is more complicated. The general setup is the same as in the single-winner case, but we are to choose a committee of a given size (i.e., a fixed-size set of candidates),[2] and evaluating the qualities of such committees is more involved. Faliszewski et al. (2017b) identified the following three idealized goals that multiwinner rules may aim at:

**Individual Excellence.** In this case we want to choose candidates that perform very well individually (e.g., that are all ranked highly by as many voters as possible). This is a natural goal if, for example, we are to choose a number of finalists of a competition based on the judgments of a group of experts. One of the prime

---

[2]In particular, we still assume that the voters rank individual candidates and not committees. The latter approach is possible as well and was studied, e.g., by Fishburn (1981a, b) in the context of multiwinner elections inspired by the Condorcet principle. Recently it is studied under the umbrella of the quickly growing area of voting in combinatorial domains (Lang and Xia 2016).

characteristics of the individual excellence goal is that if there are two similar candidates (e.g., ranked on similar positions by all the voters), then either both of them should be included in the committee or both of them should be rejected (except for the boundary cases).

**Diversity.** Here we seek a committee where each voter ranks some of its member highly, but where different committee members may be supported by possibly very different numbers of voters. For example, consider the task of choosing a set of movies that an airline would put on its transatlantic flights. It is important that every passenger finds a movie to watch, but it is not important if, say, 90% of the passengers watch a Hollywood blockbuster and the remaining 10% watch the other movies, or if every movie is watched by the same number of people. For the case of the diversity goal, if there are two similar candidates then at most one of them should be included in the committee (again, ignoring borderline cases).

**Proportionality.** When electing a collective body (such as a country's parliament or a university's senate), it is important that the chosen people represent the voters proportionally. For example, if there are two similar candidates (e.g., members of the same party for a parliamentary election) then a rule aiming at proportional representation should either select neither of them, one of them, or both of them, depending on the level o f support they receive.

In some way, the goal of proportionality is between those of individual excellence (where every committee member should receive strong support from as many voters as possible) and diversity (where as many voters as possible should be able to find a satisfying representative). Yet there are reasons to view it as a separate one. On the philosophical level, proportionality is one of the most natural goals, often required in practical applications and studied deeply in the literature (e.g., in the context of apportionment problems, Pukelsheim 2014). On a more technical level, in our recent work (Faliszewski et al. 2017a) we showed three continuous spectra of rules between the k-Borda rule (which is an archetypal example of a rule focused on individual excellence) and the Chamberlin–Courant rule (Chamberlin and Courant 1983) (which is an archetypal example of a diversity oriented one), and only one of them passed through an area of rules that could be seen as proportional (a family of rules similar to Proportional Approval Voting). Thus, even if the proportionality goal is not completely independent from the other two, the spectrum of goals that multiwinner voting rules may aim at is certainly multidimensional and finding a good rule for a given application can be quite nontrivial.

# 3 Blurring the Picture: Custom-Made Voting Rules

The space of multiwinner voting rules covers the area spanned by the three goals mentioned above, but, in fact, it is far broader and sometimes it is not even possible to say how a given rule relates to the goals of individual excellence, diversity, or proportionality. To illustrate this point we present an example of choosing merchandise via multiwinner voting (this example is motivated by our previous work, Faliszewski et al. 2016).

Our example builds on the idea of committee scoring rules. While we point the readers to our technical papers for more details (Elkind et al. 2017b; Faliszewski et al. 2016), the main idea of committee scoring rules is as follows: We assume that there is a scoring function $f$ that associates each voter $v$ and committee $S$ with a score value $f(v, S)$. The score of a committee $S$ is the sum of the scores it receives from all the voters; the committee with the highest score wins.

Now let us consider a store which needs to select $k$ items to offer to its customers, from the set $C = \{c_1, \ldots, c_m\}$ of those available on the market. The store has a community $V = (v_1, \ldots, v_n)$ of customers and for each customer it knows which items he or she finds attractive (i.e., which items he or she approves).[3] Finally, we have a scoring function that given a customer $v_i$ and a set $S$ of items that the store chooses to offer, provides the number of items that the voter would buy (we assume for simplicity that all the items are of the same type and have similar prices). This function models how the store believes its customers to behave. The store wants to find a set of $k$ items such that if it makes these items available to the customers, then the total number of items sold is as high as possible (the items from the set are put on display in the store and customers buy their copies). In other words, the store computes the winning committee under the rule defined by $f$, for the election with candidates $C$ and voters $V$. (It is worth mentioning that in this application the customers are not even aware that they participate in an election; the store simply collects its knowledge on how the customers behave and adjusts the set of items offered. This is important because it means that the store has the freedom to choose whatever voting rule it feels appropriate, without worrying if the customers would "accept" this rule.)

Now let us consider how the store might choose the function $f$ depending on the type of commodity that it sells. First, let us imagine a t-shirt store. Each customer finds some shirts attractive and we can assume that, eventually, each customer will buy all the t-shirts that he or she likes (we do not expect this to happen at a single visit to the store, but over time we can approximate the behavior of the customers this way; if a reader feels that he or she likes many shirts, but still does not buy all of them, we would argue that a person eventually buys many of those that he or she really likes a lot, but we also acknowledge that our example is very simplistic). The store would use a function $f$ such that if $S$ were the set of shirts on display and customer $v_i$ approved shirts from set $X_i$, then we would have $f(v_i, S) = |S \cap X_i|$. In other words, the store would use Multiwinner Approval Voting (see, e.g., the work of Kilgour 2010); in the world of rules based on preference rankings, an analogous rule is called Bloc. Interestingly, our recent numerical experiments on the two-dimensional Euclidean domain suggest that the Bloc rule typically selects clusters of candidates that are

---

[3]In more realistic settings we would, of course, assume that the store knows these functions in some approximate way only, but here we focus on illustrating certain simple points.

at some not-too-large distance from Borda winners (Elkind et al. 2017a).[4] Since Borda winners could be understood as representing a "rotten consensus" (or, the mainstream), the Bloc rule indeed is appealing for an apparel store: It recommends that the store should offer items that are attractive to certain groups of clients, but that do not try to meet the tastes of everyone.

In the second example we assume that the store is, in fact, a car dealership. In this case every customer $v_i$ has a set of cars $X_i$ that he or she finds appropriate for his or her needs (e.g., SUVs from a particular price range), but would only buy a single one. Thus, the dealer should use function $f$ such that $f(v_i, S) = 1$ if $X_i \cap S \neq \emptyset$ and $f(v_i, S) = 0$ otherwise. In other words, the dealer should use approval-based Chamberlin–Courant rule (Procaccia et al. 2008). For the case of this rule (although for its Borda-based variant), our experiments in the two-dimensional Euclidean domain (Elkind et al. 2017a) have shown that selected committees consist of candidates that try to cover the tastes of the voters fairly evenly.[5] This, again, confirms the intuition that car dealers should (and do) offer cars that cover as many areas of the market as possible, and try to not offer too many cars that fill the same niche.

Third, let us consider a store that sells high-end, luxurious jewelery. While the customers can afford many different jewelery items, they also want to buy exactly what they like. In our model this means that for each customer $v_i$ there is a unique item $x_i$ approved by him or her, and the store should use function $f(v_i, S) = |S \cap x_i|$. In the language of voting theory, this means that the store should use the single non-transferable voting rule (SNTV). However, SNTV is a particularly unattractive rule as it focuses on the voters' top choices only and, if there are many candidates (as in the jewelery store example), then an SNTV winning committee may be unappealing to a large majority of the voters. We interpret this as saying that in addition to the few items that the store would have on display, it should offer jewelery on an on-demand basis. Indeed, this is what jewelery stores do.

Finally, each of the stores above could use a somewhat different type of scoring function to describe its customers. For example, the t-shirt store might offer some sort of "buy two get one free" promotion and the car dealer could try to encourage its customers to replace cars more frequently or to buy a second car for the family. Such marketing strategies would need to be reflected in the scoring functions and, indeed, every store would likely require different functions, customized for its business model and particular needs.

Note that in neither of the above examples did we try to argue that we need a rule that finds individually excellent, diverse, or proportional committees. Instead, we started from the business model of each of the stores and we derived appropriate

---

[4]In this domain we assume that each candidate and each voter is a point in a two-dimensional space. The voters form preference orders by sorting the points of the candidates from the closest to the farthest with respect to the Euclidean distance from their points. In our experiment, we generated elections by drawing candidates and voters from several simple distributions (e.g., from the uniform distribution on a square).

[5]Preliminary experiments conducted for this chapter show that the approval-based variant behaves analogously.

rules. While in the car dealership example we indeed obtained a rule that is typically associated with finding diverse committees (the Chamberlin–Courant rule), it
was simply a consequence of how we expected people to buy cars. Further, for the
Bloc rule—used in the first example—it is certainly not clear what idealized goal it
achieves. On the other hand, the SNTV rule obtained in the third example can both
be seen as achieving individual excellence and diversity (Faliszewski et al. 2016).

## 4   The Problem: How to Choose Rules?

Throughout most of this chapter we argued that the space of interesting multiwinner
rules is broad and multidimensional. Indeed, we believe that we do not even understand what all these dimensions are. Some of the main consequences of this fact are
that designing custom-made rules for specific, individual applications, and arguing
that they indeed are good for these applications, is highly nontrivial. In fact, we
believe that we do not yet have the methodology needed for such tasks. For example,
the axiomatic approach would certainly not suffice on its own: Each axiom is either
satisfied or not, and with some $t$ axioms, there are only $2^t$ different "slots" to which
multiwinner rules could fall. Even if we had many axioms (and we do not believe
that we really have many), this space would still be discrete (the same problem exists
for single-winner rules, but since we believe the space of single-winner rules to be
(mostly) unidimensional, we feel that its discretization is feasible and effective).
Thus other means of evaluating multiwinner rules—possibly ones much easier to
interpret than axiomatic analysis—are truly needed.

**Acknowledgements**  Piotr Faliszewski is grateful to Hannu Nurmi, Umberto Grandi, and William
Zwicker for very useful comments they made regarding this contribution. Piotr Faliszewski's work
was supported by the National Science Centre, Poland, under project 2016/21/B/ST6/01509.

## References

Chamberlin, B., & Courant, P. (1983). Representative deliberations and representative decisions:
    Proportional representation and the Borda rule. *American Political Science Review*, 77(3), 718–
    733.
Doodle AG. http://www.doodle.com.
Elkind, E., Faliszewski, P., Laslier, J., Skowron, P., Slinko, A., & Talmon, N. (2017a). What do
    multiwinner voting rules do? An experiment over the two-dimensional euclidean domain. In
    *Proceedings of the 31st AAAI Conference on Artificial Intelligence* (pp. 494–501).
Elkind, E., Faliszewski, P., Skowron, P., & Slinko, A. (2017b). Properties of multiwinner voting
    rules. *Social Choice and Welfare*, 48(3), 599–632.
Faliszewski, P., Skowron, P., Slinko, A., & Talmon, N. (2016). Committee scoring rules: Axiomatic
    classification and hierarchy. In *Proceedings of the 25th International Joint Conference on Artifi-
    cial Intelligence* (pp. 250–256).

Faliszewski, P., Skowron, P., Slinko, A., & Talmon, N. (2017a). Multiwinner rules on paths from *k*-Borda to Chamberlin–Courant. In *Proceedings of the 26th International Joint Conference on Artificial Intelligence* (pp. 192–198).

Faliszewski, P., Skowron, P., Slinko, A., & Talmon, N. (2017b). Multiwinner voting: A new challenge for social choice theory. In U. Endriss (Ed.), *Trends in computational social choice*. AI Access Foundation.

Fishburn, P. (1981a). Majority committees. *Journal of Economic Theory, 25*(2), 255–268.

Fishburn, P. C. (1981b). An analysis of simple voting systems for electing committees. *SIAM Journal on Applied Mathematics, 41*(3), 499–502.

Kilgour, M. (2010). Approval balloting for multi-winner elections. In *Handbook on approval voting*, Chapter 6. Springer.

Lang, J., & Xia, L. (2016). Voting in combinatorial domains. In F. Brandt, V. Conitzer, U. Endriss, J. Lang, & A. D. Procaccia (Eds.), *Handbook of computational social choice*, Chapter 9. Cambridge University Press.

O'Reilly, T. (2017). What is web 2.0? Retrieved September 13, 2017 from http://www.oreilly.com/pub/a/web2/archive/what-isweb-20.html.

Procaccia, A., Rosenschein, J., & Zohar, A. (2008). On the complexity of achieving proportional representation. *Social Choice and Welfare, 30*(3), 353–362.

Pukelsheim, F. (2014). *Proportional representation: Apportionment methods and their applications.* Springer.

# On Paradoxes Afflicting Voting Procedures: Needed Knowledge Regarding Necessary and/or Sufficient Condition(s) for Their Occurrence

Dan S. Felsenthal

**Abstract** This note lists the most well-known paradoxes (or pathologies) which may afflict voting procedures designed to elect one out of several candidates and calls for future research to focus on finding the necessary and/or sufficient conditions for these paradoxes to occur under various voting procedures in order to be able to better assess the likelihood of occurrence of these paradoxes under these procedures.

I define a *voting paradox* as an undesirable outcome that a voting procedure may produce and which may be regarded at first glance, at least by some people, as surprising or counter-intuitive.

I distinguish between two types of voting paradoxes associated with a given voting procedure:

1. *"Simple" or "Straightforward" paradoxes*: These are paradoxes where the relevant data leads to "surprising" and arguably undesirable outcome.[1]
2. *"Conditional" paradoxes*: These are paradoxes where changing one relevant datum while holding constant all other relevant data leads to a "surprising" and arguably undesirable outcome.

The five best-known simple paradoxes that may afflict voting procedures designed to elect one out of two or more candidates are the following:

1. *The Condorcet Winner Paradox* (Condorcet 1785; Black 1958): A candidate $x$ is not elected despite the fact that it constitutes a "Condorcet Winner", i.e., despite the fact that $x$ is preferred by a majority of the voters over each of the other competing alternatives.

---

[1] The relevant data include, *inter alia*, the number of voters, the number of candidates, the number of candidates that must be elected, the preference ordering of every voter among the competing candidates, the amount of information voters have regarding all other voters' preference orderings, the order in which voters cast their votes if it is not simultaneous, the order in which candidates are voted upon if candidates are not voted upon simultaneously, whether voting is open or secret and the manner in which ties are to be broken.

D. S. Felsenthal (✉)
School of Political Sciences, University of Haifa, Haifa, Israel
e-mail: msdanfl@mail.huji.ac.il

© Springer Nature Switzerland AG 2019
J.-F. Laslier et al. (eds.), *The Future of Economic Design*, Studies in
Economic Design, https://doi.org/10.1007/978-3-030-18050-8_13

2. *The Absolute Majority Paradox*: This is a special case of the Condorcet Winner Paradox. A candidate, *x*, may not be elected despite the fact that s/he is the only candidate ranked first by an absolute majority of the voters.
3. *The Condorcet Loser (or Borda) Paradox* (Borda 1784; Black 1958): A candidate, *x*, may be elected despite the fact that s/he constitutes a "Condorcet Loser", i.e., despite the fact that a majority of voters prefer each of the remaining candidates to *x*.
4. *The Absolute Loser Paradox*: This is a special case of the Condorcet Loser Paradox. A candidate, *x*, may be elected despite the fact that s/he is ranked last by a majority of voters.
5. *The Pareto (or Dominated Candidate) Paradox* (Fishburn 1974): A candidate, *x*, may be elected despite the fact that *all* voters prefer another candidate, *y*, to *x*.

The eight best known "conditional" paradoxes that may afflict a voting procedure for electing a single candidate are the following:

1. *Additional Support (or Lack of Monotonicity or Negative Responsiveness) Paradox* (Smith 1973; Fishburn 1974a; Fishburn and Brams 1983; Felsenthal and Nurmi 2017): This paradox has two alternative versions, one pertaining to a fixed electorate and another pertaining to a variable electorate, a s fllows:
   If candidate *x* is elected under a given distribution of voters' preferences among the competing candidates, it is possible that, *ceteris paribus*, *x* may not be elected if some voters increase their support for *x* by moving *x* to a higher position in their preference ordering. Alternatively, if candidate *x* is not elected under a given distribution of voters' preferences among the competing candidates, it is possible that, *ceteris paribus*, *x* may be elected if some voters decrease their support for *x* by moving *x* to a lower position in their preference ordering.
   Similarly, if *x* is elected in a given electorate it is possible that, *ceteris paribus*, *x* will not be elected if additional voters join the electorate who rank *x* at the top of their preference ordering. Alternatively, if *x* is not elected in a given electorate it is possible that, *ceteris paribus*, *x* will be elected if additional voters join the electorate who rank *x* at the bottom of their preference ordering.

2. *Reinforcement (or Inconsistency or Multiple Districts) Paradox* (Young 1974): If *x* is elected in each of several disjoint electorates, it is possible that, *ceteris paribus*, *x* will not be elected if all electorates are merged into a single electorate.
3. *Truncation Paradox* (Brams 1982; Fishburn and Brams 1983): A voter may obtain a more preferable outcome if, *ceteris paribus*, s/he lists in his/her ballot only part of his/her (sincere) preference ordering among some of the competing candidates than listing his/her entire preference ordering among all the competing candidates.
4. *No-Show Paradox* (Fishburn and Brams 1983; Ray 1986; Moulin 1988; Holzman 1988/89; Pérez 1995): This is an extreme version of the Truncation Paradox. A voter may obtain a more preferable outcome if s/he decides not to participate in an election than, *ceteris paribus*, if s/he decides to participate in the election and vote sincerely for his or her top preference(s).

5. *The Twin Paradox* (Moulin 1988): This is a special version of the No-Show Paradox. Two voters having the same preference ordering may obtain a preferable outcome if, *ceteris paribus*, one of them decides not to participate in the election while the other votes sincerely.
6. *The Subset Choice Paradox* (Fishburn 1974b, c, 1977): When there are at least three candidates and candidate $x$ is the unique winner, it is possible that $x$ becomes a loser if, *ceteris paribus*, one of the other candidates is removed. In the context of individual choice theory the subset choice condition is known as 'Chernoff's condition' (1954, p. 429, postulate 4) which states that if an alternative $x$ chosen from a set $T$ is an element of a subset $S$ of $T$, then $x$ must be chosen also from $S$. This principle is called 'heritage' by Aizerman and Malishevsky (1981, p. 1033) and 'property alpha' by Sen (1970, p. 17).
7. *The Preference Inversion Paradox* (Saari and Barney 2003): If the individual preferences of each voter are inverted it is possible that, *ceteris paribus*, the (unique) original winner will still win.
8. *Lack of Path Independence Paradox* (Farquharson 1969): If the voting on the competing candidates is conducted sequentially rather than simultaneously, it is possible that candidate $x$ will be elected under a particular sequence but not, *ceteris paribus*, under an alternative sequence.

The literature contains quite a few surveys listing which of several possible voting procedures for electing a single candidate is vulnerable to which of the afore-mentioned paradoxes (see, for example, Felsenthal and Nurmi 2018). However, in order to be able to state more conclusively how likely is a given voting procedure that is susceptible to some of the afore-mentioned paradoxes to actually display these paradoxes, one must know what are the necessary and/or sufficient conditions for these paradoxes to occur under the given procedure.

Such limited knowledge exists with respect to some paradoxes and voting procedures. For example, both the Plurality and the Alternative Vote procedures are vulnerable to the Condorcet Winner paradox. However, *ceteris paribus*, if there are only three candidates then the Alternative Vote procedure[2] is more likely to elect the Condorcet winner if one exists than the Plurality procedure. This is so because a necessary and sufficient condition for a Condorcet winner (or any other candidate) to be elected under the Plurality procedure is that s/he will constitute the top preference of a plurality of the voters, whereas for a Condorcet winner to be elected under the Alternative Vote procedure when there are three candidates it is sufficient that the Condorcet winner constitutes the top preference of a plurality of the voters but this is not a necessary condition as the Condorcet winner, when one exists and there are only three candidates, may be elected under the Alternative Vote procedure even if

---

[2]This procedure is also known as Instant Runoff Voting (IRV), Hare rule, or Ranked Choice Voting. According to this procedure voters rank the candidates in order of preference. A candidate supported by a majority of first preferences is elected. Otherwise the candidate supported by the fewest first preferences is eliminated and his or her ballots are transferred to other candidates on the basis of second preferences. This process is repeated until one candidate is supported by a majority of ballots.

s/he does not constitute the top preference of the plurality of the voters. However, this is no longer true when there are more than three candidates because it is possible that with more than three candidates a Condorcet winner may not be elected under the Alternative Vote procedure even if s/he constitutes the top preference of the plurality of voters. For an example see Felsenthal (2012, Example 3.5.6.3, p. 49).

Similarly, Miller (2017) specifies the precise conditions under which variants of monotonicity failure arise in three-candidate elections under the Alternative Vote procedure.

However, a more general knowledge, as far as I know, is still lacking with respect to most voting procedures and paradoxes. Without such knowledge the assessment of the likelihood of the paradoxes to occur must be based on less reliable methods such as computer simulations, laboratory experiments, or historical election data. So in paraphrasing the title of this book, I should like to hope that the future of electoral design will focus on finding the necessary and/or sufficient conditions of the afore-mentioned voting paradoxes under as many voting procedures that are vulnerable to these paradoxes.

**Acknowledgements**  I wish to thank Felix Brandt and Piotr Faliszewski for their helpful comments.

# References

Aizerman, M. A., & Malishevsky, A. V. (1981). General theory of best variants choice: Some aspects. *IEEE Transactions on Automatic Control AC–26*, 1030–1040.

Black, D. (1958). *The theory of committees and elections*. Cambridge: Cambridge University Press.

Borda, J.–C. de (1784) [1995]. Mémoire sur les élections au scrutin, *Histoire de l'Academie Royaledes Sciences année* 1781, pp. 651–665. Paris. Reprinted and translated in I. McLean and A. B. Urken (1995), *Classics of social choice*, Ann Arbor, MI: University of Michigan Press, pp. 83–89.

Brams, S. J. (1982). The AMS nominating system is vulnerable to truncation of preferences. *Notices of the American Mathematical Society, 29,* 136–138.

Chernoff, H. (1954). Rational selection of decision functions. *Econometrica, 22,* 422–443.

de Condorcet, Marquis. (1785). *Essai sur l'application de l'analyse à la probabilité des décisions rendues à la pluralité des voix*. Paris: L'Imprimerie Royale.

Farquharson, R. (1969). *Theory of voting*. New Haven, CT: Yale University Press.

Felsenthal, D. S. (2012). Review of paradoxes afflicting procedures for electing a single candidate. In D. S. Felsenthal & M. Machover (Eds.), *Electoral systems Paradoxes, assumptions and procedures* (pp. 19–91). Heidelberg: Springer.

Felsenthal, D. S., & Nurmi, H. (2017). *Monotonicity failures afflicting procedures for electing a single candidate*. Cham, Switzerland: Springer.

Felsenthal, D. S., & Nurmi, H. (2018). *Voting procedures for electing a single candidate: Proving their (in)vulnerability to various voting paradoxes*. Cham, Switzerland: Springer.

Fishburn, P. C. (1974a). Paradoxes of voting. *American Political Science Review, 68,* 537–546.

Fishburn, P. C. (1974b). Social choice functions. *SIAM Review, 16,* 63–90.

Fishburn, P. C. (1974c). On the sum–of–ranks winner when losers are removed. *Discrete Mathematics, 8,* 25–30.

Fishburn, P. C. (1974d). Subset choice conditions and the computation of social choice sets. *The Quarterly Journal of Economics, 88,* 320–329.

Fishburn, P. C. (1977). Condorcet social choice functions. *SIAM Journal on Applied Mathematics, 33,* 469–489.

Fishburn, P. C., & Brams, S. J. (1983). Paradoxes of preferential voting. *Mathematics Magazine, 56,* 207–214.

Holzman, R. (1988/89). To vote or not to vote: What is the quota? *Discrete Applied Mathematics,* 22, 133–141.

Miller, N. R. (2017). Closeness matters: Monotonicity failure in IRV elections with three candidates. *Public Choice, 173,* 91–108.

Moulin, H. (1988). Condorcet's principle implies the No–Show paradox. *Journal of Economic Theory, 45,* 53–64.

Pérez, J. (1995). Incidence of No–Show paradoxes in Condorcet choice functions. *Investigaciones Economicas, 19,* 139–154.

Saari, D. G., & Barney, S. (2003). Consequences of reversing the preferences. *Mathematical Intelligencer, 25,* 17–31.

Sen, A. K. (1970). *Collective choice and social welfare.* San Francisco: Holden-Day.

Smith, J. H. (1973). Aggregation of preferences with variable electorate. *Econometrica, 41,* 1027–1041.

Young, H. P. (1974). An axiomatization of Borda's rule. *Journal of Economic Theory, 9,* 43–52.

# Agent-Mediated Social Choice

**Umberto Grandi**

**Abstract** Direct democracy is often proposed as a possible solution to the 21st-century problems of democracy. However, this suggestion clashes with the size and complexity of 21st-century societies, entailing an excessive cognitive burden on voters, who would have to submit informed opinions on an excessive number of issues. In this paper I argue for the development of "voting avatars", autonomous agents debating and voting on behalf of each citizen. Theoretical research from artificial intelligence, and in particular multiagent systems and computational social choice, proposes 21st-century techniques for this purpose, from the compact representation of a voter's preferences and values, to the development of voting procedures for the use of autonomous agents only.

## 1 Introduction

Computational studies of voting are mostly motivated by two intended applications: the coordination of societies of artificial agents, and the study of human collective decisions whose complexity requires the use of computational techniques. Both research directions are too often confined to theoretical studies, with unrealistic assumptions constraining their significance for real-world situations. Most practical applications of these results are therefore confined to low-stakes decisions, which are of great importance in expanding the use of algorithms in society, but are far from high-stakes choices such as political elections, referenda, or parliamentary decisions, which societies still make using old-fashioned technologies like paper ballots.

In this paper I argue in favour of conceiving "voting avatars", artificial agents that are able to act as proxies for voters in collective decisions at any level of society. Besides being an ideal test-bed for a large number of techniques developed in the field of multiagent systems and artificial intelligence in general, agent-mediated social

U. Grandi (✉)
Institut de Recherche en Informatique de Toulouse (IRIT), University of Toulouse,
Toulouse, France
e-mail: umberto.grandi@ut-capitole.fr

© Springer Nature Switzerland AG 2019
J.-F. Laslier et al. (eds.), *The Future of Economic Design*, Studies in
Economic Design, https://doi.org/10.1007/978-3-030-18050-8_14

choice may also suggest innovative solutions to low voter participation, a problem that is endemic in most practical implementations of electronic decision processes.

**From low-stakes/high-frequency to high-stakes/high-frequency social choice**. In their quest for practical applications, researchers in (computational) social choice have argued in favour of moving away from political elections, and high-stakes/low-frequency collective decisions in general, where computational techniques and studies are less relevant. Low-stakes decisions such as answering to a personalised search query, or designing group recommender systems for retailers, were instead identified as more suitable applications for their research (Boutilier and Lu 2011). A similar trend can be observed in existing platforms for electronic democracy such as Liquid-Feedback,[1] which were initially designed to be used for policy design by political parties, and evolved into decision systems for smaller committees and low-stakes decisions (Behrens et al. 2014). Many researchers, however, (including the author of this paper) became interested in social choice with the dream of having an impact on how large societies take high-stakes decisions. Technological advancements now make it possible to carry out polls and surveys among citizens in almost real-time. *Has the time come for high-stakes/high-frequency collective decision-making?*

**Electronic, direct, participative, and interactive democracy**. Electronic democracy (or e-democracy) is an umbrella term that groups several pieces of software and policies that aim at increasing citizens' participation in collective decisions by means of digital technologies. Among those, the most relevant applications for social choice theorists revolve around the development of interactive or direct forms of democracy through the design of electronic platforms. While on the one hand existing platforms tend to be rather simple, or, as mentioned in the previous paragraphs, they are restricted to low-stakes decisions, on the other hand more encompassing visions of direct democracy often seem excessively utopian in imagining an active participation of the entire electorate to collective debates and votes (Green-Armytage 2015). Real-world experiments seem to suggest the opposite: when the citizens of Madrid were asked to vote directly (and electronically) on the renovation projects of one of the most important squares in the city, only 8% of eligible voters actually took part in the voting process.[2] The use of direct (or interactive) democracy platforms for high-stakes decisions seems to produce a very low citizen participation, a phenomenon that might be considered as an instance of the well-known paradox of voting (Downs 1957), i.e., the cost of casting a single vote exceeding the expected benefit of affecting the result of the election. *Can we imagine a technological advancement that will solve the problem of low participation in direct democratic decisions?*

**Artificial intelligence and democracy**. Techniques from artificial intelligence (AI), such as the use of machine learning for user profiling and micro-targeting, have been widely used in recent political campaigns, and their effects have been widely debated by the press. However, most articles limit the future use of AI in elections to the design of centralised algorithms that take collective decisions from the collection

---

[1]http://liquidfeedback.org/.

[2]https://elpais.com/elpais/2017/02/28/inenglish/1488280371_057827.html.

of citizens' preferences and behaviours.[3] While discussing the potential misuse of data analysis techniques by central governments, a recent article by Helbing et al. (2017) claims that "If data filters and recommendation and search algorithms would be selectable and configurable by the user, we could look at problems from multiple perspectives, and we would be less prone to manipulation by distorted information". *Can we develop personalised AI techniques that help people construct and motivate their views and voting behaviour in collective decisions?*

**Agent-mediated e-commerce and computational mechanism design**. Negotiation technology, trust-building, and a vast number of other techniques developed in multi-agent systems found application in e-commerce, where human and artificial agents participate with various roles in suitably designed markets. A research field that started more than fifteen years ago (Sierra 2004; Feigenbaum et al. 2009; Dash et al. 2003) helped creating a new reality: "The anticipated agent-mediated economy is almost upon us" (Parkes 2017). *Once people get used to delegate their consumer power to artificial agents, will they be ready to delegate their citizen power as well?*

**Computational social choice and its applications**. The field of computational social choice (COMSOC) started around ten years ago when a group of researchers with common interests in computer science, economics, and political science regrouped in an international workshop,[4] and flourished up to the recent publication of a handbook (Brandt et al. 2016). Research in this field is rather theoretical, and its motivation often ambiguous between developing technologies for the coordination of artificial agents and the study of theoretical properties of voting mechanisms to be later used by humans. Unrealistic assumptions of complete knowledge or full rationality of individual agents limit the applicability of many of its results, a problem that would be less relevant in agent-mediated institutions. If the field is to succeed and prosper, it needs applications to feed its research agenda, experimenting with either real-world situations (as argued by Bouveret 2017) or agent-based technologies. A recent positive example is the creation of the Spliddit webpage (Shah 2017), which created a fruitful loop between practical applications and theoretical research. *Can we conceive of technologies enabling agent-mediated social choices, where theorems and algorithms from COMSOC research could be applied without changing their currently unrealistic assumptions?*

# 2 Voting Avatars: A Short Story

Sylvia (a human being) lives in a world where each citizen with the right to vote is paired with a *voting avatar*, an autonomous agent that is able to communicate with

---

[3] A vision already proposed by Asimov (1955) in the short story "Franchise".

[4] www.illc.uva.nl/COMSOC/workshops.html.

her and who is authenticated by a central voting authority to vote on her behalf.[5] Sylvia is the only authorised person to communicate with her voting avatar (e.g., speech recognition, fingerprint authentication, ...). The central voting authority is a democratically elected government with executive power, supported by an elected bureaucracy who sets the agenda of debates, polls, and votes to be conducted among the entire electorate. Each morning Sylvia receives the daily political agenda, with issues classified by themes and by interest: local, regional, national, and global (the world in which Sylvia lives is likely to be a world federation). Sylvia can simply ignore the message and have a good cup of coffee, as she does on most days: Her voting avatar has already been searching the internet for opinions, consulted influential avatars, and built a preliminary voting behaviour for her daily agenda. During the day, the voting avatar will follow all discussions and correlated votes, and update Sylvia's voting behaviour based on this information and on the level of strategic behaviour she set (currently she left the "strategic voting" button unchecked, like most of her friends claim to do). Her voting avatar has been training for years on a number of votes Sylvia takes directly every month, as well as by observing her conversations on social media, by reading her emails, and by having direct conversations with her every time that the avatar made a wrong or debatable decision on her behalf. Last night, for instance, Sylvia realised that the avatar suggested voting against issuing extra visas to refugees from the Mars colonies, based on a number of dubious sources that she had consulted a couple of days earlier out of curiosity. They discussed the issue for a good 5 minutes, clarifying her position on immigration, the job market, and charity (she actually found the discussion very helpful in constructing a solid view on these issues). Today Sylvia is quite interested in the debate on global freezing, and her voting avatar is proposing a vote in support of the current bill (decisions with long-term consequences involve a long series of iterated votes on improving proposals, in order to maximise consensus). Sylvia has access to a short summary of the reasons supporting the avatar's suggestions, with links to a number of articles by authors she finds reliable, extracts from email discussions she had with a friend on this topic, as well as a list of her past decisions on related issues. She notices that the coherence warning is yellow, suggesting that her vote clashes with some of the positions she defended on the energy market a couple of months ago, but not being a public figure she chooses to ignore the warning...

## 3   Multiagent Systems and Artificial Intelligence

In the past 20 years researchers in the field of multiagent systems, a research community in artificial intelligence, have been importing and adapting models from theoretical economics and political science to conceive and program societies of artificial agents (see, e.g., Shoham and Leyton-Brown 2009). Social choice mechanisms

---

[5]This section may read like a science-fiction novel, but so do many academic papers advocating for direct democracy.

have been proposed for collective decision-making in multiagent systems, stimulating research, e.g., on the computational properties of voting rules, the development of tractable procedures for strategic voting, or the development of approximation algorithms for computing the result of particularly hard rules.

Together with algorithmic game theory, computational social choice is currently one of the most well-represented research areas in conferences on multiagent systems.[6] Many of the techniques introduced by COMSOC researchers would find a prime application in the development of voting avatars for agent-mediated collective decisions, such as those described in the short story above.

**Strategic voting**. Modelling human voters as perfectly rational agents is a useful simplification for obtaining intuitive theorems, but these assumptions limit significantly the applicability of these results (take the Gibbard–Sattertwaithe Theorem as a classical example). Modelling bounded rationality is a big challenge in both AI and Economics, a problem that is absent, or less severe, in societies of artificial agents. Moreover, many of the techniques developed by COMSOC researchers for the analysis of strategic voting (see, e.g., Conitzer and Walsh 2016; Faliszewski and Rothe 2016) may find application test-beds in agent-mediated collective choices.

**Machine learning**. A voting avatar needs to be able to learn a voter's views and preferences from a set of voters' choices, that can moreover be evolving over time, and be more or less consistent with a set of existing views already present in society. This certainly is a challenging problem to frame, and one for which large personalised datasets are hard to construct. The use of machine learning techniques is still widely unexplored in social choice, and existing papers are mostly focused on creating novel voting methods (see, e.g., Xia 2013). A fruitful starting point may be the conception of decision-support systems based on past voting behaviour.

**Iterative voting**. Voting methods that have been discarded as too complex or too unintuitive for human voters could be used successfully by artificial agents that act as proxies for voters. One example is iterative voting, in which repeated elections are staged in search for a more consensual voting outcome (see, e.g., Meir 2017, for a survey). The use of voting avatars takes the burden of intensive communication away from the voter, and reinforcement learning techniques could be tested in this setting to obtain novel voting strategies and rules (Airiau et al. 2017).

**Combinatorial voting and judgment aggregation**. By making high-frequency collective decisions possible, the development of voting avatars will be confronted with a large and interconnected space of alternative choices. This would require compact representations for voters' views and preferences, such as those developed in the area of combinatorial voting (Lang and Xia 2016), as well as a detailed understanding of the counterintuitive results and paradoxical situations that arise when aggregating them, such as those settings analysed by the theory of judgment aggregation (Endriss 2016).

---

[6]See, e.g., the proceedings of the *International Conference on Autonomous Agents and Multiagent Systems, AAMAS* (www.ifaamas.org).

**Argumentation**. Political views and voters' behaviour are the result of discussions and debates, both in the social neighbourhood of a voter and in society in general. The field of (computational) argumentation is a vast enterprise ranging from the elicitation of arguments from natural language, to the identification of winning arguments in a debate between multiple agents. Theoretical research in this field (see, e.g., Rahwan and Simari 2009) is recently being complemented by technologies that can form the basis of personal assistants for the construction of political views.

**Social influence**. A voting avatar will be scouting for opinions presented in influential newspapers or expressed by a designated set of influential voting avatars. Computational-friendly models of influence need to be developed, perhaps based on the extensive research on trust and reputation in multiagent systems (Sabater and Sierra 2005). The relation between social choices and social networks have recently been investigated by a number of papers (see, e.g., Grandi 2017, for a survey).

## 4  Conclusions and Challenges

We may be quite far from a society in which collective choices of all sorts are taken every day in large numbers by means of artificial agents that vote on our behalf. However, in this paper I argue that the implementation of similar "voting avatars" is a viable solution to the lack of voter participation which affects e-democracy applications, and would moreover constitute a prime application for many of the techniques developed in computational social choice and multiagent systems in general. A first step in this direction could be the development of decision-aid systems, which help voters construct their political views and opinions on current issues, building on existing (simple) technologies such as Vote Match[7] in UK, or Stemwijzer[8] in the Netherlands.

There are, however, a number of scientific challenges that need to be tackled. First, representing a voter's view may require combining techniques from computational knowledge representation with complex models of voting behaviour coming from political science and sociology. Second, data protection and security will be a key aspect of developing artificial agents acting on behalf of humans. However, this problem is shared by virtually all electronic applications of voting. A related problem is vote buying and vote influence in general, given the availability of data about voters' views and votes that might be generated by moving to electronic platforms for collective decision-making. Solutions may be sought in cryptography algorithms, which may be easier to apply to societies of artificial agents rather than human ones (see, e.g., the recent work of Parkes et al. 2017). Third, a functional voting avatar needs to be able to explain its behaviour to the voter in an understandable and convincing way. Explanation and interpretation are among the most important challenges for the deployment of artificial intelligence techniques in society, and a

---

[7] www.votematch.org.uk.

[8] www.stemwijzer.nl.

recent survey argues in favour of importing sociological theories of explanation to create transparent and trustworthy algorithms (Miller 2019). Investigating how an artificial agent can explain simple voting behaviours in terms of the input preferences received may constitute a first step in this direction. Last, the design of voting avatars that can act as proxies for voters raises all sorts of ethical questions—from the level of autonomy of such an artificial proxy to the legal status of agent-mediated collective decisions—some of which are currently debated as general ethical issues related to the employment of artificial intelligence technologies in our societies. It turns out that voting itself might be a possible mean of developing collective views on similar ethical problems (Noothigattu et al. 2018).

**Acknowledgements** I am grateful to the participants and the speakers of the reading group on e-democracy at the University of Toulouse (www.irit.fr/~Umberto.Grandi/teaching/directdem). Many thanks to Sylvie Doutre, Piotr Faliszewski, Davide Grossi, Arianna Novaro, Ashley Piggins, Ariel Procaccia, and Marija Slavkovik for their useful comments on previous versions of this paper.

# References

Airiau, S., Grandi, U., & Perotto, F. S. (2017). Learning agents for iterative voting. In *Proceedings of the 5th International Conference on Algorithmic Decision Theory (ADT)*.

Asimov, I. (1955). Franchise. *If*.

Behrens, J., Kistner, A., Nitsche, A., & Swierczek, B. (2014). *Principles of liquid feedback*.

Boutilier, C., & Lu, T. (2011). Probabilistic and utility-theoretic models in social choice: Challenges for learning, elicitation, and manipulation. In *Proceedings of the IJCAI-2011 Workshop on Social Choice and Artificial Intelligence*.

Bouveret, S. (2017). Social choice and the web. In U. Endriss (Ed.), *Trends in computational social choice*. AI Access.

Brandt, F., Conitzer, V., Endriss, U., Lang, J., & Procaccia, A. D. (Eds.). (2016). *Handbook of computational social choice*. Cambridge: Cambridge University Press.

Conitzer, V., & Walsh, T. (2016). Barriers to manipulation in voting. In F. Brandt, V. Conitzer, U. Endriss, J. Lang & A. D. Procaccia (Eds.), *Handbook of computational social choice*. Cambridge: Cambridge University Press.

Dash, R. K., Jennings, N. R., & Parkes, D. C. (2003). Computational-mechanism design: A call to arms. *IEEE Intelligent Systems, 18*(6), 40–47.

Downs, A. (1957). *An Economic theory of democracy*. New York: Harper.

Endriss, U. (2016). Judgment aggregation. In F. Brandt, V. Conitzer, U. Endriss, J. Lang & A. D. Procaccia (Eds.), *Handbook of computational social choice*. Cambridge: Cambridge University Press.

Faliszewski, P., & Rothe, J. (2016). Control and bribery in voting. In F. Brandt, V. Conitzer, U. Endriss, J. Lang & A. D. Procaccia (Eds.), *Handbook of computational social choice*. Cambridge: Cambridge University Press.

Feigenbaum, J., Parkes, D. C., & Pennock, D. M. (2009). Computational challenges in e-commerce. *Communications of the ACM, 52*, 70–74.

Grandi, U. (2017) Social choice on social networks. In U. Endriss (Ed.), *Trends in computational social choice*. AI Access.

Green-Armytage, J. (2015). Direct voting and proxy voting. *Constitutional Political Economy, 26*(2), 190–220.

Helbing, D., Frey, B. S., Gigerenzer, G., Hafen, E., Hagner, M., Hofstetter, Y., vanden Hoven, J., Zicari, R. V., & Zwitter, A. (2017). Will democracy survive big data and artificial intelligence. *Scientific American*.

Lang, J., & Xia, L. (2016). Voting in combinatorial domains. In F. Brandt, V. Conitzer, U. Endriss, J. Lang & A. D. Procaccia (Eds.), *Handbook of computational social choice*. Cambridge: Cambridge University Press.

Meir, R. (2017). Iterative voting. In U. Endriss (Ed.), *Trends in computational social choice*. AI Access.

Miller, T. (2019). Explanation in artificial intelligence: Insights from the social sciences. *Artificial Intelligence, 267*, 1–38.

Noothigattu, R., Gaikwad, S. N. S., Awad, E., Dsouza, S., Rahwan, I., Ravikumar, P., & Procaccia, A. D. (2018). A voting-based system for ethical decision making. In *Proceedings of the 32nd AAAI Conference on Artificial Intelligence (AAAI)*.

Parkes, D. C. (2017). On AI, markets and machine learning. In *Proceedings of the 16th Conference on Autonomous Agents and MultiAgent Systems (AAMAS)*.

Parkes, D. C., Tylkin, P., & Xia, L. (2017). Thwarting vote buying through decoy ballots. In *Proceedings of the 16th Conference on Autonomous Agents and MultiAgent Systems, (AAMAS)*.

Rahwan, I., & Simari, G. R. (2009). *Argumentation in Artificial Intelligence*. Berlin: Springer.

Sabater, J., & Sierra, C. (2005). Review on computational trust and reputation models. *Artificial Intelligence Review, 24*(1), 33–60.

Shah, N. (2017). Optimal social decision making. In *Proceedings of the 16th International Conference on Autonomous Agents and MultiAgent Systems (AAMAS)*.

Shoham, Y., & Leyton-Brown, K. (2009). *Multiagent systems: Algorithmic, game-theoretic, and logical foundations*. Cambridge: Cambridge University Press.

Sierra, C. (2004). Agent-mediated electronic commerce. *Autonomous Agents and Multi-Agent Systems, 9*(3), 285–301.

Xia, L. (2013). Designing social choice mechanisms using machine learning. In *Proceedings of the 12th International Conference on Autonomous Agents and Multi-Agent Systems (AAMAS)*.

# In Praise of Nonanonymity: Nonsubstitutable Voting

Jérôme Lang

**Abstract** In voting theory, it is generally assumed that voters are substitutable with each other, a property referred to as *anonymity* (which I will also call *nonsubstitutability* because of the possible ambiguity of the former): the outcome should remain unchanged after a permutation of agents' identities. Nonsubstitutability is generally taken for granted. While it is certainly highly desirable in political elections, I try to argue here that there are a whole lot of contexts where substitutability leads to questionable outcomes, and I suggest a simple way of generalizing almost all common voting rules to nonsubstitutable settings.

In the rest of this position paper, $N = \{1, \ldots, n\}$ is a set of agents and $A$ is a finite set of alternatives. An ordinal profile is a collection $(V_1, \ldots, V_n)$ of linear orders on $A$, and in this case $rank\,(x, V_i)$ is the rank of $x$ in $V_i$ and $N(a \succ b)$ is the set of agents who prefer $a$ to $b$. An approval profile is a collection $(V_1, \ldots, V_n)$ of subsets of $A$, and in this case $App(a)$ is the set of agents who approve $a$.

## 1 Three Examples

### 1.1 Elections in Shawington

Until now, Shawington, the federal capital of the United States of Planet Mars, did not have a mayor. Now it has to elect one. As we know, Shawington is split in three ethnic communities: the Greens, the Purples, and the Blues. Each community represents one third of the population, with 100 voters each. The candidates are $a$, $b$, $c$ and $d$. The preferences of the population are as follows:

J. Lang (✉)
CNRS, PSL, LAMSADE, Université Paris-Dauphine, Paris, France
e-mail: lang@lamsade.dauphine.fr

© Springer Nature Switzerland AG 2019
J.-F. Laslier et al. (eds.), *The Future of Economic Design*, Studies in
Economic Design, https://doi.org/10.1007/978-3-030-18050-8_15

60 Greens  $a \succ b \succ c \succ d$
10 Greens  $a \succ b \succ d \succ c$
30 Greens  $a \succ c \succ d \succ b$
40 Purples $a \succ b \succ c \succ d$
30 Purples $a \succ b \succ d \succ c$
30 Purples $a \succ d \succ c \succ b$
70 Blues   $d \succ b \succ c \succ a$
30 Blues   $d \succ a \succ c \succ b$

The Shawingtonians decide to use a positional scoring rule associated with a vector $(s_1, s_2, s_3, 0)$, with $s_1 \geq s_2 \geq s_3$ and $s_1 > 0$. The scores of the 4 candidates are as follows:

$$a \quad 200s_1 + 30s_2$$
$$b \quad\quad 210s_2$$
$$c \quad 30s_2 + 230s_3$$
$$d \quad 100s_1 + 30s_2 + 70s_3$$

Since $a$ Lorenz-dominates (or stochastically dominates) all other candidates, it is always a winner (and a single winner except if $s_1 = s_2 = s_3$, in which case it is tied with $c$). Is it the best outcome? Not sure, as the cumulated scores by population are as follows (we show in boldface the minimum value in each row):

|   | Greens | Purples | Blues |
|---|---|---|---|
| a | $100s_1$ | $100s_1$ | $\mathbf{30s_2}$ |
| b | $\mathbf{70s_2}$ | $\mathbf{70s_2}$ | $\mathbf{70s_2}$ |
| c | $30s_2 + 60s_3$ | $\mathbf{70s_3}$ | $100s_3$ |
| d | $\mathbf{40s_3}$ | $30s_2 + 30s_3$ | $100s_1$ |

It seems that $b$ is a good trade-off: all populations are equally satisfied. It does not mean that *voters* are equally satisfied, though: $b$ is the worst alternative for 30 Greens, 30 Purples and 30 Blues. But each of the three communities is solidary enough so that the dissatisfaction of 30% of the community be compensated by other means; on the other hand, choosing $a$ is really unfair to the Blues, and no compensation from the other communities is possible: in other terms, on planet Mars, utility may be transferrable within a single community but not between communities.[1]

---

[1] Readers who feel uncomfortable with grouping voters by race, religion, nationality, ethnicity, or any such criterion can consider instead a similar example where professors are grouped by department when a university makes a policy decision. It may seem acceptable to be fair across departments, rather than fair across individual faculty members. Thanks to Nisarg Shah for this suggestion.

## 1.2   Family Smith Watching Television

Family Smith lives in Shawington. Every day, they watch a movie on TV. There are three channels $c_1$, $c_2$, $c_3$, and the preferences of the family members for every day of the week are as follows (we recall that Martian weeks contain three days):

|         | Earthday | Arrowday | Senday |
|---------|----------|----------|--------|
| Ann     | $c_1 \succ c_2 \succ c_3$ | $c_3 \succ c_2 \succ c_1$ | $c_1 \succ c_2 \succ c_3$ |
| Betty   | $c_1 \succ c_3 \succ c_2$ | $c_3 \succ c_2 \succ c_1$ | $c_1 \succ c_2 \succ c_3$ |
| Charles | $c_2 \succ c_3 \succ c_1$ | $c_1 \succ c_2 \succ c_3$ | $c_3 \succ c_1 \succ c_2$ |

Let us first assume that the TV programs for the whole week are known at the beginning of the week. A fair decision with a reasonable trade-off between efficiency and fairness seems to be watching $c_2$ on Earthday, $c_3$ on Arrowday, and $c_1$ on Senday.

Now, if the programs for each day are known only on the same day, planning ahead is not possible. On Earthday, $c_1$ seems a reasonable decision. Next, when choosing the program for on Arrowday, it makes sense to introduce a bias towards Charles (who had to watch a movie he did not like on Earthday) and choose $c_2$, and then $c_1$ on Senday (while choosing $c_3$ on Arrowday would likely lead to choosing $c_3$ on Senday).

These two scenarios are respectively *off-line* and *on-line* temporal voting scenarios.

Off-line temporal voting can be seen as a standard multiple election (Brams et al. 1998) and can be formulated more generally as a fair public decision making problem (Conitzer et al. 2017), with the issues being the different days of the week. On the other hand, its structure is very close to the structure of our previous example (voting by community): what plays the role of the Greens, the Purples and the Blues are here the 'the Anns' (Ann on Earthday, Ann on Arrowday and Ann on Senday), 'the Bettys' and 'the Charles'. Utility is transferrable between the three Anns (respectively, the three Bettys, the three Charles).[2]

On-line temporal voting is a very different problem that has received only little attention, one exception being (Freeman et al. 2017), who deal with it by giving at each step more weight to the voters who were less satisfied in the previous steps.

---

[2]That utility is transferrable between the three Anns, etc. may be debatable under certain circumstances: if Ann may prefer (2, 2, 2) to (3, 3, 0), i.e., she may be adverse to utility dispersion (a timewise version of risk aversion). One may also argue that if the time scale is long, agents may discount future utilities; this fits the setting, however: after scores have been transformed into utlities, we simply apply a discounting factor to them.

## 1.3   Fair Public Decision Making in Shawington

Shawington has a budget to be spent on public projects, and it wants to decide which project to build according to a *participatory budgeting* process, following other cities such as Paris.[3] Paris' participatory budgeting works as follows: each voter is allowed to support 6 projects for the whole city and 6 for her district ("arrondissement"). Shawington takes a slightly different view: it is not true that only the inhabitants of a district are concerned by the projects to be realized in the district. For instance, when choosing which facility to build in a district $d$, it is not unreasonable to give more weight to the inhabitants of district $d$, some lower weight to those who live in a district close to $d$, and yet a smaller weight to all other inhabitants. What's more, it is a good idea to be fair to the inhabitants of the various districts—again, to communities. The recent approach to fair public decision making (Freeman et al. 2017) does take fairness issues into account but does not differentiate between voters for a given project, nor considers communities of voters.

Yet other examples have to do with *epistemic social choice*: first, some agents may have more expertise on some issues than in some others; second, their opinions may often be positively correlated, for instance because they have similar sources, so that the weight of two correlated voters should count less than the sum of their weights. This idea is developed further in Ani and Nehring (2014).

## 2   Choquet Voting

The common point of Examples 1, 2 and 3 is that voters' utilities should not be simply added, but aggregated in a more subtle way. We outline here a general model for voting, which handles most of the variants of Examples 1, 2 and 3. It is based on *Choquet integrals* (see Grabisch and Labreuche 2010 for a good survey), widely known in decision under uncertainty and in multicriteria decision making, but much less in voting theory, probably because of this obsession for anonymity.

A *(normalized) capacity* on $N$ is a function $\mu : 2^N \to [0, 1]$ satisfying $\mu(\emptyset) = 0$, $\mu(N) = 1$, and monotonicity ($S \subseteq T$ implies $\mu(S) \leq \mu(T)$). Let $\vec{u} = (u_1, \ldots, u_n) \in (\mathbb{R}^+)^n$. The *Choquet integral* of $\vec{u}$ w.r.t. $\mu$ is defined by

$$C_\mu(\vec{u}) = \sum_{i=1}^{n} u_{\sigma(i)} \mu(\sigma(i)|A(i-1))$$

where

- $\sigma$ is a permutation of $N$ such that $u_{\sigma(1)} \leq \ldots \leq u_{\sigma(n)}$;
- $A(0) = \emptyset$ and for all $i \in N$, $A(i) = \{\sigma(1), \ldots, \sigma(i)\}$ and $\mu(\sigma(i)|A(i-1)) = \mu(A(i)) - \mu(A(i-1)) = \mu(A(i-1) \cup \{\sigma(i)\}) - \mu(A(i-1))$.

---

[3]https://www.paris.fr/actualites/the-participatory-budget-of-the-city-of-paris-4151.

In words, $\mu(\sigma(i)|A(i-1))$ is the marginal importance contribution of agent $\sigma(i)$ to the set of agents $\{\sigma(1),\ldots,\sigma(i-1)\}$.

If $\mu$ is additive, then $C_\mu$ is simply a weighted average. If $\mu$ is anonymous, that is, if $\mu(S)$ depends only on $|S|$, $C_\mu$ is an *ordered weighted average*. If $\mu$ is dichotomous and $\vec{u} \in \{0,1\}^n$, $C_\mu$ is a *simple game*.

Surprisingly (or perhaps not), all commonly used voting rules can be generalized to nonanonymous settings in a natural way via Choquet integrals. For approval voting, the $\mu$-approval voting rule $AV_\mu$ is simply defined as

$$AV_\mu(V) = \text{argmax}_{x \in A} \mu(App(x))$$

If $F_s$ is a positional scoring rule induced by scoring vector $s$, then for each ordinal profile $V$,

$$F_{s,\mu}(V) = \text{argmax}_{x \in A} C_\mu(s_{rank(x,V_1)}, \ldots, s_{rank(x,V_n)})$$

Rules based on iterated elimination of alternatives such as STV can also be easily generalized. For instance, for STV, the definition is as usual, replacing plurality scores by $\mu$-plurality scores.

The $\mu$-majority graph $M_{V,\mu}$ associated with $V$ contains, for each pair $(a,b)$ of alternatives, an edge $(a,b)$ if and only if $\mu(N(a \succ b)) > \mu(N(b \succ a))$. The pairwise comparison matrix associated with $V$ is defined by $W_V(a,b) = \mu(N(a \succ b)) - \mu(N(b \succ a))$. The $\mu$-majority graph and $\mu$-pairwise majority matrix being defined, rules based on the majority graph or (more generally) on the pairwise comparison matrix can be defined as usual.

If $\mu$ is anonymous and additive, that is, $\mu(S) = |S|$, then we find the classical versions of the voting rules. If $\mu$ is additive (but not necessarily anonymous), we find the versions of the rules with weighted voters. If $\mu$ is anonymous (but not necessarily additive), we find ranked-dependent versions of the rules; see Goldsmith et al. (2014), García-Lapresta and Martínez-Panero (2017) for two independent investigations of this family of rules.

Let us come back to our first introductory example (the election of the mayor of Shawington). We define the following capacity: for each $S \subseteq N$, let $S = S_G \cup S_P \cup S_B$, where $S_G$ is the set of Green voters in $S$, etc. Finally,

$$\mu'(S) = \max(|S_G|, |S_P|, |S_B|)$$

Under a scoring rule, after some simple calculations, we find that the score of an alternative is the minimum, over all three communities, of the score they obtain for this community. Take Borda as an example. As $a$ gets 300 among the Greens, 300 among the Purples and 60 among the Blues, its Borda$_\mu$ score is 60. $b$ gets 140 among the Greens, 140 among the Purples and 140 among the Blues: its Borda$_\mu$ score is 140. We let reader check that $b$ is the winner. As a slightly more sophisticated capacity, take

$$\mu'(S) = 7\max(|S_G|, |S_P|, |S_B|) + 2\,\text{med}(|S_G|, |S_P|, |S_B|) + \min(|S_G|, |S_P|, |S_B|)$$

The intuition is as follows: the least satisfied community counts 7 times more than the most satisfied community, and the one in the middle twice as much. We let the reader check that the winner is still $b$, but this time by a small margin.

We let the reader check that the other two examples can also be expressed as Choquet voting.

**Acknowledgements** Thanks to Klaus Nehring, Ali Özkes, Marcus Pivato, Nisarg Shah and Mark Wilson for useful comments and suggestions.

# References

Brams, S. J., Zwicker, W. S., & Marc Kilgour, D. (1998). The paradox of multiple elections. *Social Choice and Welfare*, *15*(2), 211–236.

Conitzer, V., Freeman, R., & Shah, N. (2017). Fair public decision making. In *Proceedings of the 2017 ACM Conference on Economics and Computation, EC '17, Cambridge, MA, USA, 26–30 June 2017*, pp. 629–646.

Freeman, R., Zahedi, S. M., & Conitzer, V. (2017). Fair and efficient social choice in dynamic settings. In *Proceedings of the Twenty-Sixth International Joint Conference on Artificial Intelligence, IJCAI 2017, Melbourne, Australia, 19–25 August 2017*, pp. 4580–4587.

García-Lapresta, J., & Martínez-Panero, M. (2017). Positional voting rules generated by aggregation functions and the role of duplication. *International Journal of Intelligent Systems*, *32*(9), 926–946.

Goldsmith, J., Lang, J., Mattei, N., & Perny, P. (2014). Voting with rank dependent scoring rules. In *Proceedings of the Twenty-Eighth AAAI Conference on Artificial Intelligence, 27–31 July 2014, Québec City, Québec, Canada*, pp. 698–704.

Grabisch, Michel, & Labreuche, Christophe. (2010). A decade of application of the choquet and sugeno integrals in multi-criteria decision aid. *Annals OR*, *175*(1), 247–286.

Guerdjikova, A., & Nehring, K. (2014). Weighing experts, weighing sources: The diversity value. Technical Report

# Realizing Epistemic Democracy

Marcus Pivato

**Abstract** Many collective decisions depend upon questions about objective facts or probabilities. Several theories in social choice and political philosophy suggest that democratic institutions can obtain accurate answers to such questions. But these theories are founded on assumptions and modelling paradigms that are both implausible and incompatible with one another. I will propose a roadmap for a more realistic and unified approach to this problem.

Suppose a group of people confronts a question for which there is an objectively correct answer. Each person has some incomplete (and perhaps partially incorrect) information regarding this question, but no one knows enough to reliably answer it. But by combining the fragmentary information of all the individuals through some voting procedure, perhaps the group can determine the correct answer with much greater reliability than any individual. This insight is the starting point of the theory of *epistemic social choice* (ESC).

The foundational results in ESC are Condorcet's (1785), Jury Theorem and Galton's (1907) Wisdom of Crowds phenomenon. In both cases, the collective decision of a group of voters (obtained through simple majority vote or simple arithmetic averaging) is a highly reliable indicator of the correct answer, even if each individual voter is quite unreliable. There is now a large literature extending these results to various voting rules and various assumptions about the probabilistic relationship between the voters' beliefs and the actual state of the world. These results generally come in three varieties. First, there are *melioristic* results, which show that under certain conditions, a particular voting rule is more reliable than any individual, and becomes more reliable as the group gets larger. Second, there are *optimality* results, which show that a certain rule yields the best possible guess about the correct answer, given the available information (e.g. that it is a *maximum likelihood estimator*).[1] Finally,

---

[1] Melioristic and optimality results are sometimes lumped together and called *nonasymptotic* ESC results.

M. Pivato (✉)
Université de Cergy-Pontoise, Cergy, France
e-mail: marcuspivato@gmail.com

there are *asymptotic* results, which show that, for very large populations, this voting rule approaches perfect accuracy.[2] Complementing these formal results is a growing interest in "epistemic democracy" in the political philosophy literature (Cohen 1986; Landemore and Elster 2012; Schwartzberg 2015). The upshot of these literatures is that democratic institutions can be surprisingly effective at aggregating socially dispersed information and arriving at the correct solutions to epistemic problems.

However, almost all of the ESC models have a major blindspot: they assume that, conditional on the state of the world, the beliefs of different voters are *independent* random variables. This is highly unrealistic: in reality, voters are exposed to common information sources (e.g. mass media, common educational institutions, common cultural background) and also influence one another's beliefs through social interactions. So we would expect their beliefs to be highly correlated. This calls into question the optimistic predictions of the ESC literature; in reality, democratic institutions may be far less epistemically competent than the results suggest. Indeed, a skeptic could suggest several recent political events as counterexamples to the putative infallibility of mass democracies.

There are a few papers in the literature which explicitly introduce correlations between voters. Unsurprisingly, the optimistic conclusions of the ESC literature generally unravel under such correlations (Nitzan and Paroush 1984; Dietrich and List 2004; Dietrich and Spiekermann 2013a, b). Nevertheless, some papers show that the *asymptotic* results survive as long as these correlations are "not too large" (Ladha 1992, 1993, 1995; Berg 1993; Peleg and Zamir 2012; Pivato 2017).[3] Likewise, some melioristic claims survive (Dietrich and Spiekermann 2013a, b). But even this more sophisticated literature is still deficient in two ways. First, it tells us nothing about the *optimality* results, which are important for modelling juries, expert committees and other small groups. Second, it employs very simple, artificial, and static models of the correlation structure between voters. In particular, it does not contain any plausible dynamical model of how the beliefs of voters are formed and changed through social interaction and deliberation.

This is problematic because a burgeoning literature argues that deliberation can improve the epistemic reliability of collective decisions (Elster 1998; Fishkin and Laslett 2003; Landemore 2013; Bächtiger et al. 2018). This literature on *deliberative democracy* shares with ESC the claim that many important collective decisions involve mainly *epistemic* questions, for which there exist objectively correct answers. But whereas most models in the ESC literature deny any correlation between voters, the literature on deliberative democracy seeks to *maximize* such correlations through dialogue. Thus, the modelling tools of ESC are completely unsuitable for formally analysing the claims of the deliberative democracy theorists.

One partial response is the small literature that has recently emerged, studying deliberation as a game of asymmetric information (Austen-Smith and Feddersen 2005, 2006, 2009; Meirowitz 2006, 2007a, b). The general message of this literature

---

[2]See Dietrich and Spiekermann (2017) for a recent review of the ESC literature. See also Pivato (2012, 2013) for reviews of the literature on optimality results in particular.

[3]See Pivato (2017) for a more detailed summary of this literature.

is that misaligned incentives will cause agents to lie or withhold information during deliberation, leading to suboptimal social decisions. (There are some exceptions to this pessimistic conclusion; see Le Quement and Yokeswaran 2015; Diemen et al. 2015). However, this literature focuses on dichotomous decisions and supermajority rules (as in Condorcet's Jury Theorem) and makes strong rationality assumptions about the voters.

Furthermore, the literatures on ESC, deliberation games and deliberative democracy all neglect the well-documented phenomena of *motivated reasoning* and *confirmation bias* in human cognition. There is abundant empirical evidence that most people do not update their beliefs like impartial Bayesian scientists when confronted with new evidence. Instead of seeking to maximize the accuracy of their beliefs, they often seek to minimize the emotional costs of repudiating previously held beliefs—especially beliefs which they identify with their peers or social group.[4] They seek out or selectively remember evidence which supports their beliefs, and avoid, forget, deny or disregard evidence which contradicts them. This effect is especially pronounced for the highly emotive questions at the center of many political controversies (Lodge and Taber 2013; Jost et al. 2013; Kahan 2016). Political polarisation grows when voters can ensconce themselves in "echo chambers" where they see only the evidence that reinforces their prejudices (Sunstein 2003, 2009). But penetrating such echo chambers is not a panacea; perversely, people sometimes adopt *more* extreme positions when confronted with evidence which disconfirms their beliefs.

Thus, if epistemic social choice theory is to be of any relevance in the design and analysis of real democratic institutions, it must employ much more detailed and realistic models of the way that voters form their beliefs, and the way that they update these beliefs in response to new information, especially information arising through deliberation and other social interactions. There is a growing literature on opinion dynamics in social networks;[5] there needs to be more contact between this literature and epistemic social choice theory. In such models, it is likely that the beliefs of different voters will end up highly correlated; furthermore, these correlations may be systematically related to voters' demographic characteristics and social affiliations. Due to their complexity, these models might be analytically intractable; we may need to resort to computer simulations.[6] The halcyon predictions of the extant literature in epistemic social choice and deliberative democracy might not survive in these more realistic models. But democratic institutions could still exhibit a surprising level of epistemic competency, if they are properly designed.

In particular, it will be necessary for epistemic social choice theorists to decide whether their central concern is voting rules or statistics. Too much of the existing literature in ESC begins with a particular voting rule, and then carefully constructs a

---

[4]I know of only two models of epistemic social choice which include such effects: Page et al. (2007) and Bednar et al. (2010).

[5]See e.g. Acemoğlu and Ozdaglar (2011) and Mossel and Tamuz (2017) for surveys of some interesting mathematical models of opinion dynamics. See Castellano et al. (2009) and Castellano (2012) for surveys of statistical physics models of opinion dynamics.

[6]See Lorenz (2007) for an introduction and literature review of computer modelling of social opinion dynamics.

stochastic model of voter belief-formation such that the original voting rule is well-adapted to extract the truth from these beliefs. In other words, it is an exercise in ex post rationalization. This is completely backwards. We should begin with a model of voter belief formation, and then design a statistical instrument to extract the truth from these beliefs as accurately as possible. The resulting instrument may be fairly complex; for example, to answer even a single yes/no question (as in Condorcet's Jury Theorem), we may need to solicit and aggregate the voters' beliefs about an entire system of logically related questions, so the process may look more like judgement aggregation.[7] The optimal statistical instrument might not resemble any known voting rule; indeed, it might not look like a voting rule at all.

On the other hand, "epistemically optimal" voting rules might sometimes turn out to be relatively simple—just not for the reasons given in the existing literature. One reason is that optimal voting rules must be robust against modelling uncertainty. We can never know the exact probabilistic relationship between voters' beliefs and the underlying reality. Furthermore, this relationship changes from one policy problem to the next. Thus, the optimal performance of a voting rule cannot depend upon precisely specified probability distributions or parameter values. Instead, the rule must be designed to perform well for a large range of parameter values and probability distributions. It is also likely that a purely theoretical approach has its limits, and we will also need to experimentally test the epistemic performance of voting rules.

When we bring these insights to the design of democratic institutions, we may need to make compromises for the sake of practicality or political expediency. The epistemically optimal voting rule might require a ballot that is too complicated or time-consuming for the average voter. The system for aggregating these ballots might be complex and opaque, so that one cannot clearly discern the link between the raw data and the resulting decision. Such opacity could erode public trust and undermine the legitimacy of collective decisions. For these reasons, it may be necessary to replace the optimal voting rule with a simpler and more transparent approximation, which is as epistemically accurate as possible given the aforementioned pragmatic constraints.

There is another mathematical theory of collective decision-making that focuses on epistemic questions, but which has evolved almost entirely independently of epistemic social choice theory; this is the theory of *probabilistic opinion pooling* (Genest and Zidek 1986; Clemen and Winkler 1999). In the ESC literature, both the beliefs expressed by voters and the collective decision take a non-probabilistic form (e.g. a selection from a finite list of alternatives, or a ranking over such a list). In contrast, the literature on opinion pooling assumes that voters' beliefs take the form of a *probability distribution* over some set of possibilities; the goal is to aggregate these beliefs into a collective belief of the same probabilistic form. The most popular aggregation methods employ a (weighted) arithmetic average or geometric average of the probabilistic beliefs of the voters.

---

[7]See List (2005), Bovens and Rabinowicz (2006), Everaere et al. (2010), Hartmann et al. (2010), Hartmann and Sprenger (2012), Bozbay et al. (2014), Ahn and Oliveros (2014) and D'Alfonso (2016) for preliminary explorations of this approach.

There is an ironic inversion between the means and ends of these two theories. Whereas the ESC literature uses probabilistic models to justify classical voting rules, the opinion pooling literature emulates the "axiomatic" approach of classical Arrovian social choice theory, to justify probability aggregation rules. The typical theorem says, "Probability aggregation rule $X$ is the only rule that satisfies axioms $A$, $B$, and $C$." But it seems odd to solve a probabilistic problem by invoking normative axioms. Arguably, it would be more appropriate to emulate the ESC literature: begin with a detailed model of *how* the voters formed their probabilistic beliefs, and from this model derive the optimal way to aggregate these beliefs.[8]

At this point, we must confront a fundamental philosophical question: what do probabilities *mean*? Do they encode objective facts about the world (as in quantum mechanics)? Do they just summarize "empirical frequencies"? Or do they merely represent the "subjective opinions" with which agents rationalize their gambling strategies?

If the probabilistic beliefs of each agent merely summarize the frequencies she observes in her empirical data, then voters should aggregate their beliefs by simply combining their datasets and then extracting frequencies from this combined data. In the simplest case (when the datasets of different agents are disjoint), the result is the *arithmetic average* of the voters' beliefs, weighted by the relative sizes of their datasets. But this frequentist interpretation is not very plausible when evaluating the probabilities of *sui generis* events such as wars, pandemics, market crashes, or technological innovations. Since uncertainty about such events plays a major role in public policy decisions, we must consider the other two interpretations of probability.

If there are objective probabilistic facts in the world, then agents can have correct or incorrect beliefs about these objective facts. In this case, determining the probability of an event is a purely epistemic problem, no different in principle than estimating the weight of an ox. Thus, it is a legitimate target for the methods of epistemic social choice theory, and the optimal probability aggregation rule depends on a detailed model of how voters acquire their beliefs. For example, suppose all voters begin with a common prior probability distribution, and each voter then observes some private information, and updates her beliefs by Bayesian conditioning. Then to aggregate the voters' beliefs, we should aggregate their private information. In the simplest case, this can be accomplished by computing the *product* of their probabilistic beliefs, multiplied by a suitable renormalization function (Dietrich 2010).

Thus, in world where there are objective probabilistic facts, the correct way to aggregate the voters' beliefs about these facts depends on specific assumptions on how they form these beliefs. On the other hand, if probabilistic opinions are purely *subjective*, as argued by de Finetti (1970), then they are more like preferences or values. No opinion is any more correct than any other. In this case, it would be absurd to apply the methods of ESC, and we are drawn to an axiomatic approach analogous to Arrovian preference aggregation.

---

[8]This is related to the *(supra)Bayesian* approach to opinion pooling; see Genest and Zidek (1986, Sect. 4) and Clemen and Winkler (1999; Sect. 2.2) for introductions.

Perhaps the truth lies somewhere between these extremes. Suppose Alice asserts, "There is 15% probability of a war in southeast Asia in the year 2045," while Bob asserts that the probability of this event is actually 13%. We might argue that no one could possibly compute such a probability with such exactitude, so these are just expressions of subjective opinion, and neither agent is any more objectively correct than the other. But if Carl then asserts that the probability of this event is exactly *zero*, while Dave asserts that it is 100%, we might regard these pronouncements as objectively false; after all, based on our best current information and analysis, surely there is *some* nontrivial chance that a military conflict will occur in 2045, and also *some* chance that it will not. This suggests that our beliefs about events in 2045 could take the form of a *set* of probabilities, or perhaps a "second-order" probability distribution *over* probabilities, or some more complex structure—and indeed, many contemporary theories of decision-making under uncertainty employ such "metaprobabilistic beliefs" (Gilboa 2009). In this case, the theory of probabilistic opinion pooling should be enlarged to a theory of *meta*probabilistic opinion pooling.

Indeed, there is some preliminary work on the pooling of metaprobabilistic beliefs, but it is still of an axiomatic nature, as befits a purely subjective interpretation (Danan et al. 2016). But in the 2045 example, it seemed that *some* metaprobabilistic beliefs were objectively false. This suggests that there are objective metaprobabilistic *facts* in the world, and in this case, it would be appropriate to apply the methods of epistemic social choice theory to glean these facts from the metaprobabilistic beliefs of the individuals in a society. An axiomatic approach would not be appropriate. Indeed, it could even be misleading.

The problems with the axiomatic approach can be illustrated by considering a problem closely related to probabilistic opinion pooling: that of *Bayesian social aggregation*. Suppose group of individuals must collectively choose a lottery from a set of lotteries. Suppose each individual has von Neumann-Morgenstern (vNM) preferences over these lotteries, and the group (considered as a collective agent) also has vNM preferences. If the group's lottery preferences satisfy the Pareto axiom with respect to the individuals' preferences, then Harsanyi (1955) showed that the collective decision must be "utilitarian", in the sense that the group's vNM utility function must be a weighted average of the vNM utility functions of the individuals. But Harsanyi assumed that the lotteries were based on *objective* probabilities that were public knowledge. If we try to repeat this exercise in a setting where each individuals has her own *subjective* probabilistic beliefs, we run into impossibility theorems (Hylland and Zeckhauser 1979; Mongin 1995; Mongin and Pivato 2015).

The culprit seems to be the Pareto axiom. Thus, Gilboa et al. (2004) proposed to weaken the Pareto axiom, so that it applies only to preferences over lotteries wherein all agents agree about the underlying probabilities. From this axiom they derive a unique social decision rule: the social utility function is an arithmetic average of the individuals' utility functions (as in Harsanyi's utilitarianism), while the social probabilistic beliefs are an arithmetic average of the individuals' probabilistic beliefs. If we regard probabilistic beliefs as purely *subjective*, then this seems reasonable enough. But if we believe that at least some probabilities are *objective*, then we have already encountered situations where the voters' beliefs should be *multiplied* rather than averaged (when agents have different private information). Thus, it seems that even the weak Pareto axiom of Gilboa, Samet and Schmeidler is too restrictive: it imposes an inappropriate method of aggregating probabilistic beliefs. To enable the group to adopt the most suitable probability aggregation in each situation, it seems that we cannot impose *any* Pareto axiom on the social preferences.[9]

The point of this example is not to call into question the Pareto axiom, which is, after all, one of the most fundamental and uncontroversial axioms in normative economics. The point is to illustrate the hazards of trying to solve an epistemic problem by applying normative axioms. Even the most innocuous of axioms can force us into incorrect conclusions. The analysis of (meta)probabilistic opinion pooling in general—and Bayesian social aggregation in particular—should start from a clear model of the relationship between the individual's beliefs and the underlying objective probabilities (if there are any), and from this model, derive the optimal aggregation rule. However, the same caveats apply here as in the rest of epistemic social choice theory; the model must contain a realistic description of how voters acquire their beliefs, and how they change these beliefs through information acquisition and social interaction.

Can democratic institutions reliably solve epistemic problems? To answer this question, axiomatic methods are unconvincing and potentially misleading. The stylised stochastic models of the existing literature in epistemic social choice theory are suggestive, but dangerously unrealistic. Future investigations must use more sophisticated mathematical methods to develop and analyze more realistic models of voter opinion formation, which explicitly describe information flows, cognitive biases, and social influences, in order to design epistemically optimal voting rules. These voting rules might appear exotic, complicated, and opaque in their operation. But this is the only way to realize the promise of epistemic democracy.

**Acknowledgements** I thank Gabriel Carroll, Franz Dietrich, Umberto Grandi, Justin Leroux, Christian List, Arianna Novaro, Kai Spiekermann, and Bill Zwicker for their very helpful comments. None of them are responsible for any errors.

# References

Acemoğlu, D., & Ozdaglar, A. (2011). Opinion dynamics and learning in social networks. *Dynamic Games and Applications, 1*(1), 3–49.

Ahn, D. S., & Oliveros, S. (2014). The Condorcet jur(ies) theorem. *Journal of Economic Theory, 150,* 841–851.

Austen-Smith, D., & Feddersen, T. (2005). Deliberation and voting rules. In Austen-Smith, D. & Duggan, J. (Eds.), *Social Choice and Strategic Decisions: Essays in honour of Jeffrey S. Banks* (pp. 269–316). Berlin, Heidelberg: Springer.

---

[9]See Mongin and Pivato (2017) for an elaboration of this argument. To be clear, we are here talking about the *ex ante* Pareto axiom (which concerns preferences over social lotteries *before* the resolution of uncertainty), rather than the *ex post* Pareto axiom (which concerns preferences over social outcomes *after* the resolution of uncertainty).

Austen-Smith, D., & Feddersen, T. J. (2006). Deliberation, preference uncertainty, and voting rules. *American Political Science Review, 100*(2), 209–217.

Austen-Smith, D., & Feddersen, T. J. (2009). Information aggregation and communication in committees. *Philosophical Transactions of the Royal Society of London B: Biological Sciences, 364*(1518), 763–769.

Bächtiger, A., Dryzek, J., Mansbridge, J., & Warren, M. (eds.). (2018). *Oxford Handbook of Deliberative Democracy*. Oxford University Press.

Bednar, J., Bramson, A., Jones-Rooy, A., & Page, S. E. (2010). Emergent cultural signatures and persistent diversity: A model of conformity and consistency. *Rationality and Society, 22*(4), 407–444.

Berg, S. (1993). Condorcet's jury theorem, dependency among jurors. *Social Choice and Welfare, 10*(1), 87–95.

Bovens, L., & Rabinowicz, W. (2006). Democratic answers to complex questions–an epistemic perspective. *Synthese, 150*(1), 131–153.

Bozbay, I., Dietrich, F., & Peters, H. (2014). Judgment aggregation in search for the truth. *Games and Economic Behavior, 87*, 571–590.

Castellano, C., Fortunato, S., & Loreto, V. (2009). Statistical physics of social dynamics. *Reviews of Modern Physics, 81*, 591.

Castellano, C. (2012). Social influence and the dynamics of opinions: The approach of statistical physics. *Managerial and Decision Economics, 33*(5–6), 311–321.

Clemen, R. T., & Winkler, R. L. (1999). Combining probability distributions from experts in risk analysis. *Risk analysis, 19*(2), 187–203.

Cohen, J. (1986). An epistemic conception of democracy. *Ethics, 97*(1), 26–38.

de Condorcet, M. (1785). Essai sur l'application de l'analyse à la probabilité des décisions rendues à la pluralité des voix, Paris.

D'Alfonso, S. (2016). Belief merging with the aim of truthlikeness. *Synthese, 193*(7), 2013–2034.

Danan, Eric, Gajdos, Thibault, Hill, Brian, & Tallon, Jean-Marc. (2016). Robust social decisions. *American Economic Review, 106*(9), 2407–2425.

de Finetti, B. (1970) [2017]. *Theory of probability*. Chichester: John Wiley & Sons, Ltd. Translated from Italian by Antonio Machi and Adrian Smith.

Deimen, I., Ketelaar, F., & Le Quement, M. T. (2015). Consistency and communication in committees. *Journal of Economic Theory, 160*, 24–35.

Dietrich, Franz. (2010). Bayesian group belief. *Social Choice and Welfare, 35*(4), 595–626.

Dietrich, Franz, & List, Christian. (2004). A model of jury decisions where all jurors have the same evidence. *Synthese, 142*(2), 175–202.

Dietrich, Franz, & Spiekermann, Kai. (2013a). Epistemic democracy with defensible premises. *Economics and Philosophy, 29*(1), 87–120.

Dietrich, Franz, & Spiekermann, Kai. (2013b). Independent opinions? On the causal foundations of belief formation and jury theorems. *Mind, 122*(487), 655–685.

Dietrich, F., & Spiekermann, K. (2017). "Jury Theorems". (in press).

Elster, J. (Ed.). (1998). *Deliberative Democracy*. Cambridge UP.

Everaere, P., Konieczny, S., & Marquis, P. (2010). August. The Epistemic View of Belief Merging: Can We Track the Truth? In *The 19th European Conference on Artificial Intelligence (ECAI'10)* (pp. 621–626).

Fishkin, J. S., & Laslett, P. (Eds.). (2003). *Debating deliberative democracy*. Blackwell.

Galton, F. (1907). Vox populi. *Nature, 75*, 450–451.

Genest, C., & Zidek, J. V. (1986). Combining probability distributions: a critique and an annotated bibliography. *Statistical Science, 1*, 114–148.

Gilboa, I., Schmeidler, D., & Samet, D. (2004). Utilitarian aggregation of beliefs and tastes. *Journal of Political Economy, 112*, 932–938.

Gilboa, I. (2009). *Theory of decision under uncertainty*. Cambridge University Press.

Harsanyi, J. C. (1955). Cardinal welfare, individualistic ethics and interpersonal comparisons of utility. *Journal of Political Economy, 63,* 309–321. Reprinted in *Essays on Ethics, Social Behavior, and Scientific Explanation.* Dordrecht: P. Reidel.

Hartmann, S., Pigozzi, G., & Sprenger, J. (2010). Reliable methods of judgement aggregation. *Journal of Logic and Computation, 20*(2), 603–617.

Hartmann, S., & Sprenger, J. (2012). Judgment aggregation and the problem of tracking the truth. *Synthese, 187*(1), 209–221.

Hylland, A., & Zeckhauser, R. (1979). The impossibility of bayesian group decision making with separate aggregation of beliefs and values. *Econometrica, 47,* 1321–1336.

Jost, J. T., Hennes, E. P.& Lavine, H. (2013). Hot political cognition: Its self-, group-, and system-serving purposes. In Carlson, D.E. (Ed.), *Oxford handbook of social cognition* 851–875. New York: Oxford University Press.

Kahan, D. M. (2016). The politically motivated reasoning paradigm, part 1: What politically motivated reasoning is and how to measure it. *Emerging Trends in the Social and Behavioral Sciences: An Interdisciplinary, Searchable, and Linkable Resource.*

Ladha, K. K. (1992). The Condorcet Jury Theorem, free speech, and correlated votes. *American Journal of Political Science, 36,* 617–634.

Ladha, K. K. (1993). Condorcet's jury theorem in light of de Finetti's theorem. Majority-rule voting with correlated votes. *Social Choice and Welfare, 10*(1), 69–85.

Ladha, K. K. (1995). Information pooling through majority-rule voting: Condorcet's jury theorem with correlated votes. *Journal of Economic Behaviour and Organization, 26,* 353–372.

Landemore, H. (2013). *Democratic reason: Politics, collective intelligence, and the rule of the many.* Princeton University Press.

Landemore, H., & Elster, J. (2012). *Collective wisdom: Principles and mechanisms.* Cambridge University Press.

List, C. (2005). Group knowledge and group rationality: A judgment aggregation perspective. *Episteme, 2*(1), 25–38.

Le Quement, M., & Yokeeswaran, V. (2015). Subgroup deliberation and voting. *Social Choice and Welfare, 45*(1), 155–186.

Lodge, Milton, & Taber, Charles. (2013). *The rationalizing voter.* New York: Cambridge University Press.

Lorenz, Jan. (2007). Continuous opinion dynamics under bounded confidence: A survey. *International Journal of Modern Physics C, 18*(12), 1819–1838.

Meirowitz, A. (2006). Designing institutions to aggregate preferences and information. *Quarterly Journal of Political Science, 1*(4), 373–392.

Meirowitz, A. (2007a). Communication and bargaining in the spatial model. *International Journal of Game Theory, 35*(2), 251.

Meirowitz, A. (2007b). In defense of exclusionary deliberation: communication and voting with private beliefs and values. *Journal of Theoretical Politics, 19*(3), 301–327.

Mongin, P. (1995). Consistent Bayesian aggregation. *Journal of Economic Theory, 66*(2), 313–351.

Mongin, P., & Pivato, M. (2015). Ranking Multidimensional Alternatives and Uncertain Prospects. *Journal of Economic Theory, 157,* 146–171.

Mongin, P., & Pivato, M. (2017). Social preference under twofold uncertainty (preprint).

Mossel, E., & Tamuz, O. (2017). Opinion exchange dynamics. *Probability Surveys, 14,* 155–204.

Nitzan, S., & Paroush, J. (1984). The significance of independent decisions in uncertain dichotomous choice situations. *Theory and Decision, 17*(1), 47–60.

Page, S. E., Sander, L., & Schneider-Mizell, C. (2007). Conformity and dissonance in generalized voter models. *Journal of Statistical Physics, 128,* 1279–1297.

Peleg, B., & Zamir, S. (2012). Extending the Condorcet jury theorem to a general dependent jury.

Pivato, M. (2012). A statistical approach to epistemic democracy. *Episteme, 9,* 115–137.

Pivato, M. (2013). Voting rules as statistical estimators. *Social Choice and Welfare, 40*(2), 581–630.

Pivato, M. (2017). Epistemic democracy with correlated voters. *Journal of Mathematical Economics, 72,* 51–69.

Schwartzberg, Melissa. (2015). Epistemic democracy and its challenges. *Annual Review of Political Science, 18,* 187–203.

Sunstein, C. R. (2003). The law of group polarization. In *Fishkin and Laslett* (pp. 80–101).

Sunstein, C. (2009). *Republic.com 2.0.* Princeton University Press.

# Who Is Afraid of Judgment Aggregation?

**Klaus Nehring**

**Abstract** There are two different fundamental conceptions of the task of social aggregation: interest aggregation and judgment aggregation. While the former aims at the fair and efficient trade-off of personal interests, the latter aims the best resolution of disagreement. Both are possible interpretations of the standard preference aggregation and preference-based voting problems. This note makes two broad points. First, the judgment aggregation conception is not yet well understood and remains undertheorized. Second, the difference between the two conceptions matters; it is a difference that makes a difference. To substantiate the first point, we argue that judgment aggregation as resolution of disagreement cannot be subsumed under statistical aggregation in the manner of Condorcet's jury theorem, and sketch a more specific articulation as "reflexive consensus under arguable disagreement". To substantiate the second point, we illustrate the difference in the context of two important voting rules, the Kemeny rule and Majority Judgment.

In this note, I will consider the normative problem of group decision by voting. A collective decision is to be taken on the basis of a profile of individual rankings over feasible alternatives. A normative criterion encoded in a voting rule selects one or more alternatives as "optimal". The criterion may do this directly, or indirectly via selecting one or more rankings, and choosing a maximal alternative based on that ranking.

This formulation accommodates (at least) two distinct conceptions of the nature of the aggregation task. I shall refer to these as the *interest aggregation* and *judgment aggregation* conceptions, respectively. These are defined by different normative aims, and are thus not merely different semantic labels for a unique well-identified normative problem of "preference aggregation". One should thus expect different salient issues and possibly different best answers (voting rules). Hence one should

I am grateful for insightful comments by Felix Brandt, Hans Gersbach, Umberto Grandi, Ashley Piggins and Clemens Puppe.

K. Nehring (✉)
University of California, Davis, CA, USA
e-mail: kdnehring@ucdavis.edu

© Springer Nature Switzerland AG 2019
J.-F. Laslier et al. (eds.), *The Future of Economic Design*, Studies in Economic Design, https://doi.org/10.1007/978-3-030-18050-8_17

expect the field of voting theory to be shaped by the polarity between the two. Yet, as of now, this polarity seems to be at best in the background. With this note, I am trying to stimulate the perception of what is missing.

## Interest Aggregation Versus Judgment Aggregation

In *interest aggregation*, voters are asked to rank alternatives in terms of personal preference or betterness ("interest"), and the aggregation determines what is best overall, aiming to trade off the interests of the voters in a fair and efficient manner. The normative study of interest aggregation can be viewed as a branch of welfare economics characterized by two features: the exclusive reliance on information on ordinal preference, and the absence of any a priori restrictions on the possible individual preferences.

While the interest aggregation conception views voters as *stakeholders* (beneficiaries), judgment aggregation views voters as (joint) *agents*. In *judgment aggregation*, each voter is asked about what "the group" should do; in the ranking framework, this means that voters are asked to compare alternatives in terms of their perceived merit or choice-worthiness in an impersonal manner. The language of referenda are often fits this conception. For example, the U.K. posed the following question to its citizens on June 23, 2016: "Should the United Kingdom remain a member of the European Union, or leave the European Union?". Note the impersonal reference to "should" instead of an appeal to personal interest as in "Do you want the U.K. ...?".

In judgment aggregation, voters *disagree,* giving different answers to the same (impersonal) question—By contrast, there is obviously no disagreement at all when different people prefer different things for themselves—So judgment aggregation is about the "best" resolution of disagreement. The precise content is to be determined by normative theory, much of it may yet to be developed. The task does not seem easy, indeed not even to be easy to bring into sharp focus, not the least both "disagreement" and "joint agency" are multi-faceted and quite slippery concepts. I'll gesture at one specific avenue to settle disagreement of a particular kind ("arguable disagreement") by "reflexive consensus".[1]

---

[1] I do not claim, of course, that the distinction between "interest aggregation" and "judgment aggregation" is novel. At the same time, from my reading of the literature, it does not seem to be really well-established or well-understood, either, be it in terms of content or relevance.

A notable example is Sen (1977) of which two commentators reminded me. Sen introduces and distinguishes these very two terms; in broad intention, the distinction Sen aims to draw is similar to, or at least consistent with, the distinction drawn here. However, Sen's elaboration and upshot are quite different, as it turns centrally on the relevant "informational basis". In particular, for Sen, interest aggregation *requires* going beyond merely ordinal information, while judgment aggregation does not. He concludes from this that Arrow's impossibility is "not disturbing" for interest aggregation, while it is "quite disturbing", or even"dismal" for judgment aggregation.

By contrast, the distinction aimed at here is a distinction given the same informational basis of ordinal preference information. Indeed, in contrast to Sen's drift of argument, Arrow's impossibility result seems a big deal for ordinal interest aggregation; further, as the discussion of the Kemeny rule below suggests, there may exist more palatable escapes from Arrowian impossibly under judgment aggregation than under interest aggregation.

## Preference Aggregation Versus Statistical Aggregation

In contrast to the polarity emphasized here, the extant literature is characterized by a different polarity, that between "preference aggregation" and "statistical judgment aggregation". Here, the term "preference" has two levels of meaning: on one level, "preference" may denote simply the formal input to the voting rule, without commitment to a particular aggregation conception. This non-specific, formal usage often fits normative discussions of voting rules in terms of their formal properties, such as the absence of presence of certain "paradoxes". On a second level, "preference" also is often understood specifically as interest as defined above.[2]

The judgment aggregation theme appears mostly as epistemic aggregation in a statistical vein, often under the name of "Condorcet jury theory". *Statistical aggregation* models assume the existence of an independent, shared standard of decision quality (often referred to as "truth" or "common value") as well as statistical assumptions on the joint distribution of true state and voters' states (the preference profile). On this basis, the group decision problem is construed as a statistical estimation problem. Since statistical aggregation is based on a different, substantially richer set of primitives than the standard framework, it employs a formally distinct set of models and gives rise to a literature that is largely distinct from the mainstream of voting theory. Indeed, much of the research in the statistical vein is conducted with two alternatives only where the standard voting problem is largely trivial.

While statistical aggregation is a species of judgment aggregation broadly conceived, it is not well-suited to address the problem of group decision as resolution of disagreement *from the inside,* that is: from the point of view of the jointly acting voters themselves. The main purpose of the statistical models is, instead, the assessment of the epistemic (truth-tracking) performance of different voting rules in different stochastic environments. Perhaps the leading theme is the "miracle of aggregation", i.e. the claim that groups can epistemically outperform their members, on average and perhaps even in comparison to the best. The logic of the miracle of aggregation is quite clear if it is about the aggregation of dispersed private information, but does it apply to the resolution of disagreement proper?[3]

In other words: can statistical aggregation models be turned into accounts of disagreement resolution? This seems difficult, for to do so, the statistical model would have to be accessible to the voters, in particular, it would have to be commonly known. Suppose that this—demanding—assumption holds. But this is not enough. In addition, it matters whether each voter accepts the validity of the statistical model?

---

[2]Analogously, in the general theory of judgment aggregation, "judgment" may be used as a non-specific, abstract term for what is to be aggregated—thus incorporating the classical preference aggregation problem as a special case at the formal level—or may be is used in a more specific sense, contrasting "judgment" with "interest", taste or desire. See, for example, Nehring and Puppe (2010) and Mongin (2018, this volume) and the references therein.

[3]Disagreement can be sharply distinguished from dispersed private information epistemically by stipulating that in situations of (pure) disagreement, voters' beliefs are commonly known; in Aumann's celebrated phrase, voters "agree to disagree" (Aumann 1976). By contrast, in the presence of private information, voters are uncertain about each others' information, hence about each others' beliefs.

If yes, why don't the voters go with the purportedly optimal (and superior) estimate according to the model rather than stick with their own belief?. If no, then how does the estimate produced by the model help to settle the disagreement? It thus stands to reason that judgment aggregation as resolution of disagreement must make do without reference to an auxiliary statistical model.

**Voting by Reflexive Consensus**. As an entry point, let us retain the assumption that the disagreement in judgments is about something independently "true" out there, but discard the appeal to any evidence about it beyond what has entered voters judgments. Let's call this "disagreement in beliefs" or *epistemic disagreement*. The judgments to be aggregated may be expressed directly as beliefs, e.g. as subjective probabilities. Or the judgments may take the form of a ranking of different policy alternatives, and disagreement in those rankings may be construed as reflecting disagreement in underlying beliefs in the presence of shared values.

Under epistemic disagreement, it is meaningful to consider the members of the group as a set of advisors to the group as the advisee. That is, voters' judgments (beliefs) can meaningfully be viewed as reasons for an advisee to serve as the basis for his own judgment (belief). Group decision by vote can then be viewed as determined by the proper *balance of advice*, understood as jointly binding on the group. To facilitate the following discussion, let's refer to "the group" in its role as an advisee as "Publius". "Publius" designates a hypothesized disembodied epistemic agent whose task is to form a judgment on the matter at solely on the basis of the group members advice. For specificity, let us focus on the case where all members are to count equally as sources of the group decision. This anonymity assumption might reflect a recognition of group members as equally reasonable in some sense, or might derive from norms of political equality.

Publius thus has to form his own judgment on the basis of the anonymized profile of members' judgments, and *nothing else*. In particular, being equipped with a *blank slate* by himself, Publius does not have statistical information or probabilistic beliefs of his own. From this perspective, a voting rule encodes, for each possible anonymized profile, the counsel derived from the profile. Normative arguments of some sort—presumably axiomatic in some way or other—are to determine the best counsel to be derived on balance. Ideally, they serve to determine Publius' best judgment uniquely (up to ties).

From the point of view of the group members, this normatively grounded best judgment represents a *reflexive consensus* in that all members can deduce its content from knowledge of the profile alone, given the normative rationale. Note the importance of the assumption of a blank slate to enable such inference.

**Arguable Disagreement**. The notion of epistemic disagreement as disagreement in beliefs about what is in fact true is common and intuitive. However, it is also rather restrictive since substantive disagreement is conceptually meaningful also when an independent standard of truth is not available. Further, what matters for the idea of "reflexive consensus" as presented here is the meaningfulness of advice-taking. Advice-taking has, I submit, a much larger ambit as truth-aptness. To underpin it, I propose—extremely sketchily—to broaden the notion of epistemic disagreement to one of "arguable disagreement" as follows. (Of course, there must be a lot of connections to a lot of work in various literatures; so I am not claiming any particular novelty here).

I shall call a judgment or disagreement in judgment "amenable to argument", or "arguable" for short, if it is, in principle, open to the giving and taking of impersonal reasons, and hence to criticism and self-criticism, to deliberation. At the same time, amenability to argument does not require that any disagreement can be definitively settled by the exchange of arguments, even under ideal conditions. Indeed, the voting problem as a problem of plural action arises precisely because agreement has not been reached at the time of group decision.

The notion of *arguable disagreement* can be illustrated by contrasting judgments on values to tastes on the one side, and to claims on moral truth on the other. "De gustibus not est disputandum"—there is no arguing about tastes—is a central part of the neo-classical understanding of welfare economics and is frequently invoked to assimilate disagreement in aims to difference in taste, and thereby to subsume it under the umbrella of preference (i.e. interest) aggregation. Preferences as tastes are "given", and it is not necessary, and may not even be meaningful, to ask why they are what they are. Value judgments are—arguably—different, at least in certain settings and conditions: they often do admit, even invite, reasoning and debate; moral and aesthetic criticism and self-criticism are meaningful activities and frequently engaged in. So they invite arguable disagreement.

If there is "arguable disagreement", presumably there is unarguable, "raw" disagreement as well? I am not sure, but leave open this conceptual possibility with the intention of allowing for situations of disagreement in which the construction of an advisee position is not meaningful. On some views, certain disagreements about judgments might be to "deep" and somehow too essentially subjective to be "arguable" in a substantive sense. Perhaps. At the same time, the holders of such judgments will rarely themselves view their own judgments as entirely arbitrary, and merely as a matter of personal taste. If so, one might say that the judgments are "*subjectively arguable*" in the sense that judges may respond to arguments presented to them, but entirely at their own discretion. "Reasons" can be given, but lack any objective force. Subjective arguability in this sense would not be adequate to meaningfully construct an advisee position in the required sense, since Publius, being characterized by a blank slate, is devoid of any subjectivity of his own, so cannot exercise any "subjective argument" on his own.

### A Difference That Makes a Difference

A paradigmatic example of reflexive consensus is given by the ranking of/choice from two alternatives. With equal numbers in favor of both, the balance of advice between the two choices is neutral. A majority of votes in favor of one alternatives tilts the balance of advice in favor of that alternative, however slightly.

The challenge of voting theory from a judgment aggregation perspective is to extend this line of argument to more than two alternatives. While this challenge must face the basic Arrowian conundrum as much as any theory of interest aggregation, the conundrum takes a rather different form, as I will try to indicate with the following observations. Thus, the difference between the two conceptions is not just a matter of "semantics" or philosophical background narrative, but a difference that makes a difference.

**Intensities**. Within an interest aggregation conception, it seems strongly desirable to capture preference *intensities*. While ordinal preferences do not provide cardinal preference information in themselves, one may attempt to estimate or impute cardinal preferences from ordinal data. This provides a highly suggestive motivation for "positional scoring rules", in which alternative scores are interpreted as imputed cardinal utilities. The Borda rule in particular captures an argument from insufficient reason that infers equal utility difference from equal rank difference. (It is seriously debatable, however, whether such imputation of utility intensities is sound. If it was, would Arrow's impossibility result have made any headlines?).

Within a judgment aggregation conception by contrast, judgment "intensities" carry much less appeal. Presumably, what matters about a judgment is its content, not what it indicates about the psychological state of the voter expressing it. Note also that if "judgment intensities" of some form might be observed, such as "intensity of conviction" indicated by assertiveness or passion of judgment expression, such intensity may well be a good reason to discount someone's judgment in determining the balance of advice, rather than to give it more weight.

**The Kemeny Rule**. The Kemeny rule selects, for any profile of rankings, the ranking(s) that minimizes the sum of distances to the individual rankings, where distance is given by the number of comparison pairs on which two rankings differ. Satisfying a local version of IIA and endowed with a deep and elegant axiomatization due to Young and Levenglick (1978), the Kemeny rule must count as a brillant response to Arrows' possibility challenge. Yet it appears to be more respected than loved, and its standing among voting rules seems rather ambivalent. Where is the obstacle?

To put it more starkly: why is the Kemeny rule (KR) not accepted more widely as *the* solution to the problem of preference aggregation? Well, in the first instance simply because "the" problem of preference aggregation is not well-defined in the absence of a clear conception of what is aimed at normatively. More specifically and less facetiously, I will argue that the normative appeal of the KR is quite strong on a judgment aggregation conception, but is much more tenuous on an interest aggregation conception.

The crux is to get a grip on what may be called "Young's dichotomy" between understanding and justifying the KR at the level of the output rankings versus at the level of the induced output choices. For example, the Kemeny rule exhibits the No Show paradox according to which a subset of voters with identical preferences may gain their top choice by not participating in the election. I shall argue that the seriousness of this and related "paradoxes" depends on the aim of the aggregation.

First, for interest aggregation, the No Show paradox is not just counterintuitive, but appears to directly undermine the very aim of interest aggregation. This is what knock-down arguments would look like—but, of course, in the Arrowian territory we have learned to be cautious about knock-down arguments, for otherwise, no one is left standing! That said, the paradox puts a burden of proof on the advocate of the KR as an interest aggregation rule.

Second, for judgment aggregation in the statistical vein, the Kemeny rule can be seen as a maximum likelihood estimator for a particular statistical model. Here, the No Show paradox may still be a bit counterintuitive, but it in no way undermines the aim of inference itself. So the KR can be considered at least *viable* for the purpose of statistical aggregation. However, from within the statistical mode of reasoning, the rationale for adopting the KR specifically becomes rather ad hoc, since the statistical inference justifying it is very sensitive to the particular statistical—and rather questionable—assumptions behind it.

Third, for judgment aggregation aiming at *reflexive consensus,* the No Show paradox again does not contradict the aim of distilling a best advice from the conflicting judgments. So, again, the KR is at least viable as a reflexive consensus rule. Furthermore, in contrast to maximum likelihood rationale, the ad-hocness of the statistical assumptions ceases to be an issue, and the Young–Levenglick axiomatics provide a strong supporting argument. Hence it is this third conception of judgment aggregation aiming at reflexive consensus from the Kemeny rule finds the strongest support.

Young himself considers a somewhat different conception of judgment aggregation aiming at *compromise.* He states that "in situations in which differences of opinion arise not from erroneous judgments but from a difference in values ... group choice is [best viewed] as an exercise in defining consensus, that is, in finding a compromise between different conflicting opinions" (Young 1996, p. 195). In the language adopted here, Young appears to identify differences in values with *unarguable disagreement.* I concur that this a legitimate modality of judgment aggregation to consider, and appreciate "compromise" as a suggestive term of what is aimed at. However, I believe that the compromise perspective must be distinguished from the consensus perspective. In particular, by renouncing "arguability", does one not also lose the advisee position underpinning the sketched argument for reflexive consensus? If so, it would appear that the compromise "among conflicting opinions" must be a compromise among the *holders of* these opinions. But would the holders of the values not care, in the first instance, about the choices themselves, rather than about coming up with a common rationale for those choices? If so, would the choice paradoxes such as the No Show paradox not rear their head again with full force?[4]

**Majority Judgment by Grading**. The above observations illustrate that, within the standard ranking framework, the lay of the land looks rather different from within the judgment aggregation and interest aggregation conceptions. The two conceptions also suggests different novel alternative informational frameworks. An intriguing example is the recent model of "Majority Judgment" proposed and worked out in impressive theoretical and empirical detail in Balinski and Laraki (2011), henceforth BL. I'll go with their grain, and discuss their model as an instance of judgment aggregation.

---

[4]Nehring and Pivato (2018a, b) elaborate the distinction between "consensus" and "compromise" further in a general judgment aggregation context.

Majority Judgment assumes that alternatives can meaningfully be evaluated by grades that capture merit in a non-comparative, "absolute" sense. For example, in artistic or sports competitions, performance may be evaluated as "mediocre", "excellent" or "outstanding" etc. The group (jury) determines the winner as the one with the highest "consensus" grade; BL propose, roughly, the median grade as the consensus grade—Note that the institution of determining a socially authoritative winner conceives of the relevant aesthetic judgment as an instance of "arguable disagreement": on the one hand, it assumes a sufficient degree of intersubjectivity to render the social construction of a 'best competitor' meaningful, on the other hand, the judgment of merit is treated as not admitting objective verification, for otherwise the evaluation by a jury rather than by objective measurement would be beside the point.

In political settings, BL propose to apply their model to elections for public office. Can political competitors be meaningfully graded? The analogy to artistic or sports competions is suggestive, but is it valid? BL couch the issue as a matter of whether a commonly understood language of grades exists, or can be brought into usage. I am not sure that this is the crux. I would rather take as a more basic question whether or not the exists a broad, perhaps vaguely defined, quality of merit such as "leadership" that the election can be viewed as being "about". The merit of positing such a focal quality arguably depends on the context. For example, in the selection of the president of a university, the CEO of a for-profit company, or perhaps even the mayor of city, it appears quite sensible. By contrast, in many contested political settings, candidates will differ in no small part in the aims they pursue; is the notion of "merit" then sufficiently well-focused to be evaluated by "grades" in some form or other?

At a meta-level, is this issue of the appropriateness of the grading frame "decidable" in some rigorous way? Is it truth-apt? I suspect not. Rather, I suspect that this issue is itself a matter of "arguable disagreement". On the whole, BL seem to believe the same; the main thrust of their argument is to ask for circumstantial evidence whether Majority Judgment appears to "work": whether voters connect to the format, whether they are able to give grades efficiently, and whether the results seem to "make sense". One may thus conclude that if Majority Judgment is to be adopted as a method of election, the adoption decision itself would need to be a matter of judgment aggregation rather than demonstrative argument. Why not, if that's what it takes?

To reiterate the more general point that a voting rule cannot be evaluated independently from the underlying normative perspective, it is worth pointing out that the same formal model can also be interpreted as an instance of interest aggregation, as is sometimes done under the name of "median utility rule". On this interpretation, social alternatives are evaluated according to the median (over voters) of their ordinal, but interpersonally comparable utilities. Such an assumption makes some sense within a theory of distributive justice such as Rawls' leximin criterion, but seems rather peculiar in a voting setting, and in any case quite different from the "common language" assumption in BL. Laslier (2012) elaborates some of the striking consequences of the median utility approach to interest aggregation.

## Summing Up

To sum up: I have attempted to make plausible my hunch that there is much unrealized potential in a fuller meeting of minds between voting theory and judgment aggregation theory. Some readers may wonder how this could be—after all, the very terms "consensus" pops up quite regularly in various strands of the literature. True enough—but I am puzzled by the fact that "consensus" doesn't seem to figure much in the big, primary picture of social choice and voting theory. In a nutshell, my conjecture is that the very problem to which "consensus" is supposed to be the answer has not been articulated sharply and forcefully enough. I have identified it—very sketchily only—as the problem of settling "arguable disagreement" by "reflexive consensus".

If and when the problem is clearly brought into view, it is likely to exhibit multiple layers, some of them novel and perhaps unexpected. The issue of framing and the role of language and judgment practice in the Majority Judgment model is but one example. Another very broad and basic issue concerns the role and significance of the equal treatment (anonymity) assumption. While its meaning and justification in the interest aggregation paradigm are quite straightforward as an expression of ethical impartiality, within the judgment aggregation paradigm they are not. Equality may be invoked as a matter of political equality of members as citizens, or as a matter of epistemic equality as judges. What would adequately justify a treatment of members as epistemic equals? In most spheres of social life, epistemic *inequality* is the salient feature, not epistemic equality. The notion of reflexive consensus as balance of advice naturally allows for different sources being weighed unequally. How should those weights be determined? Which epistemic virtues would count? "Expertise" might be adequate as a criterion when differences of opinion reflect informational asymmetries, but in circumstances of disagreement, expertise may not be enough. Qualities such as "wisdom" and "common sense" may matter as well.[5]

Not easy to get a grip! Not easy to know whether one has gotten a grip! Must we really enter the murky waters of politics and philosophy, of language and of practical judgment? Who is afraid of judgment aggregation?

# References

Aumann, R. J. (1976). Agreeing to disagree. *Annals of Statistics, 4*, 1236–1239.
Balinski, M., & Laraki, R. (2011). *Majority judgment*. Cambridge: MIT Press.
Guerdjikova, A., & Nehring, K. (2014). *Weighing experts, weighing sources. The diversity value*. Mimeo, https://editorialexpress.com/cgi-bin/conference/download.cgi?db_name=SSCWatBC&paper_id=210.
Laslier, J.-F. (2012). On choosing the alternative with the best median evaluation. *Public Economics, 153*, 269–277.

---

[5] An axiomatic model that derives source weights from a characterization of sources by attributes of epistemic quality is developed in Guerdjikova and Nehring (2014). In voting by reflexive consensus, the voters/judges are the relevant "sources of advice".

Mongin, P. (2018). The present and future of judgment aggregation theory: a law and economics perspective. In J.-F. Laslier, H. Moulin, R. Sanver & W. Zwicker (Eds.), *The future of economic design*.

Nehring, K., & Pivato, M. (2018a). Majority rule in the absence of a majority. Mimeo, http://ssrn.com/abstract=3113575.

Nehring, K., & Pivato, M. (2018b). The median rule in judgment aggregation. Mimeo, http://ssrn.com/abstract=3113562.

Nehring, K., & Puppe, C. (2010). Abstract arrowian aggregation. *Journal of Economic Theory, 145,* 467–495.

Sen, A. (1977). Social choice theory: A re-examination. *Econometrica, 45,* 53–88.

Young, H. P., & Levenglick, A. (1978). A consistent extension of Condorcet's election principle. *SIAM Journal on Applied Mathematics, 35,* 285–300.

Young, H. P. (1996). Group choice and individual judgments. In D. C. Miller (Ed.) *Perspectives on public choice - a handbook* (pp. 181–200). Cambridge, UK: Cambridge University Press.

# Why Most Politicians Don't Give a Hoot About Voting Theory?

Hannu Nurmi

**Abstract** The theory of voting has largely developed independently of the mechanism design research, but with the introduction of the concept of strategic voting the two traditions found a common ground. This happened some fifty years ago. Yet, despite the voluminous literature that has emerged since then, the impact of voting theory to the design of political institutions remains marginal. Most politicians the present writer has talked to seem politely interested in the most dramatic results achieved in the theory, but soon lose interest when the underlying assumptions are explicated. Often the assumptions are deemed too simplistic or too abstract or plainly 'out of this world'. It looks as if there is a demand for research that aims at building bridges over the wide gap that exists between the abstract social choice results and the behavioural-institutional realities characterizing political systems of today and tomorrow.

## 1 Evaluation of Institutional Performance

From its re-invention in late 1940s the social choice theory—the backbone of voting theory—has acquired somewhat negative reputation, not (of course) because of the standards of scholarship represented by its early representatives, but due to the form in which the results were expressed. The typical form is incompatibility between various choice desiderata: the more appealing the latter, the more dramatic the results establishing the incompatibility. Examples of such results abound (e.g. Arrow 1963; Gibbard 1973; Satterthwaite 1975; Sen 1970; Kelly 1978; Muller and Satterthwaite 1977, to name a few). On a more mundane level, a number of more or less comprehensive evaluations of specific voting systems have been published

The author is grateful to Felix Brandt, Piotr Faliszewiski and Gabriel Carroll for illuminating and constructive comments on an earlier version.

H. Nurmi (✉)
Department of Philosophy, Contemporary History and Political Science, University of Turku, Turku, Finland
e-mail: hnurmi@utu.fi

© Springer Nature Switzerland AG 2019
J.-F. Laslier et al. (eds.), *The Future of Economic Design*, Studies in Economic Design, https://doi.org/10.1007/978-3-030-18050-8_18

(Richelson 1979; Straffin 1980, etc.). While the former typically deal with relationships between theoretical properties, the later focus on the properties of specific
voting rules.

What makes both types of results difficult to communicate to people who are in
charge of applying, modifying and designing voting institutions is apparently the
nature of the results, on the hand, and the expectations of the practitioners, on the
other. The latter are accustomed to seeing voting systems as disagreement-settling
devices, not necessarily methods for finding the best solutions. The emphasis is in
decisiveness, not in optimization. And indeed, there is no denying that it is sometimes
far better to find a voting outcome acceptable to all than to engage in divisive conflicts. In other words, the specific other properties of voting rules may be viewed as
somewhat secondary once the decisiveness and acceptability are established. Hence,
a scholar pointing to specific choice-theoretic weaknesses of current voting rules
may be viewed as a nuisance, not a trusted advisor.

The reasons are pretty obvious: what the results suggest are possibilities, not
inevitabilities. Bizarre things can happen, but we do not encounter them all the time.
So, Condorcet winners may not be elected and Condorcet losers can be elected in
plurality voting. But is it even remotely possible in the voting body under consideration? Would the adoption of a Condorcet extension method with its demonstrated
vulnerability to the no-show paradox (Moulin 1988; Pérez 2001) make things any
better? These are relevant questions, but, due the very nature of the results, answers
are not readily available. Incompatibility or compatibility of criteria of performance
are by definition dichotomies. The same is true of possibilities: they are true or not
true. Yet, possibilities may materialize more or less often in practice, and this may
be of paramount importance for the practitioner.

## 2   From Dichotomies to Grades of Success

Given that all voting systems are plagued with some type of failure with respect
to most common intuitive choice desiderata, a natural question is: how often can
one expect such failures to occur in a given voting body? Since answers to this
depend on the kinds of preference profiles to be encountered, the focus is shifted to
classifying the profiles and determining the probabilities or relative frequencies of
failures within those classes (Gehrlein and Lepelley 2011; Merrill 1988; Tideman
and Plassmann 2012). This would seem a plausible way of answering the practitioner's query regarding the likelihood of ending up with troublesome outcomes
when adopting a given voting rule. The results of probabilistic modeling are in many
respects valuable e.g. in showing the variation of criterion violations under modifications of 'cultures', in establishing variables that increase or decrease vulnerabilities
to anomalies, in creating order of priority between rules under various cultures, etc.
Thereby a more nuanced picture of the rules and their performance emerges. The
probabilities of paradox-avoidance could even be viewed as grades of success of a
rule under a given culture.

Even though the current literature considers a far richer variety of cultures than was typically the case some thirty years ago, the probability estimates (whether analytical or relative frequencies based on simulations) suffer from a weakness that may be unsurmountable, viz. even though the culture adequately mimics the empirical electoral environment, the possibility of repeating essentially the same (or very similar) pattern or preferences is difficult to capture. In other words, the empirical elections may take place is a much narrower range of profiles than those considered in the probabilistic models. To make things even more complicated from the practical point of view, we should observe that thus far a relatively limited number of different cultures has been thoroughly studied (see Gehrlein and Lepelley 2011). Hence, the practitioner may have to consider results pertaining cultures that are not close to that of the relevant political system.

This takes the practitioner one step forward towards finding a desirable voting rule, but thorny problems remain. To wit, the success grades deal with one desideratum at a time. Suppose that system A has a better success grade than system B on criterion x, but on criterion y the ranking between the two is reversed. Making a choice calls for a ranking, or better yet, grading of criteria and rules of compensation. What one then ends up with is essentially a multiple criterion optimization problem.

There is another way of making use of softer notions in voting theory, viz. to start from degrees of preference between alternatives. This approach has been developing largely independently of the mainstream social choice theory for decades producing results (solution concepts, compatibility theorems and the like) that basically resemble those obtained in the more traditional approach (Barrett and Salles 2011; Kacprzyk et al. 2008; Pattanaik 1997; Nurmi 1981). While intuitively plausible in its flexibility with regard to representing intensity of preferences, the success of this approach hinges essentially on methods of ascertaining the degrees of preferences (fuzzy preference relations) of individuals. Yet, many practitioners would deem the starting point more plausible than the standard one. After all, it is not difficult to envision that individuals have varying degrees of preference regarding policy alternatives or candidates depending e.g. on their position in their (nonfuzzy) preference ranking. Making use of this information in voting systems would, thus, seem plausible. While the concept of fuzzy relations is often viewed as a way to model impreciseness, there is nothing fuzzy in the solutions themselves. They typically involve characteristic functions over the alternatives akin to probability distributions. There is, however, very little work on empirical methods to estimate individual degrees of preference and therefore the practical usefulness of fuzzy methodology is largely unknown.

## 3   Decision Modalities and Voting

The standard approach to voting starts from complete and transitive preference relations and the notion of 'thin' rationality which states that rational behaviour amounts to always (i.e. with probability 1) choosing x over y, when x is preferred to y and both are available. Implicitly the political choice is viewed as a decision under certainty.

This, however, flies in the face of the intuitive observation that political choices are at best choices under risk (with known probabilities over outcomes) or even under uncertainty (with unknown probabilities). Here a better match between behavioural decision theory and social choice would be desirable. It has been established by many prominent decision scientists that there are circumstances under which individual choice behaviour under risk and uncertainty systematically deviates from the basic principles of expected utility theory (Allais 1979; Kahnemann and Tversky 1979). So, if political choices are represented as lotteries with known or unknown winning probabilities, one cannot expect that the preferences are typically transitive, perhaps not even complete. How to build a model of voter that reflects the empirical (mostly experimental) results, is largely an open question. It should be added that there is nothing inherently irrational about intransitive preferences; consider e.g. the Condorcet paradox profile and a case where three roughly equally important criteria (say, linguistic skills, substantive competence, cooperation network) of goodness are substituted for voters and the candidates A, B and C are ranked ABC on criterion one, BCA on criterion two and CAB on criterion three. It would make sense (be 'rational') to prefer A t o B (since criteria one and three agree on this). Similarly, it would be sensible to prefer B to C (as criteria one and two say so), but also C to A (because of criteria two and three). It is an empirical matter how often one might encounter such settings, but they cannot be excluded a priori.

## 4   Difficult Examples

Anyone who has worked on toy examples demonstrating criterion failures (non-monotonicity, Condorcet loser choice, no-show paradox, to name a few) soon learns that some of these are relatively easy to come up with, while others are very tedious and—when eventually discovered—look very unlikely to ever occur in practice. Examples of 'difficult' profiles are Smith's (1973) demonstration that all point runoff systems violate monotonicity, Fishburn's (1977, 484) demonstration that Kemeny's rule is vulnerable to the reinforcement paradox, and Felsenthal's (2012, 74–75) example where Copeland's rule exhibits the no-show paradox. While the intuitive difficulty of constructing such demonstrations would seem to say something about their likelihood, it is not obvious how one could measure it, much less use this in evaluating voting rules.

Over the past couple of decades many mathematicians and computer scientists have focused on the computational complexity aspects of various solution concepts in voting games (for a comprehensive survey, see Brandt et al. 2016). Many fascinating results on complexity of finding winners, of (successfully) misrepresenting preferences, of forming committees with minimal misrepresentation etc. have been presented. So far no attention has, however, been paid on the difficulty or intuitive complexity of finding examples. Perhaps this is too ill-defined problem to be dealt with using algorithmic complexity concepts.

## 5　Should We Work on Our Criteria as Well?

Although the bulk of voting theory still rests of the traditional foundations in assuming complete and transitive individual preferences *cum* deterministic outcome mappings, some suggested voting rules take a different point of departure. E.g. the approval voting (Brams and Fishburn 1983; Laslier and Sanver 2010; Sanver 2010) aggregates subsets of alternatives approved by individuals, and the majority judgement (Balinski and Laraki 2010) processes grade scores submitted by the individuals into outcome rankings. The former assumes in a way less structure on the individual opinions, whereas the latter demands more than just the ranking of alternatives. Basically the approval voting is based on nominal level measurement of alternatives whereby the voters classify the alternatives into acceptable and unacceptable sets. The majority judgment, in turn, expects the voters to assign a grade to each alternative. Although both these systems can be viewed from the angle of ordinal rank-aggregation methods (e.g. Felsenthal and Machover 2008; Nurmi 1983), it can be argued that the very fact that the methods require much more or much less information about voters opinions than the more traditional procedures makes the criteria applicable to the evaluation of the traditional procedures inappropriate or even misleading. After all, criteria like Condorcet efficiency are hardly relevant in environments where individuals assign utility values to alternatives. It can be argued that the binary majority comparisons lose crucial (utility) information and are thus irrelevant. The problem is that more relevant criteria are not easily found and have not been suggested to the best of the present writer's knowledge.

## 6　To Sum Up

The practitioner asking for advice in the design of voting institutions may not feel adequately served by the results of the current theory of voting, but then again his/her expectations may be unreasonable as well. The theory is not capable of answering conclusively to the question of which voting procedure should be chosen. All procedures have serious theoretical shortcomings. All theories aim at maximum coverage or generality. In the present context this may be a vice rather than virtue since the voting procedures are to be applied in specific environments. These may imply distinctive and significant domain restrictions with respect to profiles. Hence results based on unrestricted domains may yield unnecessarily negative results. Thus, more work in classifying the specific contexts where the procedures will be applied would be most welcome.

We also need to re-examine our fundamental concepts and assumptions if for no other reason than to see how much difference small modifications make to the results to be applied. Probability models and fuzzy and rough preferences have been explored for some time with results that significantly differ from those based on standard assumptions. Admittedly these models are associated with some specific problems e.g. in coming up with reliable methods of gathering the voter input data (membership functions, rough sets, balloting probabilities). Experimental results suggest that the conditions under which the individual preferences are representable as utility functions are often violated when the choice involves risky or uncertain prospects. These results are yet to make their way to the voting theory and yet voting typically involves uncertain or at least risky alternatives.

Many incompatibility results are demonstrated by way of an example or a set of examples. It is easy to recognize differences in difficulty of constructing those examples. This does not automatically imply that incompatibilities which are easier to demonstrate would materialize more often than those that are difficult to build. Yet, this intuitive information may suggest something useful with regard to domains where problems can be expected.

Finally, relatively little attention has been paid on the importance of choice-theoretic criteria (monotonicity, consistency, Condorcet efficiency etc.) when compared with each other. The group decision context is surely a factor that determines the practical importance, but more general hierarchies of importance of criteria have not been constructed, perhaps for the plausible reason that this is a value issue on which reasonable persons may disagree.

# References

Allais, M. (1979). The foundations of positive theory of choice involving risk and a criticism of the postulates and axioms of the American school. In M. Allais & O. Hagen (Eds.), *The expected utility hypothesis and the Allais paradox*. Dordrecht: D. Reidel.

Arrow, K. J. (1963). *Social choice and individual values* (2nd ed.). New Haven: Yale University Press. (1st ed., 1951)

Balinski, M., & Laraki, R. (2010). *Majority judgment. Measuring, ranking, and electing*. Cambridge, MA: The MIT Press.

Barrett, R., & Salles, M. (2011). Social choice with fuzzy preferences. In K. J. Arrow, A. Sen & K. Suzumura (Eds.), *Handbook of social choice and welfare* (Vol. 2, pp. 367–390).

Brams, S. J., & Fishburn, P. C. (1983). *Approval voting*. Boston: Birkhäuser.

Brandt, F., Conitzer, V., Endriss, U., Lang, J., & Procaccia, A. D. (Eds.). (2016). *Handbook of computational social choice*. New York: Cambridge University Press.

Felsenthal, D. S. (2012). Review of paradoxes afflicting procedures for electing a single candidate. In D. Felsenthal & M. Machover (Eds.), *Electoral systems: Paradoxes, assumptions, and procedures* (pp. 19–91). Berlin: Springer.

Felsenthal, D. S., & Machover, M. (2008). The majority judgment voting procedure: A critical evaluation. *Homo Oeconomicus, 25*, 319–333.

Fishburn, P. C. (1977). Condorcet social choice functions. *SIAM Journal of Applied Mathematics, 33*, 469–489.

Gehrlein, W. V., & Lepelley, D. (2011). *Voting paradoxes and group coherence: The condorcet efficiency of voting rules*. Berlin: Springer.

Gibbard, A. (1973). Manipulation of voting schemes: A general result. *Econometrica, 41*, 587–601.

Kacprzyk, J., Zadrożny, S., Fedrizzi, M., & Nurmi, H. (2008). On group decision making, consensus reaching, voting, and voting paradoxes under fuzzy preferences and a fuzzy majority: A survey and a granulation perspective. In W. Pedrych, A. Skowron, & V. Kreinovich (Eds.), *Handbook of granular computing* (pp. 906–929). Chichester: Wiley.

Kahnemann, D., & Tversky, A. (1979). Prospect theory: An analysis of decision under risk. *Econometrica*, *47*, 263–291.

Kelly, J. S. (1978). *Arrow impossibility theorems*. New York: Academic Press.

Laslier, J.-F., & Sanver, M. R. (Eds.). (2010). *Handbook on approval voting*. Berlin: Springer.

Merrill, S. (1988). *Making multicandidate elections more democratic*. Princeton: Princeton University Press.

Moulin, H. (1988). Condorcet's principle implies the no-show paradox. *Journal of Economic Theory*, *45*, 53–64.

Muller, E., & Satterthwaite, M. A. (1977). The equivalence of strong positive association and strategy-proofness. *Journal of Economic Theory*, *14*, 412–418.

Nurmi, H. (1981). Approaches to collective decision making with fuzzy preference relations. *Fuzzy Sets and Systems*, *6*, 249–259.

Nurmi, H. (1983). Voting procedures: A summary analysis. *British Journal of Political Science*, *13*, 181–208.

Pattanaik, P. K. (1997). Fuzziness and the normative theory of social choice. In J. Kacprzyk, H. Nurmi, & M. Fedrizzi (Eds.), *Consensus under fuzziness*. Dordrecht: Kluwer.

Pérez, J. (2001). The strong no show paradoxes are a common flaw in Condorcet voting correspondences. *Social Choice and Welfare*, *18*, 601–616.

Richelson, J. T. (1979). A comparative analysis of social choice functions I, II, III: A summary. *Behavioral Science*, *24*, 355.

Sanver, M. R. (2010). Approval as an intrinsic part of preference. In J.-F. Laslier & M. R. Sanver (Eds.), *Handbook on approval voting* (pp. 469–481). Berlin: Springer.

Satterthwaite, M. A. (1975). Strategy-proofness and Arrow's conditions. *Journal of Economic Theory*, *10*, 187–217.

Sen, A. K. (1970). The impossibility of a Paretian liberal. *Journal of Political Economy*, *78*, 152–157.

Smith, J. H. (1973). Aggregation of preferences with variable electorate. *Econometrica*, *41*, 1027–1041.

Straffin, P. D. (1980). *Topics in the theory of voting*. Boston: Birkhäuser.

Tideman, T. N., & Plassmann, F. (2012). Modeling the outcomes of vote-casting in actual elections. In D. Felsenthal & M. Machover (Eds.), *Electoral systems: Paradoxes, assumptions, and procedures* (pp. 217–251). Berlin: Springer.

# Beyond Gibbard and Satterthwaite: Voting Manipulation Games

## Arkadii Slinko

**Abstract** The Gibbard–Satterthwaite theorem implies the existence of voters, called manipulators, who can change the election outcome in their favour by voting strategically. However, when a given preference profile admits several such manipulators, voting becomes a game played by these voters. They have to reason strategically about each other's actions.

Voting is a common method of collective decision making, which enables the participating voters to identify the best candidate for the society given their individual rankings of the candidates. However, as early as Farquharson (1969), it was noticed that for most common rules voters sometimes can misrepresent their preference and improve the outcome for themselves. In social choice this is now called a manipulation, in political science this is called tactical or strategic voting. Pattanaik (1973) conjectured that no "reasonable" voting rule is immune to manipulation. This indeed was shown independently by Gibbard (1973) and Satterthwaite (1975): if there are at least 3 candidates, then any onto, non-dictatorial voting rule admits a preference profile (a collection of voters' rankings) where some voter would be better off by submitting a ranking that differs from his truthful one. In other words, the sincere profile of preferences is not a Nash equilibrium.

The problem may be further exacerbated by the presence of a number of such voters —we will call them Gibbard–Satterthwaite manipulators, or GS-manipulators. Indeed, if several such voters—attempt to manipulate the election simultaneously in an uncoordinated fashion, the outcome may differ not just from the outcome of the truthful voting, but also from the outcome that any o f the GS-manipulators was trying to achieve, due to complex interference among the different manipulative votes. The outcome may be undesirable for all manipulators. In other word, if voters are strategic, voting becomes a simultaneous one-shot game. Here is one basic example.

Arkadii Slinko was supported by the Marsden Fund 3706352 of The Royal Society of New Zealand.

A. Slinko (✉)
The University of Auckland, Auckland, New Zealand
e-mail: a.slinko@auckland.ac.nz

© Springer Nature Switzerland AG 2019
J.-F. Laslier et al. (eds.), *The Future of Economic Design*, Studies in
Economic Design, https://doi.org/10.1007/978-3-030-18050-8_19

131

*Example 1* Suppose four people are to choose between three alternatives. Let the profile of sincere preferences be

| 1 | 2 | 3 | 4 |
|---|---|---|---|
| a | b | c | c |
| b | a | a | b |
| c | c | b | a |

and the rule used be Plurality with breaking ties in accord with the order $a > b > c$. If everybody votes sincerely, then $c$ is elected. Voters 1 and 2 are the Gibbard–Satterthwaite manipulators at this profile since voters 3 and 4 get their best possible outcome and are expected to vote sincerely. Voter 1 can make $b$ to win by voting $b > a > c$ and voter 2 can make $a$ to win by voting $a > b > c$. However, if they both try to manipulate, $c$ will remain the winner. Each of them would prefer that the other one manipulates. If, for example, for voters 1 and 2 the utility of their top preference is 2, the utility of their middle preference is 1 and the utility of their bottom preference is 0 (these are ordinal, not cardinal), these voters are playing the game which normal form is

|        | $s_2$ | $i_2$ |
|--------|-------|-------|
| $s_1$  | 0, 0  | 2, 1  |
| $i_1$  | 1, 2  | 0, 0  |

where voter 1 is the row player and voter 2 is the column player, $s_1, s_2$ are their sincere votes and $i_1, i_2$ are their manipulative votes.

There is a substantial body of research dating back to Farquharson (1969) that explores the consequences of modeling non-truthful voting as a strategic game; see, also Fishburn (1978), Moulin (1979), Feddersen et al. (1990), Myerson and Weber (1993), De Sinopoli (2000), Dhillon and Lockwood (2004), Sertel and Sanver (2004), De Sinopoli et al. (2015), Desmedt and Elkind (2010), Obraztsova et al. (2013) to mention a few. The main common feature of them is that the utility of a voter depends solely on the winner of the election. The features of the existing models that vary are:

• voters' utilities maybe ordinal or be von-Neumann-Morgenstern utilities;
• voters may use randomised strategies or not;
• voters may be fully or only boundedly rational.

The most popular frameworks so far has been the one investigated by Moulin (1981) with ordinal utilities and Myerson and Weber (1993) with von-Neumann-Morgenstern utilities. The latter model, in particular, stipulate that each voter has a utility for the election of each candidate. Both Moulin and Myerson and Weber suggested the use of Nash equilibrium as a solution concept for the analysis of voting games, however, acknowledging that sometimes this idea led to a large number of Nash equilibria.

The use of Nash equilibrium in analysis of voting games was widely criticised (see, e.g., De Sinopoli 2000) and the classical example that wanders about from one paper supporting this criticism to another is as follows.

*Example 2* Suppose each of $n \geq 3$ voters has the same preference order. Then voting for the least preferred alternative for each of them is a Nash equilibrium.

In further works many attempts have been made to weed weird equilibria out. Farquharson (1969) was aware of the problem and suggested the sophisticated voting principle: reasonable equilibria must survive iterative deletion of dominated strategies. The following methods were considered: equilibria refinements (De Sinopoli 2000), costly voting (Sinopoli and Iannantuoni 2005), truth-biased voters Obraztsova et al. (2013), generic utilities (De Sinopoli 2001; De Sinopoli et al. 2015). The problem however remains not completely solved. These attempts, to my mind, are futile. What actually Example 2 shows is that it is not the solution concept that is to blame but the implicit assumption that voters may behave irrationally. In reality, however, voters may not be fully rational but they are certainly not irrational. And it is not a coincidence that in all aforementioned papers voters' reasoning about the voting situation and other voters' possible actions was absent.

The outlined approach always assumes that voters can submit any linear order as their vote (no matter how rational it is) without any regard to appraisal. Moulin (1981) divides voters into prudent and sophisticated depending on the amount of information about other players that they have. Effectively, if you have no information about other players you vote prudently, e.g., sincere. However, if you have all the information you need for sophisticated voting but you are unable to process it, you are back to the situation of no information.

Apart from the practical difficulty of sophisticated voting, political science has observed that some voters may be *ideological* and interested in stating their preference no matter how much information they have. Some observations tell us that the number of ideological voters is significant. In the famous Florida 2000 vote, when Bush won over Gore by just 537 votes 97,488 Nader supporters voted for Nader while in such a close election every strategic voter would vote either Gore or Bush and the overwhelming majority of Nader supporters preferred Gore to Bush.

Hence another approach to analysis of voting games would be not to assume that voters miraculously always end up in one of the Nash equilibria but to take their reasoning about other players as a starting point of the analysis. Naturally we have to assume that voters are only boundedly rational and their strategies depend on the level of depth of their reasoning. These voters need a good reason to abandon their sincere preference in favour of an insincere one.

In many papers—especially those discussing indices of manipulability (see, e.g., often cited Aleskerov and Kurbanov 1999)—it is implicitly assumed that voters manipulate as soon as they discover that they are in the position to manipulate provided everyone else will vote sincerely, which means that no reasoning about actions of others is assumed. Slinko and White (2008, 2014) were the first to assume that voters make a small step in reasoning about others, namely, about other voters who have exactly the same preference order (and assuming all the rest vote sincerely). According to Slinko and White such voters reason: "I am thinking about manipulating but other voters of my type[1] must think about this too as they are in the same position as me.[2] I have to interact with them." This is illustrated in the following example.

*Example 3* Suppose four people are to choose between three alternatives. Let the profile of sincere preferences be

| 1 | 2 | 3 | 4 |
|---|---|---|---|
| a | a | b | c |
| b | b | c | b |
| c | c | a | a |

and the rule to be used is Borda with breaking ties in accord with the order $a > b > c$. If everybody votes sincerely, then $b$ is elected. Voters 1 and 2 are Gibbard–Satterthwaite manipulators. Voter 1 can make $a$ to win by voting $a > c > b$ and voter 2 can do the same. However, if they both try to manipulate, their worst alternative $c$ will become the winner. Thus, these voters are playing the game which normal form is

|        | $s_2$  | $i_2$  |
|--------|--------|--------|
| $s_1$  | 1, 1   | 2, 2   |
| $i_1$  | 2, 2   | 0, 0   |

which is a sort of an anti-coordination game.

We note that in Example 3 it is easier for the two would-be manipulators to agree on the course of action than in Example 1 since they have the same type, i.e., have identical sincere preferences, and hence no conflict of interest but the absence of communication devices prevents them from doing this. Most mis-coordinations in our framework can be classified as instances of either strategic overshooting (too many voted strategically) or strategic undershooting (too few). If mis-coordination can result in strategic voters ending up worse off than they would have been had they all just voted sincerely, we call the strategic vote *unsafe*. Slinko and White (2008, 2014) showed that under every onto and non-dictatorial social choice rule there exist circumstances where a voter has an incentive to cast a safe strategic vote. Thus they extended the Gibbard–Satterthwaite Theorem by proving that every onto and non-dictatorial social choice rule can be individually manipulated by a voter casting a safe strategic vote.

Example 3 illustrates an important point. The game has two Nash equilibria but there is no way voters 1 and 2 can derive from their votes which Nash equilibria to play. Myerson and Weber (1993) claim that opinion polls can serve as coordination device for voters which is partly true; however sometimes an aggregated information of opinion polls may be insufficient.

---

[1] I.e., voters with the same order of appreciation of candidates.

[2] Assuming anonymity.

The generalisation of the Gibbard–Satterthwaite theorem given by Slinko and White prompts us to reconsider the concept of a manipulable profile and all the business related to indices of manipulability. For example, one may ask: Are the profiles in Examples 1 and 3 really manipulable?

Assuming that all voters of types different from their own are sincere (which is most likely untrue), the voter in the model of Slinko and White is (strongly) boundedly rational and has a much simplified view of the game. When we deal with boundedly rational players we have to have it in mind that their view of the game may be different from the game itself. The more rationality we assume, the more voter's view of the game is closer to reality. The next level of rationality of the voter would be for him to consider *all* GS-manipulators (not only of his own type) as players. At the previous level of rationality voters 1 and 2 in Example 1 would consider each other as sincere but at this next level they realise that there is a game to play.

The game played by GS-manipulators was studied in Elkind et al. (2015), Grandi et al. (2017). This means voters may strategise only when they are Gibbard–Satterthwaite manipulators and they can identify all other manipulators and try to optimise their vote relative to the information about those manipulators and their potential manipulations. These games for Plurality voting rule appear to be relatively simple and, in particular, always have a Nash equilibrium. Grandi et al. (2017) identify natural conditions implying the existence of Nash equilibria for $k$-approval with $k = 2, 3$. It appeared that some additional mild rationality conditions are necessary for 2-approval voting manipulation game to have a Nash equilibria. If voters are erratic and use unsound strategies, then a Nash equilibria may not exist. For 3-approval the sufficient conditions are stronger (but it is not clear if they are necessary). However, Elkind et al. (2015), Grandi et al. (2017) showed that even the so-called minimality conditions (which require that voters resort only to minimal manipulations), fail to ensure the existence of Nash equilibria for 4-approval voting rule.

One way to increase sophistication of voters is to allow countermanipulations. This means that a voter who cannot manipulate himself can vote insincerely in order to mitigate the damage that can be done by a GS-manipulator when he manipulates. The games with participation of a countermanipulator by their nature usually do not have any Nash equilibria.

*Example 4* In a 2-by-2 game with one manipulator and one countermanipulator we do not necessarily have a Nash equilibrium. Suppose that the voting rule is 2-Approval and the profile is $V = (adcb, bdca)$ with ties broken according to $a > b > c > d$. For voter 1 switching $c$ and $d$ is a manipulation in favour of $a$. Voter 2 cannot manipulate but can countermanipulate switching $c$ and $d$ making $c$ to win in case voter 1 manipulates. But then voter 1 would be better off switching $c$ and $d$ back after which the same move will be beneficial for voter 2. The normal form for such game would be

|       | $s_2$  | $i_2$  |
|-------|--------|--------|
| $s_1$ | 2, 2   | 3, 0   |
| $i_1$ | 3, 0   | 1, 1   |

We note that any 2-by-2 game, where two manipulators play, always has a Nash equilibrium, thus a manipulator-countermanipulator 2-by-2 games are very different.

Elkind et al. (2017) modeled bounded rationality of voters differently. They also assume that the voters reason about potential actions of other voters but for modeling boundedly rational they use an adaptation of the cognitive hierarchy model. They take non-strategic (sincere) voters as those belonging to level 0. The players of level 1 give best response assuming that all other players belong to level 0, and, when this particular voter is not a Gibbard–Satthethwaite manipulator it is defined to be the sincere vote. The players of level 2 give their best responce to assuming that all other players belong to level 0 or level 1. We note that players of level 2 are already quite sophisticated. They can, for example, think of countermanipulating or they can strategically stay sincere when they can manipulate. The emphasis of the paper by Elkind et al. (2017) is on the complexity of a level 2 voter deciding whether his manipulative strategy weakly dominates his sincere strategy. They present a polynomial time algorithm for 2-approval but prove NP-hardness for 4-Approval voting rule. The case of 3-approval remains open.

The algorithmic aspects of voting games with fully rational voters have also recently received some attention (Desmedt and Elkind 2010; Xia and Conitzer 2010; Thompson et al. 2013; Obraztsova et al. 2013). Empirical analysis of Nash equilibria in plurality election has been done in Thompson et al. (2013).

## Conclusion

The study of voting manipulation games is in its infancy and it is a long way before we can get any realistic models of elections. It is clear that the assumption that voters are fully rational and that every election end up in one of the nice Nash equilibria (or one of the similar solution concepts) is not realistic. However little is known about what happens in reality. For example, it is extremely hard to estimate how many strategic voters are there in any election but the percentage of those who actually manipulated is easier to estimate; for example, Kawai and Watanabe (2013) estimate the number of such voters[3] in Japanese elections as between 2.5 and 5.5%. Moreover, Benjamin et al. (2013), show that preference misrepresentation is related to cognitive skills, and Choi et al. (2014) demonstrate that decision-making ability in laboratory experiments correlates strongly with socio-economic status and wealth. So it may be reasonable to assume that only a small fraction of voters in any election is strategic. Circumstantial evidence exists that ideological voters are always present in non-negligible numbers but there are no experimental data in this respect.

We need to further develop models of bounded rationality of voters. But primarily now we need more experimental work for better understanding of voters and their behaviour.

---

[3]They call such voters misaligned.

# References

Aleskerov, F., & Kurbanov, E. (1999). Degree of manipulability of social choice procedures. In *Current trends in economics* (pp. 13–27). Berlin: Springer.

Benjamin, D. J., Brown, S. A., & Shapiro, J. M. (2013). Who is behavioral? Cognitive ability and anomalous preferences. *Journal of the European Economic Association, 11*, 1231–1255.

Choi, S., Kariv, S., Muller, W., & Silverman, D. (2014). Who is (more) rational? *American Economic Review, 104*, 1518–1550.

De Sinopoli, F. (2000). Sophisticated voting and equilibrium refinements under plurality rule. *Social Choice and Welfare, 17*(4), 655–672.

De Sinopoli, F. (2001). On the generic finiteness of equilibrium outcomes in plurality games. *Games and Economic Behavior, 34*(2), 270–286.

De Sinopoli, F., Iannantuoni, G., & Pimienta, C. (2015). On stable outcomes of approval, plurality, and negative plurality games. *Social Choice and Welfare, 44*(4), 889–909.

Desmedt, Y., & Elkind, E. (2010). Equilibria of plurality voting with abstentions. In *Proceedings of the 11th ACM conference on Electronic Commerce (EC-2010)* (pp. 347–356).

Dhillon, A., & Lockwood, B. (2004). When are plurality rule voting games dominance-solvable? *Games and Economic Behavior, 46*, 55–75.

Elkind, E., Grandi, U., Rossi, F., & Slinko, A. (2015). Gibbard-Satterthwaite games. In *Proceedings of the Twenty-Fourth International Joint Conference on Artificial Intelligence, IJCAI 2015, Buenos Aires, Argentina, July 25–31, 2015* (pp. 533–539).

Elkind, E., Grandi, U., Rossi, F., & Slinko, A. (2017). Cognitive hierarchy and voting manipulation. arXiv preprint arXiv:1707.08598.

Farquharson, R. (1969). *Theory of voting*. New Haven: Yale University Press.

Feddersen, T. J., Sened, I., & Wright, S. G. (1990). Rational voting and candidate entry under plurality rule. *American Journal of Political Science, 34*(4), 1005–1016.

Fishburn, P. C. (1978). A strategic analysis of nonranked voting systems. *SIAM Journal on Applied Mathematics, 35*(3), 488–495.

Gibbard, A. (1973). Manipulation of voting schemes: A general result. *Econometrica, 41*(4), 587–601.

Grandi, U., Hughes, D., Rossi, F., Slinko, A. (2017). Gibbard-Satterthwaite games for k-approval voting rules. CoRR, arXiv:abs/1707.05619.

Kawai, K., & Watanabe, Y. (2013). Inferring strategic voting. *American Economic Review, 103*, 624–662.

Moulin, H. (1979). Dominance solvable voting schemes. *Econometrica, 47*, 1337–1351.

Moulin, H. (1981). Prudence versus sophistication in voting strategy. *Journal of Economic theory, 24*(3), 398–412.

Myerson, R., & Weber, R. (1993). A theory of voting equilibria. *American Political Science Review, 87*(1), 102–114.

Obraztsova, S., Markakis, E., & Thompson, D. R. M. (2013). Plurality voting with truth-biased agents. In *Proceedings of the 6th International Symposium on Algorithmic Game Theory*.

Pattanaik, P. K. (1973). On the stability of sincere voting situations. *Journal of Economic Theory, 6*(6), 558–574.

Satterthwaite, M. A. (1975). Strategy-proofness and Arrow's conditions: Existence and correspondence theorems for voting procedures and social welfare functions. *Journal of Economic Theory, 10*(2), 187–217.

Sertel, M. R., & Sanver, M. R. (2004). Strong equilibrium outcomes of voting games are the generalized Condorcet winners. *Social Choice and Welfare, 22*(2), 331–347.

Sinopoli, F. D., & Iannantuoni, G. (2005). On the generic strategic stability of Nash equilibria if voting is costly. *Economic Theory, 25*(2), 477–486.

Slinko, A., & White, S. (2008). Non-dictatorial social choice rules are safely manipulable? In *Proceedings of the 2nd International Workshop on Computational Social Choice (COMSOC-2008)*.

Slinko, A., & White, S. (2014). Is it ever safe to vote strategically? *Social Choice and Welfare*, *43*, 403–427.

Thompson, D. R. M., Lev, O., Leyton-Brown, K., & Rosenschein, J. S. (2013). Empirical analysis of plurality election equilibria. In *Proceedings of the 12th International Conference on Autonomous Agents and Multiagent Systems (AAMAS-2013)*.

Xia, L., & Conitzer, V. (2010). Stackelberg voting games: Computational aspects and paradoxes. In *Proceedings of the Twenty-Fourth AAAI Conference on Artificial Intelligence (AAAI-2010)*.

# Part III
# Algorithms and Complexity

# Credimus

Edith Hemaspaandra and Lane A. Hemaspaandra

**Abstract** We believe that economic design and computational complexity—while already important to each other—should become even more important to each other with each passing year. But for that to happen, experts in on the one hand such areas as social choice, economics, and political science and on the other hand computational complexity will have to better understand each other's worldviews. This article, written by two complexity theorists who also work in computational social choice theory, focuses on one direction of that process by presenting a brief overview of how most computational complexity theorists view the world. Although our immediate motivation is to make the lens through which complexity theorists see the world be better understood by those in the social sciences, we also feel that even within computer science it is very important for nontheoreticians to understand how theoreticians think, just as it is equally important within computer science for theoreticians to understand how nontheoreticians think.

## 1 Introduction

Predictions are cheap. Our cheap prediction is:

Economic design and computational complexity should and will in the future be even more deeply intertwined than they currently are.

E. Hemaspaandra
Department of Computer Science, Rochester Institute of Technology,
Rochester, NY 14623, USA
e-mail: eh@cs.rit.edu

L. A. Hemaspaandra (✉)
Department of Computer Science, University of Rochester,
Rochester, NY 14627, USA
URL: http://www.cs.rochester.edu/u/lane

© Springer Nature Switzerland AG 2019                                                     141
J.-F. Laslier et al. (eds.), *The Future of Economic Design*, Studies in
Economic Design, https://doi.org/10.1007/978-3-030-18050-8_20

What is a bit less cheap is working to make predictions come true. If the prediction is broad or ambitious enough, doing that is often a task beyond one paper, one lifetime, or one generation.

Nonetheless, in this article we will seek to make a small contribution toward our predication's eventual realization. In particular, as complexity theorists who have for more than a decade also been working in computational social choice theory, we have seen first-hand how deeply important computational social choice theory and computational complexity have been to each other. And the "to each other" there is not written casually. As argued in a separate paper (Hemaspaandra 2018), the benefit of the interaction of those areas has been very much a two-way street.

However, to increase the strength and quality of the interaction, and to thus reap even more benefits and insights than are currently being gained, a needed foundation will be *mutual understanding*. After all, even the different subareas of computer science have quite different views of what computer science is about, and sometimes it seems that computer scientists don't understand even each other's worldviews.

As complexity theorists, we can expertly address only one direction in which explanation is needed: trying to explain the—perhaps strange to those who are not complexity theorists—way that complexity theorists tend to view the world. We sincerely hope that the reciprocal directions will be addressed by appropriate experts from the many other disciplines whose practitioners are part of the study of economic design.

Beyond that, we also embrace and somewhat generalize a hope that for the case of (computational) social choice is expressed in the above article (Hemaspaandra 2018): We hope that in time there will be a generation of researchers who are trained through graduate programs that make students simultaneously expert in computational complexity and one of the other disciplines underpinning economic design—researchers who in a single person achieve a shared understanding of two areas. But for now, most researchers have as their core training one area, even if they do reach out to work in—or work with experts in—another area. And thus we write this article to try to make as transparent as we can within a few pages the crazy, yet (to our taste) just-right, way that complexity theorists view the world.

The remainder of this article is organized as follows. Section 2 further discusses the need for, and the importance of improving, mutual understanding. That section argues that the way that complexity theorists view the world is rarely understood even by the other areas of computer science, and that the areas of computer science themselves are separated by huge cultural gaps. Section 3 presents what we feel is the heart of how computational complexity theorists view the world, which is:

> We as complexity theorists believe that there is a landscape of beautiful mathematical richness, coherence, and elegance—waiting for researchers to perceive it better and better with the passing of time—in which problems are grouped by their computational properties.

## 2 The Need for Creed: Why Understanding Each Other Is Hard yet Needed and Important

In this section, we briefly mention some cultural chasms between complexity and social choice—and even between complexity and other areas of computer science—and suggest that shrinking or removing those chasms is important: Understanding between collaborators is of great value to the collaboration.

### 2.1 Spanning to Computational Social Choice, Economics, and Beyond

In this section, we will focus on computational social choice, as that is the particular facet of economic design that the authors are most familiar with.

This paper will not itself present the many ways that computational complexity and computational social choice have interacted positively and in ways that benefit *both* areas. As mentioned above, a separate paper (Hemaspaandra 2018) already makes that case, for example pointing out: how computational social choice has populated with problems such classes as the $\Theta_2^p$ level of the polynomial hierarchy and $NP^{PP}$; how computational social choice has provided the first natural domain in which complexity theory's search-versus-decision separation machinery could be applied; how that application itself gives insight into how to best frame the manipulative-attack definitions in computational social choice; how complexity's "join" operation has been valuable in proving the impossibility of obtaining certain impossibility results in computational social choice theory; how the study of online control yielded a completely new quantifier-alternation characterization of coNP; and much more, such as how important a role approximation, dichotomy results, and parameterized complexity have played in computational social choice.

It in fact is quite remarkable how strongly complexity has helped the study of computational social choice, and is even more remarkable—since this is the direction that might not have been apparent beforehand—how strongly computational social choice has helped the study of computational complexity theory. And most remarkable of all is that these results have usually been obtained by researchers from one side, although ones who were very interested in the other side. This clearly is a pair of areas that already is showing very strong content interactions and mutual benefits. Think of how much more waits to be seen and achieved when computational social choice theorists/social choice theorists and complexity theorists deepen their understanding of each other.

Some might worry about the "computational" in "computational social choice" above—namely, worrying that computational social choice is such a young area that no one is "native" to it. We disagree. It is true that much of the key, early work on this area—as it was emerging *as* an area with a distinct identity—was done by researchers whose training was in operations research, logic, artificial intelligence (AI), theoretical computer science, economics, social choice, political science, or mathematics. But already a generation of students, now in their 20s and 30s, has been trained whose thesis work was on computational social choice theory: researchers whose "native" area and identity is—despite the fact that their thesis advisors view themselves as at their core part of one of the older areas just listed—that they are computational social choice theory researchers. This is a very good development, and yet we are asking even more: Now that the area has its own identity, one can hope to grow researchers whose core training and identity embraces both that young area and the area of computational complexity.

## 2.2  Spanning to Other Computer-Science Areas

It is often discussed within computer science departments whether computer science is even a coherent discipline. After all, if one thinks about which other department the areas of computer science feel kinship to, for theoreticians that generally would be mathematics, for systems people that generally would be electrical and computer engineering, for symbolic AI people that often would be one of brain and cognitive sciences, linguistics, philosophy, or psychology, and for vision/robotics AI people that might be mechanical engineering, electrical and computer engineering, or visual science.

The cultural differences are also stark. For example, taking as our examples the two subareas of computer science that are most strongly represented in computational social choice research—AI and theory—we have the following contrasts in culture. Anonymous submissions at the main conferences versus submissions with the authors' identities open. Intermediate feedback and rebuttals at the main conferences versus no such round. Authors' names generally being ordered by contribution versus authors names always being listed alphabetically.[1] Large hierarchical program committees versus small almost-flat program committees. And that listing is not even mentioning the issue of the contrasting content of the areas, or their differing views on conference versus journal publication.

Almost any theoretical computer scientist will have stories of how sharply his or her perspective has differed from those of his or her nontheory colleagues, e.g., a nontheory colleague who firmly felt that 8x8 chess—not $N$x$N$ chess (Storer 1983) but actual 8x8 chess—under the standard rules (which implicitly limit the length

---

[1]One of us once asked a colleague, who at one point was the president of AAAI, whether he, upon seeing a paper with a very large number of authors with them all in alphabetical order, would really assume that Dr. Aardvark had made the largest contribution. The colleague looked back as if he'd been asked whether he really believed that $1 + 1 = 2$ and said that he of course would. In fact, in the different area within computer science known as systems, there is a running semi-joke—that excellently corresponds with reality—that one can tell how theoretical a given systems conference is by looking at what portion of its papers list the authors in alphabetical order; in fact, there is a very funny joke-paper (Appel 1992) that quantifies this—more rigorously than the earlier part of this sentence does—to prove that the POPL conference is quite theoretical.

of any game) is a great example of *asymptotic* complexity, and who advised the theoretician to go use Google to learn more about this. We suspect that nontheory computer science faculty members could write quite similar sentences—with different examples—from their own points of view, regarding the things theory faculty members say.

So even the subareas of computer science have some gaps between them as to understanding, or at least have rather large agree-to-disagree differences. Our hope is that, regarding the former, this short article may be helpful.

We mention, however, that we do not agree that anything said above shows that computer science is not a coherent discipline. To us, and in this we are merely relating an important, much loved insight that has been around in one form or another for many decades (Knuth 1997; Harel 1987), there is a unifying core to the field of computer science: algorithmic thought (and the study of algorithms). That core underpins AI, systems, and theory, and makes computer science an at least decently coherent discipline.

## 3   A Core Belief, and Its Expressions, Interpretations, and Implications

### 3.1   A Core Belief

We feel that a core view—in fact, *the* core view—of complexity theorists is the following (phrased here both as a profession of belief and as a statement of what is believed).

[**Core Belief**   We as complexity theorists believe that:] There is a landscape of beautiful mathematical richness, coherence, and elegance—waiting for researchers to perceive it better and better with the passing of time—in which problems are grouped by their computational properties.

If the subfield can be said to have a creed, this is it.

By saying that complexity theorists feel this, we don't mean to suggest that it is exclusive to them. In a less computational vein, the great mathematician Paul Erdős spoke of "The Book," which holds the most elegant proof of each mathematical theorem. He famously said, "You don't have to believe in God, but you should believe in The Book," and surely viewed as moments of true joy those when a proof so beautiful as to belong in the book was discovered. And the great computer scientist Edsger Dijkstra is traditionally credited[2] with this lovely, insightful comment:

---

[2]The quote is attributed to him in works of others as early as 1993 (Haines, 1993, p. 4), though attributing the quote to Dijkstra is disputed, as Michael Fellows published a very similar comment

> Computer Science is no more about computers than astronomy is about tele-
> scopes. — E. Dijkstra

Though different people interpret that quotation in different ways, we have always interpreted it to suggest almost precisely what our core belief is expressing. Indeed, the quotation's implicitly drawn parallel between astronomy studying the structure of the universe and computer scientists studying a similarly majestic structure is extremely powerful. And things are made even more pointed in the 1991 version by Michael Fellows, which follows the same sentiment as that of the quotation with, "There is an essential unity of mathematics and computer science."

## 3.2 The Heretics

Having read Sect. 3.1, theoretical researchers from any field may think, "Well, duh!" That is, they may think that the core belief is obvious, and wonder who could possibly think anything else.

The answer is that quite a large portion of the field computer science thinks something else. This was most famously expressed in a 1999 "Best Practices Memo" (Patterson et al. 1999) that was published in *Computing Research News*, the newsletter of a prestigious group, the Computing Research Association, of over two hundred North American organizations involved in computing research, including many universities. To this day, that memo is on the Computing Research Association's web site as a best practices memo (Patterson et al. 2017, although there certainly has been strong pushback on some of its points, see, e.g., Vardi 2009; Fortnow 2009). The most jump-off-the-page lines in that memo are these:

> ... experimentalists tend to conduct research that involves creating computational artifacts and assessing them. The ideas are embodied in the artifact, which could be a chip, circuit, computer, network, software, robot, etc. Artifacts can be compared to lab apparatus in other physical sciences or engineering in that they are a medium of experimentation. Unlike lab apparatus, however, computational artifacts embody the idea or concept as well as being a means to measure or observe it. Researchers test and measure the performance of the artifacts, evaluating their effectiveness at solving the target problem. A key research tradition is to share artifacts with other researchers to the greatest extent possible. Allowing one's colleagues to examine and use one's creation is a more intimate way of conveying one's ideas than journal publishing, and is seen to be more effective. For experimentalists conference publication is preferred to journal publication, and the premier conferences are generally more selective than the premier journals... In these and other ways experimental research is at variance with conventional academic publication traditions.

---

in a 1991 manuscript that appeared in a 1993 conference proceedings and published the identical quotation in 1993 in a *Computing Research News* article joint with Ian Parberry.

Underlying this is a worldview that is very different than that of most theoreticians. The worldview is that software systems and devices are often so complex that trying to theoretically capture their behavior and properties is hopeless, and we instead need to experiment on them to make observations. For example, that view might suggest that operating systems are so enormous and complex that we can't really capture or understand precisely their behavior.

Yet theoreticians think otherwise. Theoreticians dream of a time when essentially all programs—of any size—will have a rigorous, formally specified relationship between their inputs and their actions/outputs, and when we will seek to prove that the programs satisfy those relationships (insofar as can be done without running aground on undecidability issues). Perhaps that time will be decades or centuries away for extremely complex programs, but we believe it will come. And in fact, real progress—for example thanks to advances in automated theorem-proving/automated reasoning—has been made in the past few decades on verifying that even some quite large programming systems meet their specifications.

In brief, we don't think that because software systems are complex one can only experiment on them as if they were great mysteries; rather, we think that, precisely because they are so complex, the field should increase its efforts to formally understand them, including working on building the tools and techniques to underpin such an understanding.

To be fair to the above-quoted memo, it carefully had a very separate coverage in which it described what theoreticians do. But to many theoreticians, viewing computing systems as too complex to theoretically analyze—and more suitable for experimenting on—is far too pessimistic, at least as a long-term view.

Is our Core Belief utterly optimistic? Not purely so. It is broadly optimistic, in what it believes exists, though to be frank the landscape it is speaking of is typically more about problems and classes than about analyzing operating systems. But embracing the Core Belief does not mean that one must be delusional as to time frames. For example, in Gasarch's P versus NP poll (Gasarch 2012), only 53% percent of those polled felt that P versus NP would be resolved by the year 2100. 3% thought it would never be resolved, and 5% said they simply did not know when/if it will be resolved.

A astounding 92% of the polled theoreticians believe that it will be eventually resolved, even though currently no path for imminently resolving the question is in sight (see also the very grim possibility mentioned in the 1970s by Hartmanis and Hopcroft (1976): that the question might be independent of the axioms of set theory). Theoreticians have generally taken to heart Sir Thomas Bacon's 1605 comment from *The Advancement of Learning*:

They are ill discoverers that think there is no land, when they can see nothing but sea. — Thomas Bacon

## 3.3   Landscape and Classification

So what is this landscape that the Core Belief speaks of? And how can we bring it
into better focus?

### 3.3.1   Axes and Granularity of Classification

The landscape is one where each problem is located by its classification in terms of
various measures. What is its (asymptotic, of course) deterministic time cost? What is
its deterministic space cost? What are its nondeterministic costs? Its costs in various
probabilistic models? What about in nondeterministic models that forbid ambiguity
(i.e., that have at most one accepting path) or that polynomially bound the ambiguity?
What about in quantum computing models and biocomputing models? How well can
the problem be—in various senses—solved by heuristics or approximations? What
types of circuit families can capture the problem? What types of interactive proof
classes can capture the problem?

And that is just a quick start to listing aspects of interest. The number of interesting
dimensions along which problems can be classified is already large, and continues
to grow with time. Our landscape is not a physical one, of course, but i s a rich world
of mathematical classification.

The granularity with which we group the "locations" in this world itself is interesting. Complexity theorists typically focus on equivalence classes of problems,
linked by some type of reduction. For example, the NP-complete problems are all
those problems that are many-one, polynomial-time interreducible with the problem
of testing the satisfiability of boolean formulas. One can think of the NP-complete
sets as an extremely important feature of the landscape. Yet one can also view the
landscape with an interest in other degrees of granularity. The class of NP-Turing-
complete sets for example contains all the NP-complete sets, and may well contain
additional sets (Lutz and Mayordomo 1996), since Turing reductions are a more
powerful reduction type than many-one reductions. Going in the other direction, the
class of sets that are polynomial-time isomorphic to boolean satisfiability may well
be a strict subset of the NP-complete sets, and it is known to be a strict subset with
probability one relative to a random oracle (Kurtz et al. 1995).

Briefly put, complexity classes usually are defined by placing a bound on some
key resource, e.g., NP is the class of sets that can be accepted by polynomially time-
bounded nondeterministic computation. Complexity classes in some sense are upper
bounds on some dimension of complexity. Reductions are yardsticks by which sets
can be compared. If a set $A$ reduces to a set $B$ by some standard reduction type, we
view $A$ as being "easier or not too much harder" than $B$, with the details depending on what power the reduction itself possesses. There are now a huge number of
intensely studied reduction types, capturing such notions as, just as examples, the
amount of time or space the reduction is itself allowed to use; whether the reduction
is a single query or multiple ones and if the latter how they are used and whether

they are sequential or parallel; and to what extent the reduction itself can act non-deterministically. And completeness for complexity classes combines a class with a reduction type, identifying those sets in the class that are so powerful that every set in the class reduces to them by the given reduction type. In some sense, the completeness equivalence class of a complexity class groups together those problems, if any such problems exist (and some parts of the landscape perhaps lack complete sets Sipser 1982; Gurevich 1983; Hartmanis and Hemachandra 1988; Hemaspaandra et al. 1993), that distill the essence of the potential hardness of the class—they share the same underlying computational challenge. As such, they help complexity theorists focus on what the source of a problem's complexity is.

The joyful obsession and life's work of complexity theorists is to better understand this landscape. This often is done though classifying where important problems—or groups of problems—fall. Far more rarely yet vastly more excitingly, complexity theorists find new relationships between the different dimensions of classification, e.g., by showing that every set in the polynomial hierarchy Turing reduces to probabilistic polynomial time (Toda 1991) or by showing the class of sets having interactive proofs is precisely deterministic polynomial space (PSPACE) (Shamir 1992).

### 3.3.2 Classification Is Done for Insights Into the Landscape

The Core Belief and the previous section should hint at a truth that often is surprising to people who are not complexity theorists. That truth is that complexity theorists want to classify problems as part of the ongoing attempts to better understand the landscape of problem complexity. And in particular, we are interested in doing that even for problems where the classifications we are trying to distinguish between don't in practice differ in what they say about how quickly a problem can be solved.

For example, complexity theorists think that it is a rather big deal whether a problem—if it is an interesting one, such as about logic—is complete for double exponential time versus for example being complete for triple exponential space. This isn't because we think that complete problems for double exponential time are going to be easy to quickly solve. It is because we want to clarify where interesting problems fall in the landscape.

Looking at the other extreme, there is a huge amount of research into complexity classes (such as certain uniform circuit classes and logarithmic-space classes) all of which are contained in deterministic polynomial time. Yet to most people, deterministic polynomial time already is the promised land as to computational cost. Nonetheless, smaller classes are intensely studied, to better understand the rich world of complexities that exist there, and which problems have which complexities, although in fairness we should mention that some of this type of study is also motivated by the issue of whether the problem can or cannot be parallelized (Greenlaw et al. 1995).

But the real kicker here is that even if SAT solvers turn out to be able to do stunningly well on NP-complete problems, complexity theorists still will view the notion of NP-completeness as being of fundamental importance to the landscape. This is not because we don't care about how well heuristics can do—that too is a dimension of the landscape, and thus something on which rigorous results are important and welcome—but rather we think that the notion of NP-completeness itself is one of the greatest beauties of the landscape, and is natural and compelling in so very many ways.[3]

To take as an example one of the most beautiful examples of how profound the issue is of whether NP-complete sets belong to P, i.e., whether P = NP, we mention that a not widely known paper by Hartmanis and Yesha (1984) is in effect showing that whether humans can be perfectly replaced by machines in the task of finding and presenting particular-sized mathematical proofs of theorems—loosely put, the issue of whether humans have any chance of having any special creativity and importance in achieving mathematical proofs—can be characterized by the outcome of such basic landscape questions as whether P and NP differ, and whether P and PSPACE differ.

# 4  Conclusion

To end as we started, we believe that economic design and computational complexity should become even more important to each other with each passing year, but that an improved mutual understanding of the areas' worldviews is important in making that happen. In that spirit, this article sets out the optimistic worldview that we believe is held by most computational complexity theorists. And the most central part of that worldview is that *we as complexity theorists believe that there is a landscape of beautiful mathematical richness, coherence, and elegance—waiting for researchers to perceive it better and better with the passing of time—in which problems are grouped by their computational properties.*

That is not to say that we believe that the greatest open issues within that landscape will be resolved within our lifetimes. But we believe that—just as that landscape has already been seen to have utter surprises in what it says regarding language theory (Szelepcsényi 1988; Immerman 1988), interactive proofs (Lund et al. 1992; Shamir 1992), branching programs and safe-storage machines (Barrington 1989; Cai and Furst 1991), approximation (Arora et al. 1998), the power and lack of power of probabilistic computation (Nisan and Wigderson 1994; Impagliazzo and Wigderson 1997; Toda 1991), and much more—the landscape contains countless more surprises

---

[3]This article is not on the subject of how well heuristics can do on NP-complete problems, or the strengths and limitations of SAT solvers. On one hand, there are theoretical results showing that polynomial-time heuristics cannot have a subexponentially dense set of errors on any NP-hard problem unless the polynomial hierarchy collapses. And if someone says they have a SAT solver that works on any collection of NP problems they ever have encountered, it is interesting to point out to them that factoring numbers that are the product of two large primes can be turned into a SAT problem, and so their amazing SAT solver should be able to break RSA and make them rich... yet no one has yet been able to make that work. On the other hand, SAT solvers undeniably do perform remarkably well on a great range of data sets. For discussion of most of the issues just mentioned, and how they can be at least partially reconciled, see for example the article by Hemaspaandra and Williams (2012).

and advances that will be reached in years, in decades, and in centuries, and we believe that many of them will be in the important, rapidly growing areas at the intersection of economic design and computational complexity.

**Acknowledgements** We thank William S. Zwicker for helpful comments and suggestions. This work was done in part while on a sabbatical stay at ETH Zürich's Department of Computer Science, generously supported by that department.

# References

Appel, A. (1992). Is POPL mathematics or science? *SIGPLAN Notices*, *27*(4), 87–89.

Arora, S., Lund, C., Motwani, R., Sudan, M., & Szegedy, M. (1998). Proof verification and the hardness of approximation problems. *Journal of the ACM*, *45*(3), 501–555.

Barrington, D. (1989). Bounded-width polynomial-size branching programs recognize exactly those languages in $NC^1$. *Journal of Computer and System Sciences*, *38*(1), 150–164.

Cai, J.-Y., & Furst, M. (1991). PSPACE survives constant-width bottlenecks. *International Journal of Foundations of Computer Science*, *2*(1), 67–76.

Fortnow, L. (2009). Time for computer science to grow up. *Communications of the ACM*, *52*(8), 33–35.

Gasarch, W. (2012). The second P =? NP poll. *SIGACT News*, *43*(2), 53–77.

Greenlaw, R., Hoover, H., & Ruzzo, W. (1995). *Limits to parallel computation: P-completeness theory*. Oxford: Oxford University Press.

Gurevich, Y. (1983). Algebras of feasible functions. In *Proceedings of the 24th IEEE Symposium on Foundations of Computer Science* (pp. 210–214). IEEE Computer Society Press.

Haines, M. (1993). *Distributed runtime support for task and data management*. Ph.D. thesis, Colorado State University, Fort Colins, CO, August 1993. Available as Colordo State Univeristy Department of Computer Science Technical Report CS-93-110.

Harel, D. (1987). *Algorithmics: The spirit of computing*. Boston: Addison-Wesley.

Hartmanis, J., & Hemachandra, L. (1988). Complexity classes without machines: On complete languages for UP. *Theoretical Computer Science*, *58*(1–3), 129–142.

Hartmanis, J., & Hopcroft, J. (1976). Independence results in computer science. *SIGACT News*, *8*(4), 13–24.

Hartmanis, J., & Yesha, Y. (1984). Computation times of NP sets of different densities. *Theoretical Computer Science*, *34*(1–2), 17–32.

Hemaspaandra, L. (2018). Computational social choice and computational complexity: BFFs? In *Proceedings of the 32nd AAAI Conference on Artificial Intelligence* (pp. 7971–7977). AAAI Press.

Hemaspaandra, L., Jain, S., & Vereshchagin, N. (1993). Banishing robust Turing completeness. *International Journal of Foundations of Computer Science*, *4*(3), 245–265.

Hemaspaandra, L., & Williams, R. (2012). An atypical survey of typical-case heuristic algorithms. *SIGACT News*, *43*(4), 71–89.

Immerman, N. (1988). Nondeterministic space is closed under complementation. *SIAM Journal on Computing*, *17*(5), 935–938.

Impagliazzo, R., & Wigderson, A. (1997). P = BPP if E requires exponential circuits: Derandomizing the XOR lemma. In *Proceedings of the 29th ACM Symposium on Theory of Computing* (pp. 220–229). ACM Press.

Knuth, D. (1997). Algorithms in modern mathematics and computer science. In A. Ershov & D. Knuth (Eds.), *Algorithms in modern mathematics and computer science* (pp. 82–99). Lecture Notes in Computer Science #122. Berlin: Springer.

Kurtz, S., Mahaney, S., & Royer, J. (1995). The isomorphism conjecture fails relative to a random oracle. *Journal of the ACM, 42*(2), 401–420.

Lund, C., Fortnow, L., Karloff, H., & Nisan, N. (1992). Algebraic methods for interactive proof systems. *Journal of the ACM, 39*(4), 859–868.

Lutz, J., & Mayordomo, E. (1996). Cook versus Karp-Levin: Separating completeness notions if NP is not small. *Theoretical Computer Science, 164*(1–2), 123–140.

Nisan, N., & Wigderson, A. (1994). Hardness vs. randomness. *Journal of Computer and System Sciences, 49*(2), 149–167.

Patterson, D., Snyder, L., & Ullman, J. (1999). Best practices memo: Evaluating computer scientists and engineers for promotion and tenure. *Computing Research News, 11*(3), A–B (special insert).

Patterson, D., Snyder, L., & Ullman, J. (2017). Evaluating computer scientists and engineers for promotion and tenure. https://cra.org/resources/best-practice-memos/evaluating-computer-scientists-and-engineers-for-promotion-and-tenure/, URL verified October 31, 2017.

Shamir, A. (1992). IP = PSPACE. *Journal of the ACM, 39*(4), 869–877.

Sipser, M. (1982). On relativization and the existence of complete sets. In *Proceedings of the 9th International Colloquium on Automata, Languages, and Programming* (pp. 523–531). Lecture Notes in Computer Science #140. Berlin: Springer.

Storer, J. (1983). On the complexity of chess. *Journal of Computer and System Sciences, 27*(1), 77–100.

Szelepcsényi, R. (1988). The method of forced enumeration for nondeterministic automata. *Acta Informatica, 26*(3), 279–284.

Toda, S. (1991). PP is as hard as the polynomial-time hierarchy. *SIAM Journal on Computing, 20*(5), 865–877.

Vardi M. (2009). Conferences vs. journals in computing research. *Communications of the ACM, 52*(5), 5.

# Complexity and Simplicity in Economic Design

Noam Nisan

**Abstract** As more and more economic activity moves to the Internet, familiar economic mechanisms are being deployed at unprecedented scales of size, speed, and complexity. In many cases this new complexity becomes the defining feature of the deployed economic mechanism and the quantitative difference becomes a key qualitative one. We suggest to study this complexity and understand in which cases and to what extent it is necessary.

As more and more economic activity moves to the Internet, familiar economic mechanisms are being deployed at unprecedented scales of size, speed, and complexity. In many cases this new complexity becomes the defining feature of the deployed economic mechanism and the quantitative difference becomes a key qualitative one.

A paradigmatic example is that of an auction. Classic economic theory has studied in considerable detail the question of how to sell an indivisible item in an auction. Auctions were mathematically modeled as to study a host of issues: revenue, social welfare, risk, partial information, different formats, players' strategies, etc. As a result of much work in economic theory one may comfortably say that single item auctions are very well understood. Then enter computational platforms, and especially the Internet. While auction theory is obviously applicable to many economic scenarios on computational platforms, in these settings we often see the humble auction grow to amazing sizes and complexities. Classic examples include the FCC spectrum auctions that seek to auction off thousands of spectrum licenses worth Billions of dollars and Internet ad-auctions that sell many billions of "ad-impressions", each worth less than a penny but together giving Google, for example, its Billions of dollars of revenue. In each of these applications, the computational platform allows us to sell multiple items in new complicated and sophisticated ways, and this capability allows us to achieve unprecedented economic benefits (in various senses).

There has been much recent work in the computer science community that deals with these as well as many other instances of large or "complex" auctions or markets.

N. Nisan (✉)
School of Computer Science and Engineering, The Hebrew University of Jerusalem,
Jerusalem, Israel
e-mail: noam.nisan@gmail.com

© Springer Nature Switzerland AG 2019
J.-F. Laslier et al. (eds.), *The Future of Economic Design*, Studies in
Economic Design, https://doi.org/10.1007/978-3-030-18050-8_21

In fact, large and complex auctions are the central focus of a young and vibrant field called *Algorithmic Mechanism Design* (Noam Nisan et al. 2007). We can probably fit much of the research done in these fields in the last two decades into the following high-level template:

1. A theoretical point of departure is some well-known issue in classic economic theory or game theory (e.g. a single item auction).
2. A motivational goal is some complex scenario that seems to be addressable by some variant of this classic departure point (e.g. allocating a large number of items).
3. The research challenge is to masterfully combine the economic or game-theoretic sensibilities of the simple departure point with the size, interactions, and constraints of the complex setting.

Such a combination must deal with the various new "complexities" of the new scenario and often requires new ideas both in the game-theoretic/economic analysis and in the computational/algorithmic treatment. The results obtained may vary across several dimensions: to what extent we can recover the "good" properties of the original simple scenario, to what extent simple mechanisms are sufficient (as opposed to complex ones), which trade-offs exist, which natural special cases behave better than the general case, etc.

This note advocates the development of an organized theory along these lines. A theory that can look at a complex economic scenario and understand the crucial ways in which it differs from or is similar to simple related economic scenarios. Such a coherent theory should serve as a general framework that guides our attempts of addressing complex economic problems such as those routinely found on the Internet. Such a theory should be able to tell which parameters of the complex scenario can be "aggregated" toward a simple view, which ones must be addressed in more sophisticated and complex ways, and which ones cause unavoidable losses. A good theory of "complex economics" can guide our attempts of solving complex economic challenges in a similar way that *computational complexity theory* (Papadimitriou 2003) guides our approaches to addressing algorithmic problems: identifying crucial bottlenecks (like time or space), suggesting required compromises (like approximations or special cases), and highlighting connections between different problems (like reductions and classes).[1]

We have so far left the key notions of "complexity" and "simplicity" rather vague. We argue that there are multiple interesting meanings to these notions, and suggest that multiple ones should be addressed by the proposed theory of complexity in economics. At the most concrete level, one may study notions of complexity that are of direct interest from an economic point of view in many applications. More generally, one may hope that studying several different notions of complexity will paint a general picture of the studied landscape and will enable insights that transcend

---

[1]An existing research field often termed "Complexity Economics" applies notions and ways of thinking from "Complexity Science" (e.g. James Gleick 1997) to economic systems. The suggestion here however is to proceed in a different direction, one whose view of "complexity" is taken from theoretical computer science rather than from physics.

any particular mathematical model. At the highest level, various concrete notions of complexity should capture important aspects of the *computational complexity* of economic problems (in the usual algorithmic sense).

To demonstrate the potential of a general theory of complexity in economic settings, let us proceed with a list of examples of various notions of complexity in various economic settings that have already been studied. Not only do these demonstrate the richness of notions of complexity, but the connections between them also demonstrate the potential for a coherent theoretical framework. For each of these examples there exists a long thread of research papers, of which we will only reference a single one, often a survey or recent paper that can point to further papers.

- **Communication Complexity of Markets and Auctions**. It is well accepted in economic theory that a critical part of any market mechanism is that of communication, and that markets are rather efficient "communication devices" in classical "convex" scenarios. However, in complex scenarios such as combinatorial auctions, it turns out that one can quantify that an intractable, exponential, amount of communication is needed in order to obtain efficiency, highlighting the limitations of market mechanisms (as well as others) in such situations. Much work has been done in studying trade-offs between communication complexity, efficiency losses, and additional constraints in various scenarios (Shahar Dobzinski et al. 2014).

- **Bidding Languages**. In complex computerized settings it is very clear that eliciting the preferences of the participants is a crucial bottleneck in the path to a desired outcome. A basic approach is that of specifying a "bidding language" (Noam Nisan 2006) by which users can describe their preferences, where "stronger" languages allow smaller complexity in describing preferences but "simpler" ones may be easier to deal with algorithmically, strategically, and conceptually.

- **Queries for Preference Elicitation**. A more general approach to preference elicitation would be for the mechanism to "query" the preferences of each of the player using some natural specific set of allowed queries. This can be more general that bidding languages since the elicitation process can be adaptive. The interesting questions here concern the trade-offs between the number of types of queries and the quality of the obtainable solution. There are various types of natural queries to consider depending on the exact scenario, and for example in the context of combinatorial auctions, "value queries" and "demand queries" have received much attention (Nisan et al. 2007). At the extreme level of generality, such models turn out to capture natural notions of communication complexity.

- **Complexity of Convergence to Equilibrium**. It is commonly assumed that in strategic situations, players would reach (or at least approach) equilibrium. While this may seem to be a reasonable assumption in classical simple scenarios, the question of how (and whether) do players reach an equilibrium in complex situations is much more mysterious. Are there any realistic "learning strategies" by the participants that will lead them to a Nash equilibrium? There are various ways of quantifying the complexity of reaching equilibria (Sergiu Hart and Andreu Mas-Colell 1997), and very strong exponential complexity intractability results are known for many general models.

- **Mechanism Specification Complexity**. An economic mechanism specifies the rules for interaction between economic parties and the outcome from that interaction. How complex must we make our mechanism in order to obtain desired economic properties? For example, when auctioning a single item, Myerson shows that the very simple mechanism of a second price auction with a reserve price already ensures optimal revenue (in rather general scenarios). Unfortunately, this simplicity no longer suffices for even slight generalizations of a single-item auction, e.g. even when selling two items to a single buyer! In this case, as well as more complex ones, it turns out that there is a trade-off between the complexity of describing the mechanism and the revenue that it can ensure. There are several interesting ways of measuring the complexity of a mechanism, the simplest of which is counting the number of possible outcomes, a measure termed the "menu-size" (Sergiu Hart and Noam Nisan 2013).
- **Equilibrium Structure Complexity**. The notion of equilibrium is central in economic settings, and Many types of equilibria are considered in the economic literature: in games pure or mixed Nash equilibria, as well as "correlated" and "coarse-correlated" ones, in markets, various types based on prices, notions like stable marriage in yet other settings, etc. Some of these equilibria are always simple (e.g. pure-Nash equilibria), while others (like correlated ones) can be significantly more complex. On the other hand, complex types of equilibria may exist when simpler ones do not, and may allow improved efficiency. Imposing some simplification on the underlying economic system may simplify equilibria, perhaps at the cost of some loss of efficiency. In particular, much recent work has studied the inefficiency incurred by equilibria when multiple items are sold simultaneously but separately, as well ass possible generalizations (Tim Roughgarden and Inbal Talgam-Cohen 2015).
- **Complexities, Incentives, and Approximation**. Some of the basic tools of mechanism design, specifically, the Vickrey–Clarke–Groves payment scheme generalize to complex scenarios. However in such complex scenarios obtaining optimal solutions is often intractable due to this very complexity (whether the complexity is measured by computation, communication, queries, or other notions). While tractable *approximate* solutions are known in many cases, it turns out that these will often not "play well" with the strategic tools of mechanism design. This type of clash between complexity and strategic tools has been at the center of focus for the field of Algorithmic Mechanism Design (Noam Nisan 2014).

# References

Blumrosen, L., & Nisan, N. (2007). Combinatorial auctions (a survey). In N. Nisan, T. Roughgarden, E. Tardos, & V. Vazirani (Eds.), *Algorithmic game theory*. Cambridge: Cambridge University Press.

Dobzinski, S., Nisan, N., & Oren, S. (2014). Economic efficiency requires interaction. In *Symposium on Theory of Computing, STOC 2014, New York, NY, USA, May 31 - June 03, 2014* (pp. 233–242).

Gleick, J. (1997). *Chaos: Making a new science*. New York: Random House.

Hart, S., & Mas-Colell, A. (2013). *Simple adaptive strategies: from regret-matching to uncoupled dynamics* (Vol. 4). Singapore: World Scientific.

Hart, S., Nisan, N. (2013). The menu-size complexity of auctions. In *Proceedings of the Fourteenth ACM Conference on Electronic Commerce (EC)* (pp. 565).

Nisan, N. (2006). Combinatorial auctions. *Bidding languages*. Cambridge: MIT Press.

Nisan. N. (2014). Algorithmic mechanism design. *Handbook of game theory* (pp. 477).

Nisan, Noam, Roughgarden, Tim, Tardos, Eva, & Vazirani, Vijay V. (2007). *Algorithmic game theory*. New York: Cambridge University Press.

Papadimitriou, C. H. (2003). *Computational complexity*. New York: Wiley.

Roughgarden, T., Talgam-Cohen, I. (2015). Why prices need algorithms. In *Proceedings of the Sixteenth ACM Conference on Economics and Computation* (pp. 19–36). Providence: ACM.

# Complexity-Theoretic Barriers in Economics

Tim Roughgarden

**Abstract** We survey several unexpected connections between computational complexity and fundamental economic questions that appear unrelated to computation.

## 1 Introduction

*Computational complexity theory* is a branch of theoretical computer science that studies the amount of resources (time, space, communication, etc.) required to solve different computational problems (see e.g. Roughgarden 2010). Computational complexity has already had plenty to say about economic equilibria. Some of the most celebrated results in algorithmic game theory give strong evidence that there is no general efficient algorithm for computing various types of equilibria, including mixed Nash equilibria in finite games and market equilibria in markets with divisible goods and concave utility functions (see e.g. Chen et al. 2009b; Daskalakis et al. 2009; Chen et al. 2009a).

Recent work has established several unexpected connections between computational complexity and fundamental economic questions that seem to have little to do with computation (as opposed to the explicitly computational questions mentioned above). These new connections suggest the possibility of a rich theory with many applications. We briefly describe a few of these connections, with the hope of inspiring further work along similar lines.

## 2 Barriers for Simple Auctions

The goal of auction design is to develop auction formats that enjoy good properties when the participants act in a self-interested manner (i.e., reach an "equilibrium"). One fundamental and high-stakes example is the design of multi-item auctions for

T. Roughgarden (✉)
Department of Computer Science, Columbia University, New York, NY, USA
e-mail: tr@cs.columbia.edu

© Springer Nature Switzerland AG 2019
J.-F. Laslier et al. (eds.), *The Future of Economic Design*, Studies in
Economic Design, https://doi.org/10.1007/978-3-030-18050-8_22

selling wireless spectrum licenses, as used by the Federal Communications Commission (see e.g. Cramton et al. 2006). Many practitioners in multi-item auction design abide by the following rule of thumb: simple auctions, such as selling each item separately in parallel, can perform well if and only if there are not strong synergies across items. An example of such a synergy would be between licenses for spectrum in northern and southern California—a company looking to roll out a new statewide data plan only wants one of the licenses if it can get both. (Or for a more prosaic example, think of a left shoe and a right shoe.)

This 25-year-old rule of thumb has only been formalized over the past few years. Proving that simple auctions can work well without item synergies boils down to proving "price of anarchy (POA)" bounds, which state that all of the equilibria of a simple auction are almost as good as an optimal outcome (e.g., in terms of the social welfare achieved). The theory of "smooth games" developed in Roughgarden (2015), and its subsequent extensions (e.g. Syrgkanis and Tardos 2013), turns out to be well suited to this goal. Many good POA bounds for different auction formats are now known (see the survey paper Roughgarden et al. 2017).

Proving the converse, that simple auctions *cannot* work well when there are item synergies, seems more difficult. While a positive result only requires analyzing one simple auction format, a negative result needs to rule out *every* possible simple auction format. Fortunately, computational complexity theory has a rich toolbox for proving such impossibility results. The main result in Roughgarden (2014) uses lower bounds in communication complexity (see e.g. Kushilevitz and Nisan 1996; Roughgarden 2016) to formally prove the impossibility of good simple auctions when there are strong item synergies.

One of many intriguing open research directions is to prove impossibility results that are stronger than those implied by the communication complexity connection established in Roughgarden (2014). The POA lower bounds proved in Roughgarden (2014) rely on two properties of equilibria: guaranteed existence and efficient verifiability. Equilibria possess additional properties—can these be exploited to prove stronger impossibility results? Specifically, every equilibrium implicitly solves the best-response problem for every player. These single-agent problems are a second source of potential impossibility, orthogonal to the difficulty posed by approximating the underlying multi-agent optimization problem.

## 3   Barriers for Competitive Equilibria

One of the most basic notions in economics is that of a competitive equilibrium, which comprises prices for items that equalize supply and demand. In a market with divisible goods (milk, wheat, etc.), a competitive equilibrium exists under reasonable assumptions. Many markets with indivisible goods (houses, wireless spectrum licenses, airport landing slots, etc.), however, can fail to possess a competitive equilibrium. Recent work by Roughgarden and Talgam-Cohen (2015) uses computational complexity to explain the rampant non-existence of such equilibria.

To be more concrete, fix a set of $m$ (non-identical and indivisible) items and let $\mathcal{V}$ denote a set of valuation functions (i.e., real-valued set functions with ground set $\{1, 2, \ldots, m\}$). For example, $\mathcal{V}$ could be the set of additive valuations, unit-demand valuations, valuations satisfying the gross substitutes condition (Alexander et al. 1982), or general monotone valuations. One of the main results in Roughgarden and Talgam-Cohen (2015) proves a necessary condition for the guaranteed existence of a competitive equilibrium, which compares the computational tractability of two different problems. The *welfare-maximization problem* for $\mathcal{V}$ is: given valuations $v_1, \ldots, v_n \in \mathcal{V}$ for $n$ agents (described either succinctly or through oracle access), compute the allocation $(S_1, \ldots, S_n)$ of items to agents that maximizes the social welfare $\sum_{i=1}^{n} v_i(S_i)$. The *utility-maximization problem* for $\mathcal{V}$ is: given a single valuation $v \in \mathcal{V}$ and item prices $p_1, \ldots, p_m$, compute a utility-maximizing bundle (i.e., a bundle in $\mathrm{argmax}_S v(S) - \sum_{j \in S} p_j$). The welfare-maximization problem (with $n$ agents) is generally only harder than the utility-maximization problem (with 1 agent). The necessary condition in Roughgarden and Talgam-Cohen (2015) for the guaranteed existence of a competitive equilibrium in every market with valuations in $\mathcal{V}$ is that the welfare-maximization problem is no harder than the utility-maximization problem. By "no harder," we mean that given a subroutine that solves the utility-maximization problem, it is possible to solve also the welfare-maximization problem (with the number of invocations of the subroutine and the additional work performed bounded above by some polynomial function of $n$ and $m$).

One of many open questions in this direction is to develop analogous impossibility results for settings without prices or money, such as markets for matching (marriage, college admissions, etc.) or kidney exchange. One possible plan for proving such impossibility results is to derive a contradiction from statements of the following types: (i) Pareto optimality, which states that no one can be made better off without making someone else worse off, is computationally difficult to verify; (ii) a competitive equilibrium is computationally easy to verify (assuming that the single-agent utility-maximization problem is tractable); (iii) "first and second welfare theorems," which provide a close correspondence between competitive equilibria and Pareto-optimal outcomes.

# 4   Barriers for Tractable Characterizations

*Border's theorem* is a fundamental result about probability distributions that has important applications in auction design (Border 1991). One way to phrase Border's theorem is that it gives an explicit description of the facets of a certain polytope, which is the projection of the space of all single-item auctions onto their "interim allocation rules" with respect to a prior distribution over bidders' valuations for an item. Border's theorem is useful in auction design both for understanding the structure of revenue-optimal auctions (as one varies the prior), and in efficiently computing such an auction (given a description of the prior as input). Generalizations of Border's theorem beyond the single-item auction setting have been hard to come by (though see Alaei et al. 2012; Cai et al. 2012a; Che et al. 2013), unless one resorts to approximation (as in Cai et al. 2012a, b). Recent work by Gopalan et al. (2015) uses computational complexity theory to explain this state of affairs: under standard complexity assumptions (that the "polynomial hierarchy" does not collapse), Border's theorem cannot be extended significantly beyond single-item auctions. The techniques introduced in Gopalan et al. (2015) do not appear specific to auction design, and show the promise of ruling out tractable characterizations of many other mathematical objects. For example, in recent years there has been tremendous progress in ruling out "extended linear programming formulations" for various computational problems, including graph matching (see e.g. Rothvoß 2014). Can the techniques introduced in Gopalan et al. (2015) advance this research agenda further, under suitable complexity assumptions?

# References

Alaei, S., Fu, H., Haghpanah, N., Hartline, J. D., & Malekian, A. (2012). Bayesian optimal auctions via multi- to single-agent reduction. In *ACM Conference on Electronic Commerce, EC '12* (pp. 17).

Border, K. C. (1991). Implementation of reduced form auctions: A geometric approach. *Econometrica, 59*(4), 1175–1187.

Cai, Y., Daskalakis, C., & Matthew Weinberg, S. (2012a). An algorithmic characterization of multi-dimensional mechanisms. In *Proceedings of the 44th Symposium on Theory of Computing Conference, STOC* (pp. 459–478).

Cai, Y., Daskalakis, C., & Matthew Weinberg, S. (2012b). Optimal multi-dimensional mechanism design: Reducing revenue to welfare maximization. In *Proceedings of the 53rd Annual Symposium on Foundations of Computer Science (FOCS)* (pp. 130–139).

Che, Y.-K., Kim, J., & Mierendorff, K. (2013). Generalized reduced form auctions: A network flow approach. *Econometrica, 81*, 2487–2520.

Chen, X., Dai, D., Du, Y., & Teng, S.-H. (2009a) Settling the complexity of Arrow-Debreu equilibria in markets with additively separable utilities. In *Proceedings of the 50th Annual IEEE Symposium on Foundations of Computer Science (FOCS)* (pp. 273–282).

Chen, X., Deng, X., & Teng, S.-H. (2009b). Settling the complexity of computing two-player Nash equilibria. *Journal of the ACM, 56*(3), 14.

Cramton, P., Shoham, Y., & Steinberg, R. (Eds.). (2006). *Combinatorial auctions.* Cambridge: MIT Press.

Daskalakis, C., Goldberg, P. W., & Papadimitriou, C. H. (2009). The complexity of computing a Nash equilibrium. *SIAM Journal on Computing, 39*(1), 195–259.

Gopalan, P., Nisan, N., & Roughgarden, T. (2015) Public projects, Boolean functions, and the borders of Border's theorem. In *Proceedings of the 16th Annual ACM Conference on Economics and Computation (EC)*.

Kelso, Alexander S., & Crawford, Vincent P. (1982). Job matching, coalition formation, and gross substitutes. *Econometrica, 50*(6), 1483–1504.

Kushilevitz, E., Nisan, N. (1996). *Communication complexity.* Cambridge: Cambridge University Press.

Rothvoß, T. (2014). The matching polytope has exponential extension complexity. In *Proceedings of the 46th Annual ACM Symposium on Theory of Computing (STOC)* (pp. 263–272).

Roughgarden, T. (2010). Computing equilibria: A computational complexity perspective. *Economic Theory, 42*(1), 193–236.

Roughgarden, T. (2014). Barriers to near-optimal equilibria. In *Proceedings of the 55th Annual IEEE Symposium on Foundations of Computer Science (FOCS)* (pp. 71–80).

Roughgarden, T. (2015). Intrinsic robustness of the price of anarchy. *Journal of the ACM, 62*(5), 32.

Roughgarden. T. (2016). *Communication complexity (for algorithm designers)*. Now Publishers.

Roughgarden, T., Talgam-Cohen, I. (2015). Why prices need algorithms. In *Proceedings of the 16th Annual ACM Conference on Economics and Computation (EC)*.

Roughgarden, T., Syrgkanis, V., & Tardos, É. (2017). The price of anarchy in auctions. *Journal of Artificial Intelligence Research, 59*, 59–101.

Syrgkanis, V., Tardos, É. (2013). Composable and efficient mechanisms. In *Proceedings of the 44th Annual ACM Symposium on Theory of Computing (STOC)* (pp. 211–220).

# Inputs, Algorithms, Quality Measures: More Realistic Simulation in Social Choice

**Mark C. Wilson**

**Abstract** Much of my research deals with trying to evaluate the performance of social choice *algorithms* via simulations, which requires appropriate *inputs* and *quality measures*. All three areas offer substantial scope for improvement in the coming years. For concreteness and because of my own limited experience, I focus on the allocation of indivisible goods and on voting, although many of the ideas are more broadly applicable.

## 1 Introduction

There are hugely many possible algorithmically defined rules for voting and allocation. Why then do we see so few of them in the literature? This phenomenon is not limited to social choice—in my experience, in most application areas the number of heavily studied and implemented algorithms is fairly small. For example, there are many ways to sort a list, but quicksort, mergesort and heapsort dominate the literature and practice. Of course, in the case of sorting, there is only one really interesting performance criterion, namely optimal (average or worst case) running time, and it is (asymptotically) achieved by all the algorithms listed, while naive sorting algorithms fail to achieve optimality. However in economic design, we are typically faced with multiple competing success criteria, such as strategyproofness, efficiency, fairness, and welfare, which intuitively should lead to more algorithms that are considered viable.

Axiomatic methods are very common in economic design. An axiom is essentially a statement of the form "if the input satisfies property P, then the output satisfies property Q". For example, for the unanimity axiom for voting, P might be "all voters have the same top choice" and Q might be "the common top choice is the winner of the election". A rule satisfying an axiom leaves no room for confusion, or concerns about input distributions, since we are dealing with logic and not probability. The fact remains, however, that there are many axioms, and they usually conflict with

M. C. Wilson (✉)
University of Auckland, Auckland, New Zealand
e-mail: mc.wilson@auckland.ac.nz

© Springer Nature Switzerland AG 2019
J.-F. Laslier et al. (eds.), *The Future of Economic Design*, Studies in
Economic Design, https://doi.org/10.1007/978-3-030-18050-8_23

each other. This leads often to impossibility results if too many axioms are imposed, occasionally to characterization results if we impose exactly the right number, and frequently to tradeoffs that must be investigated when we impose even fewer. These tradeoffs necessarily use notions of probability, such as how often P implies Q, or optimization, such as how close the outcome is to an outcome in which Q holds.

For a given collection of axioms, it is unnecessarily limiting to consider only the extreme cases, namely the algorithms that satisfy at least one of the axioms. Rules satisfying some axioms often fail to satisfy others, which is not surprising. What is more surprising is how badly an "extremal" algorithm, which satisfies axiom A, fails to satisfy axiom B. For example, Serial Dictatorship is strategyproof and efficient for allocation, but overall performs relatively poorly on fairness and welfare criteria. Similarly, throwing away all items and allocating none is a fair and strategyproof procedure, but very inefficient. A large number of papers start with one of the extremal algorithms and investigate how badly it performs with respect to other axiomatic properties. We should at least aim to explore the "Pareto frontier of the space of rules" consisting of rules not dominated by any other rule. The rule we need for a given situation may fail to satisfy any of the axioms in general, but may only just fail to satisfy all of them simultaneously, while extremal rules may score 100% on one axiom but close to zero on another. I believe that failure to appreciate this fact is a major reason for the small number of algorithms in the literature.

As an aside, it is important to choose the right criteria and set of algorithms for our analysis. For example, a large number of papers have been written about minimizing the manipulability of voting rules within various classes of rules, for example positional scoring rules. But we already know that dictatorial voting rules minimize manipulability absolutely since they are strategyproof. Since dictatorial rules are considered bad, presumably, because of other axioms relating to fairness and welfare performance, any such studies should surely consider those two criteria and their relative importance. Otherwise, we cannot *a priori* exclude the possibility that a dictatorial rule performs overall worse than the rules in the class under study. Surprisingly, a large number of papers have not taken this into consideration. I published a few papers along these lines before realizing this.

Exploring the Pareto frontier of rules leads us immediately to issues of measurement. If an algorithm does not satisfy a certain axiom, then we need to measure how close it comes to doing so, and there are many more ways to do this than the literature would indicate. A general setup is to have measures $\mu_1, \ldots, \mu_k$, one for each axiom, each taking values between 0 and 1, where 1 indicates that the axiom holds for that input. For example, we might measure the fraction of agents who envy another agent under the algorithm's allocation for the given input, or simply have the indicator function which is 1 if the algorithm is strategyproof on the given input and 0 otherwise. Ideally for each input we will represent each algorithm by a $k$-tuple of values of the measures. An algorithm $A_1$ is dominated by $A_2$ if and only if $\mu_i(A_1) \leq \mu_i(A_2)$ for all $i$ and there is some $i$ for which the inequality is strict.

This is too weak a partial order to impose on our algorithms. For example, no extremal algorithm can be dominated on any input by any other algorithm, since one of the measures has value 1. Thus we may need to give weights to the various measures and consider only the weighted sum (this includes the case of weight zero, where we exclude a measure completely). But even if we do not, we can still search for rules on the frontier that are not dominated. The main problem is that the dominance relation is defined for each input separately, which is too detailed for most purposes. We typically need to relax this by considering a distribution over all inputs. But this introduces statistical notions—how to summarize all this information about distributions? The most obvious measures are the maximum, minimum, mean and median. Worst case comparisons don't tell us much—quicksort is worse than insertion sort in the worst case, but it is still much more used in practice because its expected running time is better. Best case performance is usually perfect for all algorithms. So the mean or median make more sense.

Of course, these statistics depend substantially on the input distribution. What kind of data will our algorithm be faced with? If inputs are chosen uniformly at random from all linear orders of $n$ distinct elements, then quicksort has running time in $\Theta(n \log n)$. However its worst case is quadratic. In the case of allocation rules, the Impartial Culture (where each agent independently chooses a linear preference order as above) is the easiest case. The lack of correlation between agents means that many agents have different first preferences, and even if they coincide on first preference, they have different second preferences. This makes allocation very easy (note that for voting, this is the hardest case, because every order of candidates has approximately equal support and only random variability of order $\sqrt{n}$ prevents a complete tie). Impartial Culture has value as an extreme case and is mathematically tractable, but is well known to be very unrealistic (Regenwetter et al. 2006).

Having pointed out some issues that I feel have not been adequately addressed in the literature so far, in the following sections I discuss some ideas for improvement in methodology. I assume that some kind of simulation or mathematical analysis will be used in order to compare algorithm performance. I consider the following setup. We generate several input datasets according to various distributional assumptions. Each is analysed separately. For each algorithm in our set, we compute the value of each measure $\mu_i$ when running the algorithm on the given input. These are aggregated to find the expected value with respect to the given distribution. These values $\mu_i^* := E[\mu_i]$ are taken as the coordinates in $k$-space and the algorithms compared using dominance as above. Results will have the form "Serial Dictatorship is dominated by the Boston mechanism with respect to efficiency and envy-freeness for 10 of 12 datasets."

There are other issues I have not considered here which will need to be dealt with. For example, comparing means may be reasonable in some cases, but if the distributions of a given measure for outputs of two different algorithms using the same input distribution overlap substantially and have large variance, we may need a more refined analysis.

## 2  Inputs

If we are happy to use simulation rather than proving analytic results, we can use any input distribution we like. How to stress-test our algorithm by choosing interesting and "realistic" data? For example, evaluating and comparing voting rules using only the Impartial Culture, although it has been done often, is not sufficient.

The Mallows model has been used by several authors, based on the idea of there being an underlying true ranking of alternatives $\rho$ and the probability of an agent having the preference $\pi$ being proportional to $\exp(-cd(\pi, \rho))$, for some $c > 0$, where $d$ is the Kendall tau (swap) distance. This is typically applied with independence between agents.

To get correlation between agents, we can use an urn process in which we think of a new agent as copying a randomly chosen agent. This can lead to the Impartial Anonymous Culture, which is analytically very tractable and less unrealistic than Impartial Culture, but still far from describing reality. Different values of the parameter associated with this Eggenberger-Pólya urn (we may add more than one agent at a time) lead to more general distributions which may be more realistic. I have not seen serious work on fitting such distributions to real data. Of course, obtaining real data on preferences is difficult, increasingly so as has become clear from the failure of recent electoral predictions worldwide.

Some analyses require more subtlety. For example, with a coauthor I have recently (Pritchard and Wilson 2018) investigated the performance of electoral systems. Most electoral systems involve geographic districts, and treating them as independent or merging them into a single district are both oversimplifications. We used a coupled urn model with one urn for each district, allowing for imitation both within and across districts. To my knowledge this is the most sophisticated model used in the area so far. Surely there are better ones.

We have been modelling true preferences so far. In order to understand real performance of algorithms, it would be useful to have good models of strategic behavior, so that the agents' expressed preferences can be used as inputs but the real ones used for quality measures. For simple voting rules such as plurality it is easy to model behaviour of voters. Of course, the more complicated the rule the less likely manipulation should be in practice, because of agent fears about other agents' strategic voting and simply because of the complexity of computing a strategic preference.

I hope to see a greater variety of input distributions used in the future literature.

## 3  Quality measures

At the most basic level, we can take $\mu$ to be the indicator for any axiom, so that $\mu_*$ is the probability that an input leads to an outcome satisfying the axiom. This has been widely used and is very simple.

More sophisticated measures may consider the number of agents directly affected by failure of the axiom to hold. For example, the fraction of ordered pairs $(i, j)$ of agents for which $i$ does not envy $j$'s allocation (that is, prefers it to the allocation given to $i$) measures partial envy-freeness. The fraction of agents for whom truthtelling is a dominant strategy is a measure of resistance to individual manipulation.

The next level involves measurements based on preference intensity. For example, if I have my 2nd choice and you have my 1st choice, I may envy you less than if I have my 3rd choice and you have my 1st choice. As another example, a single manipulator may find it harder to change its expressed preference if that is very different from its true preference (for example, because the latter might be partially known to other agents, and preferences may not be given secretly, or the agent may require a bigger bribe in order to submit a vote far from its sincere one).

Measures based on cost of forming coalitions are important. For example, the communication overhead of organizing a coalition to manipulate a voting rule, or trade among themselves to restore an efficient outcome, may grow rapidly with coalition size.

Implicit in the measures so far is an idea of distance from the outcome to an axiomatically perfect one. For example, in the manipulation context above, a cost function as in the Mallows model may be appropriate. There are many other possible metrics, however. As another example, consider the directed graph $G$ formed by agents, in which each agent $i$ points to the agent holding $i$'s favorite item. Gale's Top Trading Cycle (TTC) algorithm reallocates items according to cycles formed in $G$, passing to items ranked 2nd, 3rd, etc and deleting agents and items as it goes. The algorithm terminates precisely when the digraph has only trivial cycles (each agent points to itself), and that occurs if and only if the allocation is efficient. A sophisticated measure of efficiency, then, might use a metric that measures the distance from $G$ to the nearest acyclic digraph. Such a metric might take into account the number and length of cycles in the original digraph.

So far all the measures have involved finding the minimum distance (or cost) to achieving axiomatic perfection for a given input. We can also consider probabilistic measures, which may be more realistic. In practice, it may be hard to find a minimum manipulating coalition, and the prevalence of manipulators should be considered. A voting rule that can be manipulated for a given input by 5 agents, never by 4 or fewer, and by very few coalitions of size 6, may be less manipulable in practice than one for which all coaltiions of size 6 can manipulate but no smaller coalition can. A general idea is to consider the simple game on the set of agents, with a winning coalition being a subset containing a manipulating coalition. An index such as the Coleman index (or generalizations discussed in Pritchard et al. (2013)) gives a measure of how likely a randomly chosen coalition is to be winning.

A new fairness idea (to my knowledge) is having low *order bias*. Almost all deterministic allocation algorithms require a fixed order on agents. For example, Serial Dictatorship has this built into the definition, while the Boston mechanism requires ties to be broken. A completely fair algorithm would be not only fair *ex ante* (by the usual randomization of the agent order) but also *ex post*—my welfare should not suffer just because I am chosen to submit my preferences last. Serial Dictatorship is clearly extremely unfair—the expected rank of the object chosen by the last agent is much higher than 1, the rank achieved by the first player. Boston similarly is unfair to later players. Randomization during the algorithm (after player order has been chosen) might deal with this problem, but if we insist on deterministic algorithms, what can we do (see "Algorithms" below)?

In general there are so many possible measures that we ought to try to justify the ones we use. Most of the literature I have seen simply uses a naive measure or copies the choice of a previous author in the field. A bigger intellectual contribution can be made by giving axiomatic properties that we wish a measure to hold. Spurred on by a perceptive referee, my coauthors and I did this for manipulability of voting rules (Pritchard et al. 2013), and now I try to apply this general advice in all my research. Every time we want to measure "partial X", where X is an axiom, we should think about the axiomatic foundations of our measure. I hope to see much better justification of measures in the future literature.

## 4   Algorithms

In addition to the design criteria for algorithms such as computational efficiency and solution quality, simplicity of explanation to the public is also a consideration if we want our algorithm to be adopted. There are limits to this—Single Transferable Vote is unlikely to be completely understood by the public yet has been used in Australia (which, perhaps not coincidentally, has compulsory voting) for decades. Nevertheless, a coherent "story" is often helpful.

Inspired by a Christmas party game, with a coauthor I recently introduced the Yankee Swap algorithm for allocation of indivisible goods (Lo and Wilson 2017). Again spurred by a perceptive reviewer complaining of how arbitrary this seemed, we generalized this to a family of 8 algorithms, 4 of which can be profitably followed by TTC to yield 12 algorithms in total. Two of these are equivalent to our old friends Serial Dictatorship and Boston. These algorithms are all derived from the Gale-Shapley algorithm for two-sided matching by assigning fictitious preferences for items over agents, allowing these to change during the algorithm, and adopting a queue or stack discipline for agents to propose to items. The point is that a single standard algorithm for an allocation problem gave rise to 12 algorithms for a closely related problem in a coherent way. All of these algorithms have an easy interpretation in terms of a party game involving stealing gifts. Some of them appear to make good tradeoffs although none satisfies any of the standard axioms, and very likely none would have been found by concentrating only on the standard algorithms. It i s still too early to tell whether these algorithms will attract attention. One of them has good welfare properties, and another has extremely low order bias even though it is deterministic.

In voting theory, there may be many more good algorithms to be found. To mention one among several recently introduced rules, Zwicker and coauthors have introduced the *mediancenter* rule (Cervone et al. 2012) which is still little explored. The general technique of distance rationalization has been used to construct new rules with given axiomatic properties, based on given notions of consensus and distance between profiles (Elkind et al. 2015; Hadjibeyli and Wilson 2019). I am sure that appropriate choices of consensus and distance can yield some new interesting rules, which may have good overall performance even if they fail to be extremal in the sense discussed above.

# 5   Conclusion

I hope to have convinced the reader that with high probability, there are many interesting new algorithms still to be found, and researchers should be on the lookout for them, sometimes hidden in plain sight in ordinary life. This should lead in many situations to better rules being used than those currently under consideration. Much more sophisticated and better motivated measures of performance, and more thorough analysis with respect to more realistic input distributions, will be needed if these algorithms are to be fairly compared with the standard ones in the literature.

**Acknowledgements**  Thanks to Klaus Nehring for his feedback on an earlier draft.

# References

Hadjibeyli, B., & Wilson, M. C. (2019). Distance rationalization of anonymous and homogeneous voting rules. *Social Choice and Welfare*, *52*, 559–583.

Cervone, D. P., Dai, R., Gnoutcheff, D., Lanterman, G., Mackenzie, A., Morse, A., et al. (2012). Voting with rubber bands, weights, and strings. *Mathematical Social Sciences*, *64*, 11–27.

Elkind, E., Faliszewski, P., & Slinko, A. (2015). Distance rationalization of voting rules. *Social Choice and Welfare*, *45*, 345–377.

Geoffrey, P., & Wilson, M. C. (2018). Multi-district preference modelling. arXiv:10.31235/osf.io/xpb8w.

Jacky, L., & Wilson, M. C. (2017). New algorithms for matching problems, 19. arXiv:1703.04225.

Michel, R., Bernard, G., Marley, A. A. J., & Ilia, T. (2006). *Behavioral social choice: Probabilistic models, statistical inference, and applications.*, (pp. xvi+240). Cambridge: Cambridge University Press.

Pritchard, G., Reyhani, R., & Wilson, M. C. (2013). Power measures derived from the sequential query process. *Mathematical Social Sciences*, *65*, 174–180.

# Part IV
# Axiomatics

# Exit in Social Choice

**Gustavo Bergantiños**

**Abstract** In the classical social choice problem a society has to select an alternative from a given set. Once the alternative has been chosen all agents will remain members of the society. There are many situations where some agents decide to leave the society (the Brexit is a good example). We review some papers that study how to adapt some well known models of social choice for including the possibility of voluntary exit. Finally we give some list of open questions regarding this issue.

## 1 Introduction

In the classical social choice problem a society formed by a group of agents $(N)$ must choose an element of a set of alternatives $(X)$. Agents have strict preferences over alternatives. A social choice function (a rule) selects an alternative from $X$ taking into account the preferences revealed by the agents. The idea is to select the alternative representing "better" the preferences of the agents. The next three real examples can be included in this framework. The European Union deciding how to divide the budget. The United Nations deciding how to mitigate global warming. An university department deciding upon its new members.

One of the most famous results in this framework is the Gibbard–Satterthwaite theorem (see Gibbard 1973 and Satterthwaite 1975). It says that if a rule satisfies strategy-proofness (to reveal the true preference is always a weakly dominant strategy) and unanimity (if $x \in X$ is the best alternative for all agents in $N$, then the rule selects $x$), then it is dictatorial (there is an agent $i \in N$ such that the rule

I want to thank D. Berga J. Massó and A. Neme the discussions we had during several years in which we have written several papers in this topic. The author is partially supported by research grants ECO2014-52616-R and ECO2015-70119-REDT from the Spanish Ministry of Science and Competitiveness, GRC 2015/014 from "Xunta de Galicia", and 19320/PI/14 from "Fundación Séneca de la Región de Murcia".

G. Bergantiños (✉)
Universidade de Vigo, 36310 Vigo (Pontevedra), Spain
e-mail: gbergant@uvigo.es

© Springer Nature Switzerland AG 2019
J.-F. Laslier et al. (eds.), *The Future of Economic Design*, Studies in
Economic Design, https://doi.org/10.1007/978-3-030-18050-8_24

always select the best alternative for $i$). This result is considered as an impossibility theorem because it says, roughly speaking; that there is no nice rule satisfying
strategy-proofness and unanimity.

One way for avoiding such impossibility result is by assuming some structure on
the set of alternatives and then restricting the domain of preferences. We give two
examples. Barberá et al. (1991) study the case of a society choosing new members
from a set of candidates (in this case $X$ is the set of possible subsets of candidates).
They assume that preferences over candidates are separable. Under this circumstances, they characterize a set of rules, committees, satisfying strategy-proofness
and unanimity. Moulin (1980) studies the case of a society choosing the level of a
public good. He assumes that preferences are single peak. Then, he characterizes the
set of rules, median voter schemes, satisfying strategy-proofness, anonymity, and
efficiency.

## 2 Social Choice with Exit

The classical approach assumes that the composition of the society is fixed, namely it
is independent of the chosen alternative. Nevertheless, this assumption is not realistic
in many cases. We give some examples. The UK has leaved the European Union.
President Trump has announced that US will leave the Paris climate agreement.
Some professors leave their departments because they disagree with some issues (as
the recruiting policy). Thus, it seems necessary to modify the classical social choice
model if we want to study the exit. The idea is so simple. In the classical model agents
have preferences over alternatives. In the model with exit agents have preferences
over outcomes: pairs of agents and alternatives. Let us to be more precise.

The set of outcomes is the set of all pairs formed by a subset of the original society
(an element in $2^N$, the subset of the set of agents $N$ that will remain in the society)
and an alternative in the set $X$. Then, agents' preferences are defined over the set of
outcomes $2^N \times X$. Let $(S, x)$ and $(S', x')$ be two outcomes and let $i$ be some agent.
We assume that preferences satisfy two conditions. First, if $i$ belongs to at least one
of the societies, namely $i \in S \cup S'$, then $i$ has strict preferences between $(S, x)$ and
$(S', x')$. Second, if $i$ does not belong to any of the societies, namely $i \notin S \cup S'$,
then $i$ is indifferent between $(S, x)$ and $(S', x')$. The first condition is the standard
strictness requirement on preferences, which is also assumed in the classical social
choice model. The second condition is specific of this setting and says that agents
that do not belong to the final society do not care about neither its composition nor
the chosen outcome.

In this paper we are thinking in voluntary exit, agents can leave the society when
they want (this is the case of the three real examples mentioned above). This aspect
is included in the model as a property of individual rationality. If a rule selects the
outcome $(S, x)$ each agent $i \in S$ should prefer $(S, x)$ to leave the society, which
corresponds with the outcome $(\emptyset, x)$. Notice that by the second condition of the
preferences, agent $i$ is indifferent between $(\emptyset, x)$ and any outcome $(T, y)$ such that
$i \notin T$.

A natural question that arises is why we do not consider the exit case as a particular social choice problem where the set of alternatives is $2^N \times X$ instead of $X$? Under the second condition on the preferences each agent $i \in N$ is indifferent among many outcomes. Hence, the set of individual preferences is not the universal domain of preferences over the set of outcomes. Thus, the Gibbard–Satterthwaite theorem does not apply.

Bergantiños et al. (2017a) extends this theorem to the new setting. They prove that serial dictator rules are the unique rules satisfying strategy-proofness, unanimity, and outsider independent (suppose that an agent who was not a member of the society change his preference but he still remain as non member, then the outcome chosen by the rule does not change). For each order of the agents we can define a serial dictator rule as follows. The first agent selects his best alternative, and only if this agent has many indifferent alternatives at the top of his preference, the second agent in the order selects his best alternative among those declared as being at the top by the first agent. The rule proceeds similarly following the ordering of the agents. Thus, as in the classical setting, we have an impossibility result.

Besides no serial dictator rule satisfies individual rationality, which is a crucial property in this setting. Thus, we can move in two directions. First, to find rules in the general domain of preferences satisfying individual rationality (but failing some other like strategy-proofness). Second, to restrict the set of preferences and try to find rules satisfying individual rationality and other properties.

In the first direction we can mention Bergantiños et al. (2017b) where they prove that an adaptation of approval voting to this setting satisfies internal stability and other interesting properties like consistency and efficiency. Unfortunately it does not satisfy strategy-proofness.

In the second direction we can mention, for example, the following two papers. Berga et al. (2004) studies the same problem as in Barberá et al. (1991) but they consider that initial members of the society may want to exit, if they do not like the resulting new society. They show that, if preferences are separable there is a unique rule satisfying strategy-proofness, unanimity, and individual rationality. This rule is the one where candidates are chosen unanimously and no founder exits. Cantalá (2004) studies the same problem as in Moulin (1980) but assuming that agents have a reservation utility and can decide do not consume the public good whenever the utility they obtain for consuming is smaller than this reservation utility. Cantalá (2004) characterizes the set of rules satisfying strategy-proofness, efficiency, and individual rationality (in terms of the reservation utility).

# 3   Open Questions

We now discuss some open questions that could be addressed in the future.

In the general model we have imposed two conditions on preferences. The second one says that if an agents does not belong to the society, he is indifferent about the composition of the society and the alternative chosen. This condition is quite reasonable in the example of the department but not in the examples of the Brexit or the Paris climate agreement. For UK it is important which countries belong to the EU and some of the decisions taken by the EU. Thus, it makes sense to change the second condition by other conditions that could model some other cases.

Berga et al. (2004) and Cantalá (2004) study two particular cases of the general model and they find rules satisfying two nice properties as strategy-proofness and individual rationality. More particular cases of the general model could be studied.

Instead of asking for general rules satisfying nice properties we can focus in a particular procedures and then analyze what happens. For instance UK has left the EU. In other countries, some political parties are proposing to leave the EU. Thus, we can model this situation as a non-cooperative game where agents decide sequentially if they want to stay or to leave the society. This idea has been studied in Berga et al. (2006). More situations could studied. For instance we can consider a two stage game as follows. In the first stage agents decide about the alternative in $X$ following some voting procedure. In the second stage agents decide, following some protocol, to stay or to leave the society. Berga et al. (2007) apply this idea in the case of voting by committees.

Until now we have considered the case of voluntary exit, which can be considered as a part of the membership. In this general setting agents can leave the society because they do not like (voluntary exit) or because they have been expelled (even they like). Similarly, agents can remain in the society because they want or because they are forced to stay in. For instance, Jackson and Nicolò (2004) considers the case of an excludable public good where preferences of the agents depend on the level of the public good but also on the size of the set of its users. Even some of their results are impossibility results, they identify some restricted domains where it is possible to define rules satisfying strategy-proofness and efficiency. Roughly speaking, such rules decide the size of the set of users independently of the preference profile and then, decide the level of the public good following a generalized median voter scheme. Since the size of the users is independent of the preference profile it could happen that some agent is out of the society but he would like to be in.

# References

Barberà, S., Sonnenschein, H., & Zhou, L. (1991). Voting by committees. *Econometrica, 59*, 595–609.

Berga, D., Bergantiños, G., Massó, J., & Neme, A. (2004). Stability and voting by committees with exit. *Social Choice Welfare, 23*, 229–247.

Berga, D., Bergantiños, G., Massó, J., & Neme, A. (2006). On exiting after voting. *International Journal of Game Theory, 34*, 33–54.

Berga, D., Bergantiños, G., Massó, J., & Neme, A. (2007). An undominated Nash equilibrium in voting by committees with exit. *Mathematical Social Sciences, 54*, 152–175.

Bergantiños, G., Massó, J., & Neme, A. (2017a). On societies choosing social outcomes, and their membership: Strategy-proofness. *Social Choice and Welfare, 48*, 857–875.

Bergantiños, G., Massó, J., & Neme, A. (2017b). On societies choosing social outcomes, and their memberships: Internal stability and consistency. Mimeo.

Cantala, D. (2004). Choosing a level of public good when agents have an outside option. *Social Choice and Welfare, 22*, 491–514.

Gibbard, A. (1973). Manipulation of voting schemes: a general result. *Econometrica, 41*, 587–601.

Jackson, M., & Nicolò, A. (2004). The strategy-proof provision of public goods under congestion and crowding preferences. *Journal of Economic Theory, 115*, 278–308.

Moulin, H. (1980). On strategy-proofness and single-peakedness. *Public Choice, 35*, 437–455.

Satterthwaite, M. (1975). Strategy-proofness and arrow's conditions: Existence and correspondence theorems for voting procedures and social welfare functions. *Journal of Economic Theory, 10*, 187–217.

# Camels, Dummy, and the Importance of Context

Étienne Billette de Villemeur and Justin Leroux

**Abstract** Economic design problems are more successful in garnering support from practitioners when the axioms are relevant to the practical context. We use the well-known Dummy axiom as a concentrated example of several ways in which axioms can fail to be meaningful in practice. We then describe two channels through which characterization results using axioms that are not relevant to a specific context can undermine the axiomatic program, both from an internal (theoretical) standpoint and from its relationship with the practical world. Yet, a great deal of intellectual stimulation can be found in disciplining ourselves to be guided by context, despite the theorist's traditional leaning towards universality.

## 1 Introduction

The axiomatic side of the economic design program has been only partly successful in garnering the interest of practitioners of late. Some subfields have flourished, like the matching literature and its related subfields (e.g., school choice, kidney exchange and the scheduling of landing slots), while others have yet to reach their full potential (e.g, cost and surplus sharing, voting). We attribute this divergence partly to the types of axioms used to justify design recommendations. We claim that when axioms are relevant to the practical context at hand, the bridge between theory and practice is an

We are grateful to Antoinette Baujard and Yves Sprumont for stimulating conversations as well as to Jens Hougaard, Marcus Pivato, Jim Schummer, and Christian Trudeau for comments on an earlier draft.

É. Billette de Villemeur
Université de Lille and LEM-CNRS (UMR 9221), Villeneuve d'Ascq Cedex, France

J. Leroux (✉)
HEC Montréal, Montreal, Canada
e-mail: justin.leroux@hec.ca

CIRANO, Montreal, Canada

CRÉ, Montreal, Canada

© Springer Nature Switzerland AG 2019
J.-F. Laslier et al. (eds.), *The Future of Economic Design*, Studies in
Economic Design, https://doi.org/10.1007/978-3-030-18050-8_25

easy one to cross. To the contrary, when the axioms chosen ignore the context, we theorists face the danger of speaking only to ourselves.

We first clarify what we mean by an axiom being relevant to the context. We do so by illustrating the several ways in which axioms can fail to be meaningful in practice, using the well-known Dummy axiom as a concentrated example. We then describe two channels through which characterization results using axioms that are not relevant to a specific context can undermine the axiomatic program, both from an internal (theoretical) standpoint and from its relationship with the practical world. We conclude by reaffirming that plenty of interesting challenges and a great deal of intellectual stimulation can be found in disciplining ourselves to be guided by context, despite the theorist's traditional leaning towards universality.

## 2   Camels

Many works on the topic of fair division make explicit references to the Talmud. We shall break with this tradition by offering an example taken from Muslim lore, of which we give our own (shorter) version.[1]

A sheikh dies and leaves to his three sons an estate made up of 17 camels. His will stipulates that 1/2 of his estate should go to his eldest son, 1/3rd is to go to his second son and 1/9th to his youngest. The sons are puzzled, because none of these fractions amount to a whole number, and dividing camels into fractions (or selling them and sharing the proceeds) is out of the question. The matter remains unresolved until a wise man arrives, riding a camel. After hearing their plight, the wise man devises a solution. "Allow me to add my own camel to the lot. 1/2 of the total amounts to 9 camels, which the eldest shall receive. 1/3rd of the total is 6 camels, which the second son should get. Finally, 1/9th of the total is 2 camels, for the third son, and I can ride my own camel back to Medina."

The reader will have noticed that the reason why the wise man's solution works is that the shares assigned by the sheikh's will do not add up to one. But this is not what is amusing about the tale. We smile because the solution to the division problem entailed the arrival of an additional agent who was completely unrelated to the division problem. Before his arrival, the problem could not be solved. Our amusement reveals that we expect division problems to be solved among the relevant agents only, that the wise man's solution is an anomaly resulting from the fractions summing up to less than one.

Surely, no "real" division problem would exhibit the same idiosyncrasy. Yet, we claim that the way the profession frames those problems sometimes *creates* such anomalies. For instance, this happens whenever one invokes the so-called *Dummy axiom* to characterize division rules.

---

[1]For a more detailed version, see Abdul-Rauf (1996).

Informally, a division rule satisfies Dummy if it always assigns a share of zero to an agent who never has any impact on the overall amount to be allocated whatever the participation of others. Such an agent is called a *dummy agent*. At first glance, this is a very sensible property. Indeed, it could be unfair, or at least couterintuitive, to ask someone who has not ordered anything at the restaurant to pay part of the bill. Likewise, rewarding a teammate's effort when her contribution is consistently nil can only go on for so long.

Dummy is a perfectly reasonable axiom, against which we raise no objections. In fact, it can be reassuring to know that a division rule satisfies Dummy. However, the usefulness of Dummy is less clear when we invoke the axiom to *justify* a particular allocation rule, as is done through characterization theorems. This is because the proofs of these characterization theorems will then inevitably rely on the presence of dummy agents. That is, just as the story's solution required the arrival of a wise man, the characterization results require the potential existence of a non-relevant— dummy—agent, and cannot be derived in the presence of only the non-dummy agents.

Moreover, and as we argue next, the presence of dummy agents is a highly unlikely scenario.

## 3 Dummy and Other Axioms Invoking Unrealistic Scenarios

In cost-sharing problems, where each agent is being served her demand and the total cost must be divided among them, we believe that dummy agents simply cannot exist. Our skepticism originates from the fact that we have yet to encounter a reasonable framework where some agents obtain as much of a good (a resource, a service, etc.) without affecting the total cost. Is an agent who tunes in to the radio a dummy agent? Her unlimited usage certainly adds nothing to the cost of the service. But this is only true *provided the service is offered in the first place*. Otherwise, were she the only person demanding radio service, and were this service offered as a response to her demand only, the agent would certainly be responsible for a (presumably very large) increase in cost: she would not be a dummy agent. In fact, if the problem is to devise a pricing rule that shares the costs of a radio service to its future users, no-one can a priori be considered to be a dummy agent. All cost-sharing problems share this feature.

In surplus-sharing problems, the strict impossibility of dummy agents is less clear. Unlike the cost-sharing problem, where the demand of all agents must be served, the surplus-sharing setting is one where even those who contribute nothing could potentially be thought of as being relevant to the problem. For instance, a camel merchant in Saudi Arabia could potentially be considered having a claim on the profits of partners in a New York law firm, even though his actions have rigorously no effect on the law firm's profits. Taken to the extreme, the entire population of the planet (and perhaps even of others) could be seen as a priori belonging to the set of relevant agents. However, intuitively, to each division problem corresponds a population that can be readily identified. To invoke Dummy implicitly means that the practitioner who will end up applying the rule cannot be fully trusted to identify this relevant population. If so, Dummy can be thought of as a robustness requirement. Yet, contexts where the relevant population cannot be identified are the exception and not the norm.[2]

Another plausible justification of Dummy is a robustness argument of a different kind, which justifies the axiom as a continuity condition of sorts. The argument concedes that dummy agents do not actually exist, but claims that there can be agents who are *almost* dummy agents; i.e., whose impact on the total amount to be divided is very small. Accordingly, the shares of these agents should be close to zero. We find this justification unpersuasive, because the conditions that Dummy dictates "at the limit" impose restrictions on the division rule even for the profiles that actually matter. This is best seen in an example:

Consider the discrete cost-sharing problem, where $n$ agents must split the total cost of meeting their demands: $C(x_1, \ldots, x_n)$. Dummy and Additivity—the requirement that a cost-sharing rule, $\varphi$, prescribes the same cost shares when applied to total production costs as when applied separately to its additive components—together characterize the family of so-called *path-generated cost-sharing rules* (Wang 1999). While interesting, this family excludes reasonable solutions, such as sharing costs in fixed proportions—$\varphi_i(x, C) = C(x_1, \ldots, x_n) \delta_i$ with $\sum \delta_i = 1$—or even proportionally to individual demands—$\varphi_i(x, C) = \frac{C(x_1, \ldots, x_n)}{\sum x_i} \cdot x_i$. When dropping Dummy, these solutions become admissible again (Moulin and Vohra 2003). Yet, because dummy agents do not exist in practice, imposing Dummy amounts to discarding these potentially desirable solutions solely because they do not behave a certain way *outside of their actual domain of usefulness*.

The above example illustrates just how misguided the "continuity" justification of Dummy can be. Furthermore, it demonstrates that the set of admissible solutions may dramatically change when we move from a population containing almost-dummy agents to one with actual dummy agents. If continuity is a concern, we believe that continuity at a meta-level—i.e., continuity of the set of admissible rules—should take precedence over continuity of the rules themselves outside of their relevant domain of application.

*Faithfulness* is another of many axioms to convey a limiting argument of sorts, this time in the area of preference aggregation. A preference aggregation rule satisfies Faithfulness if, when applied to a population of a single agent, it returns exactly that agent's preferences. In other words, as the population shrinks to the extreme, the rule should respect the preferences of the (degenerate) population in the limit when only a single agent remains. To be sure, it would be strange for a rule to disagree with the

---

[2]Moreover, narrowing down the problem to consider only the relevant agents actually amounts to adhering to the "Dummy Consistency" axiom; i.e., the requirement that eliminating a dummy agent from the population should not affect the shares of non-dummy agents (see Moulin 2002 for a formal definition). This is very different from the Dummy axiom because discarding dummy agents from the sharing problem is not the same as including them and specifying that they should be assigned a zero share.

single member of the population whose preference it aggregates. That being said, such a scenario is not a situation we should even care about, because not only does the matter become trivial and uninteresting, it is no longer an appropriate target for the machinery of social choice theory.[3]

Dummy and Faithfulness are but two examples of what we shall refer to as *axioms invoking unrealistic scenarios*, of which there exist many others. We claim that such axioms are unhelpful for economic design and, worse, can even be downright harmful both to the quality of the recommendations and to the visibility of axiomatic economic design to practitioners.

# 4  Why Axioms Invoking Unrealistic Scenarios Can Be Harmful

We find it questionable to design sharing rules with the purpose of accommodating situations that will not occur. We now claim that characterizations involving axioms that invoke unrealistic scenarios can be harmful for at least two reasons.

First, appealing to axioms invoking unrealistic scenarios restricts the family of desirable rules for the situations we actually care about. Indeed, the very nature of axiomatic work is such that placing conditions on how a rule should behave in some scenarios has repercussions on the rule's behavior in other situations, if only indirectly. For example, requiring that a dummy agent receives a zero share will, through the mathematical structure of the problem, impose restrictions on the family of admissible rules, as seen in the above example. One may question the usefulness of restrictions that are only warranted by the desire to treat remote scenarios a certain way. Of course, what constitutes a "remote" scenario is relative to the context at hand.

Secondly, and relating once again to the importance of context, axioms invoking unrealistic scenarios can hurt the visibility of axiomatic economic design. Understandably, an administrator charged with reforming the electoral process of a nation will be unimpressed by how a rule behaves in the highly unlikely event that a single voter shows up on election day. To practitioners, rules that hinge on axioms invoking scenarios that are unrealistic—given the context—are little more than an academic curiosity. By contrast, it is quite telling that the school choice (e.g., Abdulkadiroğlu and Sönmez 2003; Abdulkadiroğlu et al. 2009; Kesten 2010), kidney exchange problems (e.g., Roth et al. 2004, 2007) and airport landing slots allocation (Schummer and Vohra 2013; Schummer and Abizada 2017) are the branches of the larger social choice program to have received the most attention recently in the world of practice. The rules that these literatures recommend typically rely *only* on axioms that

---

[3] That an aggregation rule be required to respect the ranking of the only individual whose preferences are known only makes sense if we consider that the views of a single individual can somehow summarize those of society, which is not obvious from a logical standpoint. Even so, in that case, the aggregation problem would disappear. We thank Antoinette Baujard for this observation.

adapt important concepts (like stability and fairness) to the context.[4] Practitioners can relate to these justifications and therefore find it easier to take the corresponding recommendations seriously.

## 5 Looking to the Future

To be sure, the fact that practitioners can relate to certain axioms more than to others does not make these axioms desirable per se. In fact, one could object that practitioners are not experts of the theory and thus should not be the ones to decide what axioms should and should not be used. The objection seems valid, but loses merit once we remind ourselves of the purpose of our work. As economic designers, our end goal is to solve practical economic problems. The process entails some division of labor: At the forefront are the practitioners, who witness firsthand the actual problems and who then become demanders of solutions. While they may be envisioned in an abstract way, these problems are necessarily embedded into some context. Economic designers can be seen as the *engineers* who are tasked with mustering the theoretical machinery with the aim of proposing appropriate solutions. Both appropriateness and adoption of the proposed solutions will depend on how well they are related to the context.

This analogy does not mean that we should forgo our inclination for mathematical elegance nor that we should rein in our desire for generality; however, it *does* mean that we should refrain from invoking unrealistic scenarios for the purpose of entertaining these preferences. Equivalently, we should pay explicit attention to the features of an allocation rule that matter to practitioners.[5]

It may not be easy. For example, and to come back to our initial example, refraining from using the Dummy axiom in division problems makes proofs much more difficult–yet not impossible, as proved in Moulin and Vohra (2003). Nevertheless, it also opens a vast array of possibilities. As economic designers, we should welcome contextualized problems as opportunities, but also as guides, to explore these possibilities.

---

[4]In the school-choice problem, the concept of stability adapts to the fact that schools do not have preferences per se but priority orderings over students reflecting, say, the fact that they may already have siblings enrolled in the school. The concept of "justified envy" adapts the fairness concept of envy in a similar way.

[5]We write "equivalently", because restricting the domain of application of an axiom can be seen as rewriting the axiom to accomodate the concerns of the practitioner. In this volume, Schummer and Serizawa (2019) make a very related point that the practitioner is "the expert in determining what objectives need to be met" and that the relevant axioms to consider should emanate from her concerns rather than from the desire to arrive at a characterization result.

# References

Abdul-Rauf, M. (1996). *Imam Ali Ibn Abi Talib: The First Intellectual Muslim Thinker* (p. 106). Alexandria: Al-Saadawi Publications.

Abdulkadiroğlu, A., & Sönmez, T. (2003). School choice: A mechanism design approach. *American Economic Review, 93*(3), 729–747.

Abdulkadiroğlu, A., Pathak, P. A., & Roth, A. E. (2009). Strategy-proofness versus efficiency in matching with indifferences: Redesigning the NYC high school match. *American Economic Review, 99*(5), 1954–78.

Kesten, O. (2010). School choice with consent. *Quarterly Journal of Economics, 125*(3), 1297–1348.

Moulin, H. (2002). Axiomatic cost and surplus-sharing. In K. J. Arrow, A. Sen, & K. Suzumura (Eds.), *The handbook of social choice and welfare*. Amsterdam: North-Holland.

Moulin, H., & Vohra, R. (2003). Characterization of additive cost sharing methods. *Economics Letters, 80*(3), 399–407.

Roth, A. E., Sönmez, T., & Utku Ünver, M. (2004). Kidney exchange. *Quarterly Journal of Economics, 119*(2), 457–488.

Roth, A. E., Sönmez, T., & Utku Ünver, M. (2007). Efficient kidney exchange: Coincidence of wants in a markets with compatibility-based preferences. *American Economic Review, 97*(3), 828–851.

Schummer, J., & Abizada, A. (2017). Incentives in landing slot problems. *Journal of Economic Theory, 170*, 29–55.

Schummer, J., & Serizawa, S. (2019). The Role of Characterizations in Market Design. *[This volume]*.

Schummer, J., & Vohra, R. V. (2013). Assignment of arrival slots. *American Economic Journal: Microeconomics, 5*(2), 164–185.

Wang, Y. T. (1999). The additivity and dummy axioms in the discrete cost sharing model. *Economic Letters, 64*, 187–192.

# Collective Choice Problems, Formal Representations, and Missing Information

Ashley Piggins

**Abstract** This paper discusses the appropriateness of certain formal models in social choice theory. How can we be sure that these are adequate? Does the preference for parsimonious models mean that important information, necessary to solve the social choice problem, is excluded? Two models are discussed: Sen's Paretian liberal paradox and the school choice problem. The Paretian liberal paradox teaches us the importance of motivations and context, two things that typically do not feature in social choice analysis. An alternative formulation of the school choice problem, due to Duddy, is also presented and contrasted with the original.

When studying logic for the first time, we are taught that the subject is about valid deductions (or entailments). If $P \rightarrow Q$ is true and $P$ is true, it logically follows that $Q$ is true. It does not matter what the $P$ and $Q$ actually refer to. This stripping away of content is regarded as essential. For, as Kant (1992, p. 432) put it, "Logic contains no matter at all, only form of thought".

In social choice theory, we often switch between thinking of social alternatives, $x$ and $y$, as social states (complete descriptions of all aspects of society as in Arrow's (1963, p. 17) original definition), or something resembling political candidates or parties in an election. Again, the fact that we do not need to specify whether we are in a welfare-economic or political context is regarded as a virtue. Arrow's impossibility theorem applies in both cases. The more we can remove context from a particular social choice problem, and reduce it down to what seem like natural mathematical primitives, the more general our theory is and the wider its scope. However, this begs the question of what, exactly, is a natural mathematization of a collective choice problem? How can we be sure that all of the relevant information needed to adequately solve the problem is captured by that formalism? Economists tend to favor

I would like to thank Conal Duddy, Umberto Grandi, Arianna Novaro and Bill Zwicker for their helpful comments on an earlier version of this paper.

A. Piggins (✉)
J.E. Cairnes School of Business and Economics and the Whitaker Institute,
National University of Ireland Galway, Galway, Ireland
e-mail: ashley.piggins@nuigalway.ie

J.-F. Laslier et al. (eds.), *The Future of Economic Design*, Studies in
Economic Design, https://doi.org/10.1007/978-3-030-18050-8_26

formal models that are parsimonious and grounded in *ordinalism*. This is the view that analytical representations of collective choice problems must use only ordinal preference information, with no interpersonal comparability (Sen 1979). Attempts to expand this informational basis within traditional welfare economics are described in Sen (1977).

I shall discuss two well-known literatures, the literature on rights and the litera- ture on school choice. The literature on rights has flourished since the publication of Sen's (1970) Paretian liberal paradox, and the literature on school choice has grown rapidly since the important and highly influential work of Abdulkadiroğlu and Sönmez (2003). I argue that we cannot, in general, solve the Paretian liberal paradox because the informational environment is too restrictive. This suggests that ordinalism has limits and therefore we might wish to expand the informational envi- ronment in some way (by introducing *motivations* and *context*, for example). What I have in mind is different from adding cardinal utility and interpersonal comparisons to the analysis, and we can perhaps best describe it using Sen's phrase "non-utility information".

The school choice problem is more clearly a problem of economic design than is the rights problem. A solution to a school choice problem determines a match between students and schools based on the ordinal preferences of students over schools and the ordinal priority rankings of students by schools. I shall not argue that ordinalism is an inadequate framework for solving problems of student-school matching. Instead, I present a recent argument due to Duddy (2018) that shows how we can expand the informational basis of the school choice problem while remaining in an ordinalist framework. Duddy's formulation of the problem strikes me as an improvement on the original formulation.

First of all, Sen's well-known example. There are two individuals, Prude and Lewd, and a copy of the risqué *Lady Chatterley's Lover*. There are three options: Prude reads the book ($x$), Lewd reads the book ($y$), or no one reads the book ($z$). Lewd prefers most for Prude to read the book. Sen says that Lewd would "delight" in Prude's discomfort (Sen 1970, p. 155). Given a choice between reading the book himself or no one reading it, Lewd would prefer (being a lewd) to read it. Writing $x P_l y$ to denote that Lewd strictly prefers $x$ to $y$, Lewd's preference ordering is $x P_l y P_l z$. Of course, Prude prefers most that no one reads it. However, if someone must then he would rather it be him. Prude worries that if Lewd reads it then he could become even more depraved. Therefore, Prude's ordering is $z P_p x P_p y$.

Given that the alternatives are taken to represent social states with "each state being a complete description of society including every individual's position in it" (Sen 1970, p. 152), alternatives $y$ and $z$ differ only in ways that are private to Lewd, and $x$ and $z$ differ only in ways that are private to Prude. Because of this, we might feel that Lewd's and Prude's preferences over these respective pairs should be given special status when determining the social preference. Suppose we make Lewd decisive over $\{y, z\}$. This means that $y P_l z$ implies $y P z$ and $z P_l y$ implies $z P y$. Similarly, make Prude decisive over $\{x, z\}$. If we combine this decisiveness with the weak Pareto principle then we obtain a cycle in social preferences, $y P z$, $z P x$ and $x P y$. The cycle means that it is difficult to make a social choice.

Despite the social-choice formulation of this example, there appears to be a straightforward solution to it based on the *contextual* information supplied by Sen. Critically, Sen tells us the reasons why Prude and Lewd hold the preferences that they do. Lewd is being mischievous and Prude is being puritanical. It is often said that their preferences are "meddlesome" (Blau 1975; Sen 1976). Philosophically we might wish to ignore "meddlesomeness" and focus only on the non-meddlesome preferences. Lewd prefers to read the book than not ($y P_l z$), and Prude would rather not read it ($z P_p x$). Therefore, Lewd ought to read the book and the social choice should be $y$. This solution appears to be Sen's preferred one, and it intuitively seems correct despite the fact that $y$ is Pareto inferior to $x$ (Sen 1976, p. 237).

Notice that the contextual information provided by Sen in his example is not part of the formal description of the social choice problem itself. Sen not only explicitly identifies what the $x$, $y$ and $z$ refer to, but he also describes what *motivates* Prude and Lewd to hold the preferences that they do. In economics, we typically assume that preferences are rational and well-informed. If not, then they cannot safely be used for public policy. However, we do not stipulate that preferences should be *normatively justifiable*. Prude and Lewd's preferences would possibly fail that test. Once we understand the reasons for Prude and Lewd's preferences, then we can decide which part of their preferences should "count" and which should not. The missing contextual information helps us find the right solution.

Here is another example. Professor Price and Professor Lee work together and their department is interviewing three candidates, $x$, $y$ and $z$, for a vacant position. Price has co-authored with $x$ and $z$, and Lee with $y$ and $z$. On the basis of the superior information that co-authorship typically provides, the department has assigned $\{x, z\}$ decisiveness to Price and $\{y, z\}$ decisiveness to Lee. Price ranks $z$ as the best candidate, then $x$, and finally $y$. Lee ranks $x$ best, then $y$, and finally $z$. The formal description of this social choice problem is identical to *Lady Chatterley* and the same social preference cycle obtains. However, it is clear from the context that no one is being meddlesome. Price and Lee's rankings are a sincere expression of their beliefs. *In this context*, $y$ would be a very unappealing social choice, as both Price and Lee believe that $x$ is a better candidate.

I think these examples show that solutions to social choice problems may require more information than typically features in formal descriptions of those problems. In particular, non-utility information seems to have an important role. My conjecture is that including this information into the analysis will take us beyond ordinalism, and a challenge for the future is to see how this is done.

I now discuss matching. Do ideas from the Paretian liberal have any application here? While the notion that someone can be decisive over who they are matched to is intuitively unappealing, we might feel that a person could exercise a veto over their match. This is particularly so in the "marriage problem" of Gale and Shapley (1962). However, we run immediately into an impossibility which is so trivial that I suspect it is already known. The men we denote by $a$, $b$ and $c$, the women by $d$, $e$ and $f$.

If $c$ can veto $f$, for example, then he is either matched with $d$ or $e$. It is easy to see from Table 1 that in either case, it is impossible to find a stable match.

**Table 1** Preferences

| Men | Preferences | Women | Preferences |
|-----|-------------|-------|-------------|
| $a$ | $e P_a d P_a f$ | $d$ | $a P_d b P_d c$ |
| $b$ | $d P_b e P_b f$ | $e$ | $b P_e a P_e c$ |
| $c$ | $d P_c e P_c f$ | $f$ | $a P_f b P_f c$ |

Stability is inconsistent with veto power, just as Pareto efficiency is inconsistent with decisiveness (in the Paretian liberal).

In addition, the philosophical issue raised in the previous section has a partial analog in matching theory. It is not perfectly analogous since I do not claim that we need to add non-utility information to matching theory. Rather, I wish to draw attention to work by Duddy (2018) which shows how the informational basis of the well-known school choice problem can be expanded while remaining within ordinalism.

Typically, a school choice problem is *formally* defined by five items: (1) a set $I$ of students, (2) a set $S$ of schools, (3) a list of natural numbers, each indicating the capacity of a school, (4) a list of strict preference orderings over $S$, one for each student, and (5) a list of weak priority orderings over $I$, one for each school. The total number of places available across all schools must be at least as great as the number of students, with each school having at least one available place. Note that the student preferences reflect the desirability of the schools in the eyes of the students, whereas the priority orderings of the schools are *induced* by priority criteria exogenously set by school district boards. These criteria are meant to reflect objectives of public policy. For example, a prospective student may already have a sibling at the school. If so, they potentially have higher priority at the school than another student. Having a sibling already at the school may be a priority criterion, as might living within walking distance of the school.

Now consider the following example, due to Duddy (2018). There are three students $i$, $j$ and $k$ and three schools $s_1$, $s_2$ and $s_3$. Schools $s_1$ and $s_2$ are in the Oak Hill neighborhood and $s_3$ is in Elm Hill. There is one available place at each school. All students agree that $s_1$ is the best school, followed by $s_2$ and lastly $s_3$. Student $i$ lives in Oak Hill (where $s_1$ is located) *and* has a sibling already attending $s_1$. Students $j$ and $k$ both live in Elm Hill and so one of them will have to travel to Oak Hill for their schooling.

Student $i$ will be top of $s_1$'s priority ordering by virtue of the fact that they live within walking distance and have a sibling attending the school. Students $j$ and $k$ tie for second place in this ordering. Student $i$ will also be top of $s_2$'s priority ordering because of the walking distance criterion alone. Of course, students $j$ and $k$ will be top of $s_3$'s priority ordering for the same reason. The obvious way to match students to schools is described in Table 2.

The entries refer to the probabilities of being matched to a particular school. Student $i$ ought to be sent to $s_1$ due to meeting the walking distance criterion and the sibling criterion. A fair coin toss decides which of $j$ and $k$ travels to $s_2$ for their schooling. Although they will ultimately attend the inferior Oak Hill school,

**Table 2**  A matching

|   | $s_1$ | $s_2$ | $s_3$ |
|---|---|---|---|
| $i$ | 1 | 0 | 0 |
| $j$ | 0 | $1/2$ | $1/2$ |
| $k$ | 0 | $1/2$ | $1/2$ |

**Table 3**  A matching for the second scenario

|   | $s_1$ | $s_2$ | $s_3$ |
|---|---|---|---|
| $i$ | $1/2$ | $1/2$ | 0 |
| $j$ | $1/4$ | $1/4$ | $1/2$ |
| $k$ | $1/4$ | $1/4$ | $1/2$ |

**Table 4**  An alternative matching

|   | $s_1$ | $s_2$ | $s_3$ |
|---|---|---|---|
| $i$ | $1/3$ | $2/3$ | 0 |
| $j$ | $1/3$ | $1/6$ | $1/2$ |
| $k$ | $1/3$ | $1/6$ | $1/2$ |

reserving the place at $s_1$ for student $i$ is justified by the fact that this student meets two of the priority criteria for that school. Neither $j$ nor $k$ meet any of the criteria. All of the standard mechanisms for matching students to schools will generate the expected matching in Table 2.

Duddy then invites us to reconsider the example, but this time without $i$ having a sibling at $s_1$. This small change is highly significant. Before, assigning $i$ to $s_1$ was justified on the grounds that $i$ lived within walking distance *and* had a sibling already at that school. Now the sibling is gone. Of course, it could be argued that $i$ should still be assigned to $s_1$ given that $i$ lives within walking distance, but *that objective of public policy is also perfectly met by assigning $i$ to $s_2$*. Student $i$ lives within walking distance of $s_2$ too, and so society is indifferent as to whether $i$ attends $s_1$ or $s_2$. No additional goal of public policy is served by assigning $i$ to $s_1$ rather than $s_2$. It is reasonable, then, for a coin toss to decide which of $s_1$ and $s_2$ student $i$ attends. It would be unreasonable to make $i$ attend $s_3$ because then all of the students will be attending school outside of their walk zones.

This leads to the following matching (Table 3).

Duddy presents another reasonable matching for this scenario. Each student is given an equal chance of attending the best school, $s_1$. If $i$ is not matched to $s_1$ then she is definitely assigned to $s_2$. This leads to Table 4.

Clearly, Tables 3 and 4 represent reasonable matches in the second scenario. The important point Duddy makes, however, is that when we represent the second scenario using the list of items (1) to (5) above then *the second scenario is indistinguishable from the first*. The reason is that no school's priority ordering changed when we moved from the first scenario to the second. The way that the problem is formalized means that these two very different scenarios correspond to the *same* school choice problem.

What this means is that the sensible matching in the first scenario (Table 2) must apply in the second. But, as Duddy points out, this is a very unattractive matching in that second scenario. The student from Elm Hill who travels to Oak Hill for her schooling is automatically assigned to the *inferior* Oak Hill school, with the superior Oak Hill school reserved exclusively for student $i$. There is no basis for that exclusivity in the second scenario, unlike the first. As we have noted, the policy goal of having students in their walk zone is equally met by assigning $i$ to either $s_1$ or $s_2$. The two schools are substitutable in that respect.

The philosophical point, rather like in the *Lady Chatterley* example, is that the way the problem is formalized means that normatively important information is lost. Duddy presents an alternative formulation of the school choice problem in which it is possible to distinguish between the two scenarios presented above, and he adapts the Gale-Shapley student-optimal stable mechanism to this setting. His formulation keeps items (1) to (4) of the original formulation, replaces (5) with (5'), and adds two additional items. Item (5') is a set $P$ of priority criteria, (6) is a mapping $f$ from $I \times S$ to $2^P$, the power set of $P$, and (7) is a weak ordering $\succeq$ over $2^P$ such that $A \subseteq B \rightarrow B \succeq A$.[1] The two scenarios can now be distinguished because $f(i, s_1)$ in the first scenario is different from $f(i, s_1)$ in the second scenario. I regard Duddy's work as an attempt to expand the underlying informational environment of student-school matching, and efforts along these lines are surely important to the future of economic design.

# References

Abdulkadiroğlu, A., & Sönmez, T. (2003). School choice: A mechanism design approach. *American Economic Review, 93*, 729–747.

Arrow, K. J. (1963). *Social choice and individual values*. New York: Wiley.

Blau, J. H. (1975). Liberal values and independence. *Review of Economic Studies, 42*, 395–401.

Duddy, C. (2018). The structure of priority in the school choice problem, forthcoming in Economics and Philosophy.

Gale, D., & Shapley, L. S. (1962). College admissions and the stability of marriage. *American Mathematical Monthly, 69*, 9–15.

Kant, I. (1992). *Lectures on logic* (J. Michael Young, Trans.) (Ed.). Cambridge: Cambridge University Press.

Sen, A. (1970). The impossibility of a Paretian liberal. *Journal of Political Economy, 78*, 152–157.

Sen, A. (1976). Liberty, unanimity and rights. *Economica, 43*, 217–245.

Sen, A. (1977). On weights and measures: Informational constraints in social welfare analysis. *Econometrica, 45*, 1539–1572.

Sen, A. (1979). Personal utilities and public judgements: Or what's wrong with welfare economics. *Economic Journal, 89*, 537–558.

---

[1] Umberto Grandi and Arianna Novaro point out that the power set of $P$ grows exponentially in the number of criteria. It may be computationally difficult to express such a weak ordering. However, if the number of priority criteria is low, as seems likely in school choice problems, then this would not be a problem in practice.

# Axioms Should Explain Solutions

Ariel D. Procaccia

**Abstract** In normative economics, axiomatic properties of mechanisms are often formulated by taking the viewpoint of the designer on what is desirable, rather than that of the participants. By contrast, I argue that, in real-world applications, the central role of axioms should be to help explain the mechanism's outcomes to participants. I specifically draw on my practical experience in two areas: fair division, which I view as a success story for the axiomatic approach; and voting, where this approach currently falls short.

The *axiomatic approach* advocates the design of mechanisms that provide rigorous guarantees, and, indeed, compares mechanisms based on the properties that they do, or do not, satisfy. This approach has long been a staple of economic theory, and, more recently, has guided the analysis of social computing systems (Tennenholtz and Zohar 2016).

The axioms in question typically attempt to formally capture notions of justice, fairness, efficiency, or just plain reasonableness. My view is that they are typically formulated with the *designer* in mind, that is, they are meant to convince an authority to adopt a specific mechanism. In this essay I wish to examine the idea that axioms should be formulated to help *explain* the mechanism's choices to *participants*. To demonstrate the appeal, and real-world implications, of this idea, I start by presenting a positive case study—fair division. I then discuss the practical shortcomings of the axiomatic approach to voting, and point to a way forward.

In fair division, an especially intuitive setting is that of rent division, where the goal is to fairly assign the $n$ rooms of an apartment to $n$ players, and divide the predetermined total rent between them. The preferences of the players are assumed to be *quasi-linear*, that is, the utility of a player for receiving a room that he values at $x$ for the price of $y$ is $x - y$.

From the axiomatic viewpoint, the primary question is what one means by "fairly". The gold standard of fairness axioms is (arguably) *envy-freeness*, which, in the rent division context, requires that the utility of each player for his room at its price be

A. D. Procaccia (✉)
Carnegie Mellon University, Pittsburgh, USA
e-mail: arielpro@cs.cmu.edu

© Springer Nature Switzerland AG 2019
J.-F. Laslier et al. (eds.), *The Future of Economic Design*, Studies in
Economic Design, https://doi.org/10.1007/978-3-030-18050-8_27

at least as high as his utility for any other room at the price of that room. The true power of this axiom stems from the fact that an envy-free rent division always exists, under mild assumptions (Svensson 1983; Su 1999).

By contrast, when allocating indivisible goods *without* money, envy-freeness clearly cannot be guaranteed. For example, imagine a situation with two players and a single good: whichever player receives the good will be envied by the other. This observation underlies a significant body of work on fairness axioms that *can* be guaranteed in this setting, and the design of algorithms that achieve them. My current favorite is *envy-freeness up to one good*: one player may strictly prefer the bundle of another player to his own, but it is always possible to remove a single good from the bundle of the latter player to eliminate the former player's envy. Lipton et al. (2004) have shown that such an allocation always exists, even under general (monotonic) combinatorial valuations.

The foregoing axioms play a central role in the design of the not-for-profit fair division website *Spliddit* (Goldman and Procaccia 2014), which currently provides algorithms for the division of rent, goods, credit, tasks, and fare. The website's tagline—and, indeed, its central principle—is *provably fair solutions*, that is, each of the algorithms deployed on Spliddit guarantees axiomatic notions of fairness.

This brings us to the main point of this essay: The reason I view these provable fairness guarantees as crucial is that they allow us to explain to users why a given solution is fair. Indeed, the website contains an accessible description of the relevant axioms on the page devoted to each application. More importantly, when displaying a solution to a specific problem instance, users are also shown a *personalized* explanation of why their solution satisfies the promised fairness guarantees.

For example, for rent division, Spliddit uses a polynomial-time implementation of the *maximin solution* (Alkan et al. 1991; Gal et al. 2016), which is envy free, and might display the following text:

> Why is my assignment envy free? You were assigned the room called 'Master Bedroom' for $290.00. Since you valued the room at $535.00, you gained $245.00. You valued the room called 'Basement' at $64.00. Since this room costs $554.00, you would have lost $490.00. You valued the room called '2nd Floor' at $401.00. Since this room costs $156.00, you would have also gained $245.00.

Similarly, for goods division, Spliddit employs a highly optimized implementation of the Maximum Nash Welfare solution, which guarantees an allocation that is envy free up to one good (Caragiannis et al. 2016), and displays a personalized fairness explanation to that effect.

I firmly believe that these fairness explanations make the solutions more appealing to users, and, therefore, make it more likely that users will accept them. And, while I only have anecdotal evidence to support this belief, the fact that Spliddit has had more than 124,000 users since its launch in November 2014 (as of April 2018) suggests that the approach is quite effective.

Putting my artificial intelligence (AI) hat on for a moment, the notion of *explainable AI* is all the rage now, driven by the popularity of machine learning and the opacity of most machine learning algorithms. I am often asked why machine learning cannot be used to find solutions that people perceive as fair, thereby replacing the axiomatic approach; my answer is twofold. First, there is insufficient data. Second, more pertinently, explainability is crucial, and, even if we had enough data, current machine learning algorithms would not be able to explain why a solution is fair.

Now for the bad news (although the next few paragraphs just set the stage for it). Since the publication of the book by Arrow (1951), social choice theory—which deals with aggregating individual preferences or opinions towards collective decisions— has relied heavily on the axiomatic approach. In the context of voting, specifically, several impossibility results are perhaps the most famous, but axioms are also routinely used to make the case for particular voting rules.

More concretely, the standard model of voting involves a set of voters, whose preferences over a set of alternatives are expressed as *rankings* of the alternatives. A *voting rule* (also known as a *social choice function*) takes as input a *preference profile* (the reported rankings), and returns the winning alternative. Axioms, in this context, are desirable properties of voting rules. Here are a few representative examples:

1. *Condorcet consistency:* We say that alternative $x$ beats $y$ in a pairwise comparison if a majority of voters rank $x$ above $y$. An alternative is a *Condorcet winner* if it beats every other alternative in a pairwise comparison. A voting rule is *Condorcet consistent* if it selects a Condorcet winner whenever one exists in a given preference profile.
2. *Monotonicity:* If $x$ is the winning alternative under a certain preference profile, and another profile is obtained by pushing $x$ upwards in the votes while keeping the order of all other alternatives fixed, then $x$ is also the winner under the new profile.
3. *Consistency:* If alternative $x$ is the winner under two different preference profiles, $x$ also wins under the union of the two profiles.

In terms of explainability, there is a significant difference between these axioms and, say, envy-freeness in rent division. Monotonicity and consistency tie together multiple preference profiles; explaining the outcome on any particular profile is difficult, because this would require voters to reason counterfactually (thinking about what the outcome would be if the profile was different).[1] And Condorcet consistency is very effective in explaining the outcome on a given profile—but only when a Condorcet winner exists. Consequently, while axioms do play a central role in presenting the voting systems that are deployed on several voting websites—including Pnyx, Whale, and CIVS—they are not used to explain outcomes, to the best of my knowledge.[2]

---

[1]Cailloux and Endriss (2016) develop an algorithm that automatically derives a justification for any outcome of the Borda rule, by applying both 'single-profile' and 'multi-profile' axioms to a sequence of hypothetical profiles. Their approach nicely formalizes the idea of explaining an outcome, but, in its current form, may not produce explanations that people would be able to follow.

[2]Two caveats are in order. First, the website Whale does visualize outcomes. For example, for Condorcet-based methods, the website displays the pairwise majority graph. These visualizations are useful insofar as they explain how the voting rule works, but, in my view, they do not explain its outcomes. Second, for the case of *multi-winner* elections, there are some examples of axioms that directly give rise to explainable outcomes. Notably, Aziz et al. (2017) recently developed the notion of *justified representation* for approval-based multi-winner elections, which, roughly speaking,

The shortcomings of the axiomatic approach (or, at least, the perception thereof) have motivated my work, and the work of others, on *optimization-based* approaches to voting. To be clear, implementations of rules derived from the axiomatic approach also frequently rely heavily on optimization, but here I am referring to voting rules that choose an alternative that optimizes an objective function. At the risk of sounding cryptic, let me mention that, in the context of the aggregation of subjective preferences, I am especially interested in optimizing utilitarian social welfare with respect to implicit utility functions (Boutilier 2015; Caragiannis et al. 2017). And when aggregating objective opinions, we have looked at optimizing the worst-case or expected position of the selected alternative in a set of feasible ground truth rankings (Procaccia et al. 2016; Benade et al. 2017).

These optimization-based rules are deployed on *RoboVote*, a not-for-profit voting website, which we launched in November 2016. RoboVote attempts to communicate to users the reasoning behind its calculations but, in my view, its explanations are not nearly as compelling as their counterparts on Spliddit. For example:

> Based on the medium level of disagreement in the reported votes, we estimate that voters ordered at most 2.4 pairs of alternatives incorrectly, on average. Given that, the winning alternative returned by Robovote is guaranteed to compare favorably with all but at most 1 other alternative according to the objectively correct order. No other alternative can give a better guarantee.

In fact, some RoboVote users—especially those who have used the website to make relatively high-stakes group decisions, such as selecting papers for awards at a prestigious conference—have reached out to us and asked for additional explanation. In some cases, we were lucky that the alternative selected by RoboVote happened to be a Condorcet winner. In other cases, where a Condorcet winner did not exist, we could point to an agreement between RoboVote and several simple voting rules. But no general methodology emerged from this exercise. In summary, optimization-based approaches to voting currently do *not* give rise to satisfying explanations.

Through deployment via online platforms, I believe that research in voting has the potential to change the way group decisions are made worldwide—starting from (many) small groups, and, in a utopian world, ultimately shaping global institutions. But I view the problem of explaining outcomes as a significant barrier to (at least partially) realizing this ideal. The case study of fair division suggests that axiomatically explaining solutions is an effective approach; it remains an open question whether an equivalent axiomatic framework can be developed for voting.

**Acknowledgements**  I thank Felix Brandt, Umberto Grandi, Dominik Peters, Marcus Pivato, Nisarg Shah, and Bill Zwicker for insightful feedback. This work was partially supported by NSF grants IIS-1350598, IIS-1714140, CCF-1525932, and CCF-1733556; by ONR grants N00014-16-1-3075 and N00014-17-1-2428; as well as a Sloan Research Fellowship and a Guggenheim Fellowship.

---

requires that if a sufficiently large group of voters approve the same alternative, then the winning subset must contain at least one alternative approved by some member of the group. This at least allows addressing complaints by large groups that are not represented in the outcome, by arguing that the group members themselves cannot even agree on a single alternative.

# References

Alkan, A., Demange, G., & Gale, D. (1991). Fair allocation of indivisible goods and criteria of justice. *Econometrica, 59*(4), 1023–1039.

Arrow, K. (1951). *Social choice and individual values*. Hoboken: Wiley.

Aziz, H., Brill, M., Elkind, E., Freeman, R., & Walsh, T. (2017). Justified representation in approval-based committee voting. *Social Choice and Welfare, 42*(2), 461–485.

Benade, G., Kahng, A., & Procaccia, A. D. (2017). Making right decisions based on wrong opinions. In *Proceedings of the 18th ACM Conference on Economics and Computation (EC)* (pp. 267–284).

Boutilier, C., Caragiannis, I., Haber, S., Lu, T., Procaccia, A. D., & Sheffet, O. (2015). Optimal social choice functions: A utilitarian view. *Artificial Intelligence, 227*, 190–213.

Cailloux, O., & Endriss, U. (2016). Arguing about voting rules. In *Proceedings of the 15th International Conference on Autonomous Agents and Multi-Agent Systems (AAMAS)* (pp. 287–295).

Caragiannis, I., Kurokawa, D., Moulin, H., Procaccia, A. D., Shah, N., & Wang, J. (2016). The unreasonable fairness of maximum Nash welfare. In *Proceedings of the 17th ACM Conference on Economics and Computation (EC)* (pp. 305–322).

Caragiannis, I., Nath, S., Procaccia, A. D., & Shah, N. (2017). Subset selection via implicit utilitarian voting. *Journal of Artificial Intelligence Research, 58*, 123–152.

Gal, Y., Mash, M., Procaccia, A. D., & Zick, Y. (2016). Which is the fairest (rent division) of them all? In *Proceedings of the 17th ACM Conference on Economics and Computation (EC)* (pp. 67–84).

Goldman, J., & Procaccia, A. D. (2014). Spliddit: Unleashing fair division algorithms. *SIGecom Exchanges, 13*(2), 41–46.

Lipton, R. J., Markakis, E., Mossel, E., & Saberi, A. (2004). On approximately fair allocations of indivisible goods. In *Proceedings of the 6th ACM Conference on Economics and Computation (EC)* (pp. 125–131).

Procaccia, A. D., Shah, N., & Zick, Y. (2016). Voting rules as error-correcting codes. *Artificial Intelligence, 231*, 1–16.

Su, F. E. (1999). Rental harmony: Sperner's lemma in fair division. *American Mathematical Monthly, 106*(10), 930–942.

Svensson, L.-G. (1983). Large indivisibles: An analysis with respect to price equilibrium and fairness. *Econometrica, 51*(4), 939–954.

Tennenholtz, M., & Zohar, A. (2016). The axiomatic approach and the Internet. In F. Brandt, V. Conitzer, U. Endress, J. Lang, & A. D. Procaccia (Eds.), *Handbook of computational social choice*, Chap. 18. Cambridge: Cambridge University Press.

# The Role of Characterizations
# in Market Design

James Schummer and Shigehiro Serizawa

**Abstract** The search for desirable allocation methods has long been formalized using an axiomatic approach. We contrast how this approach should be applied within the context of general, abstract analyses versus the analysis of specific market design problems. Specifically, we discuss how the role of axiomatic characterization results might differ across these two contexts. We conclude that characterization results, though informative, should not be viewed as a primary objective in market design research. We illustrate this point using a new result in an allocation problem that, abstractly, represents a hypothetical market design problem.

## 1   Introduction

Much of the literature on *mechanism design*—the study of abstract resource allocation rules—has used an axiomatic approach to describe what is achievable in the design of resource allocation rules. Even though *market design* problems are often based on such abstract mechanism design problems, they tend to include additional institutional constraints or objectives that are specific to a particular industry or application. Nevertheless, since these constraints and objectives often can be expressed axiomatically, it is not surprising that a significant amount of axiomatic analysis also takes place in the market design literature. In light of this, one should consider the degree to which axiomatic market design research should mimic the style of work that has been done in mechanism design. In particular we discuss the degree to which the market design literature should emphasize axiomatic *characterization* results as a primary research objective.

We thank Justin Leroux and William Thomson for comments.

J. Schummer (✉)
Kellogg School of Management, Northwestern University, Evanston, IL, USA
e-mail: schummer@northwestern.edu

S. Serizawa
ISER, Osaka University, Suita, Japan
e-mail: serizawa@iser.osaka.u.ac.jp

© Springer Nature Switzerland AG 2019                                          201
J.-F. Laslier et al. (eds.), *The Future of Economic Design*, Studies in
Economic Design, https://doi.org/10.1007/978-3-030-18050-8_28

Many "planners" (such as societies, firms, or non-profit institutions) at some point find themselves having to decide on a method for allocating resources or costs. A government may need to allocate public resources for private use (mineral resources, airwaves, airport capacity); a firm might need a policy for internal resource allocation; an online platform might need to decide how agents are matched with resources (or to other agents); a non-profit organization might be tasked with allocating transplant organs, refugee families, public housing, etc.

What these real-world design problems have in common at the abstract level is that the planner facing each problem is trying to achieve some *objectives* or goals, subject to various *constraints*. Where they *differ*, however, is in the details of what these objectives and constraints actually are. In terms of objectives, for instance, societies or non-profit organizations may care mostly about welfare and fairness, while firms may prioritize profit maximization. Constraints also vary across applications. A society needs to respect different forms of property rights and outside options than does a firm dealing with customers. Incentive constraints vary depending on the degree (and kind) of private information that exists in each application. Political constraints obviously vary: A society might be constrained to switch to a new resource allocation method that benefits all (or a majority of) members, while a startup company choosing a resource allocation method for the very first time may face almost no such constraints.

One could go so far as to say that these "micro level" variations in objectives and constraints across different applications is a major part of what distinguishes the field of market design from that of abstract mechanism design.[1] Therefore we believe it is natural to ask the following question.

*Given this distinction between abstract mechanism design and practical market design, to what extent should research directions differ across these two fields?*

This question is broad and worthy of discussion from many different angles; in fact other articles in this volume also touch on this question. We intend to focus on the use of axiomatic characterizations in the two literatures, and the extent to which characterizations *per se* should represent a research *objective*.

## 2 The Axiomatic Approach

### 2.1 Framework

Some real world planners who wish to improve an existing allocation rule begin merely by considering modifications to their current allocation methods. This

---

[1]The terms *market design* and *mechanism design* are not well-defined, even within the academic profession. For our purposes, we loosely consider *market design* problems to be more application-driven ones, where research questions are motivated directly from real world problems, ideally documented by some sources outside of the economics profession. We use the broad term *mechanism design* to refer to more abstract analysis of resource allocation problems, where research questions are more qualitative and abstract in nature.

approach is limiting, so to enlarge its options the planner might go further by considering rules being used by other planners, and make comparisons among them (e.g. in the context of school choice, the decision makers in one city might consider the methods used in other cities). Taking this approach one step further, a planner might use intuition and creativity to engineer a new allocation rule, showing that it improves upon the existing one, and switch to it.[2] But by using any of these approaches, even if we iterate, we are still left with the question of "Can we do better?"

The axiomatic approach attempts to answer this question directly by building the concept of "better" from the ground up. The first step is for the researcher to consider what set of objectives (or constraints) a planner might have and to formalize them as axioms; let us call this set $\mathcal{A}$. Second, the researcher searches for some or all of the rules that satisfy $\mathcal{A}$. If the researcher obtains a result, it generally falls into one of three categories:

(i) An impossibility result (no rule satisfies $\mathcal{A}$),
(ii) a characterization result (describing all of the rules satisfying $\mathcal{A}$),[3] or
(iii) a partial result (the researcher can describe some, but not all, rule(s) satisfying $\mathcal{A}$).

This classification brings us to our central question. In terms of research objectives, what is to be done in case (iii)? The researcher has answered the planner's question of "Can we do better?" in the sense of *achieving* $\mathcal{A}$, but has not completely described *all* the ways in which "better" can be accomplished. Therefore, to what extent has the researcher provided the planner with good advice at this point? Is the researcher's answer to the planner's question incomplete? These questions can be extended even to case (ii). When the set of characterized rules is "large"—which should be considered a good thing—what answer does the researcher ultimately provide for the planner? Can the researcher give even better advice?

We believe that the answers to these questions depend on the context of the research question itself, namely whether it is was asked in the context of an abstract mechanism design study or of a concrete market design problem. Therefore we consider these two contexts separately.

## 2.2   Axiomatics and Abstract Mechanism Design

When the search for "better" rules provides many choices it seems reasonable for the planner to ask for "even better" rules. A portion of the literature on abstract mechanism design seems to take this position. If we are lucky enough (in case (ii)) to have described *many* choices among those "better" rules that satisfy $\mathcal{A}$, then we can

---

[2]For example, this is a rough description of the approach used by the US Federal Aviation Administration when they developed a new landing slot reallocation procedure in the 1990's; see the historical discussion in Wambsganss (1997).

[3]One could classify (i) as a subset of (ii); this semantic distinction plays no role in our discussion.

refine the planner's search to "even better" rules by introducing additional, desirable axioms. Even if we *cannot* describe all of the "better" rules (i.e. case (iii)), then perhaps we can succeed at describing all of some set of "even better" rules, again by introducing additional, desirable axioms. In both cases this requires defining the "even better" list of requirements $\mathcal{A}' \supset \mathcal{A}$.

Since (properly chosen) axioms embody desirability, each additional axiom leads to more desirability.[4] By repeating this *find-rule-add-axiom* process the researcher might eventually reach a characterization result, but with a stronger (longer) list of axioms ($\mathcal{A}'$) than was used in the original attempt. By going further than that—by adding an additional axiom—the researcher might find an impossibility result that, with the previous characterization result, establishes a boundary between what is and is not feasible. Thus the researcher has described one boundary between how much better a rule can or cannot be, and describes the rules that achieve this.

In the abstract world of mechanism design, where stylized models are meant to capture generality rather than detail, such an approach is defensible. We wish to gain insight into which kinds of qualitative objectives can be achieved and which are incompatible. If "reasonable" objectives (embodied by $\mathcal{A}$) regularly lead us to certain classes of rules, we gain an intuition that these rules should tend to work well in the real world, even if our abstract models fail to capture every detail about that world.

A prime example of this comes from the characterization results obtained in the house reallocation model of Shapley and Scarf (1974). In their model, each agent is endowed with an indivisible object (house). A rule reallocates these objects based on the agents' preferences over them. As is well known, the Top Trading Cycle rule described by Shapley and Scarf has played a dominant role in the analysis of this model. Most famously it is characterized as selecting the unique core allocation (Roth and Postlewaite 1977). Additionally, however, other work shows that TTC is characterized by other combinations of desirable objectives, e.g., Ma (1994), Dur (2013), and Morrill (2013) among others.

Despite the abstractness of Shapley and Scarf's elegant model, these numerous results provide strong qualitative support for the use of TTC in problems where agents need to trade endowed, indivisible objects. The literature thus advises a planner facing problems with this structure to consider the use of TTC, at least as long as the planner's objectives correspond to the axioms used in any one of its characterizations. Indeed, TTC has been the foundational basis for life-saving allocation rules in the

---

[4]We emphasize the qualifier of *properly chosen*. Additional axioms chosen specifically for the mission of characterizing a class of rules add little insight; see Thomson (2017) for further discussion.

real world. In the words of Abdulkadiroğlu and Sönmez (2013), "The basic ideas of [TTC] resulted in organized kidney exchange in various parts of the world almost thirty years later, saving thousands of lives."

## 2.3   Axiomatics and Practical Market Design

Now let us consider the research process of Sect. 2.1 in the context of market design. A major difference between this context and the previous one is in the first step of the axiomatic approach: the specification of the planner's objectives $\mathcal{A}$. In the mechanism design literature, $\mathcal{A}$ is typically a literature-driven list of design objectives that are truly *axiomatic* in the lay sense of the term: they are self-evidently desirable. The concepts of, say, efficiency, fairness, or incentive compatibility are unquestionably desirable, even if there could be disagreement over the details of how to formalize these concepts in an abstract model, or over their relative importance (when incompatible).

On the other hand, the planner in a *market design* problem often has a list of objectives defined *ex ante*. It is the planner, and not the researcher, who dictates what the objectives are through his or her industry expertise, job requirements, or particular institutional constraints. In many cases, the planner already knows what objectives are desired and what constraints exist, even before the researcher arrives. For example, a planner in a school choice problem has already decided that certain students should have priority over others at a certain school; an auctioneer knows that revenue maximization is the primary objective; the F.A.A. has already decided that incentives are an important consideration in designing landing slot reassignment algorithms.[5] The planner's objectives are rarely defined by a pre-existing, abstract mechanism design literature.

This observation has implications for the question we asked in Sect. 2.1: What is to be done when the researcher, after formalizing the planner's objectives $\mathcal{A}$, finds only a partial result (case (iii))? The mechanism designer may be tempted to do what we described in Sect. 2.2 and ask for an "even better" result by reaching for a stronger list of objectives $\mathcal{A}' \supset \mathcal{A}$. A problem arises, however, when these additional requirements $(\mathcal{A}' \setminus \mathcal{A})$ are not driven by the planner's problem, but instead are literature-driven principals from the abstract mechanism design literature. When that happens, the researcher is no longer addressing that planner's specific application, but is instead mixing together what should be two *separate* research objectives: the market design research objective of solving a specific real world problem versus the mechanism design objective of describing broad generalities meant to offer qualitative insights. When the researcher does this, the takeaway from the resulting analysis is unclear: what problem is being solved? Is the researcher addressing an applied problem or

---

[5] See documentation in Wambsganss (1997) and Schummer and Vohra (2013).

attempting to make broad qualitative statements at an abstract level? To paraphrase a well known saying, "No [researcher] can serve two masters."[6]

Even if the planner's objectives $\mathcal{A}$ do not lead to a characterization result, a partial result nevertheless still answers the planner's question of how to achieve a "better" way of allocating resources, *using the planner's definition of better*. On the other hand, a characterization result artificially obtained by further strengthening the planner's objectives does *not* necessarily yield good advice to the planner, particularly when the additional axioms are literature-driven objectives not directly related to the planner's problem. An implication of this argument is that it becomes the researcher's burden to thoroughly investigate what the planner's objectives are in the first place. This typically requires more than just reading previous mechanism design literature on abstract problems. Instead researchers must talk with (non-academic) planners, or read specialized industry-focused journals outside of their normal areas of expertise.

To give an example of a problem which justifies this separation between application- and literature-driven objectives, we refer to the landing slot reassignment problem studied by Schummer and Vohra (2013) and by Schummer and Abizada (2017). The latter work designs a list of incentive compatibility axioms, showing that they are specifically tailored to the design objectives of the FAA by referencing both academic and non-academic sources. After showing that these objectives are incompatible with a strong efficiency requirement, they revert to a weaker form of efficiency that, again, is documented to be the kind of efficiency requirement that the planner (the FAA) strives to achieve. They show that these (FAA-motivated) efficiency and incentive compatibility axioms are simultaneously achievable by a class of rules based on the well-known Deferred Acceptance rule. In other words the authors have found "better" rules using the FAA's own definition of better, and documented this to be the case.

However, Schummer and Abizada's result is only a partial one (case (iii)). Could one move on to an "even better" result by imposing additional requirements? Indeed, one can look to the abstract mechanism design literature in which characterizations of the Deferred Acceptance rule already appear. For example, Kojima and Manea (2010) characterize Deferred Acceptance using a combination of normative axioms along with an *IR monotonicity* axiom. In their words, they "do not regard IR monotonicity as a normative…requirement, but as a positive comprehensive description of the deferred acceptance algorithm." (p. 635). Their use of this axiom, and the corresponding result, clearly provides qualitative insights into the workings of Deferred Acceptance (i.e. the mechanism design research objective we described earlier). On the other hand we believe it would be artificial to add such an axiom to Schummer and Abizada's list of application-driven axioms merely for the purpose of "getting a characterization." In fact, doing so would neither (i) provide broad, qualitative insights (since their model is specialized, unlike Kojima and Manea's), nor (ii) pro-

---

[6]The biblical verse, Matthew 6:24, from which we get this saying goes on to state "…You cannot serve both God and money." We let the reader decide which of those two masters best represents market design.

vide additional advice to the FAA (since Schummer and Abizada's rules have already been shown to satisfy the FAA's objectives).

## 3   A New Example

We provide a new result that further illustrates the relative desirability of characterization results in either abstract models of mechanism design or practical, specific market design problems. Though we introduce an abstract model, we ask the reader to imagine that it represents some specific, applied market design problem. We then consider which forms of axiomatic *results* might be relevant for the problem's planner. Because the point of this example is to illustrate our ideas discussed above, we minimize formality and notation, and omit proofs which are available upon request.[7]

The model concerns a rescheduling problem, where a set of $n$ agents occupy positions $\{1, 2, \ldots, n\}$ in a schedule, e.g. each position could represent a time of day when an agent receives some service at, say, a repair shop, doctor's office, etc. Each agent $i$ initially occupies position $i$ in the schedule, but has strict, single-peaked preferences over all $n$ positions, with peak denoted $p_i$. A *rescheduling rule* reassigns the agents to the $n$ positions as a function of their preferences.

The model represents situations in which agents are assigned appointments far in advance of when they learn their preferences, e.g. an agent might decide that a doctor's appointment, made months earlier, would be more convenient later in the day. Since agents are endowed with positions in the schedule before learning their preferences, the planner may wish to Pareto-efficiently reschedule the agents while still respecting the property rights they have with respect to their endowments (individual rationality) and accounting for incentives (e.g. strategy-proofness).[8]

Of course this model is a special case of the previously mentioned house reallocation model of Shapley and Scarf (1974). Therefore the TTC rule "works" for this problem: it satisfies strategy-proofness, Pareto-efficiency, and individual rationality. In fact if *all* (non-single-peaked) strict preferences were admissible in this problem, TTC would be the only rule to satisfy these three properties (Ma 1994). On the other hand, once we restrict attention to single-peaked preferences this uniqueness need no longer hold. Indeed that is the case.

For the sake of discussion, let us now imagine that this model had been inspired by some (unspecified) market design problem in which the planner also cares about a particular kind of fairness condition. Specifically we imagine a planner who wishes *to avoid creating envy among the agents*. Note that, in allocating indivisible objects,

---

[7]In fact we were recently made aware of two working papers—Bade (2017) and Liu(2018)—that consider this same model and prove some of the results we reference below. Our specific results concerning two *envy* conditions are novel, and are what we use to illustrate our main point in this chapter.

[8]The planner has precisely these objectives in the application of airport landing slot reassignment problems (see Schummer and Vohra 2013), though preferences have a different structure in that application.

envy is unavoidable: some agent typically will prefer another's position to his or her own.[9] However a planner may want to avoid creating *additional* cases in which some agent $i$ envies some agent $j$ only *after* positions are reassigned.

For example, imagine that $i$ is originally scheduled earlier than $j$ ($i < j$) and that $i$'s ideal position is earlier than his original one ($p_i < i$). At the original schedule, it is unambiguous that $i$ cannot envy $j$'s position; this follows from single-peakedness since $p_i < i < j$. We imagine a planner who would prefer to maintain that lack of envy even after $i$ and $j$ are reassigned positions, i.e. agent $i$ should prefer his new position to $j$'s new position.[10] Roughly speaking, if $i$ wants to move earlier and was already ahead of $j$, $i$ should remain "ahead" of $j$ (at least in terms of preference).

If the planner also desires the symmetric condition (replacing "earlier" with "later"), this results in the following condition.

**Definition 1** A reassignment rule *creates no new envy pairs*[11] if, whenever $p_i < i < j$ or $j < i < p_i$, agent $i$ prefers his reassigned position to that of $j$.

It is easily verified that TTC does *not* satisfy this condition.[12] Hence, despite the various ways in which TTC has been characterized on larger domains of preferences, it does not achieve the goals of our imaginary planner. Nevertheless, there exists a rule that not only *creates no new envy pairs*, but also satisfies the three standard conditions we started with above. We define the **Iterative Swapping rule** via the following algorithm.[13] The algorithm begins by taking the initial (endowment) schedule assigning agents to positions. It then iteratively adjusts the schedule, removing agents as they move into their favorite (remaining) positions.

**Iterative Swaps algorithm**

**Assign Peaks step**: Permanently assign any agent who is currently assigned to his favorite among the remaining positions, and remove that agent-position pair from the problem. Iterate this step until each remaining agent's favorite remaining position is not his current one. If any agents remain, perform the Swaps step.

**Swaps step**: Consider each remaining pair of agents, $\{i, j\}$, who are *adjacent* among remaining agents. That is, suppose they currently occupy positions $\hat{i} < \hat{j}$ respectively, and that no agent occupies any position $\hat{k}$ with $\hat{i} < \hat{k} < \hat{j}$. Swap

---

[9]The exception is when agents' peaks are distinct, so they can be simultaneously assigned their favorite positions.

[10]This concept could be motivated in a variety of ways, but to do so is not very relevant to our main point. Instead we ask the reader to simply imagine that, for whatever reason, a planner has this objective.

[11]The condition is weaker than its name suggests, because the condition only applies for certain orders of $p_i$, $i$, and $j$. See below for the stronger condition.

[12]The core property of TTC might allow $j$ to trade his endowment for any position that $i$ might find desirable, depending on its owner's preferences.

[13]Bade (2017) analyzes this rule in more detail, proving some of the results we mention below. It is also in the class of rules studied by Liu (2018).

the current positions of agents $i$ and $j$ if and only if it makes both agents better off. Repeat this step until no such Swaps can be made, then perform the Assign Peaks step.

It is easily shown that this algorithm eventually ends in some iteration of the Assign Peaks step, with each agent permanently assigned to some position. The Iterative Swapping rule is defined as the rule selecting the outcome of this algorithm, given any profile of single-peaked preferences. This rule turns out to satisfy the objectives discussed above.

**Theorem 1** *The Iterative Swapping rule satisfies strategy-proofness, efficiency, and individual rationality (see Bade 2017). Furthermore, the rule creates no new envy pairs.*

At this point we pause to reflect on what advice we can give to our imaginary planner who has precisely these objectives. The good news is that we have delivered a rule that achieves exactly what this planner has requested. What comes next? Can we ask for more?

A reasonable quest would be to strengthen this planner's notion of "no new envy pairs." The definition above is weak in that it restricts attention to situations in which agent $i$ initially prefers his endowed position to $j$'s based solely on the hypothesis that $p_i < i < j$ (or $j < i < p_i$). One could ask for a more general statement that, as long as $i$ does not envy $j$ at the original schedule, then the same should be true after reassignment *regardless* of the location of $p_i$. It turns out that such a requirement is too strong. We can demonstrate this fact even if we only require it amongst agents with identical preferences.

**Definition 2** A reassignment rule satisfies *preservation of no-envy among equals* if, for any agents $i$ and $j$ who have the same preference relation, if $i$ prefers his endowed position to $j$'s endowed position, then $i$ prefers his reassigned position to $j$'s reassigned position.

Though our rule (like TTC) fails this notion, this turns out to be a consequence of the following fact. (A simple proof is available upon request.)

**Theorem 2** *No rule satisfies strategy-proofness, efficiency, individual rationality, and preservation of no-envy among equals.*

This theorem, with the first one, provide the planner with a boundary. Together, they show the degree to which the planner can avoid the issue we described earlier: the *creation* of envy among the agents. Under the constraint of the first three axioms we considered, the planner can avoid creating envy as long as an agent $i$ prefers a new schedule position in the opposite direction of some agent $j$'s endowed position. The Iterative Swapping rule achieves this. On the other hand, this possibility breaks down when, for instance, $i$ and $j$ both prefer to move *toward* each other's endowed positions, What happens is that the other three axioms may require agent $j$ to receive a new position that agent $i$ also prefers to anything else.

To return to the main point of this article, we now evaluate the results we have. The first theorem provides a partial result, but not a characterization. If we view a result like Theorem 1 in the context of abstract mechanism design, the contribution is limited. While we have provided a rule satisfying the four axioms, it is not yet clear how it can be generalized, or what other forms of rules might also achieve those same objectives.[14]

On the other hand, if we imagine a planner facing an applied problem, and having precisely the objectives we have described, our two theorems combined give that planner significant insight: they describe the degree to which his objectives can be achieved. This is true *despite the absence of a characterization result*. If the Iterative Swapping rule could be characterized with the use of additional axioms *beyond* the requirements of the planner, such a result would provide artificial advice to the planner, even if it provides qualitative insights to the academic profession.

# 4 Summary

We have argued that, in applied market design applications, the *planner's objectives* form the exogenous primitives of the research program. The researcher must examine the relevant application (or industry) in order to determine what those objectives are, and conduct the research that follows these objectives. Once this is done, the researcher should not be tempted to add additional, abstract objectives for the sole purpose of achieving an axiomatic characterization result. This argument is strengthened by the observation that typically, *the planner is the expert* in determining what objectives need to be met, while the researcher is the expert in finding mechanisms to satisfy any such set of objectives.[15]

To illustrate our points, we have contrasted results obtained in market design models with those obtained in very related mechanism design models; Sect. 2.3 considers existing work while Sect. 3 illustrates by using new results in an abstract model that could represent a hypothetical market design problem. Despite the similarity in primitives across related models, results should vary in terms of the objectives being considered, whether they be qualitative and abstract (mechanism design) or practical and direct advice (market design). In particular, we believe that the field of market design would appeal to practitioners even more broadly than it does now if there would be an even stronger shift away from the research objective of characterizing allocation rules and further toward the objective of investigating and documenting planners' objectives across different applications.

---

[14]Indeed, we have worked a bit on this problem, discovering that the description of rules satisfying the four axioms becomes increasingly complex as $n$ grows larger.

[15]In Chapter (25), Billette de Villemeur and Leroux emphasize this point even more strongly by observing that certain axioms traditionally invoked in abstract mechanism design might be viewed as *irrelevant* depending on the applied context in which they are being invoked.

# References

Abdulkadiroğlu, A., & Che Y.-K. (2010). The role of priorities in assigning indivisible objects: A characterization of top trading cycles. Mimeo.

Abdulkadiroğlu, A., & Sönmez, T. (2013). Matching markets: Theory and practice. In D. Acemoglu, M. Arello, & E. Dekel (Eds.), *Advances in Economics and Econometrics* (Vol. 1, pp. 3–47). Cambridge.

Bade, S. (2017). Matching with single-peaked preferences. Mimeo.

Dur, U. (2013). A characterization of the top trading cycles mechanism for the school choice problem. Mimeo.

Kojima, F., & Manea, M. (2010). Axioms for deferred acceptance. *Econometrica, 78*(2), 633–653.

Liu, P. (2018). A large class of strategy-proof exchange rules with single-peaked preferences. Mimeo.

Ma, J. (1994). Strategy-proofness and the strict core in a market with indivisibilities. *International Journal of Game Theory, 23*(1), 75–83.

Morrill, T. (2013). An alternative characterization of top trading cycles. *Economic Theory, 54*(1), 181–197.

Roth, A. E., & Postlewaite, A. (1977). Weak versus strong domination in a market with indivisible goods. *Journal of Mathematical Economics, 4*(2), 131–137.

Schummer, J., & Abizada, A. (2017). Incentives in landing slot problems. *Journal of Economic Theory, 170*, 29–55.

Schummer, J., & Vohra, R. V. (2013). Assignment of arrival slots. *American Economic Journal: Microeconomics, 5*(2), 164–185.

Shapley, L., & Scarf, H. (1974). On cores and Indivisibility. *Journal of Mathematical Economics, 1*, 23–37.

Thomson, W. (2017). *Lets talk about economic design, fairness, and incentives*, book manuscript in progress.

Wambsganss, M. (1997). Collaborative decision making through dynamic information transfer. *Air Traffic Control Quarterly, 4*, 107–123.

# On the Axiomatics of Resource Allocation: Classifying Axioms and Mapping Out Promising Directions

**William Thomson**

**Abstract** The purpose of this note is to propose a two-way classification of the axioms of the theory of economic design, and to map out directions for future research that we perceive as particularly promising.

Axiomatic analysis holds a central place in economic design. It allows us to go beyond current practice and hypothetical examples. It provides explicit arguments for the use of particular allocation rules in terms of meaningful criteria of good behavior and it gives complete answers to design questions: Are those properties compatible? If yes, what rules satisfy them all?

The scope of the contexts and the richness of the requirements that have been explored keep expanding. What would the literatures on matching, queuing, school choice, organ allocation, auctions, to name a few examples, look like without the conceptual and technical apparatus that the axiomatic approach has provided?

This purpose of this note is to suggest how its impact could be increased further. We first argue for the need to organize the field of axioms and for that purpose offer a two-way classification, together with taxonomic suggestions. We then indicate several directions that we see as deserving particular attention in future axiomatic work.

I thank Shigehiro Serizawa for useful comments.

W. Thomson (✉)
Department of Economics, University of Rochester, Rochester, NY 14627, USA
e-mail: wth2@mail.rochester.edu

# 1   Organizing the Field of Axioms

What's an axiom? An **axiom** is the expression in mathematical form of the intuition we have about how a solution mapping could be required to behave in certain situations.[1]

The potential benefits of organizing the field of axioms, any field for that matter, should be obvious. Broad categories of axioms are sometimes noted in the literature and piecemeal and model-specific observations made; also, identifying logical relations between axioms is a routine part of axiomatic work. However, a comprehensive attempt at classification is lacking. Such a treatment should be useful, an inspiring precedent being the manner in which the invention of the periodic table of the elements set the research agenda of physicists for many years. We therefore begin with a nutshell summary of a two-way classification of axioms, in terms of scope and format on the one hand, and in terms of content on the other, developed in (Thomson 2019b).

To illustrate our definitions, we refer to three classes of allocation problems. In a **classical economy** a group of agents equipped with preferences defined over some commodity space have endowments of these goods, and the issue is to redistribute these endowments among them. In a **problem of fair division**, a social endowment of resources has to be distributed among agents having equal rights on it. In a **claims problems**, a group of agents have one-dimensional claims over a single resource and there is not enough of the resource to fully honor all claims.[2]

A "solution mapping", a "solution" for short, associates with each allocation problem in some class one or several allocations, interpreted as recommendations for it. We focus on single-valued solutions, which we call "rules". An agent's "assignment" is what the rule chooses for him. Depending upon the model, it could be a commodity bundle, a welfare level, a contribution expressed in resources, or effort, or money ...to some collective enterprise.

## 1.1   Axiom Scope and Format

The distinctions made below should be self-evident, and our main contribution may simply be to provide language that allows us to talk about them.[3]

   • **Universal versus model-specific axioms**. Few axioms express ideas that are meaningful in all models. Conversely, when an axiom is tailored to a specific model,

---

[1] We are not addressing here the role of axioms in other areas of economics, such as decision theory, nor in other fields, such as mathematics. For a discussion of the axiomatic method in some theory and resource allocation, see Thomson (2001).

[2] For surveys, see Thomson (2003, 2019b).

[3] Yet, not keeping them in mind may be damaging. For example, the failure to distinguish between punctual and relational axioms has been responsible for an erroneous interpretation of the consistency principle and of some results involving consistency axioms (Thomson 2012).

one can usually see in it some general principle at work. Thus, we are certainly not suggesting a partition here. Rather, one should think of axioms arranged in a spectrum from **universal** to **model-specific**.

- **Punctual versus relational axioms**. A **punctual axiom** applies to each problem separately, "point by point"; a **relational axiom** relates the choices made by a rule for problems that are related in a certain way.[4]

- **Pre-application versus post-application axioms**. In describing the relation between the problems that a relational axiom is concerned with, we may have to apply the rule to at least one of them (usually one); the rule appears on the hypothesis side. Such an axiom is a **post-application** axiom; the others are **pre-application** axioms.[5]

- **Individual-centric versus group-centric axioms**. An **individual-centric** axiom specifies how each individual's assignment should depend on a particular parameter of a problem, usually a parameter attached to that individual. A **group-centric** axiom specifies how the assignment of each member of a group, or of a group as a whole, should depend on a particular parameter, in most cases, a collective parameter.

- **Self-regarding versus other-regarding axioms**. When a model includes parameters attached to individual agents, our first inclination is usually to specify how the rule should take into account each particular agent's parameters when calculating *this agent's assignment*; we then have a **self-regarding** axiom. But we may also want to say something about how *the other agents' assignments* should depend on this parameter. This gives u s an **other-regarding** axiom.[6]

- **Monotonicity versus invariance axioms**. When the space to which some parameter of a problem belongs and the space to which the outcome chosen by a rule belong are equipped with order structures, a **monotonicity** axiom requires the order structure of the former to be reflected in the order structure of the latter. An **invariance** axiom requires that certain changes in a parameter—these changes may or may not be evaluated in some order—not be accompanied by any changes in the chosen outcome.

- **Fixed-population versus variable-population axioms**. Even when a rule is defined over a domain that contains problems involving different populations, a **fixed-population** axiom specifies how the rule should behave on a subdomain of problems in which the population of agents is unchanged. A **variable-population** axiom specifies how a rule should respond to changes in populations.

---

[4]*Efficiency* is a punctual axiom. *Resource monotonicity*, which says that if the resources available to society become more abundant, every agent should end up at least as well off as he was made initially, is a relational axiom.

[5]*Resource monotonicity* (footnote 3) is a pre-application axiom. *Consistency*, which says that if some agents leave with their assignments, and the situation is reevaluated at this point, each of the remaining agents should be assigned the same thing as initially, is a post-application axiom.

[6]The *individual endowment lower bound*, which says that each agent should find his assignment at least as desirable as his private endowment, is a self-regarding axiom. *Other-regarding endowment monotonicity*, which says that when an agent's private endowment increases, each of the others should end up at least as well off as he was made initially, is an other-regarding axiom.

## *1.2   Axiom Content*

The categories listed below are not meant to achieve a partition of the field of axioms.

• **Efficiency and related properties**. These punctual properties are at the core of modern economic thinking.[7]

• **Symmetry, anonymity, order preservation**. The first two axioms are commonly invoked as embodying minimal objectives of punctual fairness. When the parameters attached to agents belong to a space that has an economically meaningful order structure, we may require that a rule "reflect" or "respect" that order. The idea can sometimes be applied to groups of agents when the parameters attached to their members can be aggregated, or when parameters are attached to groups themselves.

• **Lower bounds and upper bounds**. Bounds can be imposed on an individual's assignment (footnote 5), or on the assignments of groups, aggregated in some way, or on the group's assignment directly. Most commonly, it is the resources that each agent owns that are used to define a lower bound on his welfare. In fact, imposing such a bound can be interpreted as a way of making operational the abstract concept of ownership. Meaningful bounds can also be based on comparisons of the economy under consideration to counterfactual economies.

Each of the general principles listed next is the "root" of many specific relational axioms.

• **Solidarity**. The principle of solidarity says that if a parameter of an economy changes and no one (or no one in a particular group), is responsible for the change, the welfare of all agents (or of all agents in that group), should be affected in the same direction. A number of monotonicity requirements are obtained as special cases, when the space to which the parameter belongs and the space to which the outcomes belong have order structures. Considerations of solidarity also often underlie other-regarding monotonicity requirements.[8]

• **Robustness with respect to changes in perspectives when evaluating a situation or a change in a situation**. There are frequently more than one way of looking at a situation, all equally valid. If they lead to outcomes that are not welfare-equivalent, adopting one of them and not the other(s) is bound to be seen as unfair by some agents. Thus, the interest of robustness requirements stating that the possible perspectives that one can take all result in the same, or in welfare-equivalent, outcomes.

• **Robustness to strategic behavior**. Strategic behavior on the part of agents may take multiple forms and to each type corresponds a robustness requirement. Misrepresentations of preferences or of some other private knowledge that agents may have, and manipulations of resources under their control (withdrawing, destroying, transferring, exaggerating …), are the underlying motivation for a number of axioms of robustness of rules, namely that individual agents or groups of agents never benefit from the behavior. A next step is the identification of properties of a rule guaranteed the existence of a "game form" so that for each economy in its domain, the set of equilibrium allocations of the resulting game (the game form augmented with the

---

[7]*Respect of unanimity* is another example.

[8]*Other-regarding endowment monotonicity* (footnote 5) is an example.

preference profile) coincides with the set of allocations the rule would select for that economy. This is the object of the vast "implementation" literature.

It would be meaningful to place in one group the first four categories of axioms just listed as they express normative or ethical concerns, and in a second group the fifth category, which has to do with strategic ones. Some axioms would belong to both groups. For example, *endowment monotonicity*[9] is clearly desirable from a normative viewpoint but it also has a strategic interpretation: if violated, an agent could gain by strategically destroying some of the resources he controls. The same comment applies to the *individual endowment lower bound*. It means "respecting" ownership but it can also be understood as providing agents the incentive to willingly participate in exchange.

• **Technical and operational axioms**. This last category stands somewhat apart from the others. Some axioms do not have a straightforward economic interpretation but are imposed for technical reasons, to allow the use of certain mathematical tools for example, or to impose some discipline on some complex class of rules. The term "operational" is also often used to designate an axiom that allows the real-world implementation of some socially meaningful objective that would be impractical otherwise. Such an axiom may concern the "engineering" or computational aspect of design, but is typically of limited normative or strategic interest. It should be noted however that ease of computation is a double-edged property: it facilitates the work of the designer but it also enables strategic agents, thereby hampering achieving the chosen social objectives.

# 2 Directions in Which Deeper Axiomatic Analysis is Needed or Seems Particularly Promising

Using the taxonomy of the previous sections, this section suggests directions that the axiomatic program would benefit from emphasizing in the future.

## 2.1 Being More Systematic in Formulating Axioms

When reading an axiomatic study, we often have the feeling that it would have benefited from their authors being more systematic in their formulation of axioms, that gaps in our knowledge would be filled faster and more thoroughly. For example, other-regarding versions of axioms are rarely explored, and group-centric versions

---

[9]This says that when an agent's private endowment increases, he should end up at least as well as was made initially.

of axioms are usually not paid much attention even though doing so could often yield some interesting insights.[10]

Let us take claims problems as an example. Because it could draw on a large body of existing axiomatic work, this literature quickly reached maturity, yet it is surprising that so many rather straightforward questions having to do with the dependence of rules on the claims vector have remained unexplored. Supposing that someone's claim increases, the first requirement that comes to mind is certainly that he not be awarded less than he was awarded initially and this self-regarding requirement has been discussed in a number of studies. However, there are multiple other ways, most of them either other-regarding or group-centric, in which a rule could be required to depend on claims: the increase in the award to an agent whose claim increases should be bounded by the amount by which his claim increases; the impact on each of the others' awards should be bounded by this amount; no one else's award should increase; the impact on that agent's award should be a monotonic function of the amount to divide; so should the impact on each of the other agents' awards; and similar bounds and monotonicity properties could be imposed when the focus is on groups of claimants (Thomson 2019a).[11]

## 2.2   Interpreting and Reinterpreting Axioms

Some axioms are sometimes unfairly perceived as being technical; conversely, economic interpretations are sometimes forced on axioms whose technical role should instead be openly acknowledged.

Technical axioms are clearly not what the focus of our investigation should be. However, upon examination, a technical axiom can often be seen to have compelling normative or strategic interpretations, sometimes both. Continuity requirements illustrate the point; they are usually thought of as technical, yet there is always some arbitrariness in the exact data of a problem and preventing small variations in these data from having a radical impact on outcomes seems normatively justified. Also, errors in measurement are unavoidable; on the other hand, errors can be corrected. It would not be right for welfare distributions to be overly sensitive to arguably irrelevant details of the situation at hand. Also, a discontinuous rule is more likely to be manipulable, so *continuity* is a step towards robustness to manipulation. (In fact, it is often a logical implication of requirements of immunity to manipulation.)

Another example is the ubiquitous requirement of *additivity* (coalitional form games, cost allocation, airport problems, claims problems). *Additivity* too is often described as a technical axiom, yet it belongs in the category of robustness axioms

---

[10]Why did it take so long for a group version of *non-bossiness* (the requirement that, if an agent's assignment does not change as a response to a change in his preferences, no one else's assignment change either), to be formulated?

[11]These axioms are related in multiple ways. This should be expected. The richer one's formulation of axioms for a model, the more likely they will be logically related.

described in Sect. 1.2: when two equally valid perspectives can be taken in evaluating a situation—taking cost allocation as an example, whether or not two projects should be looked at separately, or consolidated as one big project—it seems natural to require that these two perspectives result in the same outcome.[12]

*Consistency* is sometimes called an operational axiom, but what is it supposed to make operational? Like all invariance properties, it too belongs to the category of axioms expressing the robustness of a rule to choices of perspectives in evaluating a situation, in this case a variation in population. When some agents leave the scene with their assignments, should we stick with the initial choice for the remaining agents or should we look at the situation anew, from their perspective, and reassess their opportunities? Again, these are two equally deserving viewpoints. The axiom says that it should not make any difference which is adopted. (It also has a fairness interpretation discussed below.)

This does not mean that there is no place in the theory of economic design for technical axioms. We do want all of our axioms to have a clear economic interpretation, but when we have trouble understanding the implications of some axiom system, invoking additional technical axioms may be a useful, hopefully temporary, step towards the answers we seek. However it is counterproductive to artificially claim normative or strategic interpretations for these axioms. Consider *non-bossiness* (footnote 9). This axiom does have multiple conceptual and logical connections to various normative and strategic axioms but these connections are really quite tenuous. In spite of its name, helping prevent what we call "bossiness" in common language is not a very convincing interpretation of the axiom. Its main rationale really is that it structures a class of rules in a way that makes it manageable (Thomson 2016).

## 2.3  Parameterizing Axioms

An axiom is a yes-or-no proposition but we will argue here that there is much to gain from introducing measures of the extent to which axioms are satisfied or violated, and working with these measures.

We disqualify a rule as soon as we find one violation of an axiom that we have imposed. Of course, there are almost always other situations where violations occur and it is because our judgment is that the violations are "frequent enough" that we disqualify the rule. We rarely seek to identify all of those situations however, and most often that would be too challenging a task anyway. Usually, the most that is practical is identifying domains that are free of them, or sufficient conditions on domains for that to be the case. We propose here that parameterizing axioms would

---

[12]Consider a society simultaneously facing two allocation problems; these problems are of the same type and they can be added, for example two cost allocation problems implicating the same agent set, such as a university contracting with the same list of suppliers for two construction projects. *Additivity* says that they can be handled separately, each agent receiving two bills, one for each of them, or consolidated into one and handled as one problem.

provide sensitive tools to evaluate rules and that more effort should be expanded in this direction.

In the study of *strategy-proofness* for discrete models a range of interesting measures have been proposed many of them based on counting profiles for which things go wrong, and some work has actually been done using such measures to compare rules. For example, one can calculate the proportions of profiles of preferences at which some agent can gain by misrepresenting his preferences. One may take into account the gain that he can achieve by optimally manipulating, or the average gain that a manipulating agent can achieve by manipulation. Orders can be defined on the space of rules based on inclusion of profiles for which manipulation is possible for someone. Finally, one can compare rules in terms of the equilibrium sets of the manipulation games associated with them. This involves characterizing these sets and relating them to the set of outcomes that would be selected under truthful behavior.

When allocation space is a subset of a Euclidean space, fine quantitative measures of the extent to which an axiom is violated can usually be defined although here too, one should not expect that there would be a unique way to parameterize violations. Consider the *no-envy* axiom.[13] Calculating the minimal radial expansion to which the bundle assigned to an agent who is envious (say agent 1) of some other agent (say agent 2) should be subjected for this instance of envy to disappear is an option. Another is to use the minimal amount of some reference good that should be added to agent 1's assignment for his envy of agent 2 to disappear. Alternatively we could subject agent 2's bundle to minimal radial contractions or minimal subtractions from his consumption of a reference good for agent 1 to stop envying him (Chaudhuri 1986; Diamantaras and Thomson 1990). For a relational axiom such as *resource monotonicity*, we can measure a violation by the minimal expansion to which the bundle assigned to an agent who is hurt by the augmentation of the social endowment should be subjected for him to return to his initial welfare level (Moulin and Thomson 1988).

Once axioms are parameterized, numerical tradeoffs between them can be studied. For example, for fair division problems, a relationship has been identified between a parameter measuring the extent to which *resource monotonicity* and a parameter measuring the extent to which the *no-domination* requirement[14] can be jointly satisfied (Thomson 1987a).

For classical economies, a seminal result is that for two agents, no rule satisfies *efficiency*, the *individual-endowment lower bound*, and *strategy-proofness* (Hurwicz 1972). A parameterized version of each of these axioms has been proposed, the parameter ranging from 0 (the axiom is vacuous then) to 1 (when the standard formulation is obtained). We now know that no matter how much each of these axioms is weakened and unless it is given up altogether, an incompatibility with the other two still holds (Schummer 2004; Cho 2014). Thus, parameterizations

---

[13]This says that no agent should prefer someone else's assignment to his own.

[14]This says that the assignment of no agent should dominate commodity by commodity the assignment of anybody else.

of axioms have allowed to reach a much deeper understanding of the nature of the difficulty.

Conversely, finding ways of quantifying how well an axiom is satisfied may be worthwhile. Using the *no-envy* axiom to illustrate the point, the operations on bundles suggested above can also be used to quantify how far some agent is from envying someone else. Selections from the no-envy solution have been proposed based on measures of this type (Tadenuma and Thomson 1995). Also, having a measure of the extent to which a bound is met allows a search for rules that best satisfy the bound.

## 2.4 Beyond One-Dimensional Parameterizations

A "price of anarchy" measure of the type formulated by computer scientists is based on worst-case scenarios. Such a measure summarizes the behavior of a rule by means of a single number. We argue here that this may be too blunt of a tool and that more sensitive evaluations can sometimes be obtained by partitioning the space of problems, measuring on each component separately how well-behaved the rule is, and working with the resulting lists of answers.

To illustrate, consider the notion of a *guarantee structure* that was proposed to evaluate how bargaining solutions respond to population increases as a function of two parameters, the size of the initial population and the size of the additional population (Chun and Thomson 1989; Lensberg and Thomson 1983; Thomson 1983, Thomson 1987a). When new agents arrive without the opportunities open to society expanding, the agents initially present will typically be affected negatively as a group, but how badly they will be affected individually should be a concern. If a rule is *population monotonic*, each of the incumbents will have to make some sacrifice. How much of a sacrifice? How disruptive will this arrival be? When incumbents are allowed to express an opinion as to whether potential newcomers should be allowed in, they will want to know what it will cost them.

So for a generic agent, let us identify the minimal ratio of the final to initial numerical welfares that a rule assigns to him when such events occur. Seen positively, the resulting number can be understood as a *guarantee* to the agent that when population expands, his final welfare will reach at least a certain proportion of his initial welfare. A *guarantee structure* gives the list of guarantees to a generic agent involved in a generic problem as a function of the numbers of incumbents and newcomers. Higher guarantees are of course more desirable. These notions have recently been successfully adapted to claims problems (Harless 2017a, b). A disadvantage they have is that they only provide a partial ranking of rules. Yet, in both applications, bargaining problems and claims problems, broad rankings of rules have been obtained based on guarantee structures, and maximal elements within large classes have even been identified, and characterized.

# References

Chaudhuri, A. (1986). Some implications of an intensity measure of envy. *Social Choice and Welfare*, *3*, 255–270.

Chun, Y., & Thomson, W. (1989). Bargaining solutions and relative guarantees. *Mathematical Social Sciences*, *17*, 295–295.

Cho, W. (2014). Impossibility results for parametrized notions of efficiency and strategy-proofness in exchange economies. *Games and Economic Behavior*, *86*, 26–39.

Diamantaras, D., & Thomson, W. (1990). An extension and refinement of the no-envy concept. *Economics Letters*, *33*, 217–222.

Harless, P. (2017a). Wary of the worst: maximizing award guarantees when new claimants arrive. *Games and Economic Behavior*, *105*, 316–328.

Harless, P. (2017b). Guarantee structures for claims problems; properties and comparisons among rules. Mimeo.

Hurwicz, L. (1972). On informationally decentralized systems. In C.B. McGuire & R. Radner, (Eds.), *Decision and Organisation* (pp. 297-336). Minneapolis: University of Minnesota Press.

Lensberg, T., & Thomson, W. (1983). Guarantee structures for problems of fair division. *Mathematical Social Sciences*, *3*, 205–218.

Moulin, H., & Thomson, W. (1988). Can everyone benefit from growth? Two difficulties. *Journal of Mathematical Economics*, *17*, 339–345.

Schummer, J. (2004). Almost-dominant strategy implementation: Exchange economies. *Games and Economic Behavior*, *30*, 154–170.

Tadenuma, K., & Thomson, W. (1995). Refinements of the no-envy solution in economies with indivisible goods. *Theory and Decision*, *39*, 189–206.

Thomson, W. (1983). Collective guarantee structures. *Economics Letters*, *11*, 63–68.

Thomson, W. (1987a). Individual and collective opportunities. *International Journal of Game Theory*, *16*, 245–252.

Thomson, W. (1987b). Can everyone benefit from growth? Another impossibility. University of Rochester, Mimeo.

Thomson, W. (2001). On the axiomatic method and its recent applications to game theory and resource allocation. *Social Choice and Welfare*, *18*, 327–387.

Thomson, W. (2003). Axiomatic and game-theoretic analysis of bankruptcy and taxation problems: A survey. *Mathematical Social Sciences*, *45*, 249–297.

Thomson, W. (2012). On the axiomatics of resource allocation: Interpreting the consistency principle. *Economics and Philosophy*, *28*, 385–421.

Thomson, W. (2016). Non-bossiness. *Social Choice and Welfare*, *47*, 665–696.

Thomson, W. (forthcoming 2019a). *How to divide when there isn't enough: From Aristotle, the Talmud, and Maimonides to the axiomatics of resource allocation. Monograph of the econometric society*. Cambridge: Cambridge University Press.

Thomson, W. (2019b). *Let's Talk about Economic Design*.

# Part V
# Behavioral Aspects

# Modeling Behavior: A Critical Agenda

**Laurence Kranich**

**Abstract** The traditional approach to modeling behavioral anomalies consists of modifying the specification of agents' characteristics in order to exhibit or generate each such anomaly. Within this context, I raise two issues: (1) why agents' characteristics differ and how they are determined, and (2) the design of policies and institutions when traits are at least partially influenced by the environment. However, the main argument in the paper is that the traditional, piecemeal approach may be problematic due to interactive effects between traits or behaviors. This suggests that greater effort should be devoted to determining which traits are interrelated and to studying them jointly rather than separately. Finally, I briefly mention two alternative modeling strategies which may be more amenable to analyzing behavior in new situations.

## 1 The Status Quo

The purpose of constructing models to explain and explore human behavior is three-fold: first, to try to understand the relationship between agents' traits or characteristics and their actions[1]; second, to explore the range of behaviors consistent with such traits[2]; and, third, to extrapolate and study the implications for behavior in other settings or circumstances.[3] However, observation of the behaviors of actual individuals often conflict with the predictions of our models. Historically, in order to resolve such discrepancies and further our understanding of the phenomena, the first step has

---

[1] This requires that the model contains all essential ingredients. Otherwise, if missing factors would affect the relationship between traits and outcomes, then the mechanism is underdetermined.

[2] This serves to establish bounds for expected (and possibly not yet observed) behavior of those with such attributes. In addition, it might serve as a test of the model: if actual, observed behavior is inconsistent with the theory, then the theory is wrong.

[3] This includes prediction and forecasting.

---

L. Kranich (✉)
University at Albany, SUNY, Albany, NY, USA
e-mail: lkranich@albany.edu

© Springer Nature Switzerland AG 2019
J.-F. Laslier et al. (eds.), *The Future of Economic Design*, Studies in Economic Design, https://doi.org/10.1007/978-3-030-18050-8_30

been to modify the description of the agent in such a way as to generate the observed behavior. The presumption is that with a better, more realistic representation of the individual, we can explain a broader range of actual behavior, and thus that the model will yield better predictions in other circumstances. For example, if under standard neoclassical assumptions, agents would not engage in benevolent acts but we observe that they do,[4] then we might modify the model to allow for agents to derive some satisfaction (utility) from doing so (e.g., a so-called "warm glow") or that they are directly affected by other agents. Similar modifications might be required to exhibit spite or malice, trust, guilt, loss aversion, inequality aversion, empathy, a preference for freedom, for flexibility, etc.

Generally, our approach to treating behavioral anomalies (vis-à-vis the neoclassical standard) has been piecemeal, modifying the model to allow trust or to include enmity or to incorporate loss aversion, etc.[5] In such cases, the path from assumption to action is often direct: people engage in a behavior because they derive satisfaction from the act or the consequence. (That is not to say that indirectness is desirable in and of itself.) But while such modifications "explain" or generate a particular behavior, they are often not well-suited to explain a range of behaviors. For example, the same individual might behave benevolently at some times and malevolently at others. This suggests that the context or timing of a decision might be relevant, or that agents are at least reactive to or interactive with their environment. Thus, an individual, facing the same set of choices might choose differently depending on their mood, the menu of choices, the agents with whom they interact, the institutional framework, etc.[6]

Each such "wrinkle" presents a different and interesting agenda item. Can we accurately model the effect of mood on behavior? Does the menu of choice matter and how? Do agents reciprocate for real or perceived treatment by others? Do agents benefit from having the freedom to choose among various alternatives versus only a single alternative? Or, more generally, are they affected by the environment and/or the procedure used to allocate resources? If such factors affect the decision-maker, they might affect the decision, and we would like to understand how. Moreover, many of these issues pertain to external influences, thus providing scope to shape the social environment as well as the outcome. (More on this below.)

Within this context, one interesting question is why agents' traits differ and how they are determined. Are they inherited (nature)? Are they influenced by others or by the institutional setting (nurture)? Are they a combination of the two? If they are inherited, then there is little scope for influence.[7] Rather, the interesting issue is how they came to be and where they are heading. This suggests perhaps an evolutionary

---

[4]This assumes such acts do not serve another purely selfish motive (such as increasing tax deductions), which may be consistent with the standard model.

[5]A few examples include Brennan (1973) on malice, Kahneman and Tversky (1979) on loss aversion, Kreps (1979) on flexibility, Sen (1991) on freedom, Fehr and Schmidt (1999) on inequality aversion, Battigalli and Dufwenberg (2007) on guilt, and Anderlini and Terlizzese (2017) on trust, among many others.

[6]Or in such an interdependent environment, they might wish to proactively signal or affect the behavior of others.

[7]Other than by affecting their prevalence through selective mating.

model à la Maynard Smith (1982) in which the prevalence of traits (within society) is determined by their material success, or "fitness." But to explain current patterns of behavior using an evolutionary model presumes that the status quo is an end or absorbing state of an evolutionary process rather than simply a point on a path or trajectory. Otherwise, the analytical objective should be to characterize all paths containing the status quo, a cumbersome and likely fruitless exercise.

A more realistic view is that traits are at least partially influenced by environmental factors. If, indeed, they are the product of both nature and nurture, then a cultural transmission model à la Bisin and Verdier (2001), based on the work of Cavalli-Sforza and Feldman (1981) and Boyd and Richerson (1985), might provide an appropriate analytical framework. There, agents inherit the traits of their predecessors but are subject to further influence as a result of their interactions or exposure.[8] Similar to the evolutionary framework, one might then ask what patterns or distributions of traits/behavior are likely to emerge given that the current population determines the prevalence of successors' types as well as the likelihood of random influences?

A second question concerns active influences. If agents are affected in any way by their environment, then it might be possible to control the environment so as to affect the emergent traits. In that case, it may be possible to encourage or limit influences or to design institutions in order to affect not only the outcome of interaction but the composition and characteristics of society itself. Are there settings which foster cooperation or which are conducive to prosocial behavior versus those which encourage negative behavior? For example, people might be more inclined to contribute to charity if they are exposed to images of malnourished children or if they can personalize tragedy by associating it with a particular individual. Both of these can be and are deliberately used in order to elicit such an effect. In fact, it is highly likely that advertisers, marketers, fundraisers, lobbyists, campaigners—those whose job is to influence others—are keenly aware of the effectiveness of various methods far more than are economists.

## 2 A Criticism

The ultimate objective of the above research agenda is to model agents as realistically as possible, that is, to construct a general model which would match an agent's exhibited behavior in any and all settings and would thus be able to accurately predict their behavior in other contexts. This ever-more-general approach is demonstrated by the previous example: from selfishness, to benevolence, to reciprocity, to … However, this raises some questions.

First, this example may be idiosyncratic in that the behaviors are nested in such a way that the simpler behavior should be a special case of the more general behavior.

---

[8]Related to this issue of outside influence are the models with endogenous preferences in which agent's preferences are not given but may change in response to their interactions. (See, for example, Cervellati et al. (2010)).

Thus, selfishness should be an extreme case of the model with benevolence when the agent is insensitive to others. And benevolence should be a special case of reciprocity in which the negative trigger is not actuated. It is thus necessary that the successive models increase in generality in order to capture the more nuanced and varied behavior.

But what of other phenomena, such as trust and loss aversion? Is it necessary to construct a single model which explains both? If the phenomena are truly independent, that is, if one has no bearing on the other, then the historical approach of constructing one model to explain trust and another to explain loss aversion might be reasonable. But clearly many psychological traits or behavioral tendencies are related, such as trust and benevolence. One need only consider political party platforms to see issues and concerns linked by a world view. It is far more likely that a Democrat will agree across issues with another Democrat than with a Republican. This brings into question much of what has been done so far and what might be accomplished by the traditional, piecemeal approach. Is it logically sound to construct a model to explain a phenomenon if dependent variables are omitted? Would any results or conclusions drawn from such a model be valid? Or is the entire mode of analysis suspect, similar to partial equilibrium or the theory of the second best? This would suggest that our current practice is flawed. Instead, it calls for a cumulative approach in which all related behaviors are modeled as emerging from a single "unified theory," i.e., a general model which is capable of exhibiting/generating all such behavior.

The biggest obstacle to such an inclusive agenda is that we do not know what we do not know. Thus, even if we were to develop an aggregative model, any results obtained therein would only be reliable until the next behavioral anomaly was discovered, at which time the model would have to be expanded. (Note that this approach is reactive.) Conversely, knowing that another anomaly will eventually be identified undermines confidence in the current version. Moreover, for some issues, such as "framing effects," there might be many variations and no natural way to circumscribe and include them all.

This leads to a second, related, question. A single, "unified field theory" may be necessary, but it is clearly unattainable. Moreover, a model that is capable of explaining all such behavior must be as complex as the "real thing" and thus serves no useful purpose. Therefore, rather than consider what should be included in the model, one might ask, what could be left out? When is a model "good enough"? From this perspective, an alternative agenda is to circumscribe the (restricted) set of behaviors one would wish to model, and to develop the smallest model that is capable of exhibiting such behavior. For example, there would be no reason to model whimsical or erratic behavior since, by definition, such behavior is unpredictable. More appropriately, from a statistical point of view, agents are more likely to exhibit or encounter the most common behaviors. Thus, one might begin with those. Conversely, an "orphan drug" approach to modeling would suggest that if certain behaviors are extremely rare, then it is not worth attempting to include them. Nevertheless, in an interactive environment, the likelihood of occurrence of such behaviors is nonzero, and it may be necessary to specify some treatment, or rule-of-thumb, for how agents would behave in such unknown or unfamiliar circumstances.

Realistically, we are not going to build a fully aggregative model, nor are we going to formulate a large model which incorporates all behavior except a designated subset of independent factors. That being the case, we must recognize that even this more inclusive methodology and any results obtained therein might be flawed. One question we can and should ask is the extent to which it matters. Might such omitted variables have a significant impact on the magnitude of the effects? On the direction? A practical test might be to consider two or more obviously related behaviors and compare the implications of modeling them separately versus together. If the impact is small, this augurs well for the traditional approach. Beyond serving as a test of the overall importance of interactive effects, such comparisons might serve to identify related traits; that is, if the potential impact is significant, then such traits should be modeled jointly.

Finally, I briefly mention two alternative modeling strategies.

I have argued that a model which allows agents to care about others, to care about fairness, to care about freedom of choice, etc., is better than separate models to treat each. As mentioned earlier, this approach is reactive in that it requires that we prespecify the set of behaviors we wish to exhibit/explain. If we then wish to include another—say, we discover a new behavioral anomaly—it would be necessary to respecify the model. As such, the approach is not well-suited to explain how agents behave in new and unfamiliar situations, where they have little or no experience, that is, where we do not know what behavior to expect.

In contrast, an alternative, proactive, approach might be to assume that agents have more fundamental affective traits[9] or receptors which might be activated in various settings. Then a particular area of concern (such as for fairness) would consist of a set of triggers. This might address the question of why agents differ in their effective concerns.[10] Such a model would enable agents to engage in previously unforeseen circumstances.[11,12]

Another alternative might be to focus on the rules or procedures agents use to resolve decision problems rather than the problems or outcomes themselves. Such tools would then be applicable to a variety of (possibly unforeseen) problems. This, too, would provide a flexible model that can be adapted to any setting.

In summary, I have raised some issues that arise in modeling human behavior. Moreover, I have suggested some alternative modeling strategies which do not require that we prespecify the set of behaviors under consideration and thus may apply in more environments, including unfamiliar ones. However, the main argument I have made is that the traditional, piecemeal approach to modeling behavior may be problematic due to interactive effects, and that more effort should be devoted

---

[9]Analogous to Lancasterian characteristics.

[10]More precisely, it would shift the question of why agents have different concerns to why they have different receptors.

[11]Similar to the goal of Artificial Intelligence.

[12]Such a model might provide a "rule-of-thumb" for behavior, as mentioned earlier.

to studying the relationship *between* traits. At the very least, this would help us determine the extent of the problem with current practice.

# References

Anderlini, L., & Terlizzese, D. (2017). Equilibrium trust. *Games and Economic Behavior, 102,* 624–644.

Battigalli, P., & Dufwenberg, M. (2007). Guilt in games. *American Economic Review, 97,* 170–176.

Bisin, A., & Verdier, T. (2001). The economics of cultural transmission and the dynamics of preferences. *Journal of Economic Theory, 97,* 298–319.

Boyd, R., & Richerson, P. (1985). *Culture and the evolutionary process.* Chicago: University of Chicago Press.

Brennan, G. (1973). Pareto desirable redistribution: The case of malice and envy. *Journal of Public Economics, 2,* 173–183.

Cavalli-Sforza, L. L., & Feldman, M. (1981). *Cultural transmission and evolution: A quantitative approach.* Princeton: Princeton University Press.

Cervellati, M., Esteban, J., & Kranich, L. (2010). Work values, endogenous sentiments and redistribution. *Journal of Public Economics, 94,* 612–627.

Fehr, E., & Schmidt, K. (1999). A theory of fairness, competition and co-operation. *Quarterly Journal of Economics, 114,* 817–868.

Kahneman, D., & Tversky, A. (1979). Prospect theory: An analysis of decision under risk. *Econometrica 47,* 263–291.

Kreps, D. (1979). A representation theorem for 'Preference for flexibility'. *Econometrica, 47,* 565–577.

Maynard Smith, J. (1982). *Evolution and the theory of games.* Cambridge: Cambridge University Press.

Sen, A. (1991). Welfare, preference and freedom. *Journal of Econometrics, 50,* 15–29.

# Biological Design

Jean-François Laslier

**Abstract** The theory of games has been developed in good part by economists, but it serves as an important conceptual framework for the study of the evolution of living species. Along the same line of thought, mechanism design might inform and be informed by Biology.

By "biological design" I do not mean the question of designing new biological entities, nor some ideas about how biotech medicine might evolve in the future, I mean that ideas on the evolution of systems, currently mainly developed in theoretical biology, might be of some help for designing robust mechanisms in social sciences. The theory of games, as a mathematical theory, has been developed in good part by economists, but it also serves as an important conceptual framework for population genetics and the study of the evolution of living species. Along the same line of thought, mechanism design might also inform and be informed by Biology.

The basic, and most original, concept of Game Theory is not the rationality of action, but the interdependence of the outcomes of actions. This is why many game-theoretical concepts can be useful for non-rational agents such as animal or vegetal individuals, and species. Still, as the word "game" indicates, these concepts have been often developed by researchers having in mind rational agents. That is the case of early mathematicians like Emile Borel or John Von Neumann, and, of course, for the many game theorists working in departments of Economics and considering themselves as economists. But evolutionary game theory does not consider rational agents (at least in the common sense of the term) and is an important field of application of game theory, a field that has both taken advantage of general concepts and developed its own ones.

The first and main example is the notion of equilibrium. There, the pragmatic scientific requirements, that the natural science prefer not to give up, has led scholars not be satisfied with the modern notion of "Nash" equilibrium, because this notion embodies no requirement of approachability nor stability. Biologists had to develop their own refinements based on biological ideas; hence the study of the so-called

J.-F. Laslier (✉)
CNRS - Paris School of Economics, 48 Boulevard Jourdan, 75014 Paris, France
e-mail: jean-francois.laslier@ens.fr

© Springer Nature Switzerland AG 2019
J.-F. Laslier et al. (eds.), *The Future of Economic Design*, Studies in
Economic Design, https://doi.org/10.1007/978-3-030-18050-8_31

"evolutionary stability" (Smith 1982) and the interest in dynamics that might, or not, lead to rest points (Weibull 1995). The dynamics problem directly addresses the most important question in the theory of evolution by natural selection: where can an on-going process of survival of the fittest go, if who are the fittest has no absolute definition but depends on who is around? Clearly, this is an economic question as well, both for the theory of the long-term evolution of the economic system and for the foundation of micro-economics (Lesourne et al. 2006).

Further transfers between game theory and evolutionary biology get to more specific issues. An important one is the justification of altruism. Unlike the idea of biological evolution (which was not invented until Darwin's time), the observation of altruistic behavior is as old as can be. But it apparently clashes with the "individual selection" paradigm, in Biology as well as in Economics. So the question of altruism has become a puzzle, or at least an intriguing question, for social as well as for biological sciences (Nowak and Sigmund 2005), and it is only recently that a general notion of altruism rooted in evolutionary ideas was developed (Alger and Weibull 2013).

Another much debated point is the nature of language in an evolutionary perspective. A language can be seen as wonderful example of a mechanism in the sense of economic design: it is a set of rules that gives meaning to messages in order to reach a collective goal (mutual understanding) in a variety of situations. In the original discussion of the Stag Hunt game, Rousseau (1755) notices that coordination problems may be solved in the animal world with limited communication means:

> Il est aisé de comprendre qu'un pareil commerce n'exigeait pas un langage beaucoup plus raffiné que celui des corneilles ou des singes, qui s'attroupent à peu près de même.[1]

To what extent is the human verbal language an elaborate version of the communication mechanisms used by animals and plants, how did it start and how did it evolve? Game theory has a say on these points (Kim and Sobel 1992; Laslier 2003; Demichelis and Weibull 2008; Benz et al. 2011).

The two above-mentioned examples of cross-fertilization between game theory and the natural sciences work through conceptual, hence simple, models. This is in contrast with a remarkable feature of biological artifacts that is their complexity. The actual biological organisms, as well as the actual behavior of social animals are such that the "evolutionary stories," that explain what we observe by history, are notably complex, often tortuous and surprising, as exemplified by the famous Panda's thumb of Gould (1980).

Living entities use mechanisms that involve communication, and that are able to handle different situations. Getting closer to familiar themes in social choice, notice that animals take collective decisions, and that they do so in ways that are often akin to simple majoritarian voting, or to range voting; see for instance Walker et al. (2017) on how Lycaons vote by sneezing. The rules for collective decision may be quite involved when they include collecting information and even debating before

---

[1] "We may easily conceive that such an intercourse scarce required a more refined language than that of crows and monkeys, which flock together almost in the same manner."

deciding. Seeley (2010) describes such complex processes of information processing, and collective "democratic," decisions by honeybees, and Sumpter (2010) provides many examples of that kind throughout the animal world.

All these real-life complex mechanisms have appeared through time by evolution and selection, and some are learned and sustained by imitation. This is how they have been "designed". The fact that they endure shows their robustness, and the social mechanisms that we design should be also robust. In the future, the science of social mechanisms might benefit from looking in the direction of Biology, and the mechanisms proposed for Economic Design should be checked against these three questions: What kind of rationality and cognitive ability do they require? Can they be learned by imitation? Are the mechanisms we propose and study evolutionary reachable and stable?

# References

Alger, I., & Weibull, J. W. (2013). Homo Moralis—Preference evolution under incomplete information and assortative matching. *Econometrica, 81,* 2269–2302.

Benz, A., Ebert, C., Jäger, G., & van Rooil, R. (Eds.). (2011). *Language, games, and evolution: Trends in current research on language and game theory*. Berlin, Heidelberg: Springer.

Demichelis, S., & Weibull, J. W. (2008). Language, meaning, and games: A model of communication, coordination, and evolution. *American Economic Review, 98*(4), 1292–1311.

Gould, S. J. (1980). *The Panda's thumb*. New York: Norton.

Kim, Y.-G., & Sobel, J. (1992). An evolutionary approach to pre-play communication. *Econometrica, 63,* 1181–1194.

Laslier, J.-F. (2003). The evolutionary analysis of signal games. In P. Bourgine & J.-P. Nadal (Eds.), *Cognitive economics* (pp. 281–291). Heidelberg: Springer.

Lesourne, J., Orléan, A., & Walliser, B. (Eds.). (2006). *Evolutionary microeconomics*. Berlin, Heidelberg: Springer.

Nowak, M. A., & Sigmund, K. (2005). Evolution of indirect reciprocity. *Nature, 437,* 1293–1295.

Rousseau, J.-J. (1755). Discours sur l'origine et les fondements de l'inégalité parmi les hommes. In *Ecrits politiques*. Paris: Le livre de Poche, 1992.

Seeley, T. D. (2010). *Honeybee democracy*. Princeton, NJ: Princeton University Press.

Smith, J. M. (1982). *Evolution and the theory of games*. Cambridge, MA: Cambridge University Press.

Sumpter, D. J. T. (2010). *Collective animal behavior*. Princeton, NJ: Princeton University Press.

Walker, R. H., King, A. J., McNutt, J.W., & Jordan, N. R. (2017). Sneeze to leave: African wild dogs (*Lycaon pictus*) use variable quorum thresholds facilitated by sneezes in collective decisions. *Proceedings of the Royal Society of London B: Biological Sciences, 284.* https://doi.org/10.1098/rspb.2017.0347.

Weibull, J. (1995). *Evolutionary game theory*. Cambridge, MA: MIT Press.

# Behavioral Mechanism Design

**Debasis Mishra**

**Abstract** Behavioral economics has expanded the set of models that researchers in economic design can look at. At the same time, it presents new complications in design of mechanisms. I briefly outline some of the potential challenges in establishing foundational results in the behavioral mechanism design framework.

## 1 Choice Theory and Mechanism Design

One of the standard assumptions in mechanism design is that each agent has a single preference over the outcome space. The agent uses her preference to make decisions. When evaluating lotteries, the agent uses the expected utility criteria. For instance, the strategic voting literature assumes that each agent has a complete and transitive preference over the set of candidates. In the auction theory literature, each agent is assumed to have a quasilinear preference ordering over the outcomes—(alternative, transfer) pairs. Such assumptions are perhaps not surprising. When foundations for mechanism design theory were laid, there were already strong foundations in decision theory about such individual decision making.

Theory of mechanism design has come a long way and many of its areas seem to show signs of maturity. On the other hand, decision theory has seen a resurgence with a growing focus on providing foundation for alternate decision making behavior. In particular, there is now overwhelming evidence in psychology that in a variety of situations, choices of agents are *inconsistent* with decision making using a single preference ordering over the outcome space and expected utility. It is therefore natural that more research in economic and mechanism design is done to incorporate such decision making agents.

I am grateful to the editors, specially Jean-Francois Laslier, for inviting me to contribute to this volume and for their comments.

D. Mishra (✉)
Indian Statistical Institute, New Delhi, India
e-mail: dmishra@isid.ac.in

© Springer Nature Switzerland AG 2019
J.-F. Laslier et al. (eds.), *The Future of Economic Design*, Studies in Economic Design, https://doi.org/10.1007/978-3-030-18050-8_32

Broadly, we see two strands of literature in decision theory that aim to provide foundations for decision making of a "behavioral" agent. We can classify them as (a) **informational** foundations; (b) **preference** foundations. The informational foundation is mainly concerned with settings where outcomes consist of various lotteries and questions the expected utility theory. A prominent criteria has emerged, which is now widely used to address various departures from expected utility theory—the *max-min approach*, where the decision maker makes decisions by optimizing over multiple set of prior beliefs (Gilboa and Schmeidler 1989). Here, the multiple priors generate multiple preferences over lotteries and the decision maker evaluates them in a particular way. This approach alters the single preference approach by exploiting the uncertainty in the environment. It has already found many applications in mechanism design (Bose et al. 2006; Wolitzky 2016; Tillio et al. 2016).

There is another strand of literature that questions the single preference approach quite directly. Barring a few exceptions, this direct approach has not had many applications in mechanism design—Koszegi (2014) gives a comprehensive survey. I argue here that this approach needs more attention in future. The basic premise of this approach is that agents evaluate outcomes based on what is made available to them—simply, the *menu*. The classical approach to deal with menu dependent preferences is the *choice correspondence* approach. In particular, given an outcome space $X$ and if $\mathcal{X}$ is the set of all non-empty subsets of $X$, then each agent has a function $c : \mathcal{X} \to \mathcal{X}$ such that $c(S) \subseteq S$ for all $S \in \mathcal{X}$. In mechanism design terminology, the *type* of an agent becomes a choice correspondence—in a standard model, it was a preference ordering.

## 2  Some Specific Examples

Choice correspondences are too general to capture many kinds of decision making behavior. Even though we see various kinds of behavior of agents, there is some amount of consistency in their decision making, which naturally puts restrictions on the *type space* (and, hence, on choice correspondences). The future lies in identifying interesting type spaces (or space of choice correspondences) over which economic design will have to be studied. For many such type spaces, decision theory has already laid (axiomatic) foundations, which in turn has implications for representing them in a succinct manner. We give some examples for illustration. There is a common theme in all the examples below, where the decision maker uses some criteria to shortlist outcomes and then uses another criteria to choose an alternative.

1. **Shortlisting**. Consider an agent who has a pair of asymmetric relations ($P_1$, $P_2$) over the outcome space. It uses $P_1$ to shortlist from a set and then uses $P_2$ to make a final selection from the shortlist. Manzini and Mariotti (2007) axiomatically characterize shortlisting procedure. A model of mechanism design which consists of agents who behave as shortlisting agents, should elicit a pair of asymmetric relations from each agent. Consider a voting model where a voter first shortlists

candidates based on their *moral values* and then selects the candidate from the shortlisted candidates based on ideology of the party they belong to. We can apply shortlisting to a standard screening problem too. Mishra and Paramahamsa (2018) consider a seller who is selling an object to a (manager, delegate) pair. Their modeling of decision making by the (manager, delegate) pair is an analogue of the shortlisting procedure. They solve for the optimal contract of the seller and show that it involves an extra layer of randomization (pooling) than in the standard model. This illustrates that new insights can be gained into economic design by considering incentive constraints in such behavioral models.

2. **Multiple rationals**. An agent may have multiple rationals as in Cherepanov et al. (2013). Then, it chooses a maximal alternative using an asymmetric relation $P$ from a subset which it has picked by choosing maximal elements according to multiple asymmetric relations $(R_1, \ldots, R_k)$. This seems to make sense in some of the mechanism design problems. For instance, think of an agent as a collection of subagents—may be multiple experts or may be a bidding ring in an auction with multiple bidders. The asymmetric relations $(R_1, \ldots, R_k)$ reflect the preferences of these subagents. The agent aggregates these preferences using preference relation $P$. So, the type of the agent consists of $(P, R_1, \ldots, R_k)$.

3. **Limited attention**. Under the limited attention procedure, the decision maker chooses a subset of alternatives, called the *attention set*, from every menu of alternatives. It then chooses the best out of it using an asymmetric preference relation. The attention set is supposed to satisfy a minimality property in the sense that removing *ignored* alternatives does not change the attention set—see a formal definition in Masatlioglu et al. (2012). In a mechanism design context, consider a one-sided matching problem (a house allocation problem), where each house has two attributes (quality of construction, distance from city center). We could think of an agent evaluating a menu of houses by using limited attention—it has a threshold quality in mind and from that it picks those houses that meet the threshold (if none, then it picks the best quality house). Then, it chooses the best house in terms of distance from city center. Here, the type of an agent is captured by the limited attention sets and the asymmetric preference relation.

Inconsistent behavior of decision makers, specially across time periods, may be consequences of *multiple selves* in an agent (Fudenberg and Levine 2006). A popular way to rationalize such behavior is through self-control models (Gul and Pesendorfer 2001), where the decision making is done by optimizing over two selves in a particular manner. How does a principal design contracts for such agents?[1] The other approach is to treat the agent as less sophisticated and unable to optimize over selves—this line of work is followed in Eliaz and Spiegler (2006).

---

[1]Relevant work in this direction can be found in DellaVigna and Malmendier (2004); Esteban et al. (2007).

# 3  Questions

In the frameworks discussed above, the core economic issues and questions in mechanism design continue to be the same. First question is whether a form of revelation principle continues to hold. Saran (2011) provides some preliminary answers. In general, focusing attention to direct mechanisms in such settings is too restrictive. This is a simple consequence of the fact that what is "optimal" for the agent depends on the options offered. Indirect mechanisms may offer a variety of menus using messages that may not be captured using a direct mechanism. For instance, in an object-allocation problem if a top-trading cycle (TTC) mechanism is implemented as an indirect mechanism, the objects offered at every stage is changing for the agents. An agent which does not use a single preference order to evaluate outcomes will behave very differently in direct and indirect implementation of the TTC mechanism—see for instance Bade (2016). Is there a set of *canonical mechanisms* which is simpler to study and without loss of generality in these models?

Even the notion of incentive compatibility can be thought of in many ways (de Clippel 2014). One of the popular approaches is worth mentioning. Consider a model where the type of an agent is a choice *function* $c : \mathcal{X} \to X$, where $X$ is the set of alternatives and $\mathcal{X}$ are all the non-empty subsets of $X$. A *direct revelation mechanism* chooses an alternative for every profile of choice functions: $f(c_1, \ldots, c_n) \in X$, where $n$ is the number of agents. Now, one notion of incentive compatibility is the following: fix the profile $c_{-i}$ of agents other than agent $i$ and define $R(c_{-i}) = \{x : f(c'_i, c_{-i}) = x \text{ for some } c'_i\}$. Incentive compatibility requires for every $c_i$, we must have $c_i(R(c_{-i})) = f(c_i, c_{-i})$. de Clippel (2014) uses such notions of incentive compatibility to study various problems on full implementation—a counterpart of Maskin (1999) study using behavioral types. There are many other ways to think of incentive constraints. For instance, we can say that the agent can only make binary comparisons. So, at every profile $(c_i, c_{-i})$, agent $i$'s incentive constraint can be: $c_i(f(c_i, c_{-i}), f(c'_i, c_{-i})) = f(c_i, c_{-i})$ for all $c'_i$. Can we provide theoretical foundations for using one notion of incentive compatibility over the other?

Once the fundamental issues are resolved, we can consider specific objectives in mechanism design setting. For instance, what does efficiency mean with such behavioral agents? In the standard setting, if $n$ agents are given an outcome vector $(x_1, \ldots, x_n)$, then there should not exists another outcome vector $(y_1, \ldots, y_n)$ such that each agent $i$ prefers $y_i$ to $x_i$. However, if an agent is not making decisions with a single preference relation, this definition of efficiency is inadequate. Sugden (2004); Bernheim and Rangel (2009) propose some resolutions to this question. For instance, Bernheim and Rangel (2009) construct a new binary relation from choices of agents, where an outcome $a$ is preferred to $b$ if whenever $a$ is available in the choice set, $b$ is *not* chosen. Hence, we can say that $(x_1, \ldots, x_n)$ is preferred by $n$ agents if there is no other outcome vector $(y_1, \ldots, y_n)$ such that each $i$ prefers $y_i$ to $x_i$ using the binary relation defined by Bernheim and Rangel (2009).

How do we reconcile such definitions of efficiency with incentive compatibility? In models with quasilinearity, Groves class of mechanisms play an important role to reconcile efficiency and incentive compatibility. What is the analogue of the Groves mechanisms in settings with behavioral agents? In general, how does the structure and solution of various mechanism design problems change with behavioral agents? Does the nature of revenue maximizing single object auction change? Is efficient bilateral trading still impossible?

In matching problems, a standard desiderata is stability. In college-student admission problems, a matching is stable if no (college, student) pair can *block* the matching. But if colleges and students are not using standard preference relations, then this definition of stability is clearly inadequate. How should stability be defined in such problems? Does a stable matching still exist in such problems, and if yes, how does one find them?

Behavioral economics presents a new direction where the theory of mechanism design can be extended. By carefully thinking about issues and models in behavioral economics, we can rejuvenate economic and mechanism design. The research will also bring the field of economic design closer to reality.

# References

Bade, S. (2016). Pareto-optimal matching allocation mechanisms for boundedly rational agents. *Social Choice and Welfare, 47*, 501–510.

Bernheim, B. D., & Rangel, A. (2009). Beyond revealed preference: Choice-theoretic foundations for behavioral welfare economics. *The Quarterly Journal of Economics, 124*, 51–104.

Bose, S., Ozdenoren, E., & Pape, A. (2006). Optimal auctions with ambiguity. *Theoretical Economics, 1*, 411–438.

Cherepanov, V., Feddersen, T., & Sandroni, A. (2013). Rationalization. *Theoretical Economics, 8*, 775–800.

de Clippel, G. (2014). Behavioral implementation. *The American Economic Review, 104*, 2975–3002.

DellaVigna, S., & Malmendier, U. (2004). Contract design and self-control: Theory and evidence. *Quarterly Journal of Economics, 119*, 353–402.

Eliaz, K., & Spiegler, R. (2006). Contracting with diversely naive agents. *The Review of Economic Studies, 73*, 689–714.

Esteban, S., Miyagawa, E., & Shum, M. (2007). Nonlinear pricing with self-control preferences. *Journal of Economic theory, 135*, 306–338.

Fudenberg, D., & Levine, D. K. (2006). A dual-self model of impulse control. *The American Economic Review, 96*, 1449–1476.

Gilboa, I., & Schmeidler, D. (1989). Maxmin expected utility with non-unique prior. *Journal of mathematical economics, 18*, 141–153.

Gul, F., & Pesendorfer, W. (2001). Temptation and self-control. *Econometrica, 69*, 1403–1435.

Koszegi, B. (2014). Behavioral contract theory. *Journal of Economic Literature, 52*, 1075–1118.

Manzini, P., & Mariotti, M. (2007). Sequentially rationalizable choice. *The American Economic Review, 97*, 1824–1839.

Masatlioglu, Y., Nakajima, D., & Ozbay, E. Y. (2012). Revealed attention. *The American Economic Review, 102*, 2183.

Maskin, E. (1999). Nash equilibrium and welfare optimality. *The Review of Economic Studies, 66*, 23–38.

Mishra, D., & Paramahamsa, K. (2018). Selling to a naive agent with two rationales. Working paper, Indian Statistical Institute.

Saran, R. (2011). Menu-dependent preferences and revelation principle. *Journal of Economic Theory*, *146*, 1712–1720.

Sugden, R. (2004). The opportunity criterion: Consumer sovereignty without the assumption of coherent preferences. *The American Economic Review*, *94*, 1014–1033.

Tillio, A. D., Kos, N., & Messner, M. (2016). The design of ambiguous mechanisms. *The Review of Economic Studies*, *84*, 237–276.

Wolitzky, A. (2016). Mechanism design with maxmin agents: Theory and an application to bilateral trade. *Theoretical Economics*, *11*, 971–1004.

# Mechanism Design Theory and Economic Institutions: Some Reflections

Dilip Mookherjee

**Abstract** We highlight the shortcomings of mechanism design theory to understand key institutional design issues such as the allocation of authority, organizational structure or safeguards against abuse of power. This requires incorporation of imperfect commitment, collusion, and costs of communication and information processing.

Mechanism design theory has been remarkably successful over the past half century in many areas of economic theory, such as auctions, voting and market design, to name just a few. Yet how relevant has it been to understanding institutions such as firms, organizations or governments? They have advanced our ability to model two out of the three key components of institutions identified by Williamson (1985): asymmetric information and opportunism. But not much progress has been made on the third component: bounded rationality.

To illustrate, any theory of institutions will have to address the question of allocation of authority: who is delegated what authority, who reports to whom, what are the safeguards against 'abuse of power'. In the absence of bounded rationality, broadly interpreted, a version of the Revelation Principle applies (as exemplified in the work of Myerson 1982): the performance of any specific allocation of authority can be replicated by a comprehensive contract (administered by a court or algorithm) in which all members report their information to the administrator, await instructions on what to do, and have incentives to be honest and obedient. The central administrator would process all the information received, compute the solution to an optimization problem, and then issue instructions to all members. Underlying assumptions include ability of the Principal (organization owners) to commit to a mechanism, absence of costs of communicating or processing information, and absence of collusion among agents. In such a world, the notion of 'allocation of authority' makes no sense. Analogous to the Modigliani–Miller Theorem in corporate finance, this asserts the 'irrelevance of institutions' in classical environments of the sort usually studied in mechanism design theory *sans* bounded rationality.

D. Mookherjee (✉)
Department of Economics, Boston University, 270 Bay State Road, Boston, MA 02215, USA
e-mail: dilipm@bu.edu

© Springer Nature Switzerland AG 2019
J.-F. Laslier et al. (eds.), *The Future of Economic Design*, Studies in Economic Design, https://doi.org/10.1007/978-3-030-18050-8_33

To escape this quandary, the theory has to venture forth into difficult territory, where the assumptions underlying the Revelation Principle do not apply. The most significant of these pertain to bounded rationality, which Herbert Simon emphasized as key to understanding organizational structures. What exactly does bounded rationality mean? Two important dimensions are (i) limited capacity to store and process information, and (ii) costs of communication. The former implies limits on cognitive capacity of central administrators, which could rationalize distribution of information processing tasks to members. The latter implies the difficulty of members communicating all they know to the central administrator, which provides an alternative rationale for distributing decision-making authority to members so as to take advantage of their specialized non-communicable information. Similar arguments were made by critics of socialism in the celebrated debates of the 1930s such as Friedrich von Hayek.

However, Hayek did not draw attention to the opportunistic ('abuse of power') hazards inherent in systems with decentralization of authority. This would seem to be a key tradeoff underlying the design of allocation of authority. Studying this trade-off necessitates a theory of costs of communicating, storing and processing information. Progress was made in this respect by the literature on the 'message space size' led by Hurwicz (1960, 1977), Mount and Reiter (1974), and the literature on delay and information processing led by Bolton and Dewatripont (1994), Radner (1992, 1993) and van Zandt (1996). However, these theories abstracted from the modeling of incentives. Hence they could not be used to study the tradeoff between communication and information processing advantages of decentralization with resulting incentive problems. A few recent studies have attempted to study this tradeoff in specialized settings, but there is a long way to go. Examples are models of mechanism design that incorporate communication costs (Blumrosen et al. 2007; Fadel and Segal 2009; Kos 2012, 2013; Green and Laffont 1986; Melumad et al. 1992, 1997; Mookherjee and Tsumagari 2014), or the control of collusion (Baliga and Sjostrom 1998; Celik 2009; Dessein 2002; Faure-Grimaud et al. 2003). In my view, this is an important direction mechanism design theory needs to confront in order to make progress in studying institutions.

Another aspect of bounded rationality pertains to the inability of humans to foresee all possible future contingencies, and plan for them at some initial point of time. In such settings, a preference for 'flexibility' naturally emerges, whereby the Principal leaves open the possibility of adapting later decisions to the arrival of information as the future unfolds, rather than committing to all decisions upfront. Once the door is left open for such adaptation, the Principal would have the scope for possibly opportunistic ex post renegotiations that generate adverse ex ante incentives for agents. Many authors have argued this can give rise to a preference for delegation of authority by the Principal, or imposition of constraints on the Principal's own authority. Yet it is hard to formalize the source of the problem (i.e., unforeseen contingencies) that give rise to the commitment problem. It is equally difficult to develop a theory of how the Principal or agents can evaluate the outcomes of different allocations of authority in contingencies they cannot foresee.

There has been a profusion of recent literature in the field of organizational economics, relying mainly on theories of 'incomplete' contracts that rule out Myersonian revelation mechanisms by assumption (Aghion and Tirole 1997; Dessein 2002). The complexity of modeling bounded rationality motivate such an approach, raising questions regarding its foundations (as in the 1999 *Review of Economic Studies* symposium). Incomplete contract models rely on ad hoc restrictions on mechanisms, thereby raising concerns regarding the robustness of results to these assumptions. Very few authors succeed in modeling the underlying 'frictions' from first principles, as is the norm in mechanism design theory. This has resulted in an unfortunate divergence between the latter and contemporary models in organizational economics. The challenge is to develop models that are both derived from first principles, and tractable enough to generate interesting insights.

# References

Aghion, P., & Tirole, J. (1997). Formal and real authority in organizations. *Journal of Political Economy, 105*(1), 1–29.

Baliga, S., & Sjostrom, T. (1998). Decentralization and collusion. *Journal of Economic Theory, 83,* 196–232.

Bolton, P., & Dewatripoint, M. (1994). The firm as a communication network. *Quarterly Journal of Economics, 109*(4), 809–839.

Blumrosen, L., Nisan, N., & Segal, I. (2007). Auctions with severely bounded communication. *Journal of Artificial Intelligence Research, 28,* 233–266.

Celik, G. (2009). Mechanism design with collusive supervision. *Journal of Economic Theory, 144*(1), 69–95.

Dessein, W. (2002). Authority and communication in organizations. *Review of Economic Studies, 69,* 811–839.

Faure-Grimaud, A., Laffont, J.-J., & Martimort, D. (2003). Collusion, delegation and supervision with soft information. *Review of Economic Studies, 70,* 253–280.

Fadel, R., & Segal, I. (2009). The communication cost of selfishness. *Journal of Economic Theory, 144*(5), 1895–1920.

Green, J., & Laffont, J. (1986). Incentive theory with data compression. In W. Heller, R. Starr & D. Starrett (Eds.), *Essays in honor of Kenneth Arrow* (Vol. 3). Cambridge: Cambridge University Press.

Hurwicz, L. (1960). Optimality and informational efficiency in resource allocation processes. In K. J. Arrow, S. Karlin, & P. Suppes (Eds.), *Mathematical methods in the social sciences* (pp. 27–46). Stanford: Stanford University Press.

Hurwicz, L. (1977). On the dimensional requirements of informationally decentralized pareto-satisfactory processes. In K. J. Arrow & L. Hurwicz (Eds.), *Studies in resource allocation processes* (pp. 413–424). Cambridge: Cambridge University Press.

Kos, N. (2012). Communication and efficiency in auctions. *Games and Economic Behavior, 75,* 233–249.

Kos, N. (2013). Asking questions. *Games and Economic Behavior, 87,* 642–650.

Melumad, N., Mookherjee, D., & Reichelstein, S. (1992). A theory of responsibility centers. *Journal of Accounting and Economics, 15*(4), 445–484.

Melumad, N., Mookherjee, D., & Reichelstein, S. (1997). Contract complexity, incentives and the value of delegation. *Journal of Economic and Management Strategy, 6*(2), 257–289.

Mookherjee, Dilip, & Tsumagari, M. (2014). Mechanism design with communication constraints. *Journal of Political Economy, 122*(5), 1094–1129.

Mount, K., & Reiter, S. (1974). The informational size of message spaces. *Journal of Economic Theory, 8*(2), 161–191.

Myerson, R. (1982). Optimal coordination mechanisms in generalized principal agent problems. *Journal of Mathematical Economics, 10,* 67–81.

Radner, R. (1992). Hierarchy: The economics of managing. *Journal of Economic Literature, 30,* 1382–1415.

Radner, R. (1993). The organization of decentralized information processing. *Econometrica, 61*(5), 1109–1146.

van Zandt, T. (1996). Decentralized Information Processing in the Theory of Organizations. In M. Sertel (Ed.), *Contemporary economic development reviewed, volume 4: The enterprise and its environment.* London: Macmillan Press.

Williamson, O. (1985). *The economic institutions of capitalism.* New York: Free Press.

# How Can We Model Emotional and Behavioral Dynamics in Collective Decision Making?

Jörg Rothe

**Abstract** In common models of collective decision making, such as voting and fair division, most of the previous work has been concerned with static, complete-information settings only. As an interesting task for the future, it is proposed to model collective decision making more dynamically and by taking into account how the agents' preferences and behavior evolve over time and based on their emotions.

## 1 Introduction

I do not have answers. But I will pose a number of questions that I consider relevant for the future of economic design, in particular for collective decision making, which refers to processes where a group of individuals—each with their own incentives, preferences, intentions, and goals that often oppose each other—seek to come to a consensual decision. Focusing on two core themes of collective decision making (namely, voting and fair division of indivisible goods), I will make the case that more *dynamical models* are to be developed in these areas, and that *emotional* and *behavioral* aspects should be taken into account more deeply in these new models. While previous work on behavioral social choice by, e.g., Regenwetter et al. (2006) and Popova et al. (2013) takes a statistical perspective seeking to interpret empirical data, my focus is more on the theoretical models.

Most of the previous work in collective decision making has focused on static, complete-information settings only, which ignores the fact that many such processes in the real world proceed sequentially, evolving dynamically over time, with the involved agents having only incomplete information about the incentives, preferences, and goals of the other agents. After informally describing the basic framework of the above-mentioned core themes of collective decision making, previous attempts at more dynamical models will be described for voting in Sect. 2 and for

J. Rothe (✉)
Institut für Informatik, Heinrich-Heine-Universität Düsseldorf, 40225 Düsseldorf, Germany
e-mail: rothe@hhu.de

© Springer Nature Switzerland AG 2019
J.-F. Laslier et al. (eds.), *The Future of Economic Design*, Studies in
Economic Design, https://doi.org/10.1007/978-3-030-18050-8_34

245

fair division of indivisible goods in Sect. 3. Both sections will conclude by raising questions as to how emotional and behavioral aspects can perhaps be integrated into these models.

## 2   Voting and Computational Social Choice

An election is given by a set of candidates (or alternatives) and a list of voters whose preferences over these candidates typically are linear orders (i.e., rankings). A voting rule is then used to determine the winner(s) from each given preference profile. Plenty of voting rules have been studied in social choice theory (see, e.g., the recent book chapters by Zwicker 2016 and Baumeister and Rothe 2015), such as positional scoring rules (including plurality, veto, and Borda), Condorcet, Copeland, approval voting (which, unlike the other systems, assumes that the voters approve or disapprove of the candidates instead of ranking them), etc.

While there are tons of papers in the (computational) social choice literature on voting in static, complete-information settings, there are only a few attempts to model incomplete-information scenarios that dynamically evolve over time. Regarding the former (incomplete-information scenarios), most notable is the seminal work of Konczak and Lang (2005) on possible and necessary winners, which inspired follow-up work not only in voting (Xia and Conitzer 2011; Betzler and Dorn 2010; Chevaleyre et al. 2012; Baumeister et al. 2012; Baumeister and Rothe 2012) but also in fair division (Aziz et al. 2015b; Baumeister et al. 2017), judgment aggregation (Baumeister et al. 2015), and hedonic games (Lang et al. 2015). Their model is concerned with incomplete-information settings but does not model dynamical behavior over time. Informally speaking, the possible winner problem considers real-world scenarios where the voters do not rank all the candidates (technically, their preferences are partial instead of linear orders)—perhaps because there are just too many candidates or because they care only about their three top candidates and their three most despised candidates—and the question is whether the voters' partial orders can be extended to linear orders such that a given candidate wins. In the related necessary winner problem, the question is whether the distinguished candidate wins in all extensions of the voters' partial orders to linear ones. The possible winner problem generalizes the (unweighted) coalitional manipulation problem, while the necessary winner problem tells us exactly when we can terminate preference elicitation.

As to dynamical models, Xia and Conitzer (2010) and Desmedt and Elkind (2010) (see also the earlier work of Sloth 1993 that focused on two candidates only, the "roll-call voting game") take a game-theoretic approach and study the "Stackelberg voting game," an election in which the voters cast their votes sequentially with preferences being common knowledge, and subgame-perfect Nash equilibria are computed by backward induction. Note that this work gives a very interesting cross-link between two different fields: voting and (noncooperative) game theory. Tennenholtz (2004) also studies sequential voting, which he calls "transitive voting," focusing on an axiomatic approach. Parkes and Procaccia (2013) study sequential decision making with dynamically varying preferences via Markov decision processes.

Recently, Hemaspaandra et al. have developed models of online manipulation (Hemaspaandra et al. 2014) and online control (Hemaspaandra et al. 2017a, b) in sequential elections. In their early seminal papers, Bartholdi et al. (1989, 1992) introduced the corresponding "offline" (i.e., static, complete-information) models: Roughly, in the (offline) manipulation setting, manipulators vote strategically by not reporting their true preferences but instead casting insincere votes with the goal to make their favorite candidate win; and in the (offline) control setting, an election chair seeks to make her favorite candidate win by exerting structural changes to a given election, such as adding or deleting or partitioning either voters or candidates. In a related model of (offline) bribery due to Faliszewski et al. (2009a, b), an external agent seeks to bribe certain voters without exceeding a given budget so as to obtain her desired election outcome. An extremely fruitful body of research grew out of these initial papers, considering many further models and scenarios, such as the work of Conitzer et al. (2007) on weighted coalitional manipulation and destructive manipulation (where the goal is to prevent some candidate's victory) and our work on destructive control (Hemaspaandra et al. 2007). (For an overview, see the book chapters on manipulation in voting by Conitzer and Walsh 2016 and on control and bribery in voting by Faliszewski and Rothe 2016; see also Baumeister and Rothe 2015.)

Let us have a closer look at one of the models above: online manipulation in sequential elections (Hemaspaandra et al. 2014). Unlike in the offline manipulation model of Bartholdi et al. (1989) where all sincere votes are cast simultaneously and the manipulators have complete knowledge of all nonmanipulative voters' preferences, in our online manipulation model voters cast their votes one after the other and manipulators can see the past votes but not the future ones, so the current manipulator must decide—*right in this moment when it is her turn to vote*—what vote to cast. The manipulators have an "ideal" ranking of candidates and their goal is, in the constructive variant, to ensure victory for some given candidate $d$ or a better one in their ranking (or, in the destructive variant, to ensure that neither $d$ nor any worse candidate wins), no matter which votes the future voters will cast. This "maximin" approach is inspired by the field of online algorithms. However, even in this model that is so flexible as to allow PSPACE-completeness of the online manipulation problem for suitably constructed voting systems (Hemaspaandra et al. 2014), some things are static: e.g., the manipulators' ideal ranking of candidates does not change during the process; the order in which votes come in is fixed; it is clear from the beginning who is a manipulator and who a sincere voter.

What if a manipulator changes her mind during the process and votes sincerely instead? What if a sincere voter turns manipulative? What if there is disagreement about the manipulators' ideal ranking? What if there is uncertainty about the order in which votes are cast? Relatedly, what is the best time in a sequential election to disclose one's preferences? How can we model all these uncertainties?

The times, they are a-changin'. In these times where election campaigns worldwide become inflamed with passion and filled with hatred, emotionally heated up, boosted by the internet and social media, by fake news, by targeted campaign ads, it is time for our models to take that into account more deeply. *How can we model strong emotions in the voters' preferences?* Also, even though some may be skeptical as to whether "random last-minute voting" indeed exists, some others claim that it can be observed that many voters are still undecided right before election day. How they decide then might depend on silly things, such as the weather: If the sun is shining on election day, they may cheerfully vote for the party currently in power; if it is raining, they may ill-temperedly vote them out of office; if it is stormy, they may safely stay at home and abstain from voting. Now, the weather is known to be pretty unpredictable due to its nonlinear, chaotic behavior, and so are the voters. How can we model the dynamics of the voters' emotion-driven, chaotic behavior? How can manipulators exploit that? How can we protect elections against them?

# 3   Fair Division of Indivisible Goods

In fair division of indivisible goods (see, e.g., the book chapters by Bouveret et al. 2016 and Lang and Rothe 2015 and the references therein), we are faced with the task of assigning goods to $n$ agents; such an allocation is a partition of the set of goods into $n$ bundles (i.e., subsets) of goods, one for each agent. Goods are assumed to be indivisible, nonshareable, and of single-unit type. To have some connection with the previous section, let us focus—out of the many established models and allocation procedures—on one model that is due to Brams et al. (2003, 2005) and has also been used and extended by Aziz et al. (2015a), Baumeister et al. (2017), and Nguyen et al. (2018). In this model, agents have *ordinal* preferences over bundles of goods (rather than cardinal utilities); however, agents are assumed to rank *single goods* only so as to avoid a heavy elicitation burden and to allow specifying our problems compactly. (This assumption exacts a price from the agents: They will not be able to express preferential dependencies between goods.) Using scoring vectors (such as those used in voting, e.g., Borda—this gives another very interesting cross-link between two different fields: voting and fair division), every ordinal preference induces an additive utility function. In order to maximize the overall utility, a collective utility function is then used to aggregate the individual utilities; typically, one considers utilitarianism (i.e., the sum) and two versions of egalitarianism (i.e., min and leximin). This describes a *(scoring) allocation correspondence*, mapping any such profile of preferences to a nonempty subset of "winning" allocations. Using appropriate tie-breaking rules, a single winning allocation can be obtained.

Again, agents can have incentives to manipulate. Nguyen et al. (2018) provide a simple condition necessary and sufficient for scoring allocation correspondences to be strategy-proof. Baumeister et al. (2017) introduce and study, among others, a property called *duplication monotonicity*, which says that a cheating agent might pretend to be two agents (with identical preferences) so as to get a better share than she would have without cheating. Other ways of manipulation can be modeled as well.

Note that this model again is static and assumes full information about the agents' preferences. Not so much is known, though, about dynamical approaches to fair division. Among the rare examples are two interesting papers by Walsh (2011) and Kash et al. (2014) (note that their analysis is restricted to the setting of *divisible* goods). For allocating *indivisible* goods, Kohler and Chandrasekaran (1971) proposed an elicitation-free protocol based on picking sequences that has been studied in several follow-up papers (e.g., Aziz et al. 2015b; Bouveret and Lang 2011; Baumeister et al. 2017). To the best of our knowledge, dynamical approaches to allocation problems have not received much attention so far. One may look deeper into *online allocations* where the data of the allocation problem cannot be precisely described (e.g., the goods, or their valuation by the agents, come in one after the other). Or one may consider *randomness* where the input is generated according to a known or unknown probability distribution and the information on possible future scenarios is to be estimated. Or one may study *temporal allocations*, evolving over time, where fast and adaptive responses to environmental changes are required. In all these approaches, again, *how can we model the dynamics depending on the agents' preferences and behavior evolving based on their emotions?*

**Acknowledgements** I thank Umberto Grandi and Bill Zwicker for helpful comments. This work was supported in part by DFG grants RO 1202/14-2 and RO 1202/15-1.

# References

Aziz, H., Gaspers, S., Mackenzie, S., & Walsh, T. (2015a). Fair assignment of indivisible objects under ordinal preferences. *Artificial Intelligence*, *227*, 71–92.

Aziz, H., Walsh, T., & Xia, L. (2015b). Possible and necessary allocations via sequential mechanisms. In *Proceedings of IJCAI'15* (pp. 468–474).

Bartholdi III, J., Tovey, C., & Trick, M. (1989). The computational difficulty of manipulating an election. *Social Choice and Welfare*, *6*(3), 227–241.

Bartholdi III, J., Tovey, C., & Trick, M. (1992). How hard is it to control an election? *Mathematical and Computer Modelling*, *16*(8/9), 27–40.

Baumeister, D., Bouveret, S., Lang, J., Nguyen, N., Nguyen, T., Rothe, J., et al. (2017). Positional scoring-based allocation of indivisible goods. *Journal of Autonomous Agents and Multi-Agent Systems*, *31*(3), 628–655.

Baumeister, D., Erdélyi, G., Erdélyi, O., & Rothe, J. (2015). Complexity of manipulation and bribery in judgment aggregation for uniform premise-based quota rules. *Mathematical Social Sciences*, *76*, 19–30.

Baumeister, D., Faliszewski, P., Lang, J., & Rothe, J. (2012). Campaigns for lazy voters: Truncated ballots. In *Proceedings of AAMAS'12* (pp. 577–584). IFAAMAS.

Baumeister, D., & Rothe, J. (2012). Taking the final step to a full dichotomy of the possible winner problem in pure scoring rules. *Information Processing Letters*, *112*(5), 186–190.

Baumeister, D., & Rothe, J. (2015). Preference aggregation by voting. In J. Rothe (Ed.), *Economics and computation: An introduction to algorithmic game theory, computational social choice, and fair division*. Springer texts in business and economics. Chapter 4 (pp. 197–325). Berlin: Springer.

Betzler, N., & Dorn, B. (2010). Towards a dichotomy for the possible winner problem in elections based on scoring rules. *Journal of Computer and System Sciences*, *76*(8), 812–836.

Bouveret, S., Chevaleyre, Y., & Maudet, N. (2016). Fair allocation of indivisible goods. In F. Brandt, V. Conitzer, U. Endriss, J. Lang, & A. Procaccia (Eds.), *Handbook of computational social choice*. Chapter 12 (pp. 284–310). Cambridge: Cambridge University Press.

Bouveret, S., & Lang, J. (2011). A general elicitation-free protocol for allocating indivisible goods. In *Proceedings of IJCAI'11* (pp. 73–78).

Brams, S., Edelman, P., & Fishburn, P. (2003). Fair division of indivisible items. *Theory and Decision, 55*(2), 147–180.

Brams, S., & King, D. (2005). Efficient fair division: Help the worst off or avoid envy? *Rationality and Society, 17*(4), 387–421.

Chevaleyre, Y., Lang, J., Maudet, N., Monnot, J., & Xia, L. (2012). New candidates welcome! Possible winners with respect to the addition of new candidates. *Mathematical Social Sciences, 64*(1), 74–88.

Conitzer, V., Sandholm, T., & Lang, J. (2007). When are elections with few candidates hard to manipulate? *Journal of the ACM, 54*(3), 14.

Conitzer, V., & Walsh, T. (2016). Barriers to manipulation in voting. In F. Brandt, V. Conitzer, U. Endriss, J. Lang, & A. Procaccia (Eds.), *Handbook of computational social choice*, Chapter 6 (pp. 127–145). Cambridge: Cambridge University Press.

Desmedt Y., & Elkind, E. (2010). Equilibria of plurality voting with abstentions. In *Proceedings of ACM-EC'10* (pp. 347–356). ACM Press.

Faliszewski, P., Hemaspaandra, E., & Hemaspaandra, L. (2009a). How hard is bribery in elections? *Journal of Artificial Intelligence Research, 35*, 485–532.

Faliszewski, P., Hemaspaandra, E., Hemaspaandra, L., & Rothe, J. (2009b). Llull and Copeland voting computationally resist bribery and constructive control. *Journal of Artificial Intelligence Research, 35*, 275–341.

Faliszewski, P., & Rothe, J. (2016). Control and bribery in voting. In F. Brandt, V. Conitzer, U. Endriss, J. Lang, & A. Procaccia (Eds.), *Handbook of computational social choice*, Chapter 7 (pp. 146–168). Cambridge: Cambridge University Press.

Hemaspaandra, E., Hemaspaandra, L., & Rothe, J. (2007). Anyone but him: The complexity of precluding an alternative. *Artificial Intelligence, 171*(5–6), 255–285.

Hemaspaandra, E., Hemaspaandra, L., & Rothe, J. (2014). The complexity of online manipulation of sequential elections. *Journal of Computer and System Sciences, 80*(4), 697–710.

Hemaspaandra, E., Hemaspaandra, L., & Rothe, J. (2017a). The complexity of controlling candidate-sequential elections. *Theoretical Computer Science, 678*, 14–21.

Hemaspaandra, E., Hemaspaandra, L., & Rothe, J. (2017b). The complexity of online voter control in sequential elections. *Journal of Autonomous Agents and Multi-Agent Systems, 31*(5), 1055–1076.

Kash, I., Procaccia, A., & Shah, N. (2014). No agent left behind: Dynamic fair division of multiple resources. *Journal of Artificial Intelligence Research, 51*, 579–603.

Kohler, D., & Chandrasekaran, R. (1971). A class of sequential games. *Operations Research, 19*(2), 270–277.

Konczak, K., & Lang, J. (2005). Voting procedures with incomplete preferences. In *Proceedings of the Multidisciplinary IJCAI-05 Workshop on Advances in Preference Handling* (pp. 124–129).

Lang, J., Rey, A., Rothe, J., Schadrack, H., & Schend, L. (2015). Representing and solving hedonic games with ordinal preferences and thresholds. In *Proceedings of AAMAS'15*, pp. 1229–1237,

Lang, J., & Rothe, J. (2015). Fair division of indivisible goods. In J. Rothe (Ed.), *Economics and computation. An introduction to algorithmic game theory, computational social choice, and fair division*, Springer texts in business and economics, Chapter 8 (pp. 493–550). Berlin: Springer.

Nguyen, N., Baumeister, D., & Rothe, J. (2018). Strategy-proofness of scoring allocation correspondences for indivisible goods. *Social choice and welfare, 50*(1), 101–122.

Parkes, D., & Procaccia, A. (2013). Dynamic social choice with evolving preferences. In *Proceedings of AAAI'13*, pp. 767–773. AAAI Press.

Popova, A., Regenwetter, M., & Mattei, N. (2013). A behavioral perspective on social choice. *Annals of Mathematics and Artificial Intelligence, 68*(1–3), 5–30. In J. Goldsmith, & J. Rothe (Eds.),

*Special issue: Algorithms, approximation, and empirical studies in behavioral and computational social choice.*

Regenwetter, M., Grofman, B., Marley, A., & Tsetlin, I. (2006). *Behavioral social choice: Probabilistic models, statistical inference, and applications.* Cambridge: Cambridge University Press.

Sloth, B. (1993). The theory of voting and equilibria in noncooperative games. *Games and Economic Behavior, 5*(1), 152–169.

Tennenholtz, M. (2004). Transitive voting. In *Proceedings of ACM-EC'04* (pp. 230–231). ACM Press.

Walsh, T. (2011). Online cake cutting. In *Proceedings of ADT'11* (pp. 292–305). Berlin: Springer *LNAI #6992.*

Xia, L., & Conitzer, V. (2010). Stackelberg voting games: Computational aspects and paradoxes. In *Proceedings of the AAAI'10* (pp. 697–702). AAAI Press.

Xia, L., & Conitzer, V. (2011). Determining possible and necessary winners given partial orders. *Journal of Artificial Intelligence Research, 41*, 25–67.

Zwicker, W. (2016). Introduction to the theory of voting. In F. Brandt, V. Conitzer, U. Endriss, J. Lang, & A. Procaccia (Eds.), *Handbook of computational social choice*, Chapter 2 (pp. 23–56). Cambridge: Cambridge University Press.

# Future Design

Tatsuyoshi Saijo

**Abstract** *Future Design*, a new movement among Japanese researchers, asks the following question: what types of social systems are necessary if we are to leave future generations sustainable natural environments and sustainable societies? One such method is using an "imaginary future generation," and I overview the literature including the background of this method, the results of relevant laboratory and field experiments, and the nature of relevant practical applications in cooperation with several local governments.

## 1 Past, Present, and Future

Steffen et al. (2004, 2015) indicate that the Earth system's variables such as emission levels of $CO_2$, nitrogen oxides, and methane; ground temperatures; and the amounts of rainforest loss; have been increasing at an accelerating pace since 1950.[1] Alongside this, socio-economic variables such as population, urban population, real GDP, use of water, and flows of information have likewise been rapidly increasing. However, these data also indicate that the OECD nations have been reaping most of the results,

---

[1] Approximately 55 million years ago, there was an age called the PETM (Paleocene–Eocene Thermal Maximum) in which Earth's climate was the warmest. According to Yasunari (2018), the current CO2 emission level per year is about 10 times that of the time of the PETM.

This paper is based upon my keynote speech at the Future Earth Philippines Program Launch meeting held in Manila on the 19th of November 2018. I thank all comments and suggestions at the meeting. This research was supported by Scientific Research A (17H00980) of the Japan Society for the Promotion of Science and the Research Institute for Humanity and Nature (RIHN Project Number 14200122).

T. Saijo (✉)
Research Institute for Humanity and Nature, Kyoto, Japan
e-mail: saijo.tatsuyoshi@kochi-tech.ac.jp

Research Institute for Future Design, Kochi University of Technology, Kochi, Japan

Tokyo Foundation for Policy Research, Tokyo, Japan

© Springer Nature Switzerland AG 2019
J.-F. Laslier et al. (eds.), *The Future of Economic Design*, Studies in Economic Design, https://doi.org/10.1007/978-3-030-18050-8_35

i.e. the 'umami,' o f those changes. The above expresses what we have done until now, i.e., our "past." Steffen et al. (2004) call this phenomenon the "Great Acceleration."

An indicator of a more recent past is the relation between the Human Development Index (HDI) and Ecological Footprint (EFP) per capita.[2] The HDI's chief components are a country's average life-expectancy, level of education, and per-capita income—an indicator of a country's "well-being." Meanwhile, EFP is an indicator of sustainability, and it shows whether the given country is living within its share of global means. Developing countries have both low HDI and low EFP, while developed countries are high in both. The goal ought to be low EFP and high HDI. However, not a single country is heading in that direction.

So how could we evaluate the result of the Great Acceleration, i.e., the present? Rockström et al. (2009) evaluate the Earth system in terms of nine domains, indicating that the biochemical circulation of nitrogen and phosphorus, as well as biodiversity, have virtually gone past their tipping points. They also point out that climate change and land system changes are approaching their tipping points. Such investigation into whether the Earth system's variables are kept within a safe range for humankind is called "planetary boundary research."

Next, let us take a look at the near future. Maggio and Gaetano (2012) predict peaks in production volumes for oil, coal, and natural gas in the near future. According to this study, the amount of coal that we are projected to consume during the *first half* of the 21st century—not the whole 21st century—is about 1.7 times the amount we consumed during the entire 20th century. Furthermore, during the first half of the 21st century, we are projected to consume about 1.5 times the amount of oil and about 3 times the amount of natural gas that we consumed during the 20th century. Thus, for fossil fuel-use, the 20th century was the 'approach-run,' while the current century is the 'climax.' In October 2018 the Intergovernmental Panel on Climate Change warned that the temperature rise since pre-industrial times will exceed 1.5 °C around 2040 if we continue in this fashion.[3] In response to this, the BBC stated on October 8, 2018 that it is the "final call to save the world from 'climate catastrophe.'"[4]

As a result of the above changes the relatively stable Holocene that lasted for more than 10,000 years has ended and we have entered the Anthropocene, a new geological age wherein humankind has changed the Earth itself (see Monastersky 2015). Until now the Earth has gone through 100,000-year cycles of glacial and interglacial periods. However, Steffen et al. (2018) point out that even if we achieve the Paris Agreement's goal, there is a possibility of a future in which Earth deviates from that cycle into a new one involving a greenhouse Earth and a hothouse Earth.

Meanwhile major countries have huge levels of outstanding debt. The debt of Japan exceeds 200% of its GDP, while that of the United States exceeds 100%. For Germany, this figure is about 70%. The current generation is maintaining its wealth by using up the future generations' resources. In the case of Japan, to resolve the debt

---

[2]http://data.footprintnetwork.org/#/sustainableDevelopment?cn=all&yr=2014&type=BCpc,EFCpc.

[3]https://www.ipcc.ch/pdf/special-reports/sr15/sr15_spm_final.pdf.

[4]https://www.bbc.com/news/science-environment-45775309.

balance we would be required to raise consumer tax to about 30 or 40% and keep it that way for a hundred years; only then will the balance go down to about 70% (Hansen and Imrohoroglu 2016). It is still uncertain as to which generation would do this of its own accord.


## 2  How Did This Happen?

Why do we keep changing the future Earth's environment and taking away resources from future generations? According to Sapolsky (2012), humans possess three traits. The first is *sociality*. By the cooperation of multiple people humans have triumphed over other animals and stand at the top of the food-chain. However, sociality requires certain education and experience; it is not something that is acquired instantly. The second is *relativity*. We react not to absolute volumes, but to relative volumes of what is felt by the senses: when there is a sudden, loud noise, or if it suddenly becomes dark we naturally react—by default—to increase the chances of our own survival. In the context that we do not react unless some abrupt change occurs to an external factor, this trait may be reinterpreted as the principle of *optimality* (the behavioral principle in which in order to maximize the objective function, we differentiate it and find the point where it is zero). The third is *shortsightedness* (or *impulse* by Sapolsky 2012); it is not easy to resist eating something tasty in front of you. I would like to add a fourth human trait: *optimism*. For it is possible that humans, in order to increase the chances of their own survival, have evolved to be optimistic about the future, forgetting bad memories from the past to instead pursue immediate pleasures of the present (Sharot 2011).

We may consider these human traits as the premises upon which our society's basic frameworks, such as the market and democracy, are built. First, let us consider the market. While the market is 'an extremely good device for realizing the *short-term* desires of people,' it does not 'allocate resources in a way that takes account of future generations'—the future generations cannot participate in today's markets. Likewise, democracy is not 'a device that incorporates future generations;' it is 'a device that profits people who live *now*.' Running for a political office today to enrich a generation a hundred years later would most likely end in an election-defeat. Although sociality of some sort is clearly necessary to build social systems, these resulting devices deeply reflect relativity, shortsightedness, and optimism.

Following the Industrial Revolution various innovations took place and we began using massive amounts of fossil fuel. The social systems that we created, including the market and democracy, then fed back the human traits encouraging relativity, shortsightedness, and optimism, while weakening sociality. And these transformed human traits, in turn, transformed the content of the market, democracy, and innovations. Despite being the cause of the various future failures mentioned above, this has built a society that blindly focuses on growth. For this reason, we now need various social systems to restrain the market and democracy.

# 3   Is Transformation in Favor of a Sustainable Society Possible?

Future Earth was organized in 2012 as an international research platform for the generation of knowledge and action to accelerate radical innovation in favor of a sustainable society, and it has been active since 2015.[5] One of the basic concepts of Future Earth is *transdisciplinary* research. This is a framework in which stakeholders and scientists co-design research projects, co-produce knowledge, and co-deliver results. However, both the stakeholders and the scientists belong to the current generation and even if the outcome of their activities following the incentives is win-win, there is still the possibility that future generations will lose. Is it not the future generations, therefore, who we should include as stakeholders? And is it not the current generation's ways of thinking and acting that we ought to target for change?

From this viewpoint, "Future Design" emerged with the aim of creating human "futurability." A person exhibits *futurability* when he or she "experiences an increase in happiness as a result of deciding and acting to forego current gains in order to enrich future generations"; the design and praxis of a society generating futurability is called "Future Design" (Saijo 2018). This presents a fundamental question as to whether it is possible to feel the same happiness felt by a parent eating less to give more to his or her child by benefitting future generations to which we are unrelated by blood. The establishment of the concept of futurability reflected a concern regarding the concept of sustainable development (fulfilling the needs of the current generation without detracting from the needs of future generations) that was expounded in the Brundtland Commission's *Our Common Future*. For example, in the case of the aforementioned scenario about outstanding debt, it is impossible for the current generation to reduce the future generations' burdens without bearing a significant burden itself.

Viewed in association with the economic concept of incentive, which is based on the pursuit of self-interest, futurability may seem preposterous. However, from the viewpoint of the continuation of humankind, futurability is an important incentive. Indeed according to the conventional framework of the study of economics, participants with futurability as an incentive would be free-ridden by others, and according to evolutionary game theory, those with such an incentive would perish.

Within the framework of conventional mechanism design, the production of a social mechanism or system that would increase the number of people who have futurability may become an important task. In this paper, however, rather than using an approach of designing such a mechanism I would like to take an approach of designing a social mechanism that would *activate* futurability, according to the supposition that it is intrinsic to humans. This is based on the results of various experimental investigations. Even if we suppose humans have futurability, activating it is not easy. But is it impossible to build a new social mechanism of some form

---

[5]http://www.futureearth.org/.

that would strengthen the sociality that has been weakened under the market and democracy while weakening relativity, shortsightedness, and optimism instead?

In the past few years, Japanese researchers of various fields have begun developing new methods involving the use of an imaginary future generation (see Kobayashi 2018). I will introduce this new movement called "Future Design" below.


# 4    Towards Future Design

Let me introduce an experimental research by Kamijo et al. (2017) that became the starting point for future design research. In this study groups of three participants were composed to represent different generations, and each of these groups was given a task to hold a discussion for up to 10 min and then choose between A ($36) and B ($27), to redistribute this money among its three participants, and receive the money and leave. The experiment was designed so that if the group chooses A, A and B will each decrease by $9 for the next generation (a different, separate group of three participants), and if the group chooses B, there will neither be changes to A nor B for the next generation. Each generation was given information about the decision made by the previous generation(s) and was also made to understand that other generations would follow. The participants were paid according to the decisions that had been made by their generations. If human relativity (principle o foptimality) was activated, the choice would naturally be A. Meanwhile, in some groups, one out of the three participants were chosen and asked to represent the people of the following groups (generations) as he or she negotiated with the other two participants. The payment that this participant received would be in line, however, with the decision made with the other two on how to share the money. Let us call this participant an *imaginary future person.*

Twenty-eight percent of the groups without imaginary future persons and sixty percent of the groups with imaginary future persons chose B, confirming the effectiveness of imaginary future persons: it was found that playing the role of future generations could activate the futurability, not only of that person, but of others as well. Following this, the same experiment was conducted in Dhaka (an urban community) and a rural community in Bangladesh as well as in Kathmandu (an urban community) and a forested community in Nepal; the results confirmed that while introduction of imaginary future persons in non-urban communities is effective, it has no effect in urban communities (Shahrier et al. 2017a). Shahrier et al. (2017b) and others are also beginning to confirm effective functioning of new decision-making mechanisms that were designed based on their studies. For example, a great increase in the chance of sustainable choices was observed, not by introducing imaginary future persons, but instead by openly including a stage where the reason for the decision that was made is to be left to the next generation (Timilsina et al. 2018).

Imaginary future persons think about changes in the present from a future point of view, but it has also been confirmed that a retrospective, imaginary experience of sending advice from the present to people of the past can make that person think like an imaginary future person when thinking about future challenges. Nakagawa et al. (2018) recruited ordinary people and had them think about the future of the forests in one part of Japan (Kochi). The participants read past newspaper articles and sent imaginary advice to people in the past who were wondering which of several options they should choose. By doing so, the participants chose scenarios that support sustainable forests, just as imaginary future persons would do.

The above experimental research has also begun to spawn future design's practical application. In places such as Yahaba, Iwate Prefecture, researchers have begun the process of verifying whether an imaginary future generation created in the present—representing the interests of the future generation and given a role involved in vision design and decision-making—could negotiate the present generation into making decisions that overcome the intergenerational conflict of interests (Hara et al. 2017). The Cabinet Office had required all municipalities to produce a "long-term vision" for 2060, and taking advantage of this opportunity, a total of six workshops were held at a monthly pace from the second half of 2015 to March 2016. Four groups, each of 5–6 people, were composed of Yahaba residents, and two of these groups were asked to think about the Yahaba in 2060 from the viewpoint of the present generation and then propose policies for the present time—an ordinary workshop to draw up policies for the future. Meanwhile, the remaining two groups were asked to go directly to 2060 on a time machine and draw up policies from there.

Even after several workshops, the present generation groups still treated the current problems as tasks of the future. For example, they predicted that long waiting lists for nursery schools would remain a problem, or that there would not be enough care facilities for the elderly, even in 2060. Apparently they could not help but to view the future as an extension of the present, creating visions that focus primarily on finding solutions to current problems and producing ideas within the limits of present conditions. In contrast, the imaginary future generation's thoughts were original; raising the priority of finding solutions to complex and time-consuming tasks; freely picturing the future independently of current conditions; and taking advantage of both physical and sensory resources in their surroundings to make sustainable use of them.

Six months after the policymaking workshops, Nakagawa et al. (2017) held interviews with the locals who had been the imaginary future generation. It was then discovered that these former participants, rather than experiencing a conflict of some sort between today's 'I' and the imaginary future person 'I,' had now been viewing both from a societal perspective. Furthermore, it was discovered that they found joy in the very act of thinking as an imaginary future person, and that way of thinking had since been occurring naturally in their everyday lives. The study thus revealed the possibility of activating people's futurability by having them become imaginary future persons.

# 5  Future Tasks

Practical adoption of future design methods has begun. They have been used to deliberate on the design of the new city hall of Matsumoto, Nagano Prefecture (Nishimura et al. 2018); the introduction of renewable energy in Suita, Osaka Prefecture; future plans for private enterprises; and solutions to various problems at ministries and agencies. The future design method is similar to fractal analysis. It can be applied to local problems, and it can also be geared to face global challenges. The future design method could be applied to the UN's decision-making processes, and it also has the potential to effectively function at world leaders' congregations, such as the G7 and G20.

It also has the potential to innovate the very structure of town assemblies. For instance, let us say that a town assembly has ten members. Out of these ten, three are assigned to be future representatives. In order to become a future representative, a candidate must compete in election by making policies from the viewpoint of the town's future. At the extension of this idea is a proposal to change Japan's upper house into the "future house" and create a future ministry.

That said, although humans appear to have futurability, it is still unclear as to why it can be activated. For this reason, Aoki (2018) proposes a new field dubbed, "neuro future design." In addition, some as Kobayashi (2018) and Hiromitsu (2018) have taken the discussion into the philosophical context of John Rawls's *Theory of Justice.*

# References

Aoki, R. (2018). Neuro future design. *Trends in the Sciences, 23*(6), 64–66. (in Japanese).

Hansen, G. D., & İmrohoroğlu, S. (2016). Fiscal reform and government debt in Japan: A neoclassical perspective. *Review of Economic Dynamics, 21,* 201–224.

Hara, K., Yoshioka, R., Kuroda, M., Kurimoto S. and Saijo, T. (2017). Reconciling intergenerational conflicts with imaginary future generations - Evidence from a participatory deliberation practice in a municipality in Japan -, SDES-2017-19. Kami: Kochi University of Technology.

Hiromitsu, T. (2018). Considering the long-term fiscal problems from ethics and experiments. *Trends in the Sciences 23*(6), 24–27. (in Japanese).

Kamijo, Y., Komiya, A., Mifune, M., & Saijo, T. (2017). Negotiating with the future: Incorporating imaginary future generations into negotiations. *Sustainability Science, 12*(3), 409–420.

Kobayashi, K. (2018a). How to represent the interests of future generations now, 05 May 2018. https://voxeu.org/article/how-represent-interests-future-generations-now.

Kobayashi, K. (2018b). Three concerns on future design. *Trends in the Sciences, 23*(6), 28–30. (in Japanese).

Maggio, G., & Gaetano, C. (2012). When will oil, natural gas, and coal peak? *Fuel, 98,* 111–123.

Monastersky, R. (2015). The human age. *Nature, 519*(7542), 144.

Nakagawa, Y., Hara, K., & Saijo, T. (2017). Becoming sympathetic to the needs of future generations: A phenomenological study of participation in future design workshops, SDES-2017-4. Kami: Kochi University of Technology.

Nakagawa, Y., Kotani, K., Matsumoto, M., & Saijo, T. (2018). Intergenerational retrospective view-points and individual preferences of policies for future: A deliberative experiment for forest management. *Futures* (forthcoming).

Nishimura, N., Inoue, N., & Musha, T. (2018). Deliberation mechanism to let people speak for the future generations. *Trends in the Sciences, 23–6*, 20–23. (in Japanese).

Rockström, J., et al. (2009). A safe operating space for humanity. *Nature, 461*(7263), 472–475.

Saijo, T. (2018). Future design: Bequeathing sustainable natural environments and sustainable societies to future generations. https://researchmap.jp/?action=cv_download_main&upload_id=167703.

Sapolsky, R. M. (2012). Super humanity. *Scientific American, 307*(3), 40–43.

Shahrier, S., Kotani, K., & Saijo, T. (2017a). Intergenerational sustainability dilemma and the degree of capitalism in societies: A field experiment. *Sustainability Science, 12*(6), 957–967.

Shahrier, S., Kotani, K. and Saijo, T. (2017b). Intergenerational sustainability dilemma and a potential solution: Future ahead and back mechanism, SDES-2017-9. Kami: Kochi University of Technology.

Sharot, T. (2011). The optimism bias. *Current Biology, 21*(23), R941–R945.

Steffen, W., Sanderson, A., Tyson, P. D., et al. (2004). *Global change and the earth system: A planet under pressure.*, The IGBP Book Series Berlin: Springer.

Steffen, W., et al. (2015). Planetary boundaries: Guiding human development on a changing planet. *Science, 347*(6223), 1259855.

Steffen, W., et al. (2018). Trajectories of the earth system in the Anthropocene. *Proceedings of the National Academy of Sciences, 115*(33), 8252–8259.

Timilsina, R. R., Kotani, K., Nakagawa, Y., & Saijo, T. (2018). The notion of accountability as a resolution for intergenerational sustainability dilemma, in prep.

Yasunari, T. (2018). *Global climatology*. Tokyo: Tokyo University Press. (in Japanese).

# The Division of Scarce Resources

Christopher P. Chambers and Juan D. Moreno-Ternero

**Abstract** We explore possible future lines of research for the focal problem of dividing scarce resources. They refer to addressing dynamic aspects of these problems, their multidimensional extensions (with, possibly, the existence of mixed resources, heterogeneous preferences and negative awards), uncertainty, and the ensuing incentive aspects in the division process.

The division of scarce resources is probably one of the oldest problems in the history of economic thought. Ancient writings (e.g., Aristotle's Ethics or the Talmud) already deal with it. Nowadays, still in the aftermath of the Great Recession, the media is plagued with references to this problem. The man on the street is constantly thinking and talking about it.[1]

In (1982), Barry O'Neill published a seminal paper setting a simple model to address this problem. The model formalized a group of individuals having conflicting claims over an insufficient amount of a (perfectly divisible) good. Somewhat surprisingly, this simple model has generated a sizable literature in the last decades.[2] It has probably been one of the most explored models in economic theory, which has allowed us to have a deep understanding of it. Lessons have been derived studying this model that could be applied to many other related ones. Conversely, usual techniques in other related fields have been successfully imported to the analysis of *claims problems*.

O'Neill's model accommodates numerous real-life situations, such as the division of an insufficient estate to cover all its associated debts, the collection of a given tax from taxpayers, sharing the cost of a public facility, etc. Nevertheless, there are

---

[1] Unfortunately, this does not happen to be the case within mainstream research in Economics.
[2] See, for instance, Thomson (2003, 2015, 2017).

C. P. Chambers
Georgetown University, Washington, DC, USA
e-mail: cc1950@georgetown.edu

J. D. Moreno-Ternero (✉)
Universidad Pablo de Olavide, Sevilla, Spain
e-mail: jdmoreno@upo.es

© Springer Nature Switzerland AG 2019                                        263
J.-F. Laslier et al. (eds.), *The Future of Economic Design*, Studies in
Economic Design, https://doi.org/10.1007/978-3-030-18050-8_36

several natural aspects that the model does not capture and we conjecture that future efforts will be devoted to enrich the model in order to address them.

One is to move from the static nature of claims problems to deal with dynamic processes of rationing, which abound in real life. Think, for instance, of the allocation of public resources (collected via taxes by a central government) among the regional governments of a country with a certain degree of decentralization (which might be in charge, for instance, of the delivery of public health care or education in their regions). This process is normally undertaken every year, when the government approves the budget for the upcoming fiscal year. One could treat the problem arising each year as an independent claims problem, but it seems more natural to assume that the solution to each problem is, at least, influenced by the solution to the problems from previous years. Formally, one could consider a sequence of rationing (claims) problems (involving the same group of agents), at different periods of time, whose period-wise allocations might not only be determined by the data of the rationing problem at such period, but also by the allocations in previous periods. A plausible way to start approaching this issue would be by assuming that, at each period, the corresponding rationing problem is enriched by an index summarizing the amounts each agent obtained in the previous period. This would just be an alternative interpretation for the so-called baselines profile that Hougaard et al. (2013) consider in their model. Moreno-Ternero and Vidal-Puga (2017) go beyond that point exploring *aggregator operators* to extend rules from the standard static framework of claims problems to the dynamic setting. Two ultimate goals seem to be worth exploring. One would be to provide a dynamic rationale for some classical rationing rules.[3] Another would be to formalize new axioms reflecting ethical or operational principles for dynamic rules and, eventually, derive characterization results for some of them.

Another plausible line for future research is to consider the possibility of negative awards and, possibly, zero endowment. The former would make sense in a context of taxation, such as the one formalized by Young (1988).[4] It would represent including the possibility of subsidies while designing taxation methods. The latter gives rise to the related problem of income redistribution (e.g., Chambers and Moreno-Ternero 2018), which is receiving considerable attention lately in the media (but not in academic research). Standard notions from the literature on claims problems could be formalized here. Others would require substantial modifications. New axioms could also arise. An instance would be the requirement of stability saying that after a rule redistributes income for a given economy, the rule proposes no further redistribution for the resulting economy with the same group of agents and the new income profile.[5]

The literature has also moved to analyze multi-dimensional extensions of claims problems, in which claims refer to multiple types of assets. In some cases, the endowment to be allocated is still considered to be unidimensional and transfers among issues are feasible (e.g., Calleja et al. 2005; Ju et al. 2007). In other cases, allocations

---

[3]This was, for instance, the approach taken by Fleurbaey and Roemer (2011) for the three canonical axiomatic bargaining solutions.

[4]See also Chambers and Moreno-Ternero (2017).

[5]See Chambers and Moreno-Ternero (2018).

are multidimensional, de facto assuming that transfers among issues are not possible (e.g., Moreno-Ternero 2009; Bergantiños et al. 2010). Ju and Moreno-Ternero (2017) have recently considered a more general model allowing to explore end-state and procedural fairness in environments with conflicting (multidimensional) claims. The idea is to study, not only the issue of fair initial allocation (as determined by the solution to a claims problem), but also its influence on the final allocation after the subsequent interactions (via a decentralized exchange procedure) among claimants. In some way, the model provides a bridge between the literature on claims problems and the literature on fair allocation (e.g., Thomson 2011). It seems reasonable to argue that the bridge should be explored further in the near future.

The counterpart to claims problems is the problem of surplus sharing. In that problem, rather than facing the case of dividing scarce resources, the issue is to divide abundant resources. It seems less pervasive than the former in real life, especially in the aftermath of the Great Recession, but it is still a recurrent problem, with obvious interest. The surplus sharing problem has been less analyzed probably due to the fact that the notion of *duality*, which proved itself powerful for claims problems, cannot be meaningfully formalized in that setting. Ju and Moreno-Ternero (2018) have recently explored hybrid models in which both rationing and surplus sharing might arise. The allocation of a *mixed manna* (involving goods and bads) is also an obviously interesting problem and efforts seem to be emerging lately to analyze it from different perspectives (e.g., Bogomolnaia et al. 2017). For instance, one could think of a scenario in which agents invest on a collective start-up venture (seed money to be interpreted as claims), which can go well (thus generating a surplus sharing) or wrong (thus generating a claims problem). These hybrid problems would also bring uncertainty into the picture, another aspect that has only been marginally introduced in this literature and, in which, we presume, efforts will also be concentrated in the future.

Finally, the standard models for the division of scarce resources do not capture incentive aspects. The allocation of (scarce or mixed) manna could be thought as the first step towards analyzing the more general problem of allocating resources that are produced by the same agents involved in the allocation process. For instance, individuals may have claims on the available endowment, but this might be itself the consequence of some (unobservable) effort decisions of those same individuals. What kind of allocation rules would induce agents to exert effort? How could we reconcile a concern for compensation with a concern for efficiency (in the form of revenue maximization)? These seem to be relevant questions to be addressed in the future.

**Acknowledgements** We thank Gustavo Bergantiños for comments and suggestions. Mereno-Ternero acknowledges financial support from FEDER/Ministerio de Ciencia, Innovación y Universidades-Agencia Estatal de Investigación [Research Project ECO2017-83069-P].

# References

Bergantiños, G., Lorenzo, L., & Lorenzo-Freire, S. (2010). A characterization of the proportional rule in multi issue allocation situations. *Operations Research Letters, 38*, 17–19.

Bogomolnaia, A., Moulin, H., Sandomirskiy, F., & Yanovskaya, E. (2017). Competitive division of a mixed manna. *Econometrica, 85*, 1847–1871.

Calleja, P., Borm, P., & Hendrickx, R. (2005). Multi-issue allocation situations. *European Journal of Operational Research, 164*, 730–747.

Chambers, C. P., & Moreno-Ternero, J. D. (2017). Taxation and poverty. *Social Choice and Welfare, 48*, 153–175.

Chambers, C. P., & Moreno-Ternero, J. D. (2018) The axiomatic approach to income redistribution. Mimeo

Fleurbaey, M., & Roemer J. (2011). Judical precedent as a dynamic rationale for axiomatic bargaining theory. *Theoretical Economics, 6*, 289–310.

Hougaard, J. L., Moreno-Ternero, J. D., & Østerdal, L. P. (2013). Rationing in the presence of baselines. *Social Choice and Welfare, 40*, 1047–1066.

Ju, B.-G., Miyagawa, E., & Sakai, T. (2007). Non-manipulable division rules in claim problems and generalizations. *Journal of Economic Theory, 132*, 1–26.

Ju, B.-G., & Moreno-Ternero, J. D. (2017). Fair allocation of disputed properties. *International Economic Review, 58*, 1279–1301.

Ju, B.-G., & Moreno-Ternero, J. D. (2018) Entitlement theory of justice and end-state fairness in the allocation of economic goods. *Economics and Philosophy, 34*, 317–341.

Moreno-Ternero, J. D. (2009). The proportional rule for multi-issue bankruptcy problems. *Economics Bulletin, 29*, 474–481.

Moreno-Ternero, J. D., & Vidal-Puga, J. (2017) Aggregator operators for dynamic rationing. Mimeo.

O'Neill, B. (1982). A problem of rights arbitration from the Talmud. *Mathematical Social Sciences, 2*, 345–371.

Thomson, W. (2003). Axiomatic and game-theoretic analysis of bankruptcy and taxation problems: A survey. *Mathematical Social Sciences, 45*, 249–297.

Thomson, W. (2011). Fair allocation rules. In K. Arrow, A. Sen, K. Suzumura (Eds.), *Handbook of social choice and welfare* (Vol. 2, pp. 393–506). North Holland

Thomson, W. (2015). Axiomatic and game-theoretic analysis of bankruptcy and taxation problems: An update. *Mathematical Social Sciences, 74*, 41–59.

Thomson, W. (2017) *How to divide when there isn't enough: From Aristotle, the Talmud, and Maimonides to the axiomatics of resource allocation.* Econometric Society Monograph. Forthcoming. Cambridge, MA: Cambridge University Press.

Young, H. P. (1988). Distributive justice in taxation. *Journal of Economic Theory, 44*, 321–335.

# The Queueing Problem and Its Generalizations

## Youngsub Chun

**Abstract** A group of agents must be served in a facility. The facility can serve one agent at a time and agents incur waiting costs. The queueing problem is concerned with finding the order to serve agents and the monetary transfers they should receive. In this paper, we summarize recent developments in the queueing problem and discuss its possible generalizations to indicate the directions for future research.

## 1 Introduction

A group of agents must be served in a facility. The facility can serve one agent at a time and agents incur waiting costs. The queueing problem is concerned with finding the order to serve agents and the monetary transfers they should receive. An allocation consists of each agent's position in the queue and the amount of monetary transfer to her. An agent's assignment is a corresponding component of an allocation. We assume that an agent's waiting cost is constant per unit of time, but agents differ in their unit waiting cost. Furthermore, we assume that each agent has quasi-linear preferences, so that her utility from her assignment is equal to the amount of her monetary transfer minus her waiting cost. A rule assigns to each problem a nonempty subset of allocations.

It is not difficult to find real-life examples of the queueing situation. It occurs when agents cannot coordinate on the time when they want to have a service (long queues at the supermarket, airport, etc). Even if they can coordinate, all agents might want to have an earlier service rather than a later one. For example, researchers want

I am grateful to Gustavo Bergantiños and Bill Zwicker for their comments. This work was supported by the National Research Foundation of Korea Grant funded by the Korean Government (NRF-2016S1A3A2924944).

Y. Chun (✉)
Department of Economics, Seoul National University, Seoul 08826, Korea
e-mail: ychun@snu.ac.kr

© Springer Nature Switzerland AG 2019
J.-F. Laslier et al. (eds.), *The Future of Economic Design*, Studies in Economic Design, https://doi.org/10.1007/978-3-030-18050-8_37

to use a supercomputer or an expensive research facility; consumers want to install a new computer program in their computers.[1]

An earlier literature on the queueing problem goes back to Dolan (1978) who analyzes a strategy-proof, but not necessarily budget-balanced rule for the problem. Suijs (1996), Mitra (2001, 2002) study a strategy-proof and budget-balanced rule, now called the symmetrically balanced VCG rule.[2] On the other hand, a normative analysis on the problem has been started by Maniquet (2003), Chun (2006a).[3]

In this paper, we first summarize recent developments in the queueing problem. Then, we discuss its possible generalizations. In doing so, we hope to indicate the directions for future research.

## 2   Various Approaches to the Queueing Problem

The queueing problem can be solved by taking many different approaches. Here we give a brief sketch on the recent developments.

### 2.1   Strategic Approach

A strategic approach is interested in identifying rules satisfying strategy-proofness which requires that each agent should not have an inventive to misrepresent her unit waiting cost no matter what she believes other agents to be doing. If a rule satisfies strategy-proofness, then truthful reporting of the unit waiting cost is a weakly dominant strategy for all agents. In the queueing problem, the classic result of Holmström (1979) implies that a rule satisfies queue-efficiency, which requires that the selected queue should minimize the aggregate waiting cost, and strategy-proofness if and only if it is a VCG rule (Mitra 2001, 2002). By additionally imposing equal treatment of equals, which requires that two agents with the same waiting cost should end up with the same utilities, anonymous members of the VCG rules can be characterized (Chun et al. 2014a). This family includes the symmetrically balanced VCG rule introduced by Suijs (1996), Mitra (2001) and later characterized by Kayi and Ramaekers (2010), Hashimoto and Saitoh (2012), Chun et al. (2015), and the pivotal rules introduced and characterized by Mitra and Mutuswami (2011). Also, the pivotal rules can be characterized by strengthening strategy-proofness to weak group strategy-proofness, which requires that there does not exist a deviation which makes all deviating agents strictly better off (Mitra and Mutuswami 2011).

---

[1]For other examples of the queueing situation, see Maniquet (2003), Kayi and Ramaekers (2010), Mukherjee (2013).

[2]The family of VCG rules is due to Vickrey (1961), Clarke (1971), Groves (1973).

[3]See Chun (2016) for a survey of the literature.

## 2.2 Normative Approach

Maniquet (2003) tries to solve the queueing problem by applying the Shapley value (Shapley 1953) introduced in the cooperative game theory. To do this, first, he transforms the queueing problem into the queueing game by defining the worth of a coalition to be the minimum waiting cost incurred by its members under the optimistic assumption that they are served before the non-coalitional members. He obtains the minimal transfer rule which selects an efficient queue and transfers to each agent a half of her unit waiting cost multiplied by the number of her predecessors minus a half of the sum of the unit waiting cost of her followers.

Chun (2006a) proposes an alternative approach which defines the worth of a coalition to be the minimum waiting cost incurred by its members under the pessimistic assumption that they are served after the non-coalitional members. Even though the same Shapley value is applied, he obtains a different rule, the maximal transfer rule, which selects an efficient queue and transfers to each agent a half of the sum of the unit waiting cost of her predecessors minus a half of her unit waiting cost multiplied by the number of her followers. Other solutions developed in the cooperative game theory can be applied to the queueing problem. In fact, Chun and Hokari (2007) shows the coincidence of the nucleolus (or the prenucleolus) and the Shapley value for queueing games.

On the other hand, Chun (2006b) investigates the implications of no-envy (Foley 1967) and provides a complete characterization of rules satisfying no-envy. No-envy requires that no agent should end up with a higher utility by consuming what any other agent consumes.

## 2.3 Combining Strategic and Normative Approaches

Recently, there have been many papers which combine the strategic and the normative points of view. They characterize all rules satisfying queue-efficiency and strategy-proofness together with a normative requirement such as no-envy, egalitarian equivalence (Pazner and Schmeidler 1978), and the identical preferences lower bound (Moulin 1990, 1991). Egalitarian equivalence requires that for each problem, there should be a reference assignment such that each agent is indifferent between her assignment and the reference assignment. The identical preferences lower bound requires that each agent should be at least as well off as she would be, under queue-efficiency, budget-balance, and equal treatment of equals, if all agents had the same preferences as her. The implications of these requirements together with strategy-proofness have been investigated by Kayi and Ramaekers (2010), Chun et al. (2014b), Chun and Yengin (2017), respectively.

## 2.4 Non-cooperative Approach

Following various bargaining protocols implementing the Shapley value (Gul 1989; Hart and Mas-Colell 1996; Ju 2013; Ju and Wettstein 2009; Pérez-Castrillo and Wettstein 2001), Ju et al. (2014) provides a bargaining protocol of queueing rules in which agents can negotiate among themselves to resolve the queueing conflicts. Given a queueing situation in which agents have to form a queue in order to be served, they firstly compete with each other for a position in the queue. The winner can decide to take up the position or sell it to others. In the former case, the rest of agents proceed to compete for the remaining positions in the same manner. In the latter case, the seller proposes a queue with corresponding payments to the others which can be accepted or rejected. Depending on which position agents are going to compete for, they show that the subgame perfect equilibrium outcome of the corresponding bargaining protocol coincides with the payoff vector assigned by one of two rules for the queueing problem, either the minimum transfer rule or the maximum transfer rule, while an efficient queue is always formed in equilibrium.

# 3 Generalizations of the Queuing Problem

Even though the queueing problem has been studied from many different perspectives, its generalizations have not been studied in depth yet. Here, we discuss its possible generalizations to indicate the directions for future research.

## 3.1 Queueing Problems with an Initial Queue

So far, it is assumed that there is no initial order of agents. However, many queueing situations involve an initial order and agents are usually served on the first-come first-served basis. Although this rule is easy to implement, it may not be queue-efficient when waiting in a queue is costly for agents and agents differ in their unit waiting costs. In fact, trading queue positions are allowed in some situations to overcome the inefficiency resulted from the initial queue. This problem has been studied in Curiel et al. (1989), Gershkov and Schweinzer (2010), Chun et al. (2017).

## 3.2 Sequencing and Scheduling Problems

The sequencing problem generalizes the queueing problem by allowing each agent to have a different amount of service time, so that each agent is characterized by two parameters, a unit waiting cost and an amount of service time. Since this problem is more complicated than the queueing problem, it does not receive much attention in the literature. A few exceptions are: Suijs (1996), Chun (2011), Parikshit and Mitra (2017), among others.

On the other hand, we may consider another subclass of the sequencing problem which assumes that each agent has the identical unit waiting cost, but needs a different amount of service time. This schedule problem has been studied by Moulin (2007) who analyzes an existence of rules that do not give agents an incentive to merge or split the jobs.

## 3.3  Slot Allocation Problems

The slot allocation problem is concerned with the following situation. A group of agents must be assigned to a slot located along a line. Only one agent can be assigned to each slot. Agents differ in their unit waiting cost and the most preferred slot position, called the peak. Each agent's utility from her assignment is equal to the amount of monetary transfer minus the unit waiting cost multiplied by the distance from the peak to her assigned slot. We are interested in finding a way of assigning slots to agents and the monetary transfers they should receive. The slot allocation problem generalizes the queueing problem by allowing each agent to have a different peak. For the queueing problem, all agents have the same peak at the first slot. Chun and Park (2017) studies a special subclass of the slot allocation problem in which all agents have the identical unit waiting cost. Also, an ordinal version of this problem has been studied by Hougaard et al. (2014) and a related problem of assigning landing slots to airlines by Schummer and Vohra (2013), Schummer and Abizada (2017).

## 3.4  Other Generalizations

The queueing problem can be interpreted as a small subclass of the economies with indivisible goods[4] in which each agent has specific preferences over queue positions. Moreover, an agent's waiting cost is assumed to be constant per unit of time. It would be an interesting extension if the unit waiting cost is allowed to vary over queue positions. Also, it would be meaningful generalizations to analyze a dynamic queueing problem which allows each agent to have a different arrival time (Ghosh et al. 2014) and the queueing problem with multiple facilities which allows the facilities to handle several agents at the same time (Mitra 2005; Chun and Heo 2008; Mukherjee 2013).

---

[4]See Thomson (2013) for a survey of the literature.

# References

Chun, Y. (2006a). A pessimistic approach to the queueing problem. *Mathematical Social Sciences*, *51*, 171–181.

Chun, Y. (2006b). No-envy in queueing problems. *Economic Theory*, *29*, 151–162.

Chun, Y. (2011). Consistency and monotonicity in sequencing problems. *International Journal of Game Theory*, *40*, 29–41.

Chun, Y. (2016). *Fair queueing*. Berlin: Springer.

Chun, Y., & Heo, E. J. (2008). Queueing problems with two parallel servers. *International Journal of Economic Theory*, *4*, 299–315.

Chun, Y., & Hokari, T. (2007). On the coincidence of the Shapley value and the nucleolus in queueing problems. *Seoul Journal of Economics*, *20*(2), 223–237.

Chun, Y., Mitra, M., & Mutuswami, S. (2014a). Characterizations of pivotal mechanisms in the queueing problem. *Mathematical Social Sciences*, *72*, 62–66.

Chun, Y., Mitra, M., & Mutuswami, S. (2014b). Egalitarian equivalence and strategyproofness in the queueing problem. *Economic Theory*, *56*(2), 425–442.

Chun, Y., Mitra, M., & Mutuswami, S. (2015). A characterization of the symmetrically balanced VCG rule in the queueing problem. *Games and Economic Behavior*. https://doi.org/10.1016/j.geb.2015.04.001.

Chun, Y., Mitra, M., & Mutuswami, S. (2017). Reordering an existing queue. *Social Choice and Welfare*, *49*, 65–87.

Chun, Y., & Park, B. (2017). A graph theoretic approach to the slot allocation problem. *Social Choice and Welfare*, *48*, 133–152.

Chun, Y., & Yengin, D. (2017). Welfare lower bounds and strategy-proofness in the queueing problem. *Games and Economic Behavior*, *102*, 462–476.

Clarke, E. H. (1971). Multi-part pricing of public goods. *Public Choice*, *11*, 17–33.

Curiel, I., Pederzoli, G., & Tijs, S. (1989). Sequencing games. *European Journal of Operations Research*, *40*, 344–351.

Dolan, R. J. (1978). Incentive mechanisms for priority queuing problems. *The Bell Journal of Economics*, *9*, 421–436.

Foley, D. (1967). Resource allocation and the public sector. *Yale Economic Essays*, *7*, 45–98.

Gershkov, A., & Schweinzer, P. (2010). When queueing is better than push and shove. *International Journal of Game Theory*, *39*, 409–430.

Ghosh, S., Long, Y., & Mitra, M. (2014). *Dynamic VCG mechanisms in queueing*. New York: Mimeo.

Groves, T. (1973). Incentives in teams. *Econometrica*, *41*, 617–631.

Gul, F. (1989). Bargaining foundations of Shapely value. *Econometrica*, *57*, 81–95.

Hart, S., & Mas-Colell, A. (1996). Bargaining and value. *Econometrica*, *64*, 357–380.

Hashimoto, K., & Saitoh, H. (2012). Strategy-proof and anonymous rule in queueing problems: A relationship between equity and efficiency. *Social Choice and Welfare*, *38*, 473–480.

Holmström, B. (1979). Groves' schemes on restricted domains. *Econometrica*, *47*, 1137–1144.

Hougaard, J., Moreno-Ternero, J., & Østerdal, L. P. (2014). Assigning agents to a line. *Games and Economic Behavior*, *87*, 539–553.

Ju, Y. (2013). Efficiency and compromise: A bid-offer counteroffer mechanism with two players. *International Journal of Game Theory*, *42*, 501–520.

Ju, Y., Chun, Y., & van den Brink, R. (2014). Auctioning and selling positions: A non-cooperative approach to queueing conflicts. *Journal of Economic Theory*, *153*, 33–45.

Ju, Y., & Wettstein, D. (2009). Implementing cooperative solution concepts: A generalized bidding approach. *Economic Theory*, *39*, 307–330.

Kayi, C., & Ramaekers, E. (2010). Characterizations of Pareto-efficient, fair, and strategy-proof allocation rules in queueing problems. *Games and Economic Behavior*, *68*, 220–232.

Maniquet, F. (2003). A characterization of the Shapley value in queueing problems. *Journal of Economic Theory*, *109*, 90–103.

Mitra, M. (2001). Mechanism design in queueing problems. *Economic Theory, 17,* 277–305.

Mitra, M. (2002). Achieving the first best in sequencing problems. *Review of Economic Design, 7,* 75–91.

Mitra, M. (2005). Incomplete information and multiple machine queueing problems. *European Journal of Operational Research, 165,* 251–266.

Mitra, M., & Mutuswami, S. (2011). Group strategyproofness in queueing models. *Games and Economic Behavior, 72,* 242–254.

Moulin, H. (1990). Fair division under joint ownership: recent results and open problems. *Social Choice and Welfare, 7,* 149–170.

Moulin, H. (1991). Welfare bounds in the fair division problem. *Journal of Economic Theory, 54,* 321–337.

Moulin, H. (2007). On scheduling fees to prevent merging, splitting, and transferring of jobs. *Mathematics of Operations Research, 32,* 266–283.

Mukherjee, C. (2013). Weak group strategy-proof and queue-efficient mechanisms for the queueing problem with multiple machines. *International Journal of Game Theory, 42,* 131–163.

Parikshit, D., & Mitra, M. (2017). Incentives and justice for sequencing problems. *Economic Theory, 64,* 239–264.

Pazner, E., & Schmeidler, D. (1978). Egalitarian equivalent allocations: A new concept of equity. *Quarterly Journal of Economics, 92,* 671–687.

Pérez-Castrillo, D., & Wettstein, D. (2001). Bidding for the surplus: A non-cooperative approach to the Shapley value. *Journal of Economic Theory, 100,* 274–294.

Schummer, J., & Abizada, A. (2017). Incentives in landing slot problems. *Journal of Economic Theory, 170,* 29–55.

Schummer, J., & Vohra, R. (2013). Assignment of arrival slsots. *American Economic Journal: Microeconomics, 5*(2), 164–185.

Shapley, L. S. (1953). A value for *n*-person Games. In H. W. Kuhn & A. W. Tucker (Eds.), *Contributions to the theory of games II* (Vol. 28, pp. 307–317), Annals of Mathematical Studies. USA: Princeton University Press.

Suijs, J. (1996). On incentive compatibility and budget balancedness in public decision making. *Economic Design, 2,* 193–209.

Thomson, W. (2013). The theory of fair allocation. Book Manuscript. University of Rochester.

Vickrey, W. (1961). Counterspeculation, auctions and competitive sealed tenders. *Journal of Finance, 16,* 8–37.

# Algorithmically Driven Shared Ownership Economies

## Vincent Conitzer and Rupert Freeman

**Abstract** Resource sharing is a natural, and increasingly common, way to efficiently utilize resources. In many situations today, however, resources are owned by a single agent who has sole control over the usage of the resource. In this article, we examine the feasibility of an alternative model where ownership of resources is shared, and usage schedules are determined algorithmically by a fixed set of rules. We explore several design parameters for shared ownership algorithms, surveying existing work in each area and proposing directions for future research.

## 1 Introduction

While the broader concept of the sharing economy has had significant impact, truly joint ownership of scarce resources is fraught with difficulties. For it to have a chance of succeeding, there generally need to be clear rules. Shareholders need to have clear voting rights. Some resources may be owned by the state, and therefore in a sense are collectively owned; here, too, ideally there are clear rules for how public decisions concerning these resources are made. Going beyond ownership in the narrow sense, joint custody arrangements often involve carefully stated rules. In all these cases, disputes may still arise as ambiguities in the rules are discovered, and settling these disputes can be very costly.

The costs of drawing up precise rules and adjudicating disputes make it simply not worthwhile to pursue joint ownership arrangements in many contexts. While in principle, jointly owning a car with one's neighbor may seem like a good idea in a city where one rarely needs a car, the risk of conflict because of disagreement about the rules is likely to make one think twice. The more typical outcome will be for one of the neighbors to own the car and the other to perhaps sometimes borrow it (or rent

V. Conitzer
Microsoft University, Durham, NC 27708, USA
e-mail: conitzer@cs.duke.edu

R. Freeman (✉)
Microsoft Research, New York, NY 10011, USA
e-mail: rupert.freeman@microsoft.com

© Springer Nature Switzerland AG 2019
J.-F. Laslier et al. (eds.), *The Future of Economic Design*, Studies in
Economic Design, https://doi.org/10.1007/978-3-030-18050-8_38

it, or borrow it with a vague expectation of providing something in return at some point), for both to own a car, or for neither to own a car and both to rely on taxis, Uber/Lyft, or other modes of transportation.

However, as devices become increasingly connected and run by algorithms—i.e., as the Internet of Things takes off—shared ownership may become a more realistic alternative. The reason is that the rules of shared ownership can now be formally encoded in an algorithmic manner. When someone attempts to use the jointly owned car, the algorithm can simply check the rules to see whether that person is currently allowed to drive the car, and if the answer is "no" refuse him permission to start the car. Such rules can take an endless variety of forms, ranging from first-come first-served, to sophisticated systems based on virtual currencies for keeping track of how much of her allotment an agent has already spent. The best precise rules presumably depend on the application and the people involved; for example, the rules should be *understandable* to the people involved.

There is at least one environment where this vision is already realistic, and that is in the context of allocating shared computing resources. This allocation is naturally controlled by algorithms, making it an ideal test case for the broader vision we have laid out above. As computation becomes increasingly ubiquitous in our environment, we expect these issues to become increasingly pervasive.

# 2 Allocating Over Time

One of the most common motivations for shared ownership is that the relevant agents do not need the resource in question at every point in time, and the times at which one agent needs the resource differ from those of another. For example: today I really would like to have the car to go to a birthday party, but I do not expect to have any need for it tomorrow. A similar situation can occur for tasks: I have an important exam tomorrow so I would very much appreciate it if you could do the dishes today, and after the exam I will be able to vacuum the house. Clearly we already make such arrangements informally. But in the allocation of computing resources, we already see the use of formal protocols.

In many settings, computing resources are allocated on-the-fly in a truly online manner, such as an operating system allocating resources among different applications as demand arises. (Here, "online" is used not in the sense of being connected to a computer network, but rather in the sense of making these decisions dynamically as new information arrives, rather than solving a large scheduling problem up front from whose solution we do not expect to deviate later.) Algorithmic schedulers receive demands from different applications, and allocate resources accordingly (Lipton and Tomkins 1994; Kolen et al. 2007; Davis and Burns 2011). The demands may in turn be dependent on higher-level, human user behavior, which is difficult for the applications to accurately predict, prohibiting advance planning. In other shared ownership situations, however, we may want some flexibility to plan ahead. For instance, people with shared ownership of a car may want to schedule an hour every week to go grocery shopping, or book the car in advance to get to a particular appointment. For joint ownership of a vacation home, we may want the capacity to book the home months, or even years, in advance. In fact, this may also happen in the allocation of computing resources. Consider, for example, a researcher who is expecting a trove of data to come in from a spacecraft one week from now as it passes by a planet's moon; she may wish to reserve time on a supercomputer in advance for analyzing the data.

This consideration alone provides a rich design space. At one extreme, we can require users to express their preferences for a unit of time far in advance. This has the obvious benefit of allowing us to optimize the schedule for whatever objective we want (since at any point in time we have preference data far into the future), at the cost of requiring users to accurately predict their demand for the item in the future. At the other extreme, we can ask that users express their demand for only the unit of time immediately following the current time period. For example, users would declare at 1pm whether or not they want the car from 2–3 pm, with the car only then allocated to someone.

Both extremes are obviously impractical. We want to both allow agents to book usage in advance, and accommodate demand that appears only at the last minute. But it is not clear how to optimally design such a protocol. Currently, it is common for similar schemes (such as existing car-sharing services) to allow users to book usage for any unreserved time slot in a first-come first-served style. However, we would not want a routine, easily rescheduled shopping trip that was booked in advance to take priority over a medical emergency, say, even though the latter is impossible to predict in advance. Further, we do not want a user to *fake* a medical emergency to obtain use of the car, when in fact she simply forgot to book the car for a scheduled meeting with friends. Such strategic considerations cannot be ignored in the context of systems used by multiple self-interested agents—indeed, they feature prominently in the literature on scheduling (Tian 2013; Im and Kulkarni 2016; Chawla et al. 2017; Babaioff et al. 2017). These agents are focused on their own objectives, and generally cannot be expected to honestly reveal the importance or urgency of their request for resources, if doing so is not in their own interest. One possibility for addressing this is to implement some sort of pricing scheme, allowing usage to be prioritized along socially desirable axes such as importance, advance booking, demand, and any other factors considered important by the relevant stakeholders.

## 3   Money or No Money?

Money is a natural tool for allocating scarce resources. Users with greater need for the item are willing to pay more for its use, suggesting that the item is more likely to end up with someone who needs it more. For example, a simple scheme would run an auction that closes some time (say, a day) before the designated use period for the item. The agent that wins the auction then has the right to use the item in the specified slot. In the case that some emergency arises between the close of the auction and the designated usage period, some high 'override' price could be fixed in advance, that would allow users to capture use of the item even if they did not originally win the auction, also allowing the original winner to be compensated for the unexpected loss of the resource.

Unfortunately, it is not always clear that the greatest willingness to pay corresponds to the greatest need. A wealthy person's willingness to pay can exceed a poor person's ability to pay, even if the poor person has much greater need for the item than the wealthy person. Additionally, in many situations, there may be other disadvantages to the use of money. Distributing payments amongst a group of friends can be awkward, and if the amounts are small, often not worthwhile. Moral or legal issues can also prevent the use of money, when the shared resource is deemed to be such that every person should be granted equal access, regardless of wealth. Kidney exchange (see, among many others, (Roth et al. 2004)) is an example setting where non-monetary mechanism design has flourished due largely to the illegality of buying and selling organs, with the use of money being seen as "repugnant" (Roth 2007; Danovitch and Delmonico 2008).

For these reasons, we may be interested in the design of protocols that do not involve the use of money. This is generally a harder problem, since money allows agents to substantively back up claims that they want or need use of the item. Without money, we must introduce other incentives for the agents not to overstate their preferences, the most natural approach being that agents 'pay' for resources by sacrificing their right to the resource at some later point in time, or the right to some other resource (now or later). One way to achieve this is to introduce some *artificial* currency that can be equally distributed to agents for the sake of fairness, and treat it similarly to if it were real money (Gorokh et al. 2016, 2017). Such an approach is commonly used for course allocation in business schools (Sönmez and Ünver 2010), although poor performance of the artificial currency version of the real money mechanisms has forced some schools to move to a more sophisticated mechanism (still using artificial currency) (Budish 2011; Budish and Kessler 2014). Of course, if the artificial currency is transferable among agents, there is the risk that they will exchange it for real money, thereby bringing it back into the picture. Some approaches do not exactly introduce artificial currency, but still allow an agent to obtain extra resources now at the cost of fewer future resources. By keeping track of how much utility an agent should, in expectation, receive in the future, it is sometimes possible to achieve strong guarantees in a way that resembles the use of artificial currency (Athey and Bagwell 2001; Guo et al. 2009; Abdulkadiroğlu and Bagwell 2013). Even within a single round only, it is sometimes possible to design mechanisms for trading resources (without money) that obtain some formal approximation guarantees (Guo and Conitzer 2010; Han et al. 2011; Maya and Nisan 2012; Cole et al. 2013a, b; Cheung 2016; Amanatidis et al. 2017).

Another line of work in online resource allocation without money restricts users to reporting only binary (like/dislike) valuations, with some (randomized) tie-breaking rule in place to decide how to allocate the resource (Aleksandrov et al. 2015; Aleksandrov and Walsh 2017; Freeman et al. 2018). These schemes have the benefit of being easy to understand for users, but come at the expense of being unable to distinguish users whose need is great from those whose need is only small. There also exists literature on *dynamic fair division* (Walsh 2011; Kash et al. 2014; Friedman et al. 2017), but this work predominantly focuses on agents arriving and departing over time rather than dynamic preferences. Allowing agents to buy in and out of participation over time is of course a desirable feature, but exploiting dynamic preferences is crucial to the profitability of shared ownership.

# 4  Public Versus Private Goods

Sometimes, jointly owned items are useful to only a single agent at a time (*private* goods), while other items can be freely used by many agents simultaneously (*public* goods). For instance, the allocation of CPU time to computer applications is a private good, given that the CPU can only service one application at a time, whereas the storage of data in a shared cache is a public good, since many applications can access the data once stored (Kunjir et al. 2017). Outside of existing computer systems, we could consider a neighborhood jointly owning a maintenance robot that performs tasks such as clearing trash, repairing pot holes, and trimming trees, with the robot's tasks being collectively decided by the neighborhood.

While public goods are a particularly natural candidate for joint ownership, allowing for multiple agents to utilize a resource simultaneously and thus preventing duplicate cost and effort, they also present additional difficulties. For instance, a basic and problematic form of strategizing when goods are public is the *free-rider* problem (Samuelson 1954; Green and Laffont 1979), where an agent pretends to dislike some popular course of action in the hope that the action will proceed anyway, but they will not have to pay for it (in the sense of real or artificial currency, or future consumption). For instance, if I know that all of my neighbors are desperate for a particular pot hole to be repaired, then my best action may be to report that I would rather the robot did some other task, like removing a tree. Since all the other neighbors wanted the pot hole repaired, the robot will still perform that task (depending on the exact form of the protocol), and I get to enjoy the benefits from that, while also profiting from the fact that the protocol believes me to be unhappy with the outcome and will attempt to compensate me later.

Building on the classic work of Groves and Ledyard (1977), who propose quadratic pricing in public good economies, Lalley and Weyl (2018a) propose *Quadratic Voting* for binary collective decisions, where voters purchase votes at a price that is quadratic in the number of votes bought. Under a set of reasonable assumptions, voters buy a number of votes proportional to how much they care about the issue at hand (Posner and Weyl 2014; Weyl 2017; Lalley and Weyl 2018b). Potentially, agents could 'buy' their votes using not only real money but also artificial currency or even 'effort' (say, buying votes for the maintenance robot to pick up trash requires you to spend some time clearing trash yourself). In other situations, the free-rider problem can be avoided if it is possible to restrict access to otherwise public goods in some way, say by restricting access to a shared facility to members only, or denying access to cached data after a certain number of requests. Goods for which this is possible are known as *club goods* (Buchanan 1965). For such goods, it can be possible to prevent free-riding by disallowing access to users that report only a low utility for use of the good (Moulin 1994; Pu et al. 2016).

Another difference between public and private goods lies in how we may wish to define what a fair solution looks like. For private goods, fairness definitions can utilize the fact that each agent receives a well-defined allocation, be it in terms of what items the agent has control of, or for how much time. We may wish to guarantee each agent a proportional share of her utility for all resources combined (Steinhaus 1948), or a share that she (almost) prefers to that of every other agent (Foley 1967; Lipton et al. 2004). For public goods, no agent is ever in sole possession of an item, so asking whether an agent envies another is less natural, but fairness notions based on proportionality are still applicable (Conitzer et al. 2017), as are those based on stability with respect to coalitional deviations (the *core*) (Foley 1970; Fain et al. 2016, 2018). The Maximum Nash Welfare solution (Nash 1950), which maximizes the *product* of agents' utilities, (approximately) satisfies several of these fairness properties (Caragiannis et al. 2016; Conitzer et al. 2017), making it an appealing objective for collective decision making. In online settings, greedy algorithms for maximizing the Nash welfare provide both good approximations to the optimal Nash welfare, and satisfy desirable properties as algorithms in their own right (Freeman et al. 2017).

## 5 Communication Outside the Protocol

Often, we would expect that joint owners have the ability to communicate with each other outside of the formal allocation protocol. They may therefore be able to modify the allocation chosen by the algorithm. An agent allocated usage of the item may be able to pass use to a different agent (potentially for something in return), or to share the item with other agents (say, by carpooling, or storing data in a cache that will be useful to other applications).

While in some cases, such outside communication is desirable, in others it has the capacity to subvert the shared protocol. Secondary resale markets could easily devolve into simply having an exchange rate between artificial and real currency, effectively re-introducing money to the system. Private communication could also lead to collusion; in bidding rings, for instance, groups of buyers agree to share information prior to an auction, and then behave in a pre-determined way accordingly, sharing any profits (McAfee and McMillan 1992; Leyton-Brown et al. 2000, 2002; Marshall and Marx 2007). For sufficiently general settings, it has been shown that the space of collusion-resistant mechanisms is extremely narrow (Schummer 2000; Goldberg and Hartline 2005), but collusion-resistant mechanisms have been designed for certain special cases such as spectrum auctions (Wu et al. 2009), or when some amount of agent type verification is allowed (Penna and Ventre 2014). Resistance to such group-strategic behavior will clearly play an important role in the design of shared ownership protocols moving forward.

## 6 Who Chooses the Protocol, and What Should Be the Objective?

One clear theme emerges throughout in the study of collective decision making: that there is unlikely to be a single 'best' rule for all purposes. Thus, it is worth thinking about how a shared ownership protocol will be chosen among competing alternatives. As we have already seen, a wealthy person who wants to be able to use a shared resource at little notice may have quite different preferences over candidate protocols than a poor person with a fixed, predictable schedule.

When the resource in question is truly a publicly owned resource, such as a facility at a public park, then we may expect that the government could choose a protocol that optimizes in certain socially-desirable directions. However, when the resource is shared among a smaller group of people, then the protocol must be agreed upon by them, for example by voting over candidate protocols. This could be an onerous task, since an algorithmic protocol must be precisely defined to handle all eventualities. For informal shared ownership arrangements that we observe in practice currently, rare and minor exceptions can be handled informally with relative ease (e.g., if someone falls ill, they can have priority use of the car provided it does not happen often and they "make up for it" at some point in the future). To design protocols that take all such eventualities into account is difficult and the resulting protocols are unlikely to be generally applicable to many situations. Additionally, the space of protocols would become extremely large, possibly unmanageably so.

This perspective permits a more positive view of side deals resulting from communication outside the mechanism. Rather than handling all possible scenarios within the protocol, we may wish to allow, and even facilitate, users making informal trades that are not visible to the protocol. For instance, having established usage rights for a particular time slot, we could allow an agent to do as she wishes with those rights. Rare and exceptional cases would then be naturally handled in a way that allows for a greater degree of flexibility than would be convenient to explicitly encode into a protocol for shared ownership.

One could imagine a time where goods are sold with ready-to-use algorithms for shared usage, one of which can be activated upon purchase. While convenient, this possibility leads us to further questions. Moving from a predominantly individual ownership model to joint ownership will not benefit everyone; in particular, some manufacturers may see a decreased demand for their goods if not everyone needs a separate copy. Is it in their interest to help facilitate joint ownership by building (good) sharing protocols into their products? The transformation of society towards a true shared-ownership economy will require not only improvements in algorithmic economic design, but also buy-in from other parties, from corporations to government to individuals.

# 7 Conclusion

Algorithmic approaches for allocating shared resources among multiple self-interested parties are already being used today in the allocation of computing resources. But this may be but a harbinger of a much broader societal development. More and more resources are starting to be algorithmically controlled and connected to computer networks. Many of these resources either need to be assigned to a user at any given point in time, or decisions need to be made about their use, where multiple parties have an interest in these decisions. This makes them candidates to have their usage governed by the types of algorithms discussed in this chapter, possibly ushering in an age of a true shared-ownership economy. Many open questions remain, and by collaborating on them, computer scientists, economists and others may help to realize this vision in a desirable fashion.

**Acknowledgements** Conitzer is thankful for support from NSF under awards IIS-1527434 and CCF-1337215. Freeman is thankful for support from a Facebook PhD Fellowship.

# References

Abdulkadiroğlu, A., & Bagwell, K. (2013). Trust, reciprocity, and favors in cooperative relationships. *American Economic Journal: Microeconomics, 5*(2), 213–259.

Aleksandrov, M., & Walsh, T. (2017). Pure Nash equilibria in online fair division. In *Proceedings of the Twenty-Sixth International Joint Conference on Artificial Intelligence (IJCAI)*, Melbourne, Australia (pp. 42–48).

Aleksandrov, M., Aziz, H., Gaspers, S., & Walsh, T. (2015). Online fair division: Analysing a food bank problem. In *Proceedings of the Twenty-Fourth International Joint Conference on Artificial Intelligence (IJCAI)*, Buenos Aires, Argentina (pp. 2540–2546).

Amanatidis, G., Birmpas, G., Christodoulou, G., & Markakis, E. (2017). Truthful allocation mechanisms without payments: Characterization and implications on fairness. In *Proceedings of the Eighteenth ACM Conference on Economics and Computation (EC)*, Cambridge, MA, USA (pp. 545–562).

Athey, S., & Bagwell, K. (2001). Optimal collusion with private information. *RAND Journal of Economics, 32*(3), 428–465.

Babaioff, M., Mansour, Y., Nisan, N., Noti, G., Curino, C., Ganapathy, N., et al. (2017). ERA: A framework for economic resource allocation for the cloud. In *Proceedings of the 26th International Conference on World Wide Web Companion* (pp. 635–642).

Buchanan, J. M. (1965). An economic theory of clubs. *Economica, 32*(125), 1–14.

Budish, E., & Kessler, J. (2014). Changing the course allocation mechanism at Wharton. *Chicago Booth Research Paper, (15-08)*, 4.

Budish, E. (2011). The combinatorial assignment problem: Approximate competitive equilibrium from equal incomes. *Journal of Political Economy, 119*(6), 1061–1103.

Caragiannis, I., Kurokawa, D., Moulin, H., Procaccia, A. D., Shah, N., & Wang, J. (2016). The unreasonable fairness of maximum Nash welfare. In *Proceedings of the Seventeenth ACM Conference on Economics and Computation (EC)*, Maastricht, The Netherlands (pp. 305–322).

Chawla, S., Devanur, N., Kulkarni, J., & Niazadeh, R. (2017). Truth and regret in online scheduling. In *Proceedings of the Eighteenth ACM Conference on Economics and Computation (EC)*, Cambridge, MA, USA (pp. 423–440).

Cheung, Y. K. (2016). Better strategyproof mechanisms without payments or prior—an analytic approach. In *Proceedings of the Twenty-Fifth International Joint Conference on Artificial Intelligence (IJCAI)*, New York, NY, USA (pp. 194–200).

Cole, R., Gkatzelis, V., & Goel, G. (2013a). Mechanism design for fair division: allocating divisible items without payments. In *Proceedings of the Fourteenth ACM Conference on Electronic Commerce (EC)*, Saint Paul, MN, USA (pp. 251–268).

Cole, R., Gkatzelis, V., & Goel, G. (2013b). Positive results for mechanism design without money. In *Proceedings of the Twelfth International Conference on Autonomous Agents and Multi-agent Systems (AAMAS)*, Saint Paul, Minnesota, USA (pp. 1165–1166).

Conitzer, V., Freeman, R., & Shah, N. (2017). Fair public decision making. In *Proceedings of the 2017 ACM Conference on Economics and Computation (EC)*, Cambridge, MA, USA (pp. 629–646).

Danovitch, G. M., & Delmonico, F. L. (2008). The prohibition of kidney sales and organ markets should remain. *Current Opinion in Organ Transplantation, 13*(4), 386–394.

Davis, R. I., & Burns, A. (2011). A survey of hard real-time scheduling for multiprocessor systems. *ACM Computing Surveys (CSUR), 43*(4), 35.

Fain, B., Goel, A., & Munagala, K. (2016). The core of the participatory budgeting problem. In *Proceedings of the Twelfth International Conference on Web and Internet Economics (WINE)* (pp. 384–399).

Fain, B., Munagala, K., & Shah, N. (2018). Fair allocation of indivisible public goods. In *Proceedings of the 2018 ACM Conference on Economics and Computation (EC)*, Ithaca, NY, USA pp. 575–592.

Foley, D. K. (1970). Lindahl's solution and the core of an economy with public goods. *Econometrica: Journal of the Econometric Society*, 66–72.

Foley, D. K. (1967). Resource allocation and the public sector. *Yale Economics Essays, 7*, 45–98.

Freeman, R., Zahedi, & S. M., Conitzer, V., & Lee, B. C. (2018). Dynamic proportional sharing: A game-theoretic approach. *Proceedings of the ACM on Measurement and Analysis of Computing Systems, 2*(1), 3:1–3:36.

Freeman, R., Zahedi, S. M., & Conitzer, V. (2017). Fair and efficient social choice in dynamic settings. In *Proceedings of the Twenty Sixth International Joint Conference on Artificial Intelligence (IJCAI)*, Melbourne, Australia (pp. 4580–4587).

Friedman, E., Psomas, C.-A., & Vardi, S. (2017). Controlled dynamic fair division. In *Proceedings of the 2017 ACM Conference on Economics and Computation (EC)*, Cambridge, MA, USA (pp. 461–478).

Goldberg, A. V., & Hartline, J. D. (2005). Collusion-resistant mechanisms for single-parameter agents. In *Proceedings of the Sixteenth Annual ACM-SIAM Symposium on Discrete Algorithms (SODA)*, Vancouver, Canada (pp. 620–629).

Gorokh, A., Banerjee, S., & Iyer, K. (2016). Near-efficient allocation using artificial currency in repeated settings.

Gorokh, A., Banerjee, S., & Iyer, K. (2017). From monetary to non-monetary mechanism design via artificial currencies.

Green, J. R., & Laffont, J.-J. (1979). Incentives in public decision making.

Groves, T., & Ledyard, J. (1977). Optimal allocation of public goods: A solution to the "free rider" problem. *Econometrica: Journal of the Econometric Society, 783–809.

Guo, M., & Conitzer, V. (2010). Strategy-proof allocation of multiple items between two agents without payments or priors. In *Proceedings of the Ninth International Joint Conference on Autonomous Agents and Multi-agent Systems (AAMAS)*, Toronto, Canada (pp. 881–888).

Guo, M., Conitzer, V., & Reeves, D. M. (2009). Competitive repeated allocation without payments. In *Proceedings of the Fifth Workshop on Internet and Network Economics (WINE)*, Rome, Italy (pp. 244–255).

Han, L., Su, C., Tang, L., & Zhang, H. (2011). On strategy-proof allocation without payments or priors. In *Proceedings of the Seventh Workshop on Internet and Network Economics (WINE)*, Singapore (pp. 182–193).

Im, S., & Kulkarni, J. (2016). Fair online scheduling for selfish jobs on heterogeneous machines. In *Proceedings of the Twenty-Eighth ACM Symposium on Parallelism in Algorithms and Architectures (SPAA)*, Moneterey, CA, USA (pp. 185–194).

Kash, I., Procaccia, A. D., & Shah, N. (2014). No agent left behind: Dynamic fair division of multiple resources. *Journal of Artificial Intelligence Research, 51*, 579–603.

Kolen, A. W. J., Lenstra, J. K., Papadimitriou, C. H., & Spieksma, F. C. R. (2007). Interval scheduling: A survey. *Naval Research Logistics (NRL), 54*(5), 530–543.

Kunjir, M., Fain, B., Munagala, K., & Babu, S. (2017). ROBUS: Fair cache allocation for data-parallel workloads. In *Proceedings of the 2017 ACM International Conference on Management of Data (SIGMOD)*, Chicago, IL, USA (pp. 219–234).

Lalley, S. P., & Weyl, E. G. (2018a). Nash equilibria for quadratic voting. arXiv:1409.0264.

Lalley, S. P., & Weyl, E. G. (2018b). Quadratic voting: How mechanism design can radicalize democracy. *American Economic Association Papers and Proceedings, 1*(1).

Leyton-Brown, K., Shoham, Y., & Tennenholtz, M. (2000). Bidding clubs: institutionalized collusion in auctions. In *Proceedings of the Second ACM Conference on Electronic Commerce (EC)*, Minneapolis, MN, USA (pp. 253–259).

Leyton-Brown, K., Shoham, Y., & Tennenholtz, M. (2002). Bidding clubs in first-price auctions. In *Proceedings of the Eighteenth AAAI Conference on Artificial Intelligence (AAAI)*, Edmonton, Canada (pp. 373–378).

Lipton, R. J., & Tomkins, A. (1994). Online interval scheduling. In *Proceedings of the Fifth Annual ACM-SIAM Symposium on Discrete Algorithms (SODA)*, Arlington, VA, USA (Vol. 94, pp. 302–311).

Lipton, R. J., Markakis, E., Mossel, E., & Saberi, A. (2004). On approximately fair allocations of indivisible goods. In *Proceedings of the Fifth ACM Conference on Electronic Commerce (EC)*, New York, NY, USA (pp. 125–131). ACM.

Marshall, R. C., & Marx, L. M. (2007). Bidder collusion. *Journal of Economic Theory, 133*(1), 374–402.

Maya, A., & Nisan, N. (2012). Incentive compatible two player cake cutting. In *Proceedings of the Eighth Workshop on Internet and Network Economics (WINE)*, Liverpool, UK (pp. 170–183).

Moulin, H. (1994). Serial cost-sharing of excludable public goods. *The Review of Economic Studies, 61*(2), 305–325.

Nash, J. F. (1950). The bargaining problem. *Econometrica: Journal of the Econometric Society,* 155–162.

Penna, P., & Ventre, C. (2014). Optimal collusion-resistant mechanisms with verification. *Games and Economic Behavior, 86*, 491–509.

Posner, E. A., & Weyl, E. G. (2014). Quadratic voting as efficient corporate governance. *The University of Chicago Law Review, 81*(1), 251–272.

Preston, R. (1992). McAfee and John McMillan. Bidding rings. *The American Economic Review,* 579–599, 1992.

Pu, Q., Li, H., Zaharia, M., Ghodsi, A., & Stoica, I. (2016). FairRide: Near-optimal, fair cache sharing. In *Proceedings of the Thirteenth USENIX Symposium on Networked Systems Design and Implementation*, Santa Clara, CA, USA (pp. 393–406).

Roth, A. E. (2007). Repugnance as a constraint on markets. *The Journal of Economic Perspectives, 21*(3), 37–58.

Roth, A. E., Sönmez, T., & Utku Ünver, M. (2004). Kidney exchange. *The Quarterly Journal of Economics, 119*(2), 457–488.

Samuelson, P. A. (1954). The pure theory of public expenditure. *The Review of Economics and Statistics, 36*(4), 387–389.

Schummer, J. (2000). Manipulation through bribes. *Journal of Economic Theory, 91*(2), 180–198.

Sönmez, T., & Utku Ünver, M. (2010). Course bidding at business schools. *International Economic Review, 51*(1), 99–123.

Steinhaus, H. (1948). The problem of fair division. *Econometrica, 16*, 101–104.

Tian, Y. (2013). Strategy-proof and efficient offline interval scheduling and cake cutting. In *Proceedings of the Ninth International Conference on Web and Internet Economics (WINE)*, Cambridge, MA, USA (pp. 436–437).

Walsh, T. (2011). Online cake cutting. In *Proceedings of the Third International Conference on Algorithmic Decision Theory (ADT)*, Piscataway, NJ, USA (pp. 292–305).

Weyl, E. G. (2017). The robustness of quadratic voting. *Public Choice, 172*(1–2), 75–107.

Wu, Y., Wang, B., Liu, K. J. R., & Clancy, T. C. (2009). A scalable collusion-resistant multi-winner cognitive spectrum auction game. *IEEE Transactions on Communications, 57*(12).

# Beyond the Stand-Alone Core Conditions

Jens Leth Hougaard

**Abstract** The stand alone core conditions have played a key role in the fair allocation literature for decades and has been successfully applied in many types of models where agents share a common cost or revenue. Yet, the stand-alone core conditions are not indispensable when looking for fair ways to share. The present note provides a few examples of network models where the relevance of the stand-alone core is questionable and fairness seems to require a different approach. In a networked future, design of allocation mechanisms is therefore likely to move beyond the stand alone core.

## 1 Introduction

The stand-alone (core) conditions have played a key role in cost and surplus sharing even before the formulation of the core as a solution concept for cooperative games. For instance, such conditions were explicitly formulated by engineers involved in the Tennessee Valley project in US in the 1930s: a classic case in the cost sharing literature (Young 1994). But the principle has probably been used long before that in one way or the other. The intuition behind the stand-alone conditions is straightforward: say a group of agents are commonly liable for some cost, then the conditions state that no coalition of agents should cover more than their stand-alone cost (i.e., the cost the coalition would incur by cooperating with each other, but not with the remaining agents of the group).[1] This is equivalent to stating that no coalition of agents should subsidize the presence of the others. So allocations satisfying the stand-alone conditions sustain cooperation in the sense that no coalition of agents is better off on their own. From a social planner's viewpoint, this becomes crucial when cooperation is economically efficient.

---

[1]In case of surplus sharing the stand-alone conditions become lower bounds: no coalition of agents should get less surplus than what they could obtain standing alone.

---

J. L. Hougaard (✉)
IFRO, University of Copenhagen, Rolighedsvej 25, 1958 Frederiksberg C, Denmark
e-mail: jlh@ifro.ku.dk

© Springer Nature Switzerland AG 2019
J.-F. Laslier et al. (eds.), *The Future of Economic Design*, Studies in Economic Design, https://doi.org/10.1007/978-3-030-18050-8_39

Allocation problems related to common liability have traditionally been modeled as cooperative games, so the stand-alone conditions can be construed as the core conditions of the associated game (e.g., Young 1994; Moulin 1988, 2002; Hougaard 2009). A considerable amount of research effort has been directed towards finding classes of allocation problems for which the core of the associated cooperative game is non-empty. For such classes of problems it is natural to further characterize selections from the core. For instance, in case of agents sharing a common network resource, a leading research agenda since the 1970s has reinterpreted a series of classic problems from combinatorial optimization as cooperative games and analyzed properties of these games as well as variations of standard solutions concepts like the Shapley value and the Nucleolus (e.g., Sharkey 1995; Curiel 1997; Moulin 2013; Hougaard 2018). Many of these games turn out to be balanced (like minimum cost spanning tree games, Bird 1976) and some are even concave (like airport games, Littlechild and Thompson 1977) so the relevance of the stand-alone core conditions seems clear in such cases.[2] In fact, the stand-alone core conditions seem particularly compelling in situations where a planner has full knowledge of connection costs and there is a natural way to determine the liability of every coalition of agents. The main problem is therefore to identify an efficient network and allocate its cost such that the efficient network solution is sustained (and justified).

However, there are many important classes of games for which the core may be empty: these include, minimum cost Steiner tree games (Meggido 1978; Tamir 1991), congestion games (Quant et al. 2006), and traveling salesman games (Tamir 1989). Moreover, the coalitional liability is rarely uniquely specified. For instance, when agents within a given coalition connect to each other, are they allowed to connect via nodes controlled by agents outside the coalition? (in some cases this is possible, in others it is not, Granot and Huberman 1981); and what is the relevant cost of links that are not used in the realized network? (these costs remain hypothetical, and should they influence the way that realized costs are allocated? Bird 1976; Bergantinos and Vidal-Puga 2007). Furthermore, when networks are fixed (so we are not in a design phase trying to identify an optimal network structure) coalitions' stand-alone values are typically counterfactual: agents are part of an existing network, and hypothetical liabilities of coalitional sub-networks seem less important, at least in the short run. As such, there are many situations where the entire approach of cooperative game theory is questionable, not to mention the relevance of the stand-alone core conditions.

In the present paper we will therefore argue, with the help of a few examples, that there are plenty of good reasons to move beyond cooperative games and the associated stand-alone core conditions in many types of cost and surplus sharing problems. For many network problems the model of a cooperative game is simply not rich enough to capture all essential features of the allocation problem. Operational solutions that are more closely related to the network structure itself, with desirable normative

---

[2]See e.g., Peleg and Sudhölter (2007) for various properties of cooperative games as well as solution concepts.

properties may easily turn out to be more relevant than conventional game theoretic solution concepts, even in cases where the core is non-empty and the solution is in conflict with the stand-alone conditions.


## 2   Some Examples

A few examples will demonstrate some problems with the conventional approach of cooperative game theory. In particular, the stand-alone core conditions may force us to apply extreme allocations which in no way seem compelling, or justifiable.


### 2.1   Fixed Network, Different Connection Demands

In practice, existing networks often include redundant connections which nobody actually require, but nevertheless provide some users with with improved connectivity. The presence of such redundant connections will typically render the core empty and thereby the stand-alone conditions irrelevant (Hougaard and Moulin 2014). But, even in cases of non-redundancy the core may lead us to extreme allocations that seem hard to justify.

Consider a simple cycle with two nodes: one being a source, the other being the location of two agents, Ann and Bob. Ann demands 2-connectivity (i.e., needs two distinct connections to the source), while Bob is satisfied by any single connection to the source. Ann and Bob share the total cost of the network based on their demand for connectivity. The immediate solution seems to be that Ann should pay 2/3 of the cost while Bob should pay the remaining 1/3 based on the fact that Ann is twice as demanding as Bob.

Moulin and Laigret (2011) consider *non-redundant* problems like this one (where deleting any connection will lead to connectivity failure for at least one agent). They suggest a natural cost sharing rule, dubbed the Equal Needs rule, that splits the cost of every connection equally among users for which the connection is critical (i.e., users that will not obtain connectivity without the connection), and show that this rule is the only symmetric and cost additive rule that satisfies the stand-alone core conditions. Since cost additivity[3] (which basically implies that cost shares should be independent of the size of connection costs and only rely on the way users obtain service from the connections) is a well established axiom in the cost sharing literature, the result of Moulin and Laigret (2011) demonstrates that the stand-alone core conditions can force us towards extreme allocations. Indeed, in the simple example above, no connection is critical to Bob so using the Equal Needs rule Bob will pay nothing despite having assess to the source.

---

[3] An allocation rule $\phi$ satisfies cost additivity if, for any two cost vectors $c$ and $c'$, that $\phi(c + c') = \phi(c) + \phi(c')$.

## 2.2 *Max-Flow Problems*

Other examples relate to the class of max-flow problems. The max-flow problem is a classic in the Operations Research literature pioneered by the work of Ford and Fulkerson (1962): a directed graph has two distinguished nodes, a source and a sink; each arc has a maximum capacity; the aim is to direct as much flow as possible from source to sink given arc capacities.[4] Max-flow problems can be seen as models of integrated production, communication, or transportation systems, where the flow generates revenue: for instance, routing as many packets as possible through the network.

The allocation issue arises when agents control individual arcs and need to coordinate in order to obtain a (common) max-flow of, say, revenue, which is then subsequently distributed among the agents (Kalai and Zemel 1982a, b). Since agents have to coordinate their efforts to obtain a max-flow, it is seemingly natural to apply the tools of cooperative game theory. Indeed, for any coalition of agents we can define a coalition restricted max-flow problem by considering the sub-graph consisting of arcs controlled by coalition members only. In this way every coalition has a stand-alone value in terms of its restricted max-flow. Kalai and Zemel (1982a, b) show that the set of flow-games is equivalent to the set of totally balanced games, so at first glance the stand-alone core conditions should indeed be relevant here. In fact, finding a core allocation is simple: a *cut* is a set of arcs that reduces the max-flow to zero if deleted from the network; there exists a minimal cut with total capacity equal to the max-flow; individual capacities of a minimal cut constitutes a core selection of the flow-game. The question is whether such selections are compelling in terms of fairness (Hougaard 2018).

First we show that all allocations in the core may seem unfair. Consider a problem with four agents $\{A, B, C, D\}$ illustrated below, where e.g., "B,1" reads "agent B controls the arc with maximum capacity 1". Clearly, the max-flow is 2, and the unique minimal cut is marked by ($//$).

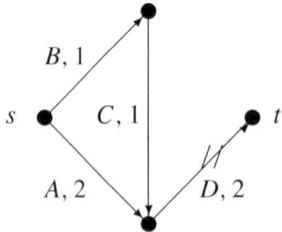

---

[4]Several algorithms exists for finding a max-flow in a given graph with given capacity constraints, see e.g., Kozen (1992).

The associated flow-game becomes:

$v(i) = 0$ for $i \in \{A, B, C, D\}$
$v(AD) = 2$, $v(ij) = 0$ otherwise,
$v(ABC) = 0$, $v(ABD) = v(ACD) = 2$, $v(BCD) = 1$,
$v(ABCD) = 2$.

The core is given by allocations;

$$\alpha(1, 0, 0, 1) + (1 - \alpha)(0, 0, 0, 2), \quad \alpha \in [0, 1]$$

guaranteeing 1 unit for agent $D$, and letting $A$ and $D$ share the remaining 1 unit, with nothing to agents $B$ and $C$. Using the minimal cut allocation we get the extreme core allocation $(0, 0, 0, 2)$. Yet, giving nothing to agents $B$ and $C$ is not compelling since one way to reach the max-flow of 2 would be to transfer 1 unit via $B$ and $C$, and 1 unit via $A$, adding up to 2 units via $D$'s arc. Indeed the Shapley value of the game (which is not a core allocation in this case) is given by,

$$x^{Sh}(v) = \left( \frac{3}{4}, \frac{1}{12}, \frac{1}{12}, \frac{13}{12} \right),$$

acknowledging that agents $B$ and $C$ ought to share the benefits, albeit with a significantly smaller amount than $A$ and $D$.

Moreover, note that if we raise the capacity of the arcs controlled by $B$ and $C$ from 1 to 2 units, we get a new flow-game $\bar{v}$ where $\bar{v}(BCD) = 2$ and $\bar{v}(S) = v(S)$ otherwise. Now, the core consists of a single allocation $(0, 0, 0, 2)$ where only $D$ receives payoff despite that fact that going from $v$ to $\bar{v}$ increases the "power" of both agents $B$ and $C$: the core reacts by an extreme punishment to $A$, but not by adding to the payoff of $B$ and $C$. Again we can compare with the more compelling Shapley value,

$$x^{Sh}(\bar{v}) = \left( \frac{1}{2}, \frac{1}{6}, \frac{1}{6}, \frac{7}{6} \right),$$

where $A$'s payoff is decreased from 0.75 to 0.5, while the payoff of the remaining agents have increased.

An arc is called *redundant* if deleting it does not change the max-flow. When the set of agents is identified by the set of arcs (i.e., each agent controls one and only one arc), an arc/agent is redundant if and only if its payoff is 0 in all core allocations (Sun and Fang 2007). In the latter part of the example above agents $A$, $B$ and $C$ are all redundant (and hence gets 0 payoff in the core), yet either $A$, or coalition $\{B, C\}$, is needed to obtain the max-flow of 2, so the core allocation is indeed extreme in this case. In other words, a requirement stating that an agent should receive a positive payoff if and only if his arc is part of *all* potential max-flows, is an axiom compatible with the core. But clearly this is much stronger than suggesting that an agent should receive a positive payoff if and only if there exists some max-flow for which his arc is involved, as indicated above for agents $B$ and $C$.

The next example shows that the cooperative game itself may not represent all issues that are crucial for fair allocation. Consider the following max-flow problem with two agents Ann and Bob.

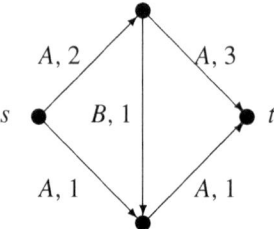

Clearly, Bob's arc can never be part of a max-flow and the associated flow-game becomes;

$v(A) = 3$, $v(B) = 0$, $v(AB) = 3$.

Now, notice that if we change the direction of Bob's arc, it can now be part of a max-flow, but the associated flow-game remains unchanged. In other words, no game theoretic solution concept can capture the change in Bob's role in the network because the associated game is too crude a representation of the situation at hand.

The final example shows that game theoretic solution concepts (here the Shapley value) may reward agents for having completely useless arcs in terms of the max-flow as in the graph below.

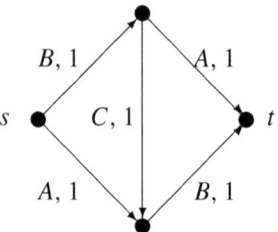

Now, there are three agents, Ann, Bob and Carl. When there are no reliability issues, Carl's arc is useless since it will never be part of a max-flow. Yet, because it may be used in a sub-optimal flow together with Bob's arcs, the associated flow-game becomes:

$v(A) = v(B) = v(C) = 0$, $v(AB) = 2, v(AC) = 0, v(BC) = 1$, $v(ABC) = 2$,

with Shapley value $(\frac{2}{3}, \frac{7}{6}, \frac{1}{6})$. So despite Carl's arc being useless he is rewarded by the Shapley value because he has one positive marginal value $v(BC) - v(B) = 1$. A much more compelling solution would be for Ann and Bob to share equally $(\frac{1}{2}, \frac{1}{2}, 0)$ in this case.

There are several alternatives to viewing the situation as a (cooperative) flow-game. Indeed, we may for example characterize each arc (agent) by some index of importance for obtaining the max-flow. Straightforward ideas include: the flow that necessarily must pass through the arc in any max-flow; or the maximum flow in any max-flow; or a weighted average of those. The max-flow can than be allocated based on such indices, e.g., proportionally.

# 3   So What?

As stressed by Moulin (2013) an empty core is no serious obstacle for the fair division approach. Here this point is taken a step further: when designing allocation schemes, the examples above have hopefully demonstrated that it is not only reasonable, but also sometimes necessary, to go beyond the stand-alone core conditions, even in cases where the core is non-empty.

As illustrated in case of max-flow problems, the model of a cooperative game is often not rich enough to capture all aspects that seems central to the way that liabilities ought to be shared (the game can be insensitive to the direction of given arcs, but the direction plays a crucial role in obtaining a max-flow and thereby creating value for the group as a whole).[5]

In a recent paper, Hougaard and Moulin (2018) study yet another aspect of this by considering limited reliability of connections: that is, connections in the network may fail at random. It is rather straightforward to define a cooperative game based on expected stand-alone costs for every coalition, but such a game does not capture that the issue of fairness becomes considerably more complicated by adding uncertainty to the problem. Now, it may matter for our conception of fairness whether we consider the situation before, or after, the resolution of uncertainties, and the relation between flexibility in demand and liability in payment becomes ambiguous: higher flexibility should arguably result in lower payment, but the more flexible agent also has a bigger chance of being served which should make her more liable, so how does allocation rules balance these opposite effects. Hougaard and Moulin (2018) suggests (and characterize) several allocation rules accounting for such effects: none of these rules guarantee core allocations (when the core is non-empty). So again the richness introduced by adding limited reliability to the model is not captured by the game in a way appropriate for fair allocation.

The future of the fair allocation approach is therefore likely to go beyond cooperative games and the stand-alone core.

**Acknowledgements** Constructive comments from Justin Leroux and Christian Trudeau are gratefully acknowledged.

---

[5]In the related context of network formation games, Jackson (2005) points at similar types of problems because a cooperative game cannot account for the fact that the same group of agents may connect in different network configurations, and these may have different value for the group.

# References

Bergantinos, G., & Vidal-Puga, J. J. (2007). A fair rule in minimum cost spanning tree problems. *Journal of Economic Theory, 137*, 326–352.

Bird, C. G. (1976). On cost allocation for a spanning tree: A game theoretic approach. *Networks, 6*, 335–350.

Curiel, I. (1997). *Cooperative game theory and applications: Cooperative games arising from combinatorial optimization problems*. Kluwer Academic Publishers.

Ford, L. R., & Fulkerson, D. R. (1962). *Flows in networks*. Princeton University Press.

Granot, D., & Huberman, G. (1981). Minimum cost spanning tree games. *Mathematical Programming, 21*, 1–18.

Hougaard, J. L. (2009). *An introduction to allocation rules*. Springer.

Hougaard, J. L. (2018). *Allocation in networks*. MIT Press.

Hougaard, J. L., & Moulin, H. (2014). Sharing the cost of redundant items. *Games and Economic Behavior, 87*, 339–352.

Hougaard, J. L., & Moulin, H. (2018). Sharing the cost of risky projects. *Economic Theory, 65*, 663–679.

Jackson, M. O. (2005). Allocation rules for network games. *Games and Economic Behavior, 51*, 128–154.

Kalai, E., & Zemel, E. (1982a). Totally balanced games and games of flow. *Mathematics of Operations Research, 7*, 476–478.

Kalai, E., & Zemel, E. (1982b). Generalized network problems yielding totally balanced games. *Operations Research, 30*, 998–1008.

Kozen, D. C. (1992). *The design and analysis of algorithms*. Springer.

Littlechild and Thompson. (1977). Aircraft landing fees: A game theoretic approach. *Bell Journal of Economics, 8*, 186–204.

Megiddo, N. (1978). Cost allocation for Steiner trees. *Networks, 8*, 1–6.

Moulin, H. (1988). *Axioms of cooperative decision making*. Cambridge University Press.

Moulin, H. (2002). Axiomatic cost and surplus sharing: In Arrow et al. (Eds.), *Handbook of social choice and welfare* (pp. 289–357).

Moulin, H. (2013). Cost sharing in networks: Some open questions. *International Game Theory Review, 15*, 1340001.

Moulin, H., & Laigret, F. (2011). Equal-need sharing of a network under connectivity constraints. *Games and Economic Behavior, 72*, 314–320.

Peleg, B., & Sudhölter, P. (2007). *Introduction to the theory of cooperative games* (2nd ed.). Kluwer.

Quant, M., Borm, P., & Reijnierse, H. (2006). Congestion network problems and related games. *European Journal of Operational Research, 172*, 919–930.

Sharkey, W. W. (1995). Network models in economics. In Ball et al. (Eds.), *Handbooks in operations research and management science* (pp. 713–765).

Sun, X., & Fang, Q. (2007). Core stability and flow games. In J. Akiyama, et al. (Eds.), *Discrete geometry, combinatorics and graph theory*. Lecture notes in computer science. Springer.

Tamir, A. (1989). On the core of the traveling salesman cost allocation game. *Operations Research Letters, 8*, 31–34.

Tamir, A. (1991). On the core of network synthesis games. *Mathematical Programming, 50*, 123–135.

Young, H. P. (1994). *Equity*. Princeton University Press.

# On "Going Unstructured" in Bargaining Experiments

Emin Karagözoğlu

**Abstract** This chapter focuses on the role of experiments in guiding theorists and policymakers working on design issues in negotiations. More specifically, our focus will be on unstructured bargaining experiments. I argue that the experimental research on bargaining should gradually shift focus from highly structured and simplified bargaining experiments to more unstructured bargaining experiments. Our supporting arguments will be based on experimental findings from bargaining and surplus sharing experiments, theoretical developments in bargaining research and behavioral economics, computerization of economic experiments, improved ability to conduct online experiments with large subject pools, and the recent developments in data science and econometrics.

Bargaining is ubiquitous. Buyer-seller negotiations, labor-management negotiations, bargaining over profits in a partnership, nuclear disarmament negotiations, bargaining over the allocation of ministries in coalition-building attempts between political parties, negotiations between supranational institutions (e.g., IMF, NATO, United Nations) and their member countries, climate change negotiations, hostage/ransom negotiations, kids negotiating with each other about who will play with a certain toy and for how long, merger and acquisition negotiations, litigation negotiations, and the list goes on. Schelling (1960) argued that "… to study the strategy of conflict is to take the view that most conflict situations are essentially bargaining situations" (Schelling 1960, p. 5). Similarly, social psychologists consider bargaining as one of the most essential social interactions (see Rubin and Brown 1975). This wide-spread presence in almost every areas of life makes institutional, procedural, or contractual details in bargaining interactions very important, as they potentially have significant efficiency and equity implications. Crawford (1982) emphasized the importance of bargaining research as follows: "… the potential welfare gains from improving the efficiency of bargaining outcomes are enormous, perhaps even greater than those that would result from a better understanding of macroeconomic policy". Along these

E. Karagözoğlu (✉)
Department of Economics, Bilkent University, 06800 Çankaya, Ankara, Turkey
e-mail: karagozoglu@bilkent.edu.tr

CESifo, Munich, Germany

© Springer Nature Switzerland AG 2019
J.-F. Laslier et al. (eds.), *The Future of Economic Design*, Studies in Economic Design, https://doi.org/10.1007/978-3-030-18050-8_40

lines, some natural questions that are potentially interesting from an institutional or contractual design perspective are:

– What is the influence of power (im)balance on the nature of agreements, and the likelihood of delays and disagreements?
– How should the roles (initiator, finalizer, proposer, responder etc.) be assigned to bargainers to achieve efficiency and/or fairness?
– Which characteristics of arbitrators or arbitration mechanisms are more conducive to reaching agreements?
– Do different contracts or bargaining protocols lead to different disagreement rates? or What happens if the bargaining parties find the procedures unfair?
– Should negotiations take place in private or in front of third parties?
– How should the production and distribution phases be integrated to maximize efficiency in bargaining problems with jointly produced surpluses?
– What is the nature of concession-making behavior? Can we design bargaining protocols that enhance mutual concessions and fast(er) agreements?
– What is the influence of communication channels on bargaining outcomes, such as frequency of (dis)agreements?

These questions and alike attracted a great deal of attention from bargaining, fair division, and mechanism design scholars for many years (some recent examples are Frechette et al. 2003, 2005; Yıldırım 2010; Rode and Le Menestrel 2011; Charness 2012; Karagözoğlu 2012; Anbarcı and Feltovich 2013, 2018; Eraslan and Merlo 2017; Bolton and Karagözoğlu 2016; Gantner and Kerschbamer 2016; Rodriguez-Lara 2016; Baranski 2017; Herbst et al. 2017; Kamm and Houba 2017).

Controlled economic experiments (lab and field), as much as theory, is instrumental in guiding institutional/contractual design (Thaler and Sunstein 2009; Falk and Heckman 2009; Riedl 2010; Bolton and Ockenfels 2012; Roth 2016). Therefore, it is important to conduct experiments that mimic the important aspects of corresponding real-world circumstances of interest and provide correct insights regarding what we should expect to observe. More than fifty years of experimental work shows that complementing theoretical work with experiments has a better chance of delivering *better* models and *better* policies/institutions (see Falk and Heckman 2009; Riedl 2010; Bolton and Ockenfels 2012; Bierbrauer et al. 2017).

This chapter focuses on the role of experiments in guiding theorists and policymakers working on design problems in negotiations. More specifically, our focus will be on unstructured bargaining experiments. In an unstructured bargaining experiment, most of the time, the experimenter only fixes the time frame (i.e., how long do the bargainers have to reach an agreement) and the disagreement payoffs. Hence, elements such as "Who makes the opening offer?", "Who concedes and when?", "Do bargainers make offers one after another?", "Who seals the deal by accepting the others' offer?", "What kind of strategies are used by bargainers?" are all endogenously determined. A s such, each of these becomes potential research questions.

I argue that the experimental research on bargaining should gradually shift focus from highly structured and simplified bargaining experiments to more unstructured bargaining experiments.[1] Our supporting arguments will be based on experimental findings from bargaining and surplus sharing experiments, theoretical developments in bargaining research and behavioral economics, computerization of economic experiments, improved ability to conduct online experiments with large subject pools, and the recent developments in data science and econometrics.

# 1  Historical Background

Nash's seminal work where he introduced the axiomatic approach to study bargaining problems (Nash 1950) had—legitimately—a long-lasting impact on the bargaining research. The axiomatic model focuses on the outcomes implied by appealing normative properties (i.e., axioms) that bargaining solutions satisfy. One interpretation for axiomatic bargaining solutions, sometimes entertained by bargaining scholars (first by Nash, himself), is as follows: an outcome proposed by an axiomatic solution concept describes (or represents) the outcome of a negotiation, where parties had, practically unlimited opportunities to discuss and bargain. Accordingly, this line of thinking did not impose restrictions on the structure of the (exogenously given) bargaining protocol or the (endogenously emerging) bargaining process. Since it was not modeled explicitly, the bargaining process was—more or less—left in the dark.

Another influential work on surplus sharing problems, around the same time was Shapley's. Shapley (1953) introduced a solution concept, named after him as the *Shapley Value*, for cooperative games. More precisely, the Shapley value assigns each cooperative game an outcome (i.e., a distribution of surplus). Given its reliance on the common-sense notion of *marginal contributions* and the set of appealing properties it satisfied, the Shapley value became a very popular solution concept in the study of cooperative bargaining games and distribution problems. Like the Nash bargaining solution, the Shapley value did not explicitly model the process that leads to the outcome it proposes.

Pretty much until early 1980s, axiomatic and cooperative approaches to bargaining dominated the field; and this—naturally—influenced the type of experimental research conducted on bargaining. Consequently, most of the bargaining experiments conducted during these three decades were unstructured bargaining experiments or characteristic function game experiments. Some well-known examples are Nydegger and Owen (1975), Rapoport et al. (1977), Murnighan and Roth (1979), and Roth and Malouf (1979).[2] Experimental studies in this era mostly aimed at testing the empirical validity of the Nash bargaining solution (or alternatives such as the Kalai-Smorodinsky solution, the egalitarian solution etc.), frequently used axioms in the

---

[1]Güth (2012) also mentions the revival of unstructured bargaining experiments as one of the future directions.

[2]It can be argued that Chamberlin (1948) and Smith (1962) were also early examples of unstructured bargaining experiments. The former utilized bilateral bargaining to study the efficiency of markets and the latter utilized double auctions for the same reason. We do not describe them in detail here since they were not directly connected to the theoretical literature on bargaining.

axiomatic study of bargaining (e.g., Pareto optimality, symmetry, scale invariance, contraction independence, individual monotonicity, etc.), or the Shapley value, in simple experimental setups. Possibly, since the bargaining process was not the focus of the prominent theoretical models, it did not have a priority in the experimental work either.[3]

It would not be an exaggeration to say that the year 1982 witnessed two of the most influential articles on bargaining. Rubinstein's (noncooperative game) theoretical analysis of the alternating-offers bargaining game and Güth, Schmittberger, and Schwarze's experimental investigation of the ultimatum game.[4]

Rubinstein studied a two-player, inifinite horizon bargaining game where every period, one player makes a proposal and the other responds (accept or reject), and the players' roles alternate every period. This is a well-defined extensive-form game, and thanks to the developments in game theory in 1970s—more precisely, the subgame perfect Nash equilibrium in this case—its equilibrium could be characterized. A striking finding is that, in the corresponding equilibrium, under standard rationality assumptions, the proposer in the first period makes an offer that the responder prefers not to reject, and the bargaining game ends with an immediate agreement. Hence, from an ex-ante perspective the game studied had the potential to offer insights on the bargaining process, but the unique equilibrium did not leave much room for that since there was only one offer and no give-and-take, no haggling etc.

Güth and co-authors, on the other hand, introduced and experimentally studied possibly one of the simplest extensive-form bargaining games. The results of this study led to a plethora of research, which later produced new lines of research such as social or other-regarding preferences. However, for this chapter, the importance of Güth and co-authors' work is its simplicity. If bargaining is a movie, the ultimatum game is possibly one scene.

The alternating offers bargaining game and the ultimatum game soon became the most popular models in bargaining theory and bargaining experiments literatures, respectively. Since there were game theoretical models where strategic considerations, actions, and the process were explicitly incorporated, unstructured bargaining experiments went out of fashion. It was extremely complicated—almost impossible—to model the highly complex strategy spaces present in such bargaining games. Noncooperative game theory was experiencing significant progress, and the profession favored bargaining games that could be formally modeled as a strategic game and for which, equilibrium predictions could be given.

---

[3]Some exceptions are studies by Roth and coauthors, who also analyzed process variables and the work of social psychologists who were always interested in the process.

[4]Ståhl (1972) introduced the alternating offers bargaining model but did not provide an equilibrium characterization, which is possibly why Rubinstein's work made a larger impact.

## 2   Economists' Outcome-Oriented Mindset

Another factor that contributed to the rise of highly structured bargaining experiments (e.g., ultimatum game, Nash demand game, finite horizon alternating-offers bargaining game experiments) was economists' primary focus on outcomes rather than processes. The economics profession, traditionally, adopts an outcome-oriented mindset. This can be traced back to Friedman's "instrumentalism" (Friedman 1953), or even to the predominantly used inductive methodology for reaching general/universal theories in economics. It is fair to say that economists do not completely deny the fact that the process may matter, but it's rather that given the potential complexity (or sometimes even the impossibility) of modelling the process, and their primary interest in outcomes over processes, the latter is usually sacrificed. For instance, game theorists acknowledge the fact that fine details of a game possibly matter; but given that those fine details are not observable in many situations (or even if they are observable, models incorporating them may not be tractable), they focus on models/theories that provide predictions that are as valid as possible for a variety of processes.

Being one of the most studied topics in game theory and economics, bargaining—not surprisingly—was not an exception. The axiomatic model was, by construction, outcome-oriented. Nash's strategic model of bargaining (Nash 1953) captured the essential characteristics of bargaining situations (i.e., aligned interests on reaching an agreement and conflicting interests on the agreement to be reached). Furthermore, Binmore (2007) argued that the Nash's strategic bargaining game can be thought as describing a situation where players who can commit to their demands are rushing to place their demand first on the table; and Skyrms (1996) interpreted it as the "bottom line" stage of bargaining after all haggling was done. Nonetheless, it was a static model without much detail or depth regarding the bargaining process. Later, Ståhl and Rubinstein explicitly modeled bargaining process as an alternating offer exchanges. Nevertheless, the equilibrium of that game did not have much to say about the process. Finally, Güth and co-authors' ultimatum game was a major simplification of real-world bargaining situations. Consequently, when unstructured bargaining experiments were more popular, the process data was not completely exploited, and when non-cooperative game theory dominated the field, they became unpopular.

## 3   Why Should We Care About the Bargaining Process More?

I Argue that There Are at Least Two Reasons to Care About the Bargaining Process

1. "Why" or "how" questions should also be of interest to economists. Hence, even when two different processes lead to the same outcome, we should be investigating these processes and understand why they lead to the same outcome.

Otherwise, we're leaving a big black box behind us; and as a result, our understanding of human behavior is rather incomplete.

2. Experimental work on bargaining as well as fair division consistently show that the process matters a lot (Farrell and Gibbons 1989; Sebenius 1992; Gächter and Riedl 2005; Charness 2012; Güth 2012; Crusius et al. 2012; Karagözoğlu and Riedl 2015; Bolton and Karagözoğlu 2016; Camerer et al. 2018). Different processes/procedures/mechanisms can indeed lead to drastically different outcomes. Moreover, people may have strong preferences even among different procedures that lead to the same physical/material outcomes (e.g., procedural justice).

From an applied/policy-oriented perspective, the second reason is possibly more relevant. For instance, when endogenous sequencing in offer-making, strategic waiting between offers, verbal messages sent to provide supporting arguments for offers made or to reduce the social distance between bargainers (e.g., chit chat), etc. are important in explaining bargaining outcomes, institutions that are designed without taking these into consideration, based on simple, structured models of bargaining may not offer desirable outcomes. The experimental literature on bargaining games provides evidence that bargaining behavior and outcomes differ across structured and unstructured bargaining games. For instance, Roth (1985, 1987) and Forsythe et al. (1991) report that unstructured bargaining institution facilitate bargainers' focus on a small number of focal points (induced by fairness judgments) and make them more likely to emerge as agreement outcomes. Charness (2012) argues that communication typically leads to socially more efficient agreements in bargaining environments. Bolton and Karagözoğlu (2016) report differences between an unstructured bargaining game and an ultimatum game in that the soft power brought by asymmetric reference points is more completely exploited in the ultimatum game. Anbarcı and Feltovich (2018) report that bargainers' behavior more fully reflect changes in disagreement points in the unstructured bargaining institution than in the Nash demand game.

## 4    What Can the Data from Unstructured Bargaining Experiments Tell Us?

As I mentioned at the beginning, the unstructured protocol, which closely resembles many real-world bargaining situations provide a very rich data set to the researcher. For instance, the researcher can now study the identities of the bargainers who make the opening proposals and whether that matters or not, when is the opening proposal made, what is the time difference between the first proposals of two bargainers, how much time do bargainers spend to make an offer or a counter offer, are the concessions reciprocal, are there turning points (is it monotonic convergence to an agreement or are oscillations present), which verbal messages make numerical offers more acceptable, what is the role of persuasiveness, how does the power imbalances influence the bargaining atmosphere, which "type" of bargaining processes lead to "which" type of bargaining outcomes, which robust negotiation styles emerge, how does the social distance evolve in the course of negotiation, and what is its effect on the bargaining outcome, and so on. The answers to these questions have implications for the design of negotiation institutions/rules.

## 5   Favorable Developments

All the arguments for more unstructured bargaining experiments I mentioned until now are *demand-side* arguments (i.e., we need to know more about the bargaining process). On the other hand, there are favorable developments on the *supply-side* as well. In particular, developments that may lead to a revival of unstructured bargaining experiments are:

- *Developments in theory:* We have more sophisticated theoretical models, which are equipped with behavioral regularities compared to two-three decades ago. They allow us to test hypotheses regarding bargaining process. We have more theoretical models of bargaining with protocol-free predictions as well, which can deliver predictions about timing of concessions, timing of agreements, and disagreements, and as such they can act as benchmark models to use in experiments. These developments in theory make it easier for us to resort to more unstructured bargaining experiments.
- *Ability to run large-scale, online experiments:* One potential disadvantage of the unstructured bargaining data is noise. There are potentially many factors in action, which requires relatively larger data sets for clean statistical/econometric analysis. That said, compared to 10-20 years ago, a fast and economic access to large subject pools is much more possible thanks to online platforms such as Amazon's M-Turk. Nowadays, one can collect data from hundreds of subjects in an hour or two, something almost completely impossible back in the 1990s.
- *Computerized experiments and anonymity:* Back in 70s and 80s (and part of 90s), most experiments were conducted as pen-and-paper experiments; and it was very difficult to keep anonymity. Most unstructured bargaining experiments at the time were face-to-face, which lead to disproportionately many 50-50 agreements. The dominant prevalence of 50-50 agreements due to lack of anonymity and further practical problems (e.g., someone should carry and deliver offers and responses between players) prevents a fruitful/effective use of unstructured bargaining protocols to study bargaining behavior. The computerization of experiments and the development of software for running economic experiments helped researchers to keep anonymity in their experiments, easily implement the unstructured bargaining protocol (without needing someone to carry offers and responses for each bargaining pair) and, arguably, make these experiments more interesting/attractive.
- *Large data sets from online trading platforms:* A while ago, e-bay introduced an alternative platform to its usual auction format: best-offers. Best-offer platform is essentially a bargaining platform. Millions of exchanges are made via the best-offer platform (Backus et al. 2018). Although the bargaining protocol is fairly

structured, it still shares some common aspects with more unstructured protocols such as the option to send text messages and flexible offer-making time horizon. Platforms like this offer a great opportunity to bargaining scholars to study the influence of various process variables on bargaining outcomes.

– *Developments in Data Science and Econometrics:* As one may easily notice, by browsing some pages of a scientific magazine, newspaper, or an academic journal, we are going through what's called "Big Data Revolution". Data analysis toolkits in data science, statistics, information theory, and econometrics and—may be more importantly—computing technology have developed immensely, which allow researchers to handle extremely large data sets from the analysis of which we can learn a lot about human behavior. In the age of big data, investigating bargaining process is a natural way to go (see Camerer et al. 2018 for the first study in that direction).

# 6 Recap

To summarize: Simple, compact bargaining games served the profession greatly and the results they produced led to major developments in theory building and policy making. However, it could be that we already learnt the maximum we can from them. I think that it is time to *go unstructured* in conducting bargaining experiments. It is the time for the revival of unstructured bargaining experiments after long decades. They more closely resemble real-world bargaining encounters. That is not a necessarily sufficient reason alone, but we accumulated enough evidence that the process does matter a lot in many circumstances. We developed behavioral theories that can capture some of these observations. We also have sophisticated theoretical models that produce protocol-free predictions, which can constitute null hypotheses for experimental research. We have much more opportunities to reach big data sets compared to 90s (or even early 2000s), and we have the econometric toolbox and the computing technology to analyze them. The benefits of understanding the bargaining process is greater than ever and costs are much less than they were in the past.

**Acknowledgements** I would like to thank Nick Feltovich, Werner Güth, and Jean-François Laslier for helpful comments.

# References

Anbarcı, N., & Feltovich, N. (2013). How sensitive are bargaining outcomes to changes in disagreement payoffs? *Experimental Economics, 16,* 560–596.

Anbarcı, N., & Feltovich, N. (2018). How fully do people exploit their bargaining position? The effects of bargaining institution and the 50-50 norm. *Journal of Economic Behavior & Organization, 145,* 320–334.

Backus, M., Blake, T., Larsen, B., & Tadelis, S. (2018) Sequential bargaining in the field: Evidence from millions of online bargaining interactions. *NBER Working Paper #24306*.

Baranski, A. (2017). Redistribution and pre-distribution: An experimental study on the timing of profit-sharing negotiations and production. *Maastricht University Working Paper*.

Bierbrauer, F., Ockenfels, A., Pollak, A., & Rückert, D. (2017). Robust mechanism design and social preferences. *Journal of Public Economics, 149*, 59–80.

Binmore, K. (2007). *Playing for real: A text on game theory*. Oxford: Oxford University Press.

Bolton, G. E., & Karagözoğlu, E. (2016). On the influence of hard leverage in a soft leverage bargaining game: The importance of credible claims. *Games and Economic Behavior, 99*, 164–179.

Bolton, G. E., & Ockenfels, A. (2012). Behavioral economic engineering. *Journal of Economic Psychology, 33*, 665–676.

Camerer, C., Nave, G., & Smith, A. (2018) Dynamic unstructured bargaining with private information: theory, experiment, and outcome prediction via machine learning. *Management Science*, forthcoming.

Chamberlin, E. (1948). An experimental imperfect market. *Journal of Political Economy, 56*, 95–108.

Charness, G. (2012). Communication in bargaining experiments. In G. Bolton & R. Croson (Eds.), *The oxford handbook of economic conflict resolution*, Chap. 2 (pp. 7–18). New York: Oxford University Press.

Crawford, V. P. (1982). A theory of disagreement in bargaining. *Econometrica, 50*, 607–638.

Crusius, J., van Horen, F., & Mussweiler, T. (2012). Why process matters: A social cognition perspective on economic behavior. *Journal of Economic Psychology, 33*, 677–685.

Eraslan, H., & Merlo, A. (2017). Some unpleasant bargaining arithmetic? *Journal of Economic Theory, 171*, 293–315.

Falk, A., & Heckman, J. (2009). Lab experiments are a major source of knowledge in the social sciences. *Science, 326*, 535–538.

Farrell, J., & Gibbons, R. (1989). Cheap talk can matter in bargaining. *Journal of Economic Theory, 48*, 221–237.

Frechette, G., Kagel, J., & Lehrer, S. (2003). Bargaining in legislatures: An experimental investigation of open versus closed amendment rules. *American Political Science Review, 97*, 221–232.

Frechette, G., Kagel, J., & Morelli, M. (2005). Nominal bargaining power, selection protocol, and discounting in legislative bargaining. *Journal of Public Economics, 89*, 1497–1517.

Friedman, M. (1953). The methodology of positive economics. In M. Friedman (Eds.), *Essays in positive economics*. Chicago: University of Chicago Press.

Forsythe, R., Kennan, J., & Sopher, B. (1991). An experimental analysis of strikes in bargaining games with one-sided private information. *American Economic Review, 81*, 253–278.

Gantner, A., & Kerschbamer, R. (2016). Fairness and efficiency in a subjective claims problem. *Journal of Economic Behavior & Organization, 131*, 21–36.

Gächter, S., & Riedl, A. (2005). Moral property rights in bargaining with infeasible claims. *Management Science, 51*, 249–263.

Güth, W. (2012). Bargaining and negotiations: What should experimentalists explore more thoroughly? In G. Bolton & R. Croson (Eds.), *The oxford handbook of economic conflict resolution*, Chap. 17 (pp. 241–253). New York: Oxford University Press.

Güth, W., Schmittberger, R., & Schwarze, B. (1982). An experimental analysis of ultimatum bargaining. *Journal of Economic Behaviour and Organization, 3*, 367–388.

Harsanyi, J. C. (1956). Approaches to the bargaining problem before and after the Theory of Games: a critical discussion of Zeuthen's, Hicks', and Nash's theories. *Econometrica, 24*, 144–157.

Herbst, L., Konrad, K. A., & Morath, F. (2017). Balance of power and the propensity of conflict. *Games and Economic Behavior, 103*, 168–184.

Houba, H. & Kamm, A. (2017) A bargaining experiment with asymmetric institutions and preferences. *Tinbergen Institute Discussion Paper*.

Karagözoğlu, E. (2012). Bargaining games with joint production. In G. Bolton & R. Croson (Eds.), *The oxford handbook of economic conflict resolution*, Chap. 24 (pp. 359–372). New York: Oxford University Press.

Karagözoğlu, E., & Riedl, A. (2015). Performance information, production uncertainty, and subjective entitlements in bargaining. *Management Science, 61,* 2611–2626.

Murnighan, J. K., & Roth, A. E. (1979). Large group bargaining in a characteristic function game. *Journal of Conflict Resolution, 22,* 299–317.

Nash, J. F. (1950). The bargaining problem. *Econometrica, 18,* 155–162.

Nash, J. F. (1953). Two-person cooperative games. *Econometrica, 21,* 128–140.

Nydegger, R., & Owen, G. (1975). Two person bargaining: An experimental test of Nash axioms. *International Journal of Game Theory, 3,* 239–249.

Rapoport, A., Frenkel, O., & Perner, J. (1977). Experiments with cooperative $2 \times 2$ games. *Theory and Decision, 8,* 67–92.

Riedl, A. (2010). Behavioral and experimental economics do inform public policy. *FinanzArchiv: Public Finance Analysis, 66,* 65–95.

Rode, J., & Le Menestrel, M. (2011). The influence of decision power on distributive fairness. *Journal of Economic Behavior and Organization, 79,* 246–265.

Rodriguez-Lara, I. (2016). Equity and bargaining power in ultimatum games. *Journal of Economic Behavior & Organization, 130,* 144–165.

Roth, A. E. (1985). Toward a focal point theory of bargaining. In A. E. Roth (Ed.), *Game Theoretical Models of Bargaining* (pp. 259–268). Cambridge, UK: Cambridge University Press.

Roth, A. E. (1987). Bargaining phenomena and bargaining theory. In A. E. Roth (Ed.), *Laboratory experimentation in economics: six points of view* (pp. 14–41). Cambridge, UK: Cambridge University Press.

Roth, A. E. (2016). Experiments in market design. In J. H. Kagel & A. E. Roth *The handbook of experimental economics* (Vol. 2, pp. 290–346). Princeton: Princeton University Press

Roth, A. E., & Malouf, M. K. (1979). Game theoretic models and the role of information in bargining. *Pscyhological Review, 86,* 574–594.

Rubin, J. Z., & Brown, B. R. (1975). *The social psychology of bargaining and negotiation*. New York: Academic.

Rubinstein, A. (1982). Perfect equilibrium in a bargaining model. *Econometrica, 50,* 97–110.

Schelling, T. (1960). *The strategy of conflict*. Cambridge, MA: Harvard University Press.

Sebenius, J. K. (1992). Negotiation analysis: A characterization and review. *Management Science, 38,* 18–38.

Shapley, L. (1953). A value for n-person games. In H. W. Kuhn & W. A. Tucker (Eds.), *Contributions to the theory of games*, vol. II. Annals of Mathematical Studies (Vol. 28, pp. 307–317). Princeton University Press, Princeton, New Jersey.

Skyrms, B. (1996). *Evolution of the social contract*. Cambridge, UK: Cambridge University Press.

Smith, V. (1962). An experimental study of competitive market behavior. *Journal of Political Economy, 70,* 111–137.

Ståhl, I. (1972). *Bargaining Theory*. Stockholm School of Economics: Mimeo.

Thaler, R. H., & Sunstein, C. R. (2009). *Nudge: Improving decisions about health, wealth, and happiness*. London: Penguin Books.

Yıldırım, H. (2010). Distribution of surplus in sequential bargaining with endogenous recognition. *Public Choice, 142,* 41–57.

# When Are Operations Research Algorithms Useful in Fair Division Problems?

Christian Trudeau

**Abstract** In many fair division problems, there is an underlying operations research problem to determine how to organize the coalition of agents and compute the value created or the cost generated by the various groups. Through examples, this chapter discusses the possibility of using the algorithms used in the underlying operations research problem to answer fair division questions.

Many interesting economic problems come from classic operations research problems. In many non-transferable utility problems, for instance the stable marriage problem, the two problems are basically the same: finding stable matches is an operations research problem, and the famous deferred acceptance algorithm solves the problem. But in transferable utility problems, there is usually a tension between the general objectives of the OR problem (typically setting up the project so that it creates as much value as possible) and the objectives of the fair division problem. For example, matching, queueing and source connection problems are well-studied as operations research problems, and the focus is on how to efficiently organize the agents. The questions of interest to economists, for instance constructing transfers among agents or dividing the value created, usually come after these first objectives are achieved.

While operations research solutions are often built to answer economics questions, for example in helping to find the set of solutions satisfying No Envy, we are interested here in economic problems built on operations research problems. Then, the question of how to solve the optimization problem is anterior to the economic questions, and a natural question that arises is if the methods provided by operations research can be used by economists to answer the questions they are interested in. If, for example in a cost sharing problem based on an operations research problem, we can share the cost at the same time as we solve the optimization problem, we have obtained a method that is intuitive and much easier to explain. Given that the cost sharing method/mechanism proposed by the economists will likely try to differentiate agents by considering how they affect the operations research problem, simultaneously solving the problem and assigning cost/value to agents is very natural and appealing.

C. Trudeau (✉)
Department of Economics, University of Windsor,
401 Sunset Avenue, Windsor, ON, Canada
e-mail: trudeauc@uwindsor.ca

© Springer Nature Switzerland AG 2019                                      305
J.-F. Laslier et al. (eds.), *The Future of Economic Design*, Studies in
Economic Design, https://doi.org/10.1007/978-3-030-18050-8_41

A theoretically elegant solution provided by the economist might lose some of its appeal if it cannot be easily computed in a context where the optimization problem is simple. Unfortunately, not all operations research problems allow for an easy use of such algorithms by economists. This brief chapter will discuss some successful and unsuccessful cases and speculate on what makes an algorithm more likely to be useful to economists interested in the fairness questions. In essence, the use of OR algorithms will be useful if it improves our understanding of the problem and allows for a clearer and more intuitive presentation of the solutions. It will be feasible to use such algorithms if at each step there is a natural interpretation of the relevant fairness issues. Naturally, a polynomial-time algorithm seems crucial. Otherwise, the complexity of the algorithm will transfer to the fair division mechanism. However, the most efficient algorithm is not necessarily the most appropriate for the economists. How the various algorithms perform on the aforementioned criterias is as important, and different algorithms solving the same problem might lead to different economic solutions.

To fix ideas, we consider the family of source connection problems, as such problems can take many different forms, from very simple to very complex, solved by all sort of OR algorithms. The problem can be described as follows: we have a set of agents and a source, all located on nodes of the network. Agents require a connection to the source to obtain a good or service, with the quantity to provide to each agent potentially differing. Connecting these nodes requires building costly edges. Each edge has a cost function, with cost depending on the flow going through the edges. We obtain many variants of the problems by restricting the class of admissible cost functions.

One clear difficulty in going from the operations research problem to a cost sharing problem is to reconcile the global objective of the operations research problem (building the network at the cheapest total cost) with the more local objective of the economics problem (for example assigning fair shares to neighbours that are in line with their differences and similarities). Unsurprisingly the best examples we have of uses of OR algorithms in economics are adaptations of greedy algorithms, that are such that the global problem is solved by a series of local operations. One of the simplest source connection problems, the minimum cost spanning tree (mcst) problem, fits the bill. In this problem in which there is a fixed cost but no marginal cost on any edge, not only there is always a tree among the set of optimal networks, but we can also build it by adding one edge at a time, with algorithms proposed by Boruvka (1926), Kruskal (1956) and Prim (1957). In Prim's algorithm, at each step we add the cheapest edge that allows to connect an agent to the source (directly or through a previously connected agent). Bird (1976) proposed a simple method which consists in assigning, at each step, the cost of the added edge to the connected agent. Not only is the method simple, allowing to share cost at the same time as we build the network, but it yields a core allocation. Even though the allocation method has its faults (often unfair, problems when we have multiple optimal trees, non-continuity with respect to cost), it is so simple that any discussion of the mcst problem starts with it, as one would invariably discuss how the minimal cost is obtained. More involved are the obligation rules (Tijs et al. 2006). Based on Kruskal's algorithm which at each step adds the cheapest edge that does not create a cycle, the obligation rules divide the cost of the added edge among the agents that benefit from its addition. The large family of obligation rules includes the folk solution (Feltkamp et al. 1994; Bergantinos and Vidal-Puga 2007), probably the most studied method in the mcst literature. The folk solution has many interpretations, and is usually computed by modifying the cost structure in a certain way (by building an irreducible matrix, in which the costs of unused edges are reduced as much as possible without changing the total cost to connect the grand coalition) before taking the Shapley value of the resulting stand-alone game. As an obligation rule, the folk solution is defined by using the fact that at each step the added edge connects two connected components. If a connected component already contains the source, it does not pay anything for the added edge. If neither connected components contain the source, the added edge is shared in proportion to the number of agents a component gets access to. While the Shapley value definition is elegant, the connection with the OR algorithm allows to explain the method simply and efficiently, especially to a crowd not familiar with the Shapley value. It has also led to the discovery of new properties satisfied by the method and to connections with other solutions. Interestingly, Bergantinos and Vidal-Puga (2011) discuss how fairness principles should apply when we add an edge, as they compare Kruskal's and Boruvka's algorithms.

The mcst problem supposes that edges are non-directed (i.e. the cost is the same if the flow is from $i$ to $j$ or in the opposite direction). If we remove that assumption, we obtain the minimum cost arborescence problem, and the algorithm to solve the resulting OR problem is not greedy anymore, but still solvable in polynomial time. The algorithm can be loosely described as selecting for each agent his cheapest incoming edge, and combining agents that have selected edges that create a cycle, repeating the process until all agents or groups of agents select edges that form a network connecting all groups to the source. A procedure similar to the obligation rules version of the folk solution also yields core allocations. See Dutta and Mishra (2012) and Bahel and Trudeau (2017).

An interesting case in which we obtain no satisfactory links with the OR problem is the general convex network flow problems, in which all cost functions are convex. There is a well-known algorithm to solve the problem (the negative cycle-cancelling algorithm), but not only is that algorithm not greedy, it involves correcting an existing network. To do so, we start from an existing network and compute the cost of sending one unit on each edge and each direction and we look for cycles with negative costs, which indicates cost-reducing opportunities. We have then moved from building a network from scratch, at each step assigning cost shares to agents helped by the addition of the edges, to improving an existing network. One could start with the very inefficient network in which each agent is directly connected to the source, for which it is natural to assign to each agent its cost to connect directly to the source. Conceptually, we could reduce the cost of the network in steps by removing negative cycles, and by simultaneously assigning the cost savings to the agents in that cycle. But there is no clear way on how to proceed: which negative cycles to choose first, and how to share the savings. In particular, there are no easy ways to obtain a core allocation, even though the core is known to always be non-empty (Quant et al. 2006).

Even if no solution can be directly built from the OR algorithms, there might be a lot to learn from these algorithms. Fair division problems require providing constraints on how shares vary when parameters of the problem vary, say the demand of an agent or the cost structure. Understanding how these changes affect the OR algorithms can be very useful in determining if the constraints can be met and in finding methods that meet them. For many problems, in particular for core stability questions, we have to understand not only how these changes affect the problem for the grand coalition, but also for all of its subsets.

The fairness questions applied to dynamic settings, considerably more complex than in static settings, are good examples of problems in which the described approach might bear fruits. Imagine a setting in which how we share resources today affects how much resources we have tomorrow. Finding (or approximating) optimal uses of the resources will involve algorithms, which can help us understand the fairness implications. See the Chapter "The Division of Scarce Resources" by Chambers and Moreno-Ternero in this volume for further discussions of such dynamic problems.

**Acknowledgements** Christian Trudeau acknowledges financial support by the Social Sciences and Humanities Research Council of Canada [grant number 435-2014-0140]. Marcelo Arbex, Gustavo Bergantinos, Jens Hougaard, Justin Leroux, Juan Moreno-Ternero, Sang-Chul Suh and Yuntong Wang provided helpful feedback.

# References

Bahel, E., & Trudeau, C. (2017). Minimum incoming cost rules for arborescences. *Social Choice and Welfare, 49,* 287–314.

Bergantinos, G., & Vidal-Puga, J. (2007). A fair rule in minimum cost spanning tree problems. *Journal of Economic Theory, 137,* 326–352.

Bergantinos, G., & Vidal-Puga, J. (2011). The folk solution and borukva's algorithm in minimum cost spanning tree problems. *Discrete Applied Mathematics, 159,* 1279–1283.

Bird, C. (1976). On cost allocation for a spanning tree: A game theoretic approach. *Networks, 6,* 335–350.

Boruvka, O. (1926). About a certain minimal problem. Praca Moravske Prirodovedecke Spolecnosti(in Czech) 3, 37–58.

Dutta, B., & Mishra, D. (2012). Minimum cost arborescences. *Games and Economic Behavior, 74,* 120–143.

Feltkamp, V., Tijs, S., & Muto, S. (1994). *On the irreducible core and the equal remaining obligations rule of minimum cost spanning extension problems.* Discussion Paper 94106, Tilburg University Center.

Kruskal, J. (1956). On the shortest spanning subtree of a graph and the traveling salesman problem. *Proceedings of the American Mathematical Society, 7,* 48–50.

Prim, R. C. (1957). Shortest connection networks and some generalizations. *Bell System Technical Journal, 36,* 1389–1401.

Quant, M., Borm, P., & Reijnierse, H. (2006). Congestion network problems and related games. *European Journal of Operational Research, 172,* 919–930.

Tijs, S., Branzei, R., Moretti, S., & Norde, H. (2006). Obligation rules for msct situations and their monotonicity properties. *European Journal of Operational Reearch, 175,* 121–134.

# Part VII
# Implementation

# From the Revelation Principle to the Principles of Revelation

Olivier Bochet

**Abstract** In this short piece, I discuss a recent topic in the mechanism design litera-ture regarding the performance of strategy-proof rules. From the revelation principle, all strategy-proof rules are equivalent but it has been found that some perform better than others in practice. In line with some of the recent literature, I advocate for a more thorough study and the addition of protective criterion on top of strategy-proofness.

A decision rule $f$, is strategy-proof if truth-telling is a weakly dominant strategy in the direct revelation mechanism associated to $f$. In other words, the revelation principle asserts that if a rule is implementable in dominant strategy, then it must be strategy-proof. This requirement has many appeals. In particular, it is parsimonious regarding the (common) knowledge and rationality assumptions made with respect to agents' behavior. Complicated forecasts à la Nash are not required and, in fact, agents do not need to necessarily pay attention to what others may do.

Following the negative message conveyed by the Gibbard-Satterthwaite theo-rem, it would have been easy to just turn off the light and move on to mechanism design problems which make more stringent assumptions on agents' behavior—e.g. Bayesian mechanism design. The vast literature on the design of strategy-proof rules shows us that the latter choice would have been wrong. In contrast to the dictatorship result of Gibbard-Satterthwaite, the quest for normatively appealing strategy-proof rules has been, to say the least, a successful one. I would not want to offend anyone by omitting to cite some relevant papers, and my task here is not to do a thorough literature review. Instead I would like to open a discussion regarding a possible over-emphasis of us, practitioners of the field, on the message delivered by the revelation principle.

Indeed, it seems that we have been, for a long time, content with coming up with new models/domains where one can provide interesting characterization theorems of strategy-proof rules. But have we, in the process, asked ourselves whether all these rules work well in practice? Despite their normative appeal, is it possible that some

O. Bochet
Division of Social Science, New York University Abu Dhabi, Abu Dhabi,
United Arab Emirates
e-mail: olivier.bochet@nyu.edu

© Springer Nature Switzerland AG 2019
J.-F. Laslier et al. (eds.), *The Future of Economic Design*, Studies in
Economic Design, https://doi.org/10.1007/978-3-030-18050-8_42

rules in fact perform poorly when put to the real world test? From the point of view of the revelation principle, one probably does not have to ask such questions. Because of the minimal rationality assumption required to identify a dominant strategy, in practice if such strategies exist agents should be able to identify them. Proceeding with such a conclusion may in fact very well be wrong.

Going back to the revelation principle, recall that it only states that strategy-proofness is a necessary condition for implementation in dominant strategies, it is typically not sufficient. However, even equipped with both a necessary and sufficient condition—typically, a variation of the well-known non-bossiness condition in addition to the requirement of strategy-proofness—a decision rule may fail to perform adequately. By the latter, I mean that agents may fail to identify their dominant strategies and as a result they may get stuck with outcomes that differ from the decision rule under the true preference profile.

Our task as practitioners of the field is to understand where these failures may come from. Are these generated by the lack of strictly dominant strategy? After all strategy-proofness is a requirement in terms of weakly dominant strategies, so there may exist alternative dominant strategies which fare as well as truth-telling. Are there behavioral issues, e.g. some forms of bounded rationality or cognitive limitations that would stand in the way for agents to identify what the "right" play actually is? To me, this calls for a shift from the revelation principle to a study of the "principles of revelation".

It is not until recently that we have started to pay attention to such questions. From my point of view, the first two seminal papers which study this issue are Saijo et al. (2007) and Cason et al. (2006). The former is a theory piece while the latter is experimental and testing the findings of the former. In Saijo et al. (2007), the authors acknowledge the possible lack of "robustness" of some of our dearest strategy-proof rules.[1] Their key observation is that many strategy-proof rules are plagued with Nash equilibria whose outcomes differ from the decision rule under the true preference profile. Behaviorally, it is not entirely clear what may drive agents to wonder away from dominant strategy play and stuck at misreports with the wrong outcomes. Cason et al. (2006) however confirms that decision rules which suffer from this deficit perform badly—with a rate of dominant strategy play that is in the neighborhood of 50%, while those which pass the test have a drastically (and statistically significant) higher rate of dominant strategy play.[2] Saijo et al. (2007) introduced the protective criterion called *secure implementation* which stands as double implementation in both dominant strategies and Nash equilibria. A new condition called the rectangular property is at the heart of Saijo et al. (2007) characterization. Unfortunately this condition is so strong that many of our dearest strategy-proof rules fail to satisfy it. Another possible way out of the negative conclusion of secure implementation is to try to understand the sources or reasons for the failure of the rectangular property. Bochet and Sakai

---

[1]For instance, the top trading cycle, most sequential allotment rules including the uniform rule, most of the generalized median voting rules.

[2]I find it surprising that the results of Cason et al. (2006) have not received the attention they deserved.

(2010) study the Sprumont model Sprumont (1991). While only priority rules are secure implementable, they show that none of the bad Nash equilibria obtained under the uniform rule are coalition-proof. More recently, Bochet and Tumennasan (2017a) show that for all sequential allotment rules (Barbera et al. (1997), the set of Nash equilibrium allocations is a complete lattice whose supremum is the outcome under truth-telling, i.e. truth-telling Pareto dominates all preference reports that deliver an allocation that differ from the one under truth-telling. Surprisingly, even with strong properties, the few experimental sessions I ran testing the uniform rule showed an average rate of dominant strategy play that is also in the neighborhood of 55%. But in line with the findings of the previous two papers, I also found that the average rate of Nash equilibrium play was in the vicinity of of 80% and all Nash equilibria delivered the uniform rule outcome.

Another way out? If there are some known behavioral features of agents' behavior, the failure of the rectangular property may be moot. For instance, suppose that agents exhibit some preference for honesty (Dutta and Sen (2012)), e.g. in the face of indifference they would rather report truthfully. Then the only necessary and sufficient condition for secure implementation is strategy-proofness as the rectangular property is vacuously satisfied (Saporiti (2014)).

As a matter of caution (and in line with the Wilson doctrine), I think we ought to emphasize robustness properties of strategy-proof rules—or of their implementing mechanisms—rather than rely on behavioral assumptions which may be difficult to check or guarantee in practice. In addition, I believe that theory and experiments should go hand-in-hand. In that respect, Li (2017) beautifully uses the two approaches. Li's starting point is the difference observed between play of the second price auction versus its sequential version, the ascending price auction. Both are theoretically outcome equivalent while in practice the ascending price auction outperforms the second price auction. Li's notion of *obvious strategy-proofness* is a restriction imposed on the mechanism. Strikingly a decision rule is obviously strategy-proof implementable if even cognitively limited agents can recognize their strategies as weakly dominant. Here the features of sequentiality are essential compared to the static second price auction. Li delivered an important result, which in my opinion deserved all the praised and attention it has received. While Li's path seem to be different than the one followed by Saijo et al. (2007), in fact the two are convergent towards the study of the principles of revelation. Both aim at understanding what separates strategy-proof rules that work well in practice, from those that do not. Both aim at imposing robustness requirement either on strategy-proof rules themselves, or on their possible implementing mechanisms.

I would like to conclude by coming back one last time to the revelation principle. Once again, little is said regarding the "attractiveness" of truth-telling. From the revelation principle, we know that collectively reporting truthfully is the right decision. But what happens when agents wonder away from truth-telling, may be by mistake, may be because of other more behavioral reasons? Does a strategy-proof rule promotes incentives to revert to truth-telling? In Bochet and Tumennasan (2017b), we introduced a new property that is tailor-made to this discussion. We say that a rule is *resilient* if, whenever agents wonder away from truth-telling there always exists at least one agent who would find it profitable to revert to truth-telling. While there is obviously no dynamic in the play of a direct revelation mechanism, such a robustness property makes truth-telling to be more salient and thus may prevent agents from trying out reports that differ from truth-telling. We find that resilience turns out to be equivalent to requiring secure implementation, thereby putting Saijo et al. (2007) under a new light. We also study a group version of the resilience property and show that it is equivalent to a (new) coalitional version of secure implementation. A by-product of the latter result is to show that the failure of the rectangular property observed for most strategy-proof rules is typically not so severe: all strategy-proof and non-bossy in welfare rules turn out to be coalitional secure. This echoes the result of Bochet and Tumennasan (2017a) on the lack of robustness and credibility of the Nash equilibria that differ from the truth-telling outcome in the Sprumont model.

Finally, a new paper by Schummer and Velez (2017) study issue of sequential revelation. Pick a strategy-proof rule and imagine that the planner now asks agents to report their preference relation in turn (i.e. via a perfect information mechanism). There is an obvious leakage of information at play. They show that sequential elicitation can lead to equilibria where agents actually have a strict disincentive to truthfully report. Their contribution is then to identify conditions (on the cartesian product of priors) that eradicate this problem.

The aim of this short piece was to argue that the quest for satisfactory strategy-proof decision rules does not end at their characterization. On the latter, the quest relied on the revelation principle and has been much more successful than initially anticipated. We should now move to a systematic study of what we may call the principles of revelation. This is probably a misnomer, I apologize. I, however, hope to have convinced the readers of its importance. While my co-author Norov Tumennasan and myself intend to continue promoting this question—I am sure that the same goes for the authors cited in this piece—I hope that others will also join.

# References

Barbera, S., Jackson, M., & Neme, A. (1997). Strategy-proof allotment rules. *Games and Economic Behavior, 18*(1), 1–21.

Bochet, O., & Sakai, T. (2010). Secure implementation in allotment economies. *Games and Economic Behavior, 68*(1), 35–49.

Bochet, O., & Tumennasan, N. (2017a). Dominance of truthtelling, implementation and the lattice structure of nash equilibria. Technical report.

Bochet, O., & Tumennasan, N. (2017b). Prevalence of truthtelling and implementation. Technical report.

Cason, T. N., Saijo, T., Sjöström, T., & Yamato, T. (2006). Secure implementation experiments: Do strategy-proof mechanisms really work? *Games and Economic Behavior, 57*(2), 206–235.

Dutta, B., & Sen, A. (2012). Nash implementation with partially honest individuals. *Games and Economic Behavior, 74*, 154–169.

Li, S. (2017). Obviously strategy-proof mechanisms. *American Economic Review, 107*(11), 3257–87.

Saijo, T., Sjöström, T., & Yamato, T. (2007). Secure implementation. *Theoretical Economics*, *2*(3), 203–229.

Saporiti, A. (2014). Securely implementable social choice rules with partially honest agents. *Journal of Economic Theory*, *154*, 216–228.

Schummer, J., & Velez, R. (2017). Sequential preference revelation in incomplete information settings. Technical report.

Sprumont, Y. (1991). The division problem with single-peaked preferences: A characterization of the uniform allocation rule. *Econometrica*, *59*(2), 509–519.

# Complete Information Implementation and Behavioral Economics

## Bhaskar Dutta

**Abstract** Behavioural economics emphasizes the importance of (sometimes) departing from the neoclassical paradigm of choice behaviour being determined by preference maximisation. This note outlines ways in which Nash implementation theory
can be married with behavioural economics.

The classic problem of mechanism design under complete information is concerned
with a scenario where the planner's goals, represented as a *social choice correspondence* $F : \Theta \rightarrow X$, depend upon the state of the world. Here $\Theta$ is the set of states of
the world while $X$ is the set of feasible social alternatives. The state of the world is
common knowledge amongst $N$, the set of individuals in the society, but is not known
to the planner. The planner cannot simply ask the agent to announce the state of the
world since the agents' preferences may not coincide with that of the planner. So,
the planner has to design a mechanism that will induce agents to correctly reveal the
state of the world. A mechanism along with a state of the world describes a complete
information game. In a seminal paper, Maskin (1999) focussed on mechanisms where
agents choose messages simultaneously. He used the conventional game-theoretic
notion of *Nash equilibrium*, and provided an almost complete characterization of the
class of social choice correspondences that are *implementable* in Nash equilibrium.[1]

However, there is now considerable empirical evidence (both in the real world
as well as in experimental laboratories)suggesting that individuals do not quite conform to the neoclassical paradigm of maximisation of a preference ordering. This
body of evidence has given rise to the recent literature in Behavioral economics which
focuses on departures from classical notions of choice behaviour determined by preference maximisation. Choice behavior can be influenced by a variety of phenomena
such as menu dependence, framing, temptation and self-control. In multi-person

---

[1]Implementability of $F$ in any equilibrium notion $E$ means that there is a mechanism such that
the equilibrium correspondence of $E$ for the game induced by the mechanism and any state
$\theta \in \Theta$ coincides with $F(\theta)$.

B. Dutta (✉)
Ashoka University and University of Warwick,
Sonipat, India
e-mail: b.dutta@warwick.ac.uk

© Springer Nature Switzerland AG 2019
J.-F. Laslier et al. (eds.), *The Future of Economic Design*, Studies in
Economic Design, https://doi.org/10.1007/978-3-030-18050-8_43

settings, considerations of *reciprocity* mean that one agent's choice behavior or optimal action(s) may be influenced by other agents' actions. These phenomena then mean that choice behavior may no longer be rationalizable by means of a preference ordering defined over $X$.

Obviously, implementation theory has to change if it is to keep up with developments in behavioral economics. In an important paper, de Clippel (2014) takes a big step in incorporating such behavioral concerns into complete information implementation. He assumes that each individual's choice behavior is described by a choice correspondence $C_i(-, \theta)$ which describes the set of alternatives that individual $i$ would choose from any set $S \subseteq X$. This explicitly allows for menu dependence since the choice out of a set $S$ need not have any relationship with choice out of a set $T$ in any state of the world. de Clippel goes on to derive necessary and sufficient conditions for Nash implementability in this setting. But, these conditions are far from a complete characterization, at least partly because of the generality of his framework—no restriction is imposed at all on the choice correspondences $\{C_i\}_{i \in N}$.

However, a common approach in behavioral economics is to impose *some* restrictions in the form of axioms on individual choice behavior.[2] For example, one axiom -WWARP—is a weakening of the well-known Weak axiom of Revealed Preference that is necessary for full rationalizability of choice behavior. Manzini and Mariotti (2007) use WWARP and another relatively weak axiom to characterize a class of choice correspondences that can be described as short-list method. Given any set $S$, the individual first shortlists candidates according to one criterion and then chooses one candidate out of the set of candidates according to a (common) preference ordering defined over $X$. Thus, choice behavior follows a well-defined procedure providing some structure to the choice correspondences $\{C_i\}_{i \in N}$. This suggests the possibility of using the de Clippel framework, but borrowing from the axiomatic approach in behavioral economics to develop a theory of behavioral implementation.

Another interesting and unexplored area again arises from the empirical observation that individual preferences may be endogenous and may in fact be influenced by the institutional mechanism itself. Thaler and Sunnstein's *Nudge Theory* discusses how a kind of "psychological manipulation" may be used to alter human behaviour to induce agents to make appropriate choices. *Nudge units* have actually been employed in many countries to improve public administration. An interesting example is the application of Nudge theory to pension savings. In the "Save More Tomorrow" used by several companies in the U.S.A. to increase pension savings, an individual commits to allocating a share of future salary increases to savings. Since joining the programme is voluntary and individuals can opt out in the future, individual freedom of choice is not affected.

It is also important to realise that a mechanism which may seem appropriate if individual preferences are *fixed* may make matters worse because of endogeneity of preferences. Bowles and Polania-Reyes discuss some literature which describes how incentives can alter individuals' social preferences for adhering to social norms of doing the right thing—the introduction of overtime allowances resulting in shorter

---

[2]See, for instance, Manzini and Mariotti (2007) or Cherepanov et al. (2013).

hours worked being an example. More generally, the provision of incentives may actually be counterproductive.

These examples suggest that in a variety of contexts, it may be inappropriate to follow the standard design approach of constructing a mechanism which would operate on *fixed* individual preferences to produce desirable outcomes in equilibrium.

Let $\mathcal{M}$ denote the set of all mechanisms. If mechanisms change preferences, then one way to proceed is define a mapping $\phi : \Theta \times \mathcal{M} \to \Theta$. Thus, $\phi(\theta, \mu)$ denotes the state which arises or is induced if $\theta$ is the original state and $\mu$ is the mechanism chosen by the planner. Let $F$ be the planner's desired social choice correspondence. Then, the following definition seems appropriate.

**Definition 1** The social choice correspondence $F$ is implementable in Nash equilibrium if there is a mechanism $\mu$ such that for all $\theta \in \Theta$, $F(\theta) = NE((\mu, \phi(\theta, \mu)))$.

A tricky issue is how to decide on the appropriate specification of the mapping $\phi$. The right specification may well depend on the finer details of the specific context that is being sought to be analysed. While this may make it difficult to get general results, the approach itself seems an extremely fruitful area of research.

**Acknowledgements** I am grateful to Jean-Francois Laslier for useful comments on an earlier draft.

# References

Bowles, S., & Polania-Reyes, S. (2012). Economic incentives and social preferences. *Journal of Economic Literature*, *50*, 368–425.

Cherepanov, V., Feddersen, T., & Sandroni, A. (2013). Rationalization. *Theoretical Economics, 8*, 775–800.

de Clippel, G. (2014). Behavioral implementation. *American Economic Review, 104*, 2975–3002.

Manzini, P., & Mariotti, M. (2007). Sequentially rationalizable choice. *American Economic Review, 97*, 1824–1839.

Maskin, E. (1999). Nash equilibrium and welfare optimality. *Review of Economic Studies, 66*, 23–38.

Thaler, R., & Sunnstein, C. R. (2018). *Nudge : Improving decisions about health*. Wealth and Happiness: Yale University Press.

# Robust Mechanism Design: An Analogy-Based Expectation Equilibrium Perspective

Philippe Jehiel

**Abstract** In this short essay, I revisit the idea of robust mechanism design explicitly taking into account that it may be hard to access the beliefs held by agents in previous plays of the mechanism. I propose modelling such an environment using the apparatus of the analogy-based expectation equilibrium with payoff-relevant analogy partitions. Such an approach allows to move away from impossibility results that arise with ex post implementation.

Mechanism design provides the conceptual framework to analyze what can be implemented in terms of collective action or assignment of allocations when economic agents possess private information. It has sometimes been criticized on the ground that the optimal design so derived generally relies on fine details of the environment such as how information is distributed among agents and what agents know about others' information. As voiced by Wilson (1987), a worry is that a mechanism that would be optimal for a specific environment need not be even approximately good for a different environment. To the extent that the environment is not known with great precision, that is a serious concern.

Such considerations have led researchers to propose a stronger notion of implementation that would be robust to how beliefs and higher order beliefs of agents about others' information are defined (see in particular Bergemann and Morris (2005)). In quasi-linear environments, robust implementation boils down to ex post implementation. That is, agents should find it optimal to follow their course of action even assuming they know the information of others no matter what this information may be. Such an equivalence arises because one possible belief is that the information of others is concentrated on one realization, and varying this realization yields that robust implementation must imply ex post implementation (the converse part follows a simple logic as well). Ex post implementation covers implementation in dominant strategies that was considered earlier in social choice theory (see the vast literature on strategyproofness), but it is more general in interdependent value contexts in which

P. Jehiel (✉)
Paris School of Economics (Ecole des Ponts, ParisTech) and University College London,
Paris, France
e-mail: jehiel@pse.ens.fr

© Springer Nature Switzerland AG 2019                                                          321
J.-F. Laslier et al. (eds.), *The Future of Economic Design*, Studies in
Economic Design, https://doi.org/10.1007/978-3-030-18050-8_44

the payoff attached by an agent to the various alternatives may be directly affected by the information held by others.

While ex post implementation has obviously desirable properties, a serious limitation is that in rich enough environments it is too restrictive. More precisely, in interdependent value contexts, when there are at least two alternatives, and the private information held by agents has at least two dimensions, no social choice rule is ex post implementable for generic specifications of the preferences in quasi-linear environments (see Jehiel et al. (2006)). Given this impossibility result, what are the possible routes?

First, one can restrict the environment so that ex post implementation gets feasible, which typically requires reducing the private information to be one-dimensional in interdependent value contexts. I find this too restrictive and not satisfactory from a conceptual viewpoint. Second, one can require robustness with respect to a smaller set of beliefs as opposed to the universal set of beliefs. Unfortunately, as observed in Jehiel et al. (2012), allowing only for beliefs in a neighborhood of those beliefs arising from a common prior environment would not allow to get away from the impossibility result, thereby leaving little hope for this approach to work with great generality, unless one is willing to impose tight constraints on the admissible beliefs. Third, one can allow the implementation mapping to depend on the beliefs held by agents, thereby relaxing the notion of robust implementation. This is certainly a valid route, but no decisive progress has been made along these lines as far as I know.

I suggest below another route that incorporates elements of bounded rationality in how agents process past observed behaviors as a function of the environment in an attempt to form expectations about others' behaviors in the current interaction. While not yet developed, I would suggest this is a promising route to make further progress on robust mechanism design.

A compelling motivation for the study of robust mechanism design as pioneered by Bergemann and Morris is that it is probably hard (sometimes even impossible) to know what the beliefs of agents were when decisions were made when looking at past implementation problems. For example, in standard first-price auctions, both the analyst and the economic agents themselves may observe the bids and possibly the value of the good to the various bidders after the auction has taken place, but they do not always have access to what bidders knew when they submitted their bids in the first place. If agents form their view about how other agents behave based on such coarse feedback about previous plays of the same mechanism, they may not be able to form rational expectations as usually considered in economic theory. Instead, I would suggest that when not observing beliefs from past plays, agents form their expectations about others' behaviors by aggregating what they observe from past interactions over all situations that share the same payoff-relevant information but may differ in terms of the beliefs held by agents. In steady state, agents would thus behave as in the analogy-based expectation equilibrium (Jehiel 2005; Jehiel and Koessler 2008) in which agents aggregating behaviors in such a manner would be viewed as using the payoff-relevant analogy partition as defined in Jehiel and Koessler (2008). Of course, for those agents who observe the beliefs as well as the payoff-relevant information, they may rely on finer analogy partitions. From this perspective, robustness may now require implementation for all analogy partitions that would be finer than the payoff-relevant analogy partition. Even if such robustness cannot be achieved, one can always analyze the profile of what is implemented as a function of the analogy partition.

Such a view differs from previous approaches to robust mechanism design in a number of ways. First, the description of the environment remains Bayesian. That is, how beliefs, higher order beliefs but also analogy partitions are distributed across agents takes itself a probabilistic form assumed to be objectively given by nature. Second, the possible beliefs entertained by agents in this approach are constrained to be derived from the analogy-based expectations for some analogy partition assumed to be finer than the payoff-relevant analogy partition. This puts a lot of structure on the possible beliefs, which may facilitate robust implementation so defined. For example, if we are in an environment in which the profile of payoffs for all players and all alternatives fully determines the beliefs of all agents, then all equilibria must coincide with the Nash Bayes equilibria, since the profile of payoffs is then a sufficient statistic for the profile of types and it is thus enough to correctly infer the behavior of agents. By contrast, if the environment may allow for the same profile of preferences but different beliefs, equilibria need not coincide with the Nash Bayes equilibria. For example, if we are in a private value first-price auction environment in which sometimes bidders fully observe the entire profile of valuations for all participants and sometimes observe only their own valuation, the payoff-relevant analogy partition would lead to aggregate the bidding behavior in the two scenarios, which would typically move away from the Nash Bayes equilibrium.

It may be useful at this stage to introduce a little bit of formalism, which may help clarify some features of the approach. Following (Bergemann and Morris 2005), let me consider an $n$-agent environment and let me refer to the type of agent $i$ -assumed to be known to agent $i$ only- as $t_i = (\theta_i, \pi_i) \in T_i$ where $\theta_i \in \Theta_i$ refers to the payoff type of agent $i$ and $\pi_i \in \Delta(T_{-i})$ refers to the belief of $i$ about other agents' types.[1] Payoffs depend only on the profile of payoff types $\theta$, not the belief types $\pi$. That is, in a game $G$, with action profile $s = (s_i)_i$, $i$'s payoff can be written as a sole function of $s$ and the profile of payoff types $(\theta_i, \theta_{-i})$: $u_i^G(s_i, s_{-i}; \theta_i, \theta_{-i})$. Ex post equilibrium $(s_i(t_i))_i$ of $G$ requires that the strategy of $i$ depends only on the payoff type. That is, with a slight abuse of notation, referring to $s_i(\theta_i)$ as the function describing how agent $i$'s action depends on the payoff-relevant part of the type of $i$, ex-post equilibrium requires that for all $\theta_i, \theta_{-i}, s_i'$ :

$$u_i^G(s_i', s_{-i}(\theta_{-i}); \theta_i, \theta_{-i}) \leq u_i^G(s_i(\theta_i), s_{-i}(\theta_{-i}); \theta_i, \theta_{-i}).$$

Consider by contrast the analogy-based expectation equilibrium scenario in which agent $i$ relies on the payoff-relevant analogy partition. Such a setting assumes that there is an objective distribution of $t$ viewed as being decided by nature. It also considers that after the previous plays of the same game $G$, player $i$ observes

---

[1] Observe that such a recursive formulation allows to model the idea of infinite hierarchies of beliefs that the various agents may have about each other's information.

$\theta = (\theta_j)_{j=1}^n$ and $s = (s_j)_{j=1}^n$ but not $\pi = (\pi_j)_{j=1}^n$. Formally, the analogy partition of player $i$ is then defined by the collection of analogy classes

$$A_i(\theta) = \left\{ t' = (\theta_j', \pi_j')_{j=1}^n \mid \theta_j' = \theta_j \text{ for all } j \right\}$$

parameterized by $\theta = (\theta_j)_{j=1}^n$. Given the strategy $\sigma_{-i} : T_{-i} \to \Delta S_{-i}$ of agents $-i$, one can compute the aggregate distribution of $\sigma_{-i}$ in every $A_i(\theta)$. Such an aggregate distribution over $\Delta S_{-i}$ is denoted by $\overline{\sigma}_{-i}(A_i(\theta))$ and shapes the steady state belief held by agent $i$ about how the strategy of $-i$ depends on the profile of types $t$. More precisely, an analogy-based expectation equilibrium (ABEE) is such that agent $i$ with type $t_i = (\theta_i, \pi_i)$ plays $\sigma_i(t_i)$ required to be a best-response to the conjecture that agent $-i$ plays according to $\overline{\sigma}_{-i}(A_i(\theta))$ whenever $\theta = (\theta_i, \theta_{-i})$ prevails. Observe that within this approach, the belief of agent $i$ may affect agent $i$'s choice of strategy but only through the first-order belief (i.e. the part of $\pi_i$ that concerns $\theta_{-i}$), since agents $-i$ behaviors are conjectured by agent $i$ to depend only on $\theta = (\theta_i, \theta_{-i})$.

In environments in which the profile of payoff-relevant types $\theta$ fully determines $t$, then ABEE coincides with a Nash Bayes equilibrium given that the aggregation procedure would lead player $i$ to have the correct belief about the strategy of other players. Moreover, when an ex post equilibrium exists, it is also an ABEE with payoff-relevant analogy partition and in fact no matter what analogy partition finer than the payoff-relevant analogy partition that players may use (this follows because the aggregation procedure would then lead to aggregate identical behaviors and thus agents would have correct expectations). But, there may be ABEE which are not ex post equilibria. Given that an ABEE always exists in finite environments (see Jehiel and Koessler (2008)), a clear advantage of the approach is to avoid the inexistence issues that may arise when considering ex post equilibria.

In the above, I have assumed that the profile of payoff-relevant types $\theta = (\theta_i)_i$ is accessible after the play of the game $G$. It may be argued that a more natural assumption is that the payoffs of agents in the various possible alternatives are observed instead of $\theta$. When the profile of payoffs determines uniquely $\theta$, this makes no difference. But, when multiple $\theta$ correspond to the same profile of preferences, this would lead to a different definition of ABEE. To give an illustration, consider a two bidder auction setup in which $i$ observes $(x_i, w_i)$ and $i$'s valuation for the good is $v_i = x_i + w_i + w_{-i}$. There are obviously multiple $(x_i, w_i), (x_{-i}, w_{-i})$ leading to the same profile of valuations $v = (v_i, v_{-i})$. ABEE with payoff-relevant analogy partition would require that bids be aggregated over all $(x, w)$ corresponding to the same $v$, which would be different from the Nash Bayes equilibrium in contrast to the case in which the analogy classes coincide with $(x, w)$ as previously considered. This comment is illustrative of a more general observation. In the formalism of Bergemann and Morris, there is not a unique way of decomposing the type into a payoff-relevant component and a belief component. I would suggest decomposing the type in such a way that the profile of payoff-relevant types allows to uniquely recover the profile of payoffs in the various available alternatives.

As already mentioned, the above proposed approach needs to be further developed. For example, it may be useful to analyze under what conditions the full rent extraction result of Crémer and McLean (1988) can be relaxed under this approach or under what conditions one can implement desirable outcomes even without any information about how analogy partitions are distributed in the population. This should be the subject of future research.

I would conclude with one more comment. The above proposed approach is meant to represent situations in which a given mechanism would be played many times while (some) agents would have access only to the payoff-relevant information from past plays. This is a different context than one in which agents would not have seen any past play of the mechanism. In such a case, implementation in dominant strategy if feasible seems to be more appropriate (of course, when ex post implementation is not feasible, implementation in dominant strategy is not feasible either), or if one suspects that agents are not good at identifying (weakly) dominant strategies by introspection one may favor obviously strategy-proof mechanisms as introduced by Li (2017) (and extended by Zhang and Levin (2017)).

**Acknowledgements** I thank Jean-François Laslier for useful comments and the ERC (grant no 742816) for funding.

# References

Bergemann, D., & Morris, S. (2005). Robust mechanism design. *Econometrica, 73*(6), 1771–1813.

Crémer, J., & McLean, R. (1988). Full extraction of the surplus in bayesian and dominant strategy auctions. *Econometrica, 56,* 1247–1257.

Jehiel, P. (2005). Analogy-based expectation equilibrium. *Journal of Economic Theory, 123,* 81–104.

Jehiel, P., & Koessler, F. (2008). Revisiting games of incomplete information with analogy based expectations. *Games and Economic Behavior, 62,* 533–557.

Jehiel, P., Meyer-ter-Vehn, M., & Moldovanu, B. (2012). Locally robust implementation and its limits. *Journal of Economic Theory, 147,* 2439–2452.

Jehiel, P., Meyer-ter-Vehn, M., Moldovanu, B., & Zame, W. R. (2006). The limits of ex-post implementation. *Econometrica, 74,* 585–610.

Li, S. (2017). Obviously strategy-proof mechanisms. *American Economic Review* (Forthcoming).

Wilson, R. (1987). Game-theoretic analyses of trading processes. In T. Bewley (Ed.), *Advances in economic theory: Fifth world congress* (Chap. 2, pp. 33–70). Cambridge: Cambridge University Press.

Zhang, L., & Levin, D. (2017). *Partition obvious preference and mechanism design: Theory and experiment* (Working Paper).

# Continuous Mechanism Design

Ruben Juarez

**Abstract** Despite the accumulating literature on mechanism design over the last decades, the equilibrium outcome of most mechanisms described to date fails to be robust to small changes in the specifications and parameters of the model. Lacking such robustness, the designer may fail to implement his desired outcome. Thus, here we argue that new mechanisms should be more resistant to physical, modeling and behavioral misspecifications.

Over the last decades, the field of mechanism design has shown that incentives can achieve desirable objectives when players are rational. This has been validated in a variety of applications ranging from auctions and cost-sharing to matching and voting. The natural engineering approach of the field to construct such mechanisms requires the designer to make strong assumptions about the preferences and the behavior of the agents, in particular. However, errors can and do occur. The designer might be incorrect about the parameters applied to the model, the preferences of the agents, or their behaviors. Even if the designer was initially correct, agents may change their preferences or alter their behavior over time. Traditional mechanism design is not immune to these changes, including those that may be subtle. Therefore, there is a need in the field to design robust mechanisms that would, in effect, be error-proof.

While it might not be realistic to design completely resilient mechanisms that work for every possible change, our goal is to study the subtle changes that might happen due to misestimation of the parameters, preferences, or behavior of the agents to minimize potential errors. We posit that such *small changes should not have a large impact on the outcome (equilibrium) of the mechanism*. To formalize this, continuity requires that sufficiently small changes in the input of a mechanism (parameters, preferences, etc.) result in arbitrarily small changes in the output (payoff of the agents, the objective of the planner, etc.). Continuity in the outcome of the mechanism is a desirable property, as the parameters, preferences, and behavior of the agents are hard to precisely estimate, and small misestimations should not have a significant impact on the outcome of the mechanism. The lack of continuity in the mechanism

R. Juarez (✉)
Department of Economics, University of Hawaii, Honolulu, USA
e-mail: rubenj@hawaii.edu

© Springer Nature Switzerland AG 2019                                         327
J.-F. Laslier et al. (eds.), *The Future of Economic Design*, Studies in
Economic Design, https://doi.org/10.1007/978-3-030-18050-8_45

has implications for efficiency and fairness, both of which are central objectives in mechanism design.

It is important to clarify that we focus here on outcome continuous mechanisms, which take into account the final allocation to the agents at equilibrium. Because the (presumably Nash) behavior of the agents determines the allocation of the agents, this continuity property is substantially stronger than traditional continuity requirements and has received little attention in the literature.[1]

The study of continuous mechanism design fits within the extensive literature of robust (prior-free) mechanism design, where the intent is to design mechanisms that are robust for an uninformed planner about the characteristics of agents participating in the mechanism—e.g., when agents know each other and may collude (Moulin 1999; Moulin and Shenker 2001; Juarez 2013) or when the planner does not have a priori knowledge about the utility of the agents (Moulin 2008; Juarez 2008; Mehta et al. 2007).[2] Unlike most of the work previously done in robust mechanism design, we propose to examine and study the impact of small changes to the resiliency of the model.[3]

The complexity of engineering resilient mechanisms creates difficult challenges for the designer. To illustrate this, we focus on specific mechanism design settings applied to cost-sharing problems, however, similar examples can be extended to more general mechanism design settings where money is available.

# 1  Continuity with Respect to the Parameters in the Model

Consider two agents, 1 and 2, interested in connecting from the source $s$ to $t_1$ and $t_2$, respectively (see Fig. 1). The agents simultaneously choose paths in the network to connect their unique source to their unique sink, and the choice of paths of all the agents leads to the network that must be constructed and/or maintained. A mechanism splits the total cost of the network formed between the two agents who care about minimizing their cost-share.

The Shapley mechanism distributes the cost of the link selected equally among the agents. This mechanism is considered fair, is easy to implement (decentralized), and has been widely studied in the literature (Anshelevich et al. 2004; Chen et al. 2008; Epstein et al. 2007, 2009; Fiat et al. 2006). Unfortunately, the equilibrium of the Shapley mechanism is not continuous when simple measurement errors in the

---

[1] An exception is the literature on payoff-continuity when there is incomplete information in a game, see Kajii and Morris (1997, 1998), Rothschild (2005). Unfortunately, such a literature have had limited applicability to specific mechanism design problems.

[2] Due to space limitations, we omit most of the extensive literature about the worst-case analysis from the fields of Economics and Computer Sciences.

[3] A natural open question is to understand to what degree local changes lead to global properties. A recent strand of literature has focused on whether local strategy-proofness implies global strategy-proofness (Carroll 2012; Sato 2013; Archer and Kleinberg 2014).

**Fig. 1** Network with two
agents at the common source
$s$ and two different sinks $t_1$
and $t_2$

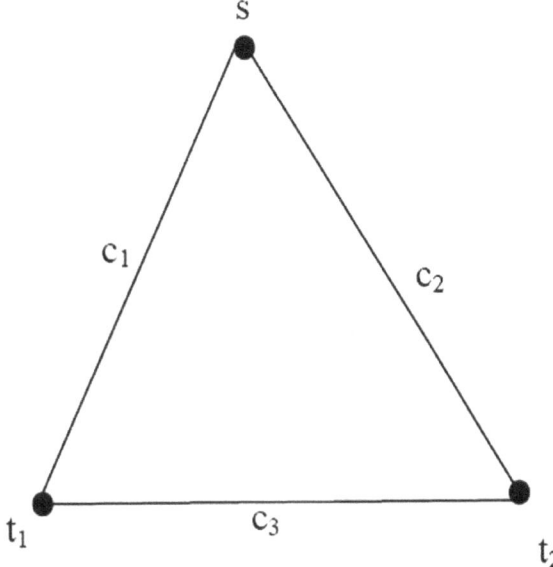

cost of the links occur. Indeed, consider the case of a symmetric problem where the
costs are $c_1 = c_2 = c > 0$ and $c_3 = \frac{c}{2} + \epsilon$, where $\epsilon$ is a small error.

When $\epsilon > 0$, the unique Nash equilibrium requires each agent to connect directly,
select $st_1$ and $st_2$, and pay a cost equal to $c$. However, when the error term $\epsilon$ is negative,
$\epsilon < 0$, there are exactly two Nash equilibriums $(st_1, st_1t_2)$ and $(st_2t_1, st_2)$, both of
which require the agents to coordinate. The payoffs to the agents are $(\frac{c}{2}, c + \epsilon)$ or
$(c + \epsilon, \frac{c}{2})$, respectively.

This discontinuity of the Nash equilibrium is highly undesirable in many dimensions. First, the payments of the agents substantially change, creating highly asymmetric payoffs when the entire problem is symmetric. Second, the strategies of the
agents widely differ depending on whether $\epsilon$ is positive or negative. For a positive $\epsilon$,
the agents' strategies are simple since they do not require coordination. In such a case,
agents pay the cost of connecting directly to their link. However, when $\epsilon$ is negative,
agents need to coordinate on the route selected, thus creating added complexity in
the strategies.

While there are some recent characterizations regarding the continuity of the Nash
equilibrium in a variety of cost-sharing models, mainly providing highly restrictive
mechanisms that only depend on the aggregate cost and disregard the demand of the
agents,[4] more work is needed to address the role of payoff-continuity in more general

---

[4]See, Juarez and Kumar (2013) for payoff-continuity in a routing model, Juarez et al. (2018) for
payoff-continuity in a sequential sharing model, and Juarez et al. (2018) for continuity in a time
allocation model. Hougaard and Tvede (2012, 2015) also characterize cost minimizing networks
when the announcement rule can change.

cost-sharing and allocation problems. The study of continuity beyond payoffs, for instance by focusing on "similar" strategies, warrants further examination.

## 2  Continuity with Respect to the Preferences of the Agents

Most economic theory assumes that individual behavior is driven by rational decisions, with a few theories challenging this approach. We believe that many failures of the theory of mechanism design occur not due to a rationality assumption, but due to a design failure, such as eliciting appropriate information from the agents. Indeed, we find two severely under-explored issues that can better link the theory with the practice of mechanism design.

**Preference Representation**: Humans are complex and so are their preferences. Representing preferences of individuals is particularly difficult in cost-sharing problems, where agents might have preferences over a continuum of options (e.g., the division of an infinitely divisible resource among agents). It is the designer's responsibility to elicit information from the agents that approximate their complex preferences, for instance by eliciting simple information [e.g., a one parameter function Moulin (2017), Goldberg and Hartline (2005)] or a more accurately description of their taste.[5]

A mechanism that solicits complex preferences from the agents will likely produce large changes in the final allocation of the agents at equilibrium (especially when agents have difficulty estimating their own preference), which in turn might create even more complexities to predict the equilibrium. Therefore, a trade-off for the designer arises: soliciting simple information from the agents that is more stable to changes yet potentially farther away from the planner's objectives (e.g., efficiency, fairness, etc.) versus soliciting more complex information with less predictive power yet closer to the planner's objectives when the prediction actually works.

**Externalities**: Perhaps one of the largest under-explored issues in mechanism design occurs when agents experience externalities. Such externalities arise when agents care not only about their own utility (i.e., allocation, cost-share, etc.) but also about that of others. Indeed, in the presence of social networks, agents may experience a satisfaction (or dissatisfaction) when their close friends (or enemies) are better off. A particular case of externalities occurs when agents experience envy, i.e., the preference of someone else's allocation to his own. However, in general, the externalities experienced by agents are hard to estimate and model into the utility function.

The study of externalities presents a challenge for the designer, who might be uninformed about the social network and its magnitude in the utility of the agents. Therefore, there is a need to understand how new mechanisms will change in the

---

[5]There is also a large literature on what preferences are in the social sciences, and more importantly, the oversimplification of preferences and behavior in Economics.

presence of externalities, and identify more resilient mechanisms that work well under such circumstances.[6]

# 3  Is the Nash Equilibrium the Right Equilibrium?

The predictive power of the Nash equilibrium has been widely tested, especially in the experimental literature, results of which have been mixed. The success of prediction is specific to every problem and highly dependent on the type of equilibrium that it generates; e.g., there are positive results for games that generate a strong Nash equilibrium, and even some games with Pareto dominant equilibria. The accuracy of the Nash equilibrium is also dependent on several other exogenous factors, such as the framing of the problem or the type of agents who are playing.

As theorists, our research should be informed by the experimental literature as an aid to refine our models to more closely resemble what has been observed in real-world settings. This is particularly crucial in the mechanism design literature, where the planner often has a menu of mechanisms, each generating equilibrium(s) with different properties, some of which may or may not appear in reality.

Specifically, consider the example represented by Fig. 1 when $\epsilon > 0$. The Shapley mechanism produces a unique inefficient equilibrium where each agent pays $c$, for a total cost of $2c$, whereas the cheapest way to connect the agents costs $\frac{3c}{2} + \epsilon$. A simpler cost-allocating mechanism that divides the total cost of the network formed in proportion to the cheapest connection of their source-sink (in this case $c$ for each agent) generates two equivalent equilibrium, where each agent pays $\frac{3c}{4} + \frac{\epsilon}{2}$. Similarly to Shapley, this mechanism might produce multiple equilibria, but all of them will be Pareto ranked. Thus, if the Nash equilibrium is a good predictor of the outcome implemented by the mechanism, the least inefficient equilibrium stands out as a reasonable selection that all the agents prefer Juarez and Kumar (2013). This produces a higher predictive power than Shapley, even when the costs $c_1, c_2$ and $c_3$ change slightly.

# 4  Conclusion

Mechanism design has been a major component in the economic analysis of institutions and markets, especially when the designer has incomplete information about the participants. While there has been a recent expansion in the field with interest in robust mechanisms that work well when the planner has little information about the

---

[6]This happens even in the most simplistic model of resource sharing of a fixed unit of a good, where Juarez and Kumar (2018) shows the difficulty to tackling this problem when the planner is uninformed about the social network and its effect on the utility function. It characterizes a class of strategy-proof resource allocating mechanisms.

agents, the field can be improved by identifying the right trade-offs that account for physical, modeling and behavioral errors, some of which might be alleviated at the design stage and by better linking the theory with more practical studies.

# References

Anshelevich, E., Dasgupta, A., Kleinberg, L., Tardos, E., Wexler, T., & Roughgarden, T. (2004). The price of stability for network design with fair cost allocation. In *45th Annual IEEE Symposium on Foundations of Computer Science (FOCS)*, pp. 59–73.

Archer, A., & Kleinberg, R. (2014). Truthful germs are contagious: a local-to-global characterization of truthfulness. *Games and Economic Behavior, 86*, 340–366.

Carroll, G. (2012). When are local incentive constraints sufficient? *Econometrica, 80*(2), 661–686.

Chen, H. L., Roughgarden, T., & Valiant, G. (2008). Designing networks with good equilibria. *Proceedings of the 19th Annual ACM-SIAM Symposium on Discrete Algorithms.*

Epstein, A., Feldman, M., & Mansour, Y. (2007). *Strong equilibrium in cost sharing connection games*. San Diego: ACM conference on Electronic Commerce.

Epstein, A., Feldman, M., & Mansour, Y. (2009). Equilibrium in cost sharing connection games. *Games and Economic Behavior, 67*, 51–68.

Fiat, A., Kaplan, H., Levy, M., Olonetsky, S., & Shabo, R. (2006). On the price of stability for designing undirected networks with fair cost allocations. *ICALP06.*

Goldberg, A. V., & Hartline, J. D. (2005). Collusion-resistant mechanisms for single-parameter agents. In *Proceedings of the Sixteenth Annual ACM-SIAM Symposium on Discrete algorithms*, pp. 620–629. Society for Industrial and Applied Mathematics.

Hougaard, J. L., & Tvede, M. (2012). Truth-telling and nash equilibria in minimum cost spanning tree models. *European Journal of Operational Research, 222*(3), 566–570.

Hougaard, J. L., & Tvede, M. (2015). Minimum cost connection networks: Truth-telling and implementation. *Journal of Economic Theory, 157*, 76–99.

Juarez, R. (2008). The worst absolute surplus loss in the problem of commons: Random priority versus average cost. *Economic Theory, 34*(1), 69–84.

Juarez, R. (2013). Group strategyproof cost sharing: The role of indifferences. *Games and Economic Behavior, 82*, 218–239.

Juarez, R., Ko, C. Y., & Jingyi, X. (2018). Sharing sequential values in a network. *Journal of Economic Theory, 177*, 734–779.

Juarez, R., & Kumar, R. (2013). Implementing efficient graphs in connection networks. *Economic Theory, 54*(2), 359–403.

Juarez, R. & Kumar, R. (2018). Dividing a dollar in a network.

Juarez, R., Nitta, K., & Vargas, M. (2018). Profit-sharing and efficient time allocation (Technical Report). Mimeo: University of Hawaii.

Kajii, A., & Morris, S. (1997). The robustness of equilibria to incomplete information. *Econometrica: Journal of the Econometric Society*, pp 1283–1309.

Kajii, A. & Morris, S. (1998). Payoff continuity in incomplete information games. *Journal of Economic Theory, 82*(1), 267–276.

Mehta, A., Roughgarden, T., & Sundararajan, M. (2007). Beyond moulin mechanisms. In *Proceedings of the 8th ACM Conference on Electronic Commerce*, pp. 1–10. ACM.

Moulin, H. (1999). Incremental cost sharing: Characterization by coalition strategy-proofness. *Social Choice and Welfare, 16*(2), 279–320.

Moulin, H. (2008). The price of anarchy of serial, average and incremental cost sharing. *Economic Theory, 36*(3), 379–405.

Moulin, H. (2017). One-dimensional mechanism design. *Theoretical Economics, 12*(2), 587–619.

Moulin, H., & Shenker, S. (2001). Strategyproof sharing of submodular costs: Budget balance versus efficiency. *Economic Theory, 18*(3), 511–533.

Rothschild, C. G. (2005). Payoff continuity in incomplete information games: A comment. *Journal of Economic Theory, 120*(2), 270–274.

Sato, S. (2013). A sufficient condition for the equivalence of strategy-proofness and nonmanipulability by preferences adjacent to the sincere one. *Journal of Economic Theory, 148*(1), 259–278.

# Decision and Information Disclosure Mechanism Design

Claudio Mezzetti

**Abstract** This chapter argues that there are important settings in which the designer may be able to control both the social outcomes and the information privately flowing to the agents. In such settings there is greater scope for implementing desirable social outcomes.

From Arrow and Gibbard–Satterthwaite to Myerson–Satterthwaite, many of the most celebrated results in mechanism design are impossibility theorems. All of them require that the set of preferences agents might have, the preference domain, be sufficiently rich; a recurring trend in the literature has been to look at restricted domains where "good" social decisions can be implemented. In this chapter, I will argue that a way in which the preference domain could be restricted is by controlling the agents' information flow. This avenue has been relatively less explored, yet I think it has substantial practical importance, beyond the narrow area in which it has been studied.

While the standard assumption in the mechanism design literature is that they know their preferences, agents often need information about the "state of the world" in order to decide how to rank different alternatives. For example, one would want to know the details of a projected new development in the city center before having a firm opinion in favor or against it. In such circumstances, what could be accomplished if the "designer" controls both the information flowing to the agents and the process leading to a decision? Intuitively, controlling information as well as the decision process introduces a new trade-off. Less information is bad, as it prevents the designer to perfectly tailor the decision to the agents' preferences, but it is also good, as it reduces the manipulation opportunities of the agents. How much mileage may one get out of designing decision procedures and information flows at the same time?

Inspired by the work on Bayesian persuasion by Kamenica and Gentzkow (2011), in recent years there has been a flourishing of interest in information design. The approach of this literature can be quickly summarized by one of the opening sentences in the survey by Bergemann and Morris (2017): the "information designer ... can

C. Mezzetti (✉)
School of Economics, University of Queensland, Brisbane, Australia
e-mail: c.mezzetti@uq.edu.au

© Springer Nature Switzerland AG 2019                                                      335
J.-F. Laslier et al. (eds.), *The Future of Economic Design*, Studies in
Economic Design, https://doi.org/10.1007/978-3-030-18050-8_46

commit to providing information about the states to the players to serve her ends, but has no ability to change the mechanism". There is also a literature on agents deciding how much information to acquire before the designer selects a mechanism; see Bergemann and Välimäki (2007) for a survey. In this chapter I will focus on the smaller literature in which the designer can choose both the information the agents have and the decision mechanism. As far as I know, most of this literature has studied surplus extraction by a seller of a single item facing buyers with quasi-linear utility. I will thus briefly review this literature, but the reader should keep in mind that the main goal of this chapter is to suggest that the approach that I will call *decision and information disclosure mechanism design* could be profitably applied to many more problems.

The papers in the literature on surplus extraction by a seller typically assume independent private values, allow buyers to have some private information and let the seller control the additional information that buyers may receive. I will maintain the private value assumption, but to simplify matters further and focus on one of the main trade-offs, I will assume that buyers have no prior private information. They only know the distribution from which the value is drawn and hence they know its expected value. The second assumption, which is standard, is that the information disclosed by the seller to a buyer is privately observed by the buyer in question and only by her. The typical application is that the seller decides how much buyers can test the value of the item for sale (e.g., an asset) but does not observe the results of the tests. The main trade-off at work is that disclosing more information makes it feasible to achieve more efficient outcomes and hence, indirectly, to extract more surplus, but at the same time puts buyers in a position to extract more information rents from the seller.

The main goal of this literature is to study the ways in which judicious disclosure of information helps surplus extraction. To see the interplay between information disclosure and incentives, I will use a simple example.

*Example 1* There is a seller of an indivisible good facing $n$ buyers. The seller's cost of producing the good is zero. Buyer i's valuation is $v_i \in [0, 1]$. It is common knowledge that $v_i$ is drawn from a distribution $F$ with positive density in the support $[0, 1]$.

*Single Buyer.* Consider first the case of a single buyer. If the value $v_i$ is known to the buyer, but not the seller, then the revenue maximizing mechanism is to charge the price at which virtual value is zero; that is the price that solves $p - \frac{1-F(p)}{f(p)} = 0$, as long as the virtual value is increasing in the value.

If the value $v_i$ is unknown to all, then the revenue maximizing mechanism is to charge a price equal to the expected value of the item to the buyer $p = \mathbb{E}_F[v_i]$, as long as the buyer buys when indifferent.

Now suppose the seller may decide how much information the buyer should privately obtain, together with the selling mechanism. It is immediate to see that no information disclosure is efficient, and allows the seller to extract the entire ex-ante surplus.

The elementary lesson of this example with a single buyer is that disclosing private information interferes with surplus extraction as an information rent must be given to the buyer.

*Two or More Buyers.* Suppose now that there are at least two buyers. If the value $v_i$ is known to each buyer $i$, but not the seller, then the revenue maximizing mechanism is Myerson's (1981) optimal auction, which in this symmetric example can be implemented by any standard auction (first-price or second-price sealed-bid, ascending or descending open-bid) with a reserve price equal to the price at which virtual value is zero; that is, the price that solves $p - \frac{1-F(p)}{f(p)} = 0$, again as long as the virtual value is increasing in the value.

As in the case of a single buyer, if the value $v_i$ is unknown to the buyer, then the revenue maximizing decision mechanism is to randomly assign the item to one of the buyers and to charge a price equal to the expected value of the item, $p = \mathbb{E}_F[v]$, again as long as the buyers buy when indifferent.

Note that now disclosing no information mechanism is not ex post efficient. Depending on the distribution $F$ and on how many buyers there are, the seller may prefer to disclose full private information to no disclosure of information, or vice versa.

Bergemann and Pesendorfer (2007) showed that neither policy is optimal. They considered the following class of decision and information disclosure mechanisms. First the seller chooses the information each buyer gets. Then, based on this information, buyers participate in an allocation mechanism (which, by the revelation principle, can be a direct revelation mechanism without loss of generality). As in classical mechanism design, the seller is committed to follow the decision and disclosure mechanism. An important assumption is that buyers must be willing to participate once they have discovered their private information (i.e., the individual rationality constraint must hold at this interim stage).

Even with just two buyers, the optimal decision and information disclosure mechanism can be complex, depending on the shape of the distribution function $F$, but a simple characterization obtains if $F$ is uniform. In that case, the optimal mechanism in the class considered is to randomly select one of the buyers and disclose no information to her, while the information disclosed to the other buyer is a partitional information structure with two partitions: the buyer discovers whether her value is above or below $\mathbb{E}_F[v]$. The pricing decision mechanism is to charge a price $p'$ to the better informed buyer equal to $\mathbb{E}_F[v \mid v \geq \mathbb{E}_F[v]]$, the expected value of the item conditional on the value being in the top interval of the binary information partition. If the better informed buyer rejects the offer (because she has discovered that her value is in the bottom interval of the partition), then the item is offered to the uninformed buyer at price $p^U = \mathbb{E}_F[v]$.[1]

Esö and Szentes (2007) allowed the seller to charge buyers a fee before disclosing information. Thus, in their approach the participation constraint only needs to

---

[1] Bergemann and Pesendorfer (2007) showed that with general distributions and $n > 1$ buyers, the optimal information disclosure policy involves finite, interval, partitions and that the information partitions of different buyers are different - asymmetric - even if the prior $F$ is the same for all buyers.

hold at the ex-ante stage, before any information is disclosed. For Example 1 with $n > 1$ buyers, their results imply that full information disclosure is optimal for the seller.[2,3] Each buyer is charged a participation fee equal to what they expect to gain from participating in the allocation mechanism after they have been revealed their true values. The allocation mechanism is efficient. Hence the seller extracts all the surplus. Specifically, the allocation mechanism is a VCG mechanism, or a second price auction. Let $v_{(1)}$ and $v_{(2)}$ be the first and second order statistic, respectively. The expected revenue from the VCG is $\mathbb{E}_F[v_{(2)}]$. The expected payoff of the winner is thus $\mathbb{E}_F[v_{(1)} - v_{(2)}]$, as the $n$ buyers are a priori equally likely to be the highest value bidder, the participation fee can be set to be equal to $\frac{1}{n}\mathbb{E}_F[v_{(1)} - v_{(2)}]$. Thus expected revenue is $\mathbb{E}_F[v_{(1)}]$, which is higher than the expected revenue under the optimal mechanism in Bergemann and Pesendorfer (2007). Note that after the information disclosure stage of the mechanism (and before the second-price auction stage) a low value buyer regrets participating and would rather quit the mechanism if she could avoid paying the participation fee.

Bergemann and Wambach (2015) showed that the participation (or individual rationality) constraint can be strengthened if the seller uses a *dynamic decision and information disclosure mechanism*. The mechanism they propose is an ascending clock auction, with the clock price $p$ starting at zero and linearly increasing. When the price is $p$ the only information privately disclosed to a buyer is whether her value for the item is above or below $p$. The only choice a buyer has is when to drop out, irrevocably, from the mechanism. When there is only one buyer left that is active, the clock stops, information disclosure stops, and a decision is made. The object is awarded to the last buyer who has remained active after all the others have dropped out (with a coin flip deciding the winner if the last active buyers drop out at the same price). The winner of the object pays a price that depends on the clock price at which the last buyer dropped out and the clock stopped. If $p^D$ is such a stopping price, the winner pays her expected value for the item, conditional on this value being above $p^D$: $\mathbb{E}_F[v \mid v > p^D]$. It is immediate to see that such a dynamic decision and information disclosure mechanism yields the seller the same expected payoff as the optimal mechanism of Esö and Szentes, but contrary to their mechanism at no point does any agent regret participating. The participation constraint holds up to the end of the mechanism (it may only be violated if there is an additional ex post stage at which the winning buyer observes her true value); that is, the mechanism satisfies posterior individual rationality.

---

[2] Indeed, even with one buyer, once the buyer has been charged $\mathbb{E}[v_i]$, the seller could let her observe her true value.

[3] Esö and Szentes (2007) study a more general problem in which buyers initially have some private information and the seller may disclose additional, "orthogonal" information. They show that optimal surplus extraction requires that the seller fully discloses all the orthogonal information. Krähmer and Strausz (2015) prove that full information disclosure is no longer optimal if the ex-ante information of the buyers is discrete rather than continuous. Li and Shi (2017) prove that if a more general class of "direct disclosure" policies is allowed, then discriminatory, as opposed to full, disclosure could be optimal.

Gottardi and Mezzetti (2019) have applied the decision and information disclosure mechanism design approach to Myerson and Satterhwaite (1983) classic study of bargaining under incomplete information. The typical application is a seller who has acquired an asset and contemplates selling it to a buyer, with both needing to run tests to discover how much they value the asset. Gottardi and Mezzetti (2019) showed that if one restricts attention to the class of static mechanisms studied by Bergemann and Pesendorfer (2007), then one can find a decision and information disclosure mechanism with a dominant strategy equilibrium in which the expected gains from trade are higher than in the Bayesian mechanism that maximizes gains from trade when buyer and seller are fully informed. In the most efficient Bayesian mechanism under full information there is too little trade; sometimes trade does not take place even though the buyer values the item more than the seller. In the mechanism proposed by Gottardi and Mezzetti (2019), the gain from restricting the agents' information is to permit completion of some of the most efficient trades that are lost in the incentive efficient mechanism under full information. The cost is that it is sometimes the case that an inefficient trade takes place, but the loss in such cases is small and thus the expected cost of restricting information is more than offset by the expected gain. Gottardi and Mezzetti (2019) also show that when buyer and seller have no prior information it is possible to design a dynamic decision and information disclosure mechanism that is ex post efficient; more precisely, iterative elimination of weakly dominated strategies yield the ex post efficient outcome that trade takes place if and only if the buyer's value is above the seller's cost.

To convey my view that many interesting results are waiting to be uncovered in settings in which the designer is able to control the flow of information to the agents, I now conclude with a completely different example.

*Example 2* There are three alternatives $\{x, y, z\}$ and three agents. The domain of preferences is the set of all possible strict rankings, but agents may not know their preferences. Consider the social choice function $C$ which selects the alternative that is ranked at the top by at least two agents and if each agent has a different alternative as the most preferred, then $C$ selects alternative $x$.

By the Gibbard–Satterthwaite theorem the social choice function $C$ is not strategy-proof when agents are fully informed about their ranking, and hence it cannot be implemented in dominant strategy equilibrium. Note that $C$ only needs information about the top ranked alternative of each agent. Suppose that it is possible to control the information flow to the agents, so that they only learn which alternative is at the top of their ranking. As before, suppose this information is learned privately. There is an issue (perhaps a delicate issue) about how an agent should evaluate her preferences about the two alternatives over which she does not know her ranking. To simplify the example and make the point quickly, suppose the agent acts as if she is indifferent between them. Then truthfully reporting her top alternative is a dominant strategy in the mechanism that asks agents to vote for one alternative, and the social choice function $C$ can be implemented when the designer has full control over the information flow to the agents. Clearly, it will not always be realistic to assume such a level of control, but there are also many instances when control is available, as agents need information to make up their mind about the different alternatives.

# References

Bergemann, D., & Juuso, V. (2007). Information in mechanism design. In R. Blundell, W. Newey & T. Persson (Eds.), *Proceedings of the 9th World Congress of the Econometric Society* (pp. 186–221). Cambridge: Cambridge University Press.

Bergemann, D., & Pesendorfer, M. (2007). Information structures in optimal auctions. *Journal of Economic Theory, 137,* 580–609.

Bergemann, D., & Wambach, A. (2015). Sequential information disclosure in auctions. *Journal of Economic Theory, 159,* 1074–1095.

Bergemann, D., & Morris, S. (2017). *Information design: A unified perspective.* mimeo.

Esö, P., & Szentes, B. (2007). Optimal information disclosure in auctions and the handicap auction. *Review of Economic Studies, 74,* 705–731.

Gottardi, P., & Mezzetti C. (2019). *What do mediators do? An information and bargaining design point of view.* Mimeo.

Kamenica, E., & Gentzkow, M. (2011). Bayesian Persuasion. *American Economic Review, 101,* 2590–2615.

Krähmer, D., & Strausz, R. (2015). Ex post information rents in sequential screening. *Games and Economic Behavior, 90,* 257–273.

Li, H., & Shi X. (2017). *Discriminatory information disclosure.* mimeo.

Myerson, R. (1981). Optimal auction design. *Mathematics of Operations Research, 6,* 58–73.

Myerson, R., & Satterhwaite, M. (1983). Efficient mechanisms for bilateral trading. *Journal of Economic Theory, 28,* 265–281.

# Towards Transparent Mechanisms

## Matías Núñez

**Abstract** How easy is it for an agent to understand the consequences of his behavior when confronted with a mechanism? In an transparent mechanism, an agent should theoretically have almost no difficulties in doing so. Similarly, an experimental subject should find more intuitive a transparent mechanism than a non-transparent one. The current work argues that transparency should be a natural development of mechanism design in the future years. After reviewing some basic reasons underlying the interest of the transparency notion, several applications where transparent mechanisms can be of particular appeal are discussed.

Economic Design is a vast field concerned with the formal analysis of institutions. It aims to predict the behavior of agents interacting through different institutions by applying rational choice techniques. It hence allows both the practitioner and the researcher to compare different structures to organize elections, to design markets, to match individuals, to rank institutions, etc.. The toolkit of the economic designer is large; among others, strategic and normative approaches are often the basic methods when determining the quality of an institution. Recent years have also seen the development of two new main methods: experimental methods (in the lab to test game theoretical predictions and in the field to get a grasp of the implications in mass settings) and the algorithmic approach emerged from the interest of the computer science community in economic design.

Through these tools, the scholar literature has achieved to identify several important features of institutions such as their efficiency, their economic sustainability, their representativeness, etc.. In my view, the next years of mechanism design will add to this list a novel and important concept: the one of institutional transparency. The idea of transparency is based on the cognitive ability of each player to

M. Núñez (✉)
CNRS & Université Paris Dauphine, Paris, France
e-mail: matias.nunez@dauphine.fr

© Springer Nature Switzerland AG 2019                                               341
J.-F. Laslier et al. (eds.), *The Future of Economic Design*, Studies in
Economic Design, https://doi.org/10.1007/978-3-030-18050-8_47

understand the consequences of his actions. The more transparent[1] a mechanism, the less cognitively complex for an agent to compute his best responses. The transparency is often informally advocated by arguing that certain mechanisms are particularly intuitive to use and/or that the experimental results fit well the theoretical prediction. Yet, we lack some formal test or some conditions that ensure the transparency of a mechanism.I suspect that a potentially interesting development of the field is the development of (i) a formal definition of the transparency of institutions and (ii) practical transparency tests that help the economic designer in experimental settings to check whether the experimental subjects understand well the mechanism at hand. The next pages summarize my views of these two points and justify why we should pay interest to transparent mechanisms.

# 1   Beyond Direct Mechanisms

In order to properly define the notion of transparency I consider of interest, some comment is in order: one should not think that the transparent mechanisms are a particular class of the direct ones, in which the message space coincides with the type space. More precisely, a transparent mechanism need not be a direct one nor a direct one must be particularly transparent. A direct mechanism can aggregate the information in the agents' types in a rather cumbersome manner while an indirect mechanism can be very easy to understand while ignoring the players' types. In order to clarify the main idea behind the transparency of a mechanism, I now introduce two examples in the electoral context. Examples with a similar spirit can be found in a rather large number of economic environments such as auctions, matching and fair division among others.

*Example 1* Consider an election with two candidates *a* and *b* and two different voting rules. The two rules request each voter to vote for one of the candidates and ties are broken alphabetically. The first voting rule, Majority voting, selects the candidate with the most votes. The second rule selects the candidate with the most votes when the number of votes is even and the candidate with the least votes when this number is odd. It seems obvious that Majority Voting is more transparent than the second rule since (i) the message space is equal and (ii) the aggregation of the ballots is more straightforward. In this case, the more transparent mechanism (Majority voting) is more appealing.

---

[1]Milgrom (2010) introduces the notion of simplification of mechanisms, according to which a mechanism $\phi$ is simpler than a mechanism $\beta$ if the strategy space of $\phi$ is contained on the one of $\beta$, and $\phi$ does not generate additional equilibria. The ideas of simplicity and transparency are hence somewhat related. Li (2017) also proposes a notion of obvious strategy-proof mechanisms according to which each agent finds obvious recognize that misreporting his type leads to no gain. The idea of transparency is logically independent from Li (2017)'s one. For instance, a game in which the optimal misreport is intuitive (such as the average game as studied by Yamamura and Kawasaki 2013) is arguably transparent while clearly failing to be strategy-proof.

*Example 2* Consider the following voting rule: $k$-Evaluative Voting, with $k$ being a positive integer. With this rule, the scores assigned to each candidate vary from 0 up to $k$, the winner being the candidate with the most votes. When $k = 1$, the rule is simply Approval voting. Approval Voting is more transparent than Evaluative Voting since ($i$) the message space is smaller and ($ii$) the aggregation of the ballots is equal. In this case, the theoretical literature tends to consider both rules as equivalent since strategic voters have an incentive to overstate their vote. In this case, the more transparent mechanism seems to be preferable.

Intuitively, it need not be always the case that the more transparent, the better the mechanism. There should be some optimal level of transparency, that is, the mechanism should be transparent enough to ensure efficiency and while not perturbing the strategic decisions of the agents through unnecessary complicated options or complex computations.

The next example shows that in order to compare the transparency of two mechanisms, one should consider the overall structure of the mechanism and not focus exclusively on the aggregation of the messages (Example 1) nor on the message space (Example 2).

*Example 3* In an election held under Plurality with a Runoff, voters vote twice. In the first round, they vote for one of the candidates. The pair of candidates with the most votes go to the runoff or second round, the rest of them are removed from the election. In the second round, the voters vote for one of the surviving candidates and the majoritarian candidate wins the election. It seems rather normal to state that this rule is less transparent or simple than the standard Plurality rule. Yet, to compare this rule with Approval voting seems far from obvious.

How can we then control for the transparency of a mechanism?

First, in a experimental setting, a very intuitive manner to detect transparency appears. One can ask the agents whether they understand the mechanism, after a short presentation of its main features. Of course, this question is rather general and there might be over/underconfidence biases. To avoid them, one needs to build an objective measure of the ability of the subjects to discern the features of the mechanism. While I see difficult to obtain a universal test, one could easily design a series of examples for each particular mechanism in which the agents should guess the correct outcome and/or the consequences of their actions. In short, the experimental subjects could be evaluated on their proficiency of the mechanism. The share of subjects passing the test could be understood as a measure of the transparency of the institution. If a large enough share of them pass the test, one is inclined to think that the institution is fairly transparent.

Second, from a theoretical perspective, the possibilities are quite large. I suspect however that combining the ideas of hierarchy of beliefs and bounded rationality with the theory of implementation is a promising venue for this strand of research. Indeed, the theory of implementation has identified the objective functions we can expect to implement with rational players. However, the methods used by the implementation literature tend to be highly abstract and are often criticized by their complexity. The canonical example of an abstract mechanism is the integer game, the building block of the Nash implementation results. These games are conceived by their generality and elegance rather than for their practicality. A first step in this direction could be to compare two mechanisms by the sets of the best responses of agents with different levels of bounded rationality. It follows that a mechanism that gives a unique best response for each agent and each level of rationality seems to be the most desirable one from this perspective. However, one should also take into account the size of the message space and whether only certain type of agents are able to correctly guess their best response (as in the beauty-contest games).

## 2 Reasons for Transparency

In the usual economic models, there is no room for defining transparency.[2] Implicit in the classical framework is the following observation: a rational agent can understand *any institution independently of how complex it is.* However, it is often the case that agents differ through their understanding of the "rules of the game". The reasons abound; it can be either through education, cultural transmission or through private information. It could also be that the different players have different cognitive skills and this generates these levels of understanding. In other words, the bounded rationality of the agents explains why some agents understand well how an institution works whereas some of them do not. Moreover, it could be even the case that the agents face some uncertainty over the functioning of the institution and hence cannot even understand which are their best responses.

Since the goal of the economic designer is to predict the functioning of institutions, focusing on transparent institutions seems to be a key step. In transparent institutions, it is reasonable to expect that players behave as the canonical rational agent. The more transparent an institution, the closer the experimental and the theoretical results should be. This last point is important, since, as has been emphasized in the last years, we are far from a precise prediction of the agents' strategic behavior. The experimental literature has shown that strategic agents might fail to recognize some strategies as dominant. It can also be the case that agents that feel that their action has no impact in the final outcome behave at random. These two effects imply that the aggregate behavior might be very far from the one prescribed by the theoretical analysis. For instance, Van der Straeten et al. (2010) among others suggest that voters are neither fully sincere nor fully strategic: they tend to follow very simple patterns of strategic behavior.

---

[2]In computer science, the notion of complexity could be understood as a measure of the difficulty faced the agents when checking the result of the election. Yet, complexity does not seem to be conceived to measure the cognitive difficulty of understanding the rules of a game form. The concept of explainability in artificial intelligence claims that "axioms should be formulated to help explain the mechanism's choices to participants" (Procaccia 2018 in this same volume). While following a different approach, explainability pursues the same goal as transparency as it intends to ensure that agents do not face any ambiguity over the consequences of their acts when making decisions.

# 3 Applications of Transparent Mechanisms

I now detail three areas in which the notion of transparency seems to be relevant.[3]

**Sequential mechanisms**. For instance, if a voting rule involves several steps, it seems hard to verify that voters clearly anticipate the consequences of their actions. In a complex institution, an agent might fail to understand the incentives and, in turn, this lack of understanding might decrease the institutional efficiency. This is why an important desiderata of a sequential mechanism is that there are as few stages as possible so that backwards induction is relatively straightforward to execute as argued by Binmore et al. (2002) and De Clippel et al. (2014). When focusing on dynamic institutions (i.e. with several steps), one should privilege those where backward induction is less demanding, with as few steps as possible. Understanding how the number o f steps affects the degree of transparency seems to be of relevance.

**Participation**. Most voting rules are flawed with the No-show paradox in the sense that it might be rational not to vote in an election. More generally, a high degree of complexity of a voting rule can decrease the participation rate since agents might feel that opaque institutions are less representative. An interesting question is whether the voters understand and anticipate the No-show paradox, and whether we can avoid these concerns by constructing new and transparent voting rules.

**Single-Peaked Domain**. In order to clarify the role of transparent mechanisms in the single-peaked domain, consider Moulin's generalized median rules. In this domain, each agent has an ideal point, the peak; the further away the outcome is from the agent's peak, the lower his utility. The simplest one of the generalized median rules, the median rule, simply selects the median of the peaks of the agents (see Arribillaga and Massó 2016 and Moulin 2017 for recent work on this area). This rule is strategy-proof, unanimous and efficient and hence gathers most of the desiderata required in the literature. Therefore, if agents play their dominant strategy, the outcome is the median of the true peaks. However, if every agent announces the same value $x$, this generates an (Nash) equilibrium since no unilateral deviation modifies the outcome. This in turn implies that any point can be obtained as an equilibrium outcome. In other words, focusing on equilibrium outcomes can be very different than focusing on the outcomes reached when the agents choose their dominant strategy. This observation leads (Cason et al. 2006) to propose his notion of secure implementation that combines both ideas. Unfortunately, without economic transfers, secure implementation leads mostly to impossibility results. These impossibility results are obtained by focusing on direct mechanisms. Hence, a possible solution to this impossibilities could be to focus on indirect and transparent mechanisms.

**Acknowledgements** I would like to thank Haris Aziz, Peter Biro and Umberto Grandi for their comments that significantly upgraded the quality of this work.

---

[3]In the field of matching theory, concerns over the opacity of university admissions system have been identified, as for instance, the French APB platform. See Biro (2018) in this same volume for a contribution exploring matching markets and recent advances on this point.

# References

Arribillaga, R. P., & Massó, J. (2016). Comparing generalized median voter schemes according to their manipulability. *Theoretical Economics*, *11*(2), 547–586.

Binmore, K., McCarthy, J., Ponti, G., Samuelson, L., & Shaked, A. (2002). A backward induction experiment. *Journal of Economic theory*, *104*(1), 48–88.

Biro, P. (2018). Engineering design in matching markets. In J.-F. Laslier, H. Moulin, M. R. Sanver, & W. S. Zwicker (Eds.), *Future of economic design*. Studies in economic design. Switzerland: Springer.

Cason, T. N., Saijo, T., Sjöström, T., & Yamato, T. (2006). Secure implementation experiments: Do strategy-proof mechanisms really work? *Games and Economic Behavior*, *57*(2), 206–235.

De Clippel, G., Eliaz, K., & Knight, B. (2014). On the selection of arbitrators. *The American Economic Review*, *104*(11), 3434–3458.

Li, S. (2017). Obviously strategy-proof mechanisms. *American Economic Review*, *107*(11), 3257–87.

Milgrom, P. (2010). Simplified mechanisms with an application to sponsored-search auctions. *Games and Economic Behavior*, *70*(1), 62–70.

Moulin, H. (2017). One-dimensional mechanism design. *Theoretical Economics*, *12*(2), 587–619.

Procaccia, A. (2018). Axioms should explain solutions. In J.-F. Laslier, H. Moulin, M. R. Sanver, & W. S. Zwicker (Eds.), *Future of economic design*. Studies in economic design. Switzerland: Springer.

Van der Straeten, K., Laslier, J.-F., Sauger, N., & Blais, A. (2010). Strategic, sincere, and heuristic voting under four election rules: An experimental study. *Social Choice and Welfare*, *35*(3), 435–472.

Yamamura, H., & Kawasaki, R. (2013). Generalized average rules as stable nash mechanisms to implement generalized median rules. *Social Choice and Welfare*, *40*(3), 815–832.

# Some Open Questions on Manipulability and Implementation

**Hans Peters**

**Abstract** We discuss a few partially open and related questions on manipulability of social choice functions and implementation of social choice correspondences. In particular, self-optimality of a preference profile for a given social choice function as a game form means that it is a Nash equilibrium in the game where it is the true preference profile. Self-implementation of a social choice correspondence means that it can be implemented by a social choice function which is a selection from it.

## 1 Introduction

Social choice theory has much in common with the theory of economic design, in particular the design of multi-person mechanisms for choosing from a set of possible alternatives. Although social choice theorists also study existing decision (e.g., political voting) mechanisms or procedures, often the ultimate goal is to design procedures that have 'good' properties, where 'good' may mean quite different things.

Some of the celebrated classical results in social choice theory concern, basically, extreme cases. Arrow's classical result (1951, 1963) says that it is impossible to obtain, in a reasonable way, a social preference (ranking) of a set of alternatives, based on individual preferences, such that the social preference between any two alternatives $x$ and $y$ depends only on the individual preference between $x$ and $y$. An individual naive voter $j$ might find this condition ('independence of irrelevant alternatives') not desirable per se: if $j$ slightly prefers $y$ over $x$ and socially $x$ is only slightly preferred over $y$, then $j$ might expect $y$ to become socially preferred over $x$ if $j$'s preference of $y$ over $x$ becomes much stronger. Such reasoning includes cardinal preference arguments which are absent from Arrow's model, but nevertheless quite natural in reality. A good account of the implications of Arrow's result for the theory of democracy and welfare economics can be found in van den Doel and van Velthoven (1993)—one of my first acquaintances with social choice theory.

H. Peters (✉)
Maastricht University, Maastricht, The Netherlands
e-mail: h.peters@maastrichtuniversity.nl

© Springer Nature Switzerland AG 2019
J.-F. Laslier et al. (eds.), *The Future of Economic Design*, Studies in
Economic Design, https://doi.org/10.1007/978-3-030-18050-8_48

347

The classical result of Gibbard (1973) and Satterthwaite (1975) states that there is no reasonable social choice function where it is always optimal to report one's true preference. Assuming that every individual is indeed capable of establishing its own true preference, this condition ('strategy-proofness') is very desirable from the point of view of simplicity: there is no need to form beliefs about the preferences, true or reported, of others. Moreover, ex ante good properties of the mechanism (social choice function) are maintained ex post if individuals report truthfully. Still, strategic voting is in general not regarded as something that is wrong by definition and, moreover, the theorem says nothing about the degree of manipulability. Interesting, in this respect, is a 'quantitative' version of the Gibbard-Satterthwaite result for three alternatives which establishes a relation between the degree of manipulability and the closeness to dictatorship, see Friedgut et al. (2011).

In this note I will discuss some potentially interesting and partially open questions concentrated on manipulability and, related, implementation. Unless stated otherwise, the background model is the standard social choice model with a finite set of individuals $N$ who have strict preferences (linear orderings) on a finite set of alternatives. A social choice function [correspondence] assigns to each profile of preferences an alternative [nonempty set of alternatives].

## 2 Self-Optimality

The strategy-proofness or non-manipulability condition can also be formulated as follows. Consider the game form (or mechanism) in which each individual's strategy space is the set of all preferences over the set of alternatives, and the outcome function is a given social choice function. In other words, each individual 'plays' (reports) a preference, and the outcome is chosen by the social choice function applied to the 'strategy profile' (preference profile) resulting this way. By adding a profile of ('true') preferences to this game form a strategic (ordinal, in the standard model) game arises. Call a preference profile *self-optimal* if it is a Nash equilibrium in the game with that preference profile as true preferences. Then a social choice function is strategy-proof exactly if every preference profile is self-optimal. The Gibbard-Satterthwaite result says that for every social choice function with range at least three and which is non-dictatorial on this range – that is, there is no individual such that always that individual's best alternative within the range is assigned – there is at least one preference profile which is not self-optimal. This gives rise to a number of potentially interesting questions, not all of which have been completely answered so far.

One question is to find (non-dictatorial) social choice functions for which the 'probability' of a non-self-optimal preference profile is as small as possible. For instance, one can count the number such preference profiles and try to minimize it. This is basically the 'minimal manipulability' approach as initiated by Kelly (1988).

A number of results has been obtained, of which I mention only a few.[1] In Maus et al. (2007) it is established that under the additional assumptions of top-onliness (meaning that only the top alternatives of the preferences matter), surjectivity and anonymity, the number of self-optimal preference profiles is maximal for social choice functions that are 'unanimous with status quo': that is, they assign a prefixed alternative unless there is unanimous agreement on some other alternative. In fact, decision making in the European union is a noteworthy example of this. In a similar setting, Veselova (2016) considers what happens if the number of individuals increases. Clearly, generally speaking one expects the probability of a non-self-optimal preference profile to go to zero in that case, simply because the influence of any single individual vanishes. It then becomes interesting to study manipulation by larger groups of individuals: this is what may happen in reality (e.g., political voting), even though individuals do not communicate directly, as it is natural to assume that like-minded individuals expect each other to vote in the same way, in particular to strategically vote (manipulate) in the same way. One may also impose restrictions on the set of preferences and on the collections of (non-)self-optimal profiles, as in Arribillaga and Massó (2016).

Another potentially interesting approach could originate from the point of view of the mechanism designer. Following the admittedly strong but usual assumptions that the designer knows nothing about the preferences of the individuals but assumes that they play best replies and in particular a Nash equilibrium, and the individuals know each other's preferences, the designer can check whether a reported preference profile is self-optimal: if it is not, then this can be regarded as evidence of manipulation, and the designer may for instance maintain the status quo or impose some predestined unattractive alternative. This is somewhat related to the result in Maus et al. (2007) mentioned above: there the mechanism imposes a prefixed alternative—which in particular could be the status quo—unless the individuals report unanimously, where the latter is a quite special self-optimal preference profile. In a model quite different from the standard social choice model, namely in the very specific context of the Nash bargaining model, this is known to result in the individuals reporting linear utility functions (see Sobel 1981; Peters 1992).

# 3   Implementation

Next, I would like to concentrate on the issue of implementation, which again is related to the Gibbard-Satterthwaite result. A game form implements a social choice correspondence in x-equilibrium (e.g., Nash equilibrium, strong equilibrium, ...) if for every profile of true preferences the x-equilibrium alternatives of the resulting game coincide with the alternatives assigned to that preference profile by the social choice correspondence. A strategy-proof social choice function with range at least

---

[1]This is not a literature survey and I apologize beforehand to all those authors whose work could equally well have been mentioned.

three is dictatorial on its range according to the Gibbard-Satterthwaite result, and this implies that it is implementable in Nash equilibrium, e.g. by the game form above with strategy spaces equal to the set of preferences and the given social choice function as outcome function. In general, a sufficient condition for implementability in Nash equilibrium is Maskin monotonicity (see Maskin 1999).

The fundamental idea behind implementation is the following. A social choice correspondence can be viewed as a central rule or procedure to make decisions, having some nice and desirable properties. It may, however, be difficult or impossible to impose it, both for legal reasons and, more importantly, because of lacking or unverifiable information about the preferences of individuals. Thus, one looks for a game, or rather game form or mechanism (e.g., a set of practical laws or rules) which implements the given social choice correspondence in some x-equilibrium. There is an extensive literature on implementation over the past forty years; for an overview up to 2001 see Jackson (2001).

Game forms used for implementing social choice correspondences tend to be quite complicated. Typically, they may ask individuals to report complete strategy profiles, and use integers or similar devices as part of strategies in order to endow individuals or coalitions (depending on the equilibrium concept) with sufficient possibilities to deviate, thus avoiding equilibria with outcomes that are not contained in the set assigned by the social choice correspondence. Indeed, the 'complexity of the message space' (which is an alternative expression for the space of strategies) is an issue that has gathered quite some attention in the literature (see already Mount and Reiter 1974).

In this note I would like to draw attention to implementing game forms or mechanisms which look quite 'natural'—I will not formally use this word since it already has different meanings in this context (e.g., Saijo et al. 1996). Given the point of departure that one wishes to impose a 'desirable' social choice correspondence in a decentralized way, it would be convenient both for simplicity and in order to convince individuals to accept the game form (the rules of the game), if the social choice correspondence would implement itself. In what follows, I will give an exact definition of this idea.

## 4   Self-Implementation

A social choice function $F$ is a *selection* from a social choice correspondence $H$ if $F(R^N) \in H(R^N)$ for every preference profile $R^N$ of the set of individuals $N$. A social choice function $F$ can be regarded as a game form in the way explained earlier, resulting in a(n ordinal) game $(F, R^N)$ for every (true) preference profile $R^N$. We call social choice correspondence $H$ *self-implementable* in x-equilibrium if it is implementable in x-equilibrium by a selection $F$ from it.[2]

---

[2]The term 'self-implementation' is not new. See Remark 4.2.7 in Peleg (2002).

For instance, suppose that $H$ is implementable in Nash equilibrium by the selection $F$. Then each individual simply reports a preference, every Nash equilibrium of this game results in a desirable alternative (i.e., an alternative assigned by $H$ to the true preference profile), and every desirable alternative can be obtained this way. For the rest of this note I will concentrate on two questions: which social choice correspondences are self-implementable in Nash equilibrium? and which are self-implementable in strong equilibrium? I will start with the latter question since a substantial answer to this question can be derived from the existing literature.

## 5  Self-Implementation in Strong Equilibrium

Suppose that social choice function $F$ is a selection from choice correspondence $H$ and implements $H$ in strong equilibrium. In other words, for every (true) preference profile $R^N$ we have

$$H(R^N) = \{F(Q^N) \mid Q^N \text{ is a strong equilibrium of } (F, R^N)\}.$$

Recall that $Q^N$ is a strong equilibrium of $(F, R^N)$ if no coalition of individuals can deviate from $Q^N$ such that all members of the coalition are better off.

We say that the social choice function $F$ satisfies *no-veto-power* (see Maskin 1999; Jackson 2001) if no individual can on its own exclude any alternative in the range of $F$ from being chosen. More formally, if $x$ is an alternative in the range of $F$ and $j \in N$, then there is no preference $R^j$ such that $F(Q^N) \neq x$ whenever $Q^j = R^j$.

Suppose that the number of alternatives $m$ is smaller than or equal to the number of individuals $n$ plus one. Attach weights $\beta(x)$ (natural numbers) to the alternatives $x \in A$ such that the total weight is equal to $n + 1$. Given these weights a social choice correspondence $M_\beta$ can be defined, based on so-called feasible elimination (Peleg 1978). This works as follows. For a given preference profile, look for an alternative $x$ that occurs at bottom for at least $\beta(x)$ individuals. Delete that alternative from the profile, and also $\beta(x)$ individuals who have that alternative at bottom. Repeat this procedure until only one alternative is left. This results in a social choice correspondence $M_\beta$. The following result can be derived from combining several sources in the literature, starting with Peleg (1978). Here, I just refer to Peleg and Peters (2010), which covers most of this material, and to the self-contained proof in Peleg and Peters (2018).

**Theorem** *Let $m \leq n + 1$. Then for any weight function $\beta$ any selection $F$ from $M_\beta$ implements $M_\beta$ in strong equilibrium. Conversely, let $H$ be a social choice correspondence implementable in strong equilibrium by a selection $F$ which is anonymous and satisfies no-veto-power. Then $H = M_\beta$ for some weight function $\beta$.*

The condition that the number of individuals be at least the number of alternatives minus one is fulfilled in many applications—in particular political elections, and no-veto-power and anonymity are natural conditions in such applications. Thus, the above Theorem provides a rather complete answer to the question which social choice correspondences are self-implementable in strong equilibrium if the number of individuals is not very small. It does not cover situations where the number of individuals is small compared to the number of alternatives: for instance when a company of three has to choose from the extensive list of wines in a restaurant. That question is still open.

## 6 Self-Implementation in Nash Equilibrium

Which social choice correspondences $H$ are self-implementable in Nash equilibrium? This question is largely open, although I have some partial answers for the (natural) case where $H$ is anonymous and assigns $\{x\}$ if all individuals have $x$ on top. If there are two individuals then this is impossible (this already follows from Hurwicz and Schmeidler 1978.) If there are at least four individuals and at least three alternatives then the top correspondence, assigning exactly those alternatives that appear on top at least once, is self-implementable in Nash equilibrium. These and further results will appear in future work (Mukherjee and Peters 2018).

## 7 Concluding

The literature on (non)manipulability and implementation in economic mechanism design and in particular social choice theory provides many answers to what is theoretically possible. In reality, mechanisms are often relatively simple, and then questions arise about what is feasible given this simplicity. In this note I have gathered a few questions which seem to be interesting and, to the best of my knowledge, thus far unanswered.

## References

Arribillaga, R. P., & Massó, J. (2016). Comparing generalized median voter schemes according to their manipulability. *Theoretical Economics, 11*, 547–586.
Arrow, K. J. (1951, 1963). *Social choice and individual values*. New York: Wiley.
Friedgut, E., Kalai, G., Keller, N., & Nisan, N. (2011). A quantitative version of the Gibbard–Satterthwaite theorem for three alternatives. *SIAM Journal on Computing, 40*, 934–952.
Gibbard, A. (1973). Manipulation of voting schemes: A general result. *Econometrica, 41*, 587–602.
Hurwicz, L., & Schmeidler, D. (1978). Construction of outcome functions guaranteeing existence and Pareto optimality of Nash equilibria. *Econometrica, 46*, 1447–1474.
Jackson, M. O. (2001). A crash course in implementation theory. *Social Choice and Welfare, 18*, 655–708.

Kelly, J. S. (1988). Minimal manipulability and local strategy-proofness. *Social Choice and Welfare, 5*, 81–85.

Maskin, E. (1999). Nash equilibrium and welfare optimality. *Review of Economic Studies, 66*, 23–38.

Maus, S., Peters, H., & Storcken, T. (2007). Anonymous voting and minimal manipulability. *Journal of Economic Theory, 135*, 533–544.

Mount, K., & Reiter, S. (1974). The informational size of message spaces. *Journal of Economic Theory, 8*, 161–192.

Mukherjee, S., & Peters, H. (2018). On self-implementation in Nash equilibrium. Working paper, To appear

Peleg, B. (1978). Consistent voting systems. *Econometrica, 46*, 153–161.

Peleg, B. (2002). Game-theoretic analysis of voting in committees. In K. J. Arrow, A. K. Sen, & K. Suzumura (Eds.) *Handbook of social choice and welfare* (Chapter 8) (Vol. 1). North-Holland: Amsterdam.

Peleg, B., & Peters, H. (2010). *Strategic social choice: Stable representations of constitutions.* Heidelberg: Springer.

Peleg, B. & Peters, H. (2018). Self-implementation of social choice correspondences in strong equilibrium. GSBE Research Memorandum 18/005, Maastricht.

Peters, H. (1992). Self-optimality and efficiency in utility distortion games. *Games and Economic Behavior, 4*, 252–260.

Saijo, T., Tatamitani, I., & Yamato, T. (1996). Natural implementation with a simple punishment. *The Japanese Economic Review, 47*, 170–185.

Satterthwaite, M. (1975). Strategy-proofness and arrow's conditions: Existence and correspondence theorems for voting procedures and social welfare functions. *Journal of Economic Theory, 10*, 187–217.

Sobel, J. (1981). Distortion of utilities and the bargaining problem. *Econometrica, 49*, 597–620.

van den Doel, H., & van Velthoven, B. (1993). *Democracy and welfare economics.* Cambridge: Cambridge University Press.

Veselova, Y. (2016). The differences between manipulability indices in the IC and IANC models. *Social Choice and Welfare, 46*, 609–638.

# Some Issues in Mechanism Design Theory

Arunava Sen

**Abstract** The paper proposes some directions for research in mechanism design theory. It is well-known that the set of incentive-compatible social choice functions defined over a domain of ordinal preferences expands when randomization is permitted. It is proposed to characterize the set of random incentive-compatible social choice functions via the properties of the extreme points of this set. Problems with quasi-linear preferences and monetary transfers are also discussed.

One of the reasons underlying the extraordinary richness of mechanism design theory is that the structure of incentive-compatible social choice functions depends intricately on the restrictions placed on the set of possible preferences (or domain restrictions). Different problems require different methods of attack and answers vary widely depending on the formulation of the question. For instance, the same incentive-compatibility axiom has different consequences in voting, matching and auction design environments. Nevertheless, some general features do emerge in the theory. One of them is that possibility results are greatly enhanced if the designer has the option of *randomization* or allowing for *monetary compensation*.[1] However, the precise ways in which these devices improve the prospects for the existence of well-behaved incentive-compatibility, is not yet well-understood.

Let $A$ and $N$ denote finite sets of alternatives (or candidates) and agents (or voters) respectively with $|A|, |N| \geq 2$. A preference ordering is a ranking of the elements of $A$. A domain $\mathcal{D}$ is a collection of orderings. Every voting/allocation problem has a specific preference domain. For instance, the domain may be complete as in the Gibbard-Satterthwaite setting or single-peaked as in several political economy models and so on.[2] A random voting rule is a map $\varphi : \mathcal{D}^{|N|} \rightarrow \mathcal{L}(A)$ where $\mathcal{L}(A)$ is the set of probability distributions over $A$. A voting rule is a random voting rule that always

---

[1] In an interesting parallel, randomization and monetary compensation are also the two basic methods to achieve fairness in the allocation of indivisible objects (Young 1995).

[2] In the matching model, an alternative is an $|N|$-tuple where an agent is indifferent between two alternatives whenever her component of the allocation is unchanged.

---

A. Sen (✉)
Indian Statistical Institute, New Delhi, India
e-mail: asen@isid.ac.in

© Springer Nature Switzerland AG 2019                                          355
J.-F. Laslier et al. (eds.), *The Future of Economic Design*, Studies in
Economic Design, https://doi.org/10.1007/978-3-030-18050-8_49

outputs a degenerate probability distribution. A voting rule is *incentive-compatible* if the probability distribution under truth-telling (weakly) first-order stochastically dominates the probability distribution under any preference misrepresentation by an agent.[3] Another property of random voting rules that will be imposed is a weak form of efficiency called *unanimity*. This requires that an alternative first-ranked by all agents (if such an alternative exists) be chosen with certainty.

Two observations about the set of incentive-compatible random voting rules satisfying unanimity are important. The first is that it is a convex set. The second is that deterministic incentive-compatible rules satisfying unanimity are extreme points of this set. An obvious question is whether *all* extreme points of this set are deterministic. If the answer to this question is yes for domain $\mathcal{D}$, it will be said to satisfy the deterministic extreme point (or DEP) property. If a domain has the DEP property, *every* incentive-compatible voting rule satisfying unanimity can be obtained by choosing a deterministic incentive-compatible voting rule satisfying unanimity according to a fixed probability distribution. Randomization is "helpful" if and only if the relevant domain does not satisfy the DEP property. Although the discussion above is set in a voting model, the notion of a DEP domain can be formulated in models with monetary transfers—for instance in a auction domain with quasi-linear preferences. The same issues are pertinent in these models as well.

A few specific domains are known to be DEP domains such as the complete domain of strict preferences (Gibbard 1977), the single-peaked domain (Peters et al. 2014; Pycia and Unver 2015) and the multi-component domain with lexicographically separable preferences (Chatterji et al. 2012). However, not all domains are DEP domains as Chatterji et al. (2014) show. However there are no general results on DEP domains. Existing results rely heavily on the structure of deterministic strategy-proof social choice functions and do not provide any insight into the general problem. An even more challenging and perhaps more important problem is to determine all the extreme points in non-DEP domains. Answers to these questions are relevant for the finding solutions to "optimal second-best" random mechanism design problems.

A question of a similar flavour arises in models where monetary compensation is permitted and utilities are quasi-linear. As before, let $\mathcal{D}$ be a collection of orderings over $A$. For every $P_i \in \mathcal{D}$ let $v(P_i)$ be a utility representation of $P_i$ and let $\mathcal{V}(\mathcal{D})$ denote the set of all possible representations of preferences in $\mathcal{D}$. The utility of agent $i$ when alternative $a$ is selected and payment $p_i$ is imposed on her, is given by $v_i(a) - p_i$ where $v_i \in \mathcal{V}(\mathcal{D})$. The mechanism design problem is to find incentive-compatible allocation and payment functions $(f, P_1, \ldots, p_n)$ where $f : \mathcal{V}(\mathcal{D})^n \to A$ and $p_i : \mathcal{V}(\mathcal{D})^n \to \Re$. For example, $A$ could be possible locations of a public good (or public bad) with the preferences being single-peaked (resp. single-dipped). The departure from the standard location model formulation is that agents can now be charged, which seems quite natural. A characterization of incentive-compatible allocation functions in this setting is important but probably difficult. In particular, all incentive-compatible social choice functions where no transfers are made at any

---

[3] This follows Gibbard (1977). Other definitions of incentive-compatibility can be formulated and have been used.

profile remain incentive-compatible. However several additional allocation rules are incentive compatible—for instance, the efficient rule using VCG payments. A beautiful solution to the problem exists in the form of the *Affine Maximizer Theorem* (Roberts 1979) for the case where $\mathcal{D}$ is the complete domain. Very little is known in the case of other domains although Mishra et al. (2014) provide a less direct characterization in the single-peaked case in terms of the *cycle-monotonicity* condition of Rochet (1987).

# References

Chatterji, S., Roy, S., & Sen, A. (2012). The structure of strategy-proof random social choice functions over product domains and separable preferences. *Journal of Mathematical Economics, 48*, 353–366.

Chatterji, S., Sen, A., & Zeng, H. (2014). Random dictatorship domains. *Games and Economic Behavior, 86*, 212–236.

Gibbard, A. (1977). Manipulation of voting schemes that mix voting with chance. *Econometrica, 45*, 665–681.

Mishra, D., Pramanik, A., & Roy, S. (2014). A multidimensional mechanism design in single peaked type spaces. *Journal of Economic Theory, 153*, 103–116.

Peters, H., Roy, S., Sen, A., & Storcken, T. (2014). Probabilistic strategy-proof rules over single-peaked domains. *Journal of Mathematical Economics, 52*, 123–127.

Pycia, M., & Ünver, U. (2015). Decomposing random mechanisms. *Journal of Mathematical Economics, 61*, 21–33.

Roberts, K. W. W. (1979). The characterization of implementable choice rules. In J.-J. Laffont (Ed.), *Aggregation and revelation of preferences* (pp. 321–348). North Holland Publishing.

Rochet, J.-C. (1987). A necessary and sufficient condition for rationalizability in a quasi-linear context. *Journal of Mathematical Economics, 16*(2), 191–200.

Young, P. (1995). *Equity in theory and practice*. Princeton University Press.

# Implementation with Boundedly Rational Agents

**Kemal Yildiz**

**Abstract** A common underlying assumption in mechanism design literature is the rationality of the agents. That is, agents take actions as to maximize their preferences. However, ample evidence in marketing, psychology and behavioral economics shows that people's choices are not always consistent with the maximization of a preference relation. This inconsistency is of concern to principals, policymakers, institutional designers who aim to implement their goals by designing mechanisms for "real" agents who may not be rational, but may follow simple heuristics, rules of thumb in taking actions. The relevance and importance of taking bounded rationality into account when designing mechanisms is attracting greater attention especially since Thaler and Sunstein's famous book (Thaler, Sunstein, Nudge: improving decisions about health, wealth, and happiness, Yale University Press, 2008). However, only recently have we been presented with formal mechanism design models that incorporate bounded rationality. Here, we first present a brief overview and classification of these models. Then, we formally present a class of mechanisms, introduced by Koray, Yildiz (J Econ Theory, 176:479–502, 2018), that incorporates bounded rationality into the implementation problem.

## 1 Introduction

We first give an overview of the implementation problem. The social objective to be implemented—mostly represented by a social choice rule—is assumed to have been unanimously agreed upon in the society. Moreover, the objective to be implemented relates the social desirability of an outcome to the current societal preferences and other relevant parameters in a precise manner. It is exactly this ambitious aim, which

I thank Gabriel Carroll for his comments. I gratefully acknowledge the hospitality of New York University, Department of Economics during his visit in 2017–2018 academic year, and the support from the Scientific and Research Council of Turkey (TUBITAK) under grant number 1059B191601712.

K. Yildiz (✉)
Department of Economics, Bilkent University, Ankara, Turkey
e-mail: kemal.yildiz@bilkent.edu.tr

© Springer Nature Switzerland AG 2019
J.-F. Laslier et al. (eds.), *The Future of Economic Design*, Studies in Economic Design, https://doi.org/10.1007/978-3-030-18050-8_50

gives rise to the main problem concerning implementation, namely the designer lacks the ability to observe the individuals' actual preferences. To deal with this problem, Hurwicz (1972) introduced the notion of a mechanism. In the past three decades, many successful results have been obtained in identifying the set of social choice rules implementable according to widely used game theoretic equilibrium notions. In all these studies a common assumption is the rationality of the agents.

On the other hand, a recent and growing body of research, boundedly rational choice theory, seeks to explain documented choice behavior that a rational agent does not exhibit. A boundedly rational choice procedure, which an agent follows to make a choice, might contain a component that can be interpreted as the welfare preference of the agent, as well as components like an alternative that serves as a *status quo*, a *list* that provides an ordering to encounter the outcomes, a *second relation* used to shortlist some of the outcomes for a final choice, or an *attention filter* that provides the set of outcomes that grabs the attention of the decision maker at a first glance.

In mechanism design literature, bounded rationality is typically incorporated as to create an additional challenge, in which the designed mechanisms should implement the desired goals even if agents fail to behave rationally. Put differently, the designed mechanism should be robust to bounded rationality. However, as recently exemplified by Glazer and Rubinstein (2012), the designer can nudge the agents as to benefit from the bounded rationality of the agents and implement the goals that he cannot implement with rational agents via using classical tools such as costly screening or monetary transfers. We observe that the bulk of research in mechanism design with boundedly rational agents can be classified as either following, what we call, the robustness approach or the nudging approach. First let us briefly describe these two related, but different approaches.

**Robustness approach**: In this approach the designer should design a robust mechanism that would implement his goals even if the agents fail to behave rationally. Hence, the bounded rationality of the agents creates an additional challenge for the designer.

**Nudging approach**: In the nudging approach the mechanism designer implements his goals by taking the advantage of agent's bounded rationality, and implements the goals that he may not implement with rational agents. The nudging approach facilitates the implementation problem, since the designer has the additional power to nudge the agents by designing the external conditions such as the default action, the shortlisting rationale or the list that guides the agents in choosing their actions. Hence, nudging approach creates the possibility of implementation even if the designer does not have the traditional tools for implementation.

To best of our knowledge the first model that incorporates bounded rationality into mechanism design is offered by Eliaz (2002), who formulated an implementation model that allows faulty agents. In classical implementation framework preferences have two roles. First, given a social choice rule, a preference realization determines the set of acceptable outcomes. Second, given a mechanism, we assume that each agent takes an action so to maximize his preference relation. In fault tolerant implementation model of Eliaz, agents are endowed with preference relations, which are used to determine the set of acceptable outcomes as usual. However, not all the agents take actions so to maximize their preferences. The agents who do not act according to their preferences are called faulty. In this setting, the aim of the principal is to implement the set of acceptable outcomes regardless of the number and the identity of these faulty agents. This model perfectly exemplifies what we name as the robustness approach since the designer should design a mechanism for implementation that would tolerate arbitrary behavior of faulty agents.

In a rather recent paper, de Clippel (2014) provides a general framework, called *behavioral implementation*. Given a class of boundedly rational choice procedures, it is assumed that each agent is endowed with a choice procedure that belongs to this class. This leads to two departures from the classical setup. First, a social choice rule determines a set of acceptable outcomes for each choice function profile. Second, given a mechanism, each agent chooses an action according to his realized choice function. As in Eliaz (2002), the task of the principal is to implement the social goals given that each agent follows a procedure that belongs to the given class. Hence, this model is also more in line with the robustness approach.

In contrast to these two models, Glazer and Rubinstein (2012) propose a model in which the principal does not have the traditional tools for implementation, such as costly screening or monetary incentives, yet he can still implement his goals by nudging, that is by taking the advantage of agent's bounded rationality. Glazer and Rubinstein present a persuasion situation as a leader-follower relation. In this model, the persuasion rule and its frame is determined by the leader. The actions that the follower shows attention are restricted depending on how the persuasion rule is framed by the leader.

As complementary to the main model, Koray and Yildiz (2018) formulate a general strategic framework that incorporates bounded rationality into the implementation problem in line with the nudging approach. In this framework, the design object is called a deviation-constrained mechanism (dc-mechanism). In a classical mechanism, an agent can deviate from a joint strategy by choosing any strategy from his strategy set, as he is independent of the joint strategy from which the deviation is to be made. On this front, a dc-mechanism is different from a classical mechanism. Namely, deviation strategies of each agent are constrained depending on the joint strategy from which the deviation is to be made. Next, we introduce some necessary notation, and then formally present implementation via dc-mechanisms.

## 2 Implementation via Deviation-Constrained Mechanisms

We use $A$ to denote a nonempty, finite set of alternatives, and $N$ a nonempty finite set of agents. Each nonempty subset of $N$ is called a **coalition**, and denoted generically by $K$. Given $A$ and $N$, for each $i \in N$, $u_i$ denotes the **preference relation**—a complete, transitive, antisymmetric binary relation—on $A$ of agent $i$. For each distinct pair $a, b \in A$, $a\, u_i\, b$ denotes that "$i$ prefers $a$ to $b$". A **preference profile** $u = (u_1, \ldots, u_n)$. The collection of all preference profiles is denoted by $\mathcal{P}$. A **social**

**choice rule** (SCR) $F$ maps each preference profile into a nonempty subset of $A$, i.e. $F : \mathcal{P} \to 2^A \setminus \{\emptyset\}$. An SCR $F$ is **unanimous** if for each $a \in A$ we have $F(u) = \{a\}$ whenever every agent in the society prefers $a$ to all other alternatives.

Formally, a **dc-mechanism** is a triple $(M, \mathcal{D}, g)$. As usual, the joint strategy space $M = \Pi_{i \in N} M_i$, where $M_i$ stands for the strategy set of agent $i$. The outcome function $g$ maps every joint strategy to an alternative, i.e. $g : M \to A$. For each agent $i$, a constraint function $\mathcal{D}_i$ maps each joint strategy $m$ to a subset of $M_i$, i.e. $\mathcal{D}_i : M \to M_i$. In a dc-mechanism, if an agent $i$ would deviate from strategy $m$, he is constrained to choosing his strategy from $\mathcal{D}_i(m)$. Given a preference profile $u$, a joint strategy $m$ is an equilibrium of the dc-mechanism $(M, \mathcal{D}, g)$ at $u$ if for each $i \in N$ and $m_i' \in \mathcal{D}_i(m)$, $g(m) \; u_i \; g(m_i', m_{-i})$. We denote the equilibria of $(M, \mathcal{D}, g)$ at $u$ by $\mathbf{E}(M, \mathcal{D}, g, u)$.

**Definition 1** An SCR $F$ is implementable via a dc-mechanism $(M, \mathcal{D}, g)$ if for each $u \in \mathcal{P}$, $F(u) = g(\mathbf{E}(M, \mathcal{D}, g, u))$.

In classical implementation via mechanisms, Maskin (1985) shows that every Nash-implementable SCR is monotonic, and monotonicity combined with *no veto power* condition is sufficient for Nash implementability in the presence of at least three agents. It follows from the findings of Koray and Yildiz (2018) that Maskin monotonicity is both necessary and sufficient for a unanimous SCR to be implementable via a dc-mechanism in the presence of at least three alternatives. Before proceeding to the definition of Maskin monotonicity, let us introduce some useful notation. The **lower contour set of** $u_i$ **with respect to** $a \in A$, denoted by $\mathrm{L}(u_i, a)$, is the set of alternatives to which $a$ is preferred by agent $i$, i.e. $\mathrm{L}(u_i, a) = \{b \in A : a \; u_i \; b\}$. An SCR $F$ is **Maskin monotonic** if for each $u^1, u^2 \in \mathcal{P}$ and $a \in F(u^1)$, we have $a \in F(u^2)$ whenever for every $i \in N$, $\mathrm{L}(u_i^1, a) \subseteq \mathrm{L}(u_i^2, a)$.

**Proposition 1** *In the presence of at least three agents, a unanimous SCR $F$ is implementable via a dc-mechanism if and only if $F$ is Maskin monotonic.*

*Proof* Directly follows from Propositions 1 and 6 of Koray and Yildiz (2018). ■

Specific forms of deviation constraints have been considered previously in the mechanism design literature. For example, Hurwicz et al. (1995) assume that an agent can claim to have any subset of his true endowment but can not claim a larger set of endowments in a pure exchange economy setting. In the model of Glazer and Rubinstein (2012) that is discussed earlier, the persuasion rule and its frame is determined by the leader, and, as in a dc-mechanism, the strategies that the follower can choose are restricted depending on how the persuasion rule is framed by the leader. Implementation with dc-mechanisms differs from the behavioral implementation model of de Clippel (2014) in two main directions. First, de Clippel assumes that an SCR aggregates a choice function profile. In contrast, here an SCR aggregates a preference profile as usual. Secondly, in both models, given a mechanism, each agent can be boundedly rational in choosing an action. However, in dc-mechanisms it is assumed that the attention filter of the agents are also subject to design. Thus, one can implement social choice rules that can be implemented neither in the classical setup, nor in the setup of de Clippel.

# 3 Final Remarks

From the bounded rationality perspective, implementation via dc-mechanisms proposes a general framework in which agents may exhibit limited attention—in the vein of Masatlioglu et al. (2012)—to their available strategies. In this framework the designer can additionally design the actions that each agent shows attention whenever he considers a deviation from a given joint action. Assuming that the designer can shape the actions that grab the attention of each agent in any arbitrary way can be too general and unrealistic. However, Proposition 1 shows that even this unrealistic freedom of designing the deviation constraints does not bring much in terms of implementability. This is because deviation constraints should be both stringent enough to keep a particular strategy when it is an equilibrium at a preference profile, and also permissive enough to eliminate the same strategy when it is not an equilibrium at another preference profile.

As for the future studies, rather specific forms of dc-mechanisms can be formulated in which agents follow different boundedly rational choice procedures. In these models, the designer can be endowed with the additional power of nudging or completely designing the primitives such as a status quo, a list, or an attention filter that are relevant for the strategic choice of the agents. It would be an interesting theoretical exercise to see which SCRs are implementable via these models.

# References

de Clippel, G. (2014). Behavioral implementation. *The American Economic Review, 104*(10), 2975–3002.

Eliaz, K. (2002). Fault tolerant implementation. *The Review of Economic Studies, 69*(3), 589–610.

Glazer, J., & Rubinstein, A. (2012). A model of persuasion with boundedly rational agents. *Journal of Political Economy, 120*(6), 1057–1082.

Hurwicz, L. (1972). On informationally decentralized systems. *Decision and organization*. Amsterdam: North Holland.

Hurwicz, L., Maskin, E., & Postlewaite, A. (1995). Feasible Nash implementation of social choice rules when the designer does not know endowments or production sets. In *The Economics of Informational Decentralization: Complexity, Efficiency, and Stability* (pp. 367–433). Springer.

Koray, S., & Yildiz, K. (2018). Implementation via rights structures. *Journal of Economic Theory, 176*, 479–502.

Maskin, E. (1985). The theory of implementation in Nash equilibrium: A survey. In Hurwicz, L., Scheidler, D., & Sonnenschein, H. (Eds.), *Social goals and social organization*. Cambridge: Cambridge University Press.

Masatlioglu, Y., Nakajima, D., & Ozbay, E. Y. (2012). Revealed attention. *The American Economic Review, 102*(5), 2183–2205.

Thaler, R. H., & Sunstein, C. R. (2008). *Nudge: Improving decisions about health, wealth, and happiness*. Yale University Press.

# Part VIII
# Interpersonal Relations

# More About Social Relations

## Marc Fleurbaey

**Abstract** This paper makes a plea for incorporating the study of social relations into the theory of fair allocation, in order to better take account of the importance of the quality of social interactions in people's lives and in comprehensive theories of justice. This involves in particular modelling power and status in a more concrete way than has been done in the existing economic literature on voting power or social influence.

## 1 Social Interactions in Economics and Game Theory

There has always been a complex relationship between economics and sociology. Sociologists tend to see economists through the lens of homo economicus, and think that they, themselves, put a much greater emphasis on power relations in their own analysis of social interactions (e.g., Coleman 1994). Yet, economists have been interested in social interactions and in power, and substantial developments on these topics are owed to economists and game theorists. The study of strategic interactions (i.e., all of game theory), the measurement of power in simple games (e.g., Felsenthal and Machover 1998), the analysis of imitation effects (e.g., Durlauf 2001) as well as status seeking (e.g., Frank 1985), the study of network structures (e.g., Jackson 2008), the incorporation of identity concerns (e.g., Akerlof and Kranton 2000), and many similar themes, have been thoroughly pursued, illuminating, and influential. Empirical studies of the determinants of subjective well-being have found that employment and marital status, occupational prestige, and the quality of community life, have a strong impact (Layard 2005) and this clearly undermines the idea that ordinary people are selfish materialists.

M. Fleurbaey (✉)
Princeton University, Princeton, NJ 08544, USA
e-mail: mfleurba@princeton.edu; marc.fleurbaey@gmail.com

© Springer Nature Switzerland AG 2019
J.-F. Laslier et al. (eds.), *The Future of Economic Design*, Studies in
Economic Design, https://doi.org/10.1007/978-3-030-18050-8_51

367

## 2   The Theory of Fairness Focuses on Resources

But the field I know best, the theory of social choice and fair allocation (summarized in the Handbooks of Social Choice and Welfare edited by Arrow, Sen and Suzumura 1999–2011), has not done much to incorporate social interactions. The initial models of the theory of social choice, as developed in Arrow's (1951) seminal work, contain either purely abstract options (the "alternatives"), or a consumer model with ordinary commodities and standard consumer preferences. The models of fair allocation, as developed in the works of Kolm (1972) and Moulin and Thomson (1997) focus on the allocation of various types of resources, private or public goods, divisible or discrete goods, without or with production, but never directly feature social interactions. The models on which I have focused (as summarized in Fleurbaey 2008; Fleurbaey and Maniquet 2011) include non-transferable goods and (transferable) money, or production of a private good with unequal skills, but again, nothing that can represent social relations in an interesting way.

This is somewhat frustrating because if one thought of this theory as the formal branch of the broad class of theories of justice, one would think that social interactions should be quite important.

## 3   Is Philosophy Any Better?

As a matter of fact, a quick look at the state of philosophical theories of justice, an important source of inspiration for social choice and fairness, reveals that philosophers are often satisfied with a rather homo-economicus-like approach to social phenomena. Take Ronald Dworkin (2000), for instance. His theory of equality of resources is fleshed out in terms of a hypothetical insurance market in which individuals would buy insurance even for disadvantages that are not typically insurable in real-life markets, and the actual mechanism for redistributing resources would try to mimic the outcome of this hypothetical market. This is a dry market-equilibrium approach, and there is nothing specific about social relations in his theory. John Rawls' (1971) theory is more complex, since he puts basic liberties at the top of the principles of justice, and also thinks of certain "goods" as including "prerogatives and powers" as well as "the social bases of self-respect." By and large, though, his approach remains centered on resources, and theorized in terms of "goods." Interestingly, even Michael Walzer (1983), who extensively wrote about power and social relations, also conceives of the "spheres" of his approach as delineated by the specific "good" that they distribute—power being one of them.

There are philosophers, such as Elizabeth Anderson (1999) and Samuel Scheffler (2005), who really want to depart from the commodity framing, and articulate their theory in terms of social relations being respectful of the basic equal dignity of every member of society (see also Wolff 1998). They somewhat overshoot, since they reject the idea that distributions of goods are central to social justice, and want to define social justice in terms of social relations. That seems excessive, as the distribution of resources cannot be completely ignored or treated only derivatively. Moreover, their description of social justice in terms of just social interactions remains rather vague and abstract. It is hard to avoid reading their approach as focusing on the distribution of specific relational goods, which is precisely the reading they seek to avert.

# 4 Why the Fuss?

Why should we worry about social relations, in the context of social choice and fair allocation? To cut short a long story, the current predicament of societies nowadays bears the mark of the legacy of deep and entrenched unfairness against women, workers, and non-white ethnic groups. This unfairness has not been only a matter of resource distribution, but also, and perhaps primarily, a matter of status and recognition, as well as control and empowerment. It is embarrassing that our core theories have little specific to say about these historical forms of unfairness, and little to say about how to resolve them.

One might argue that this is not really necessary, since some models of social choice are sufficiently abstract to encompass any social issue that can be deemed relevant, and some models of fair allocation can be reinterpreted in terms of functionings (Sen 1992) rather than commodities. This argument would be convincing if the abstract models contained all the relevant ethical considerations, but this is dubious. The models of fair allocation are more concrete than the abstract social choice model, precisely because one could not express key notions of fairness such as no-envy or egalitarian equivalence, lower bounds and solidarity requirements, without the economic structure that fair allocation models display. And among the models of fair allocation, some are able to introduce considerations that are specific to public goods or to production and unequal skills, again because they incorporate the required economic structure. Therefore, it seems most likely that ethical notions bearing on dignity, control, power, status, recognition, and so on, need the concrete structure of similar models to be studied in terms of social orderings or allocation rules.

Another objection would say that it is not really necessary to worry about the quality of social interactions, since one can simply assume that fairness requires such interactions to be of a certain kind, i.e., with sufficiently equal respect and dignity for all. Obviously, for instance, fairness excludes treating people differently on the basis of race, and one can read the usual models in which race is absent as implicitly embodying the idea that race is irrelevant for resource allocation in a fair society. But this objection assumes that the object of analysis is allocation rules that select fair and efficient allocations and procedures. It does not hold when the purpose is to construct social orderings over imperfect allocations and procedures—and this comparative approach to fairness is more relevant to the study of real-life situations and policies, as argued in Fleurbaey and Maniquet (2011).

# 5   What Modeling Strategy?

Once these objections are set aside, it seems that the structure of the suitable models remains to be imagined, because existing models of social relations focus on very specific contexts that seem to fail to simultaneously capture the main dimensions of real-life situations in which individuals interact. For instance, models in which power indices are studied contain simple games in which simple sets of options (often binary) are considered, and seem insufficient to capture, for instance, the power relations between employer and employee in a firm, or between husband and wife in a household. Models of social interactions (imitation or status seeking) typically focus on one action variable, and introduce a social comparison term in agents' utility, leading to endogenous loops creating interesting multiplier effects, or prisoner's dilemma situations; again, such unidimensionality appears too limiting for a study of fairness in social relations.

A Beckerian variation around the consumer model (in reference to Becker 1974) seems a possible direction. Imagine that each individual interacts with others in a specific network structure defining who interacts with whom in a particular way. For instance, there is a market game in which the individual interacts with production firms, a household game in which the interaction is restricted to household members, a workplace game in which the interaction is restricted to colleagues and managers, and so on. Each game, in addition to having special members, has special rules and specific outcomes. For instance, the household game involves work and production, but the output is directly consumed by the members, whereas the production game in the firm involves selling the output and sharing the proceeds. One can also add a layer in which the memberships of the various games become endogenous, people being introduced or expelled, welcomed or refused.

# 6   Power and Status in Games

It remains to imagine how power and control, status and similar aspects of social interactions would appear in such games. Most of these interactions can be described as about activities that have to be coordinated with other people, and distribution of resources. Power and control are then described in terms of influence over the choice of activities and the distribution. For instance, a typical wage earner negotiates a wage when entering a firm, then obeys orders about tasks, sometimes gets to choose some tasks if belonging to management, and receives the agreed reward. In contrast, a coop member gets a voice and vote about the general choice of activities in the firm, as well as about the ex post distribution of resources between earnings and investment.

In addition to holding a certain degree of power, status (in the sense of standing, inclusion and recognition rather than simply in terms of ranking in a pecking order, as in the status-seeking economic literature) can come in two forms, either by granting or blocking access to certain activities (or even to certain groups and associations), or by flowing from taking up and succeeding in certain activities.

People have preferences over activities and resources that derive from more fundamental preferences over comfort, control, status, and genuine taste for various activities. They may also have preferences over who they interact with in these activities. Individuals will also adopt strategies that are coordinated across spheres of interaction. For instance, they may seek to obtain better status or more resources at work in order to have a better family life, and conversely.

# 7 Stylized Facts and Tractability

In building such models, the challenge is not to find their basic structure, apparently, but to find sufficiently stylized models that capture the relevant dimensions of the problems without falling into intractable complexity.

For some purposes, for instance, to measure inequalities or unfairness in terms of the distribution of individual situations measured both in terms of resources and in terms of quality of social interactions, it might be sufficient to rely on reduced-form descriptions of the interaction games, simply relating the inputs brought by individuals (their degree of control, their initial status, their resources) to the outputs they receive in the same dimensions after the interaction has taken place. Such an approach is pursued in Fleurbaey (2017), in a companion paper that focuses on theories of social justice and measures of inequality.

# 8 A Pioneering Example

Hupkau and Maniquet (2017) provide an interesting example of the type of analysis envisioned here. Consider the interaction between welfare recipients and social agencies. Social agencies may prefer to help needy people of a suitable type (e.g., those who care about their children, who do not drink…), but finding out about people's type is not so easy. If the bad-type people can mimic the good-type people at a cost that is not prohibitive, the following collapse of the social policy may happen. Good-type people may find it outrageous to be suspected of being of the bad type, and may abhor the interaction with social workers in which this suspicion poisons the relationship. Therefore they may prefer not to take up the opportunity to receive this benefit. But then, in equilibrium, social agencies learn that most of the applicants are of the bad type, and are no longer willing to implement the policy since the target population is not there. In this case everyone suffers from the collapse of the policy.

Replacing this policy by an unconditional policy that accepts the bad types would produce a Pareto improvement in the sense that all the types would now benefit from the policy. Of course, this policy is more expensive than a policy that excludes the bad type and delivers the benefit to the good types, but such a policy is not feasible.

This example relies on the particular mechanism that the good types are ashamed or feel insulted when the interaction with social workers is intrusive or involves a suspicion that they might be of the bad type. This is an "identity" mechanism (protecting one's identify as a good type) akin to other analyses in Akerlof and Kranton (2000). But there are many ways in which the interaction can be undesirable in the context of aid conditionality. Another problem with conditionality in income support is that it involves delays and costs due to monitoring, as well as uncertainty which reduces the ex ante value of applying, for the potential recipients. And beside those who abhor the suspicion of being of the bad type, there are also those who easily adopt a sheepish attitude of subservience and somehow reinforce the power imbalance between the social workers and the recipients.

Hupkau and Maniquet's work is very good at showing that the consequences of people caring about the quality of the social interaction can have consequences that are material, not just subjective. The productive sector is also plagued by similar phenomena. There are many reasons to think that productive firms in which social interactions are bad are less efficient, due to lack of trust, insufficient information flow, dampened motivation, and so on (Lazear 1995).

# 9   Conclusion

Whether social relations will become more prominent in welfare economics and economic design in the future may depend on the possibility to transform the vague ideas of this paper into rigorous concepts amenable to empirical measurement. Therefore one should probably think of moving at the same time on the empirical side and the theoretical side. Good measures and telling facts can motivate theoretical development, while good theory is needed to understand available data and design better data sets.

**Acknowledgements** This paper has benefited from conversations with F. Maniquet.

# References

Akerlof, G. A., & Kranton, R. E. (2000). Economics and identity. *Quarterly Journal of Economics, 115,* 715–753.
Anderson, E. (1999). What is the point of equality? *Ethics, 109,* 287–337.
Arrow, K. J. (1951). *Social choice and individual values.* New York: Wiley.
Arrow, K. J., Sen, A. K., & Suzumura, K. (1999–2011), *Handbook of social choice and welfare* (Vol. 2). Amsterdam: North-Holland.
Becker, G. S. (1974). A theory of social interactions. *Journal of Political Economy, 82,* 1063–1093.
Coleman, J. S. (1994). *Foundations of social theory.* Cambridge, Mass.: Belknap Press.
Durlauf, S. N. (2001). A framework for the study of individual behavior and social interactions. *Sociological Methodology, 31,* 47–87.
Dworkin, R. (2000). *Sovereign virtue.* Cambridge, Mass.: Harvard University Press.

Felsenthal, D. S., & Machover, M. (1998). *The measurement of voting power*. Cheltenham: Edward Elgar.

Fleurbaey, M. (2008). *Fairness, responsibility, and welfare*. Oxford: Oxford University Press.

Fleurbaey M. (2017). Inequalities, social justice, and the web of social interactions. Mimeo.

Fleurbaey, M., & Maniquet, F. (2011). *A theory of fairness and social welfare*. Cambridge: Cambridge University Press.

Frank, R. (1985). *Choosing the right pond: Human behavior and the quest for status*. Oxford: Oxford University Press.

Hupkau, C., & Maniquet, F. (forthcoming 2017). Identity, non-take-up and welfare conditionality. *Journal of Economic Behavior and Organization*.

Jackson, M. O. (2008). *Social and economic networks*. Princeton: Princeton University Press.

Kolm, S. C. (1972). Justice et équité, Paris: Ed. du CNRS.

Layard, R. (2005). *Happiness: Lessons from a new science*. London: Allen Lane.

Lazear, E. P. (1995). *Personnel economics*. Cambridge, Ma.: MIT Press.

Moulin, H., & Thomson, W. (1997). Axiomatic analysis of resource allocation problems. In K. J. Arrow, A. K. Sen, & K. Suzumura (Eds.), *Social choice re-examined* (Vol. 1). London: Macmillan and New-York: St. Martin's Press.

Rawls, J. (1971). *A theory of justice*. Cambridge, Mass.: Harvard University Press.

Scheffler, S. (2005). Choice, circumstance, and the value of equality. *Politics, Philosophy and Economics, 4*, 5–28.

Sen, A. K. (1992). *Inequality reexamined*. Oxford: Clarendon Press.

Walzer, M. (1983). *Spheres of justice: A Defense of Pluralism and Equality*. New York: Basic Books.

Wolff, J. (1998). Fairness, respect, and the egalitarian ethos. *Philosophy & Public Affairs, 27*, 97–122.

# Collective Time Preferences

**Anke Gerber**

**Abstract** This note discusses a possibility result for collective intertemporal decision-making. For the special case of timed outcomes and an odd number of individuals, we show that individual preferences over discount factors are single-peaked. Consequently, simple majority voting over the collective discount factor defines a transitive social preference relation on the set of discount factors $(0, 1)$ and the voting rule that assigns to any profile of individual discount factors the unique Condorcet winner is coalitionally strategy-proof.

## 1 Introduction

In this note we will draw attention to a specific problem in the field of economic design that is of utmost importance for practical collective decision-making and yet has not been explored to the full extent so far: How should a group of individuals take a collective decision on a common intertemporal consumption stream? Needless to say, the question is highly relevant since many, if not most decision-making problems involve some trade-off between earlier and later rewards. This includes consumption and saving decisions by households, taxation and spending decisions by governments and, last not least, decisions on environmental protection and climate policy by an international community of states.

Early on researchers have realized the specific difficulties of aggregating heterogeneous time-preferences (Marglin 1963; Feldstein 1964). For private consumption streams, Gollier and Zeckhauser (2005) and Zuber (2011) have shown that a representative agent or social planner will have time-inconsistent preferences as soon as there is some heterogeneity in individual discount rates.

The recent paper by Jackson and Yariv (2015) seems to support this pessimistic view on the aggregation of intertemporal preferences. The authors consider common consumption streams and demonstrate that every Pareto efficient collective utility

A. Gerber (✉)
Department of Economics, Universität Hamburg,
Von-Melle-Park 5, 20146 Hamburg, Germany
e-mail: anke.gerber@uni-hamburg.de

© Springer Nature Switzerland AG 2019                                                          375
J.-F. Laslier et al. (eds.), *The Future of Economic Design*, Studies in
Economic Design, https://doi.org/10.1007/978-3-030-18050-8_52

function must either violate time-consistency or non-dictatorship in the presence of heterogeneity in individual time preferences. While this result can be seen in the tradition of social choice theory with its numerous impossibility results there is an important difference, e.g. to Arrow's impossibility theorem (Arrow 1951, 1963): The dictatorship property implied by efficiency and time-consistency in Jackson and Yariv (2015) is only local, i.e. the dictator, whose utility function coincides with the collective utility function, may change with the preference profile. The local dictatorship property has been pointed out by the authors in passing (Jackson and Yariv 2015, p. 161), but in our view it has not received the recognition it deserves. After all, as long as the local dictator is chosen by a normatively sound procedure, there is nothing objectionable about a collective utility function that coincides with one individual's utility function at every utility profile. Hence, it seems promising to switch attention from finding a consensus in every intertemporal decision situation to the election of a local dictator at every utility profile.

For illustration we consider the special case where individuals have discounted utility functions over timed outcomes[1] and only differ with respect to their time-preferences, i.e. their discount factors. Then, under some natural assumptions, individual preferences over discount factors are single-peaked which guarantees the existence of a Condorcet winner and the coalitional strategy-proofness of simple majority voting over the collective discount factor. Thus, while the collective time preference is the time preference of one voter (the median) it is derived from a decision-making procedure that has appealing normative properties.

## 2   Voting on Collective Time Preferences

Consider a set of individuals $N = \{1, \ldots, n\}$ who have to take a collective decision over timed outcomes $(x, t)$, where payoff $x \in \mathbb{R}_+$ is received at time $t \in \mathbb{R}_+$.[2] Time $t = 0$ corresponds to the present day, when the decision is being taken. The study of preferences over timed outcomes goes back to Fishburn and Rubinstein (1982) and many experimental studies have used timed outcomes to elicit individual time preferences (see Frederick et al. 2002).

For simplicity we assume that $n$ is odd. Every individual $i$ has a continuous preference ordering over the set of timed outcomes $\mathbb{R}_+^2$ which is represented by a discounted utility function

$$U_i(x, t) = \delta_i^t u(x) \text{ for all } (x, t) \in \mathbb{R}_+^2, \tag{1}$$

where $\delta_i \in (0, 1)$ and $u : \mathbb{R}_+ \to \mathbb{R}$ is a continuous and increasing instantaneous utility function. We normalize $u$ such that $u(0) = 0$. Thus, we assume that all individuals have the same instantaneous utility function, but we allow for heterogeneous discount

---

[1] A timed outcome delivers a nonzero payoff at one point in time only.

[2] $\mathbb{R}_+$ denotes the set of nonnegative real numbers.

factors $\delta_i$. From (1) it is immediate that every $i$'s utility function over timed outcomes is *stationary*, i.e. for all $(x, t), (y, s) \in \mathbb{R}_+^2$,[3,4]

$$U_i(x, t) = U_i(y, s) \iff U_i(x, t + \Delta) = U_i(y, s + \Delta) \text{ for all } \Delta \geq 0.$$

Given our assumption that every individual has the same instantaneous utility function, the profile of individual preferences is uniquely characterized by the profile of discount factors $D = (\delta_1, \ldots, \delta_n)$. Consider the collective utility function $V :$ $(0, 1)^n \times \mathbb{R}_+^2$ given by

$$V[D](x, t) = \delta_M^t u(x) \text{ for all } D \in (0, 1)^n \text{ and } (x, t) \in \mathbb{R}_+^2,$$

where $\delta_M$ is the median discount factor at $D$. Since $V[D]$ is the median's utility function it follows that $V$ satisfies Pareto efficiency[5] and stationarity. Note that the median is a local dictator at the given preference profile. We will now show that $\delta_M$ is the Condorcet winner under simple majority voting over the collective discount factor and hence is determined by an aggregation procedure that has normatively appealing properties.

We first derive individual preferences $\succsim_i$ over collective discount factors $\delta \in (0, 1)$. It seems natural to assume that $i$ weakly prefers $\delta$ over $\hat{\delta}$, if the preference over two timed outcomes under $\delta$ coincides with $i$'s true preference under $\delta_i$, whenever the latter coincides with the preference under $\hat{\delta}$. Formally, let $\delta, \hat{\delta} \in (0, 1)$. Then $\delta \succsim_i \hat{\delta}$, if the following condition is satisfied:

For all $(x, t), (y, s) \in \mathbb{R}_+^2$,

$$\delta_i^t u(x) \geq \delta_i^s u(y) \text{ and } \hat{\delta}^t u(x) \geq \hat{\delta}^s u(y)$$
$$\text{implies that} \quad \delta^t u(x) \geq \delta^s u(y). \tag{2}$$

Note that by (2) the preference relation $\succsim_i$ is complete and transitive on each subinterval $(0, \delta_i]$ and $[\delta_i, 1)$. Moreover, if $0 < \hat{\delta} \leq \delta \leq \delta_i$ or $\delta_i \leq \delta \leq \hat{\delta} < 1$, then $\delta \succsim_i \hat{\delta}$ and the preference is strict, i.e. $\delta \succ_i \hat{\delta}$, whenever $0 < \hat{\delta} < \delta \leq \delta_i$ or $\delta_i \leq \delta < \hat{\delta} < 1$.[6]

It remains to consider the case where $\delta < \delta_i < \hat{\delta}$. Since $u$ is continuous and $u(0) = 0$, there exist timed outcomes $(x, t), (x', t'), (y, s), (y', s') \in \mathbb{R}_+^2$ such that

---

[3]Note that stationarity is equivalent to time consistency in Jackson and Yariv (2015) if consumption streams have nonzero payoffs in one period only.

[4]If decision time can vary and $U_i^\tau(x, t) = \delta_i^{t-\tau} u(x)$ represents $i$'s preferences over timed outcomes $(x, t)$ with $t \geq \tau$, where $\tau \geq 0$ is decision time, then $i$'s preferences also satisfy *time-invariance* (preferences are independent of a common delay in decision time and payoff times) and *time-consistency* (preferences over timed outcomes are independent of decision time).

[5]A collective utility function $V$ satisfies *Pareto efficiency* if for all $D = (\delta_1, \ldots, \delta_n)$ and all $(x, t), (y, s) \in \mathbb{R}_+^2$, $\delta_i^t u(x) > \delta_i^s u(y)$ for all $i = 1, \ldots, n$, implies that $V[D](x, t) > V[D](y, s)$.

[6]$\delta \succ_i \hat{\delta}$ if $\delta \succsim_i \hat{\delta}$ and not $\hat{\delta} \succsim_i \delta$.

$$\delta^{t-s} < \frac{u(y)}{u(x)} \le \delta_i^{t-s} < \hat{\delta}^{t-s} \tag{3}$$

and

$$\delta^{t'-s'} < \delta_i^{t'-s'} \le \frac{u(y')}{u(x')} < \hat{\delta}^{t'-s'}, \tag{4}$$

i.e. from (2) we can neither conclude that $\delta \succsim_i \hat{\delta}$ nor that $\hat{\delta} \succsim_i \delta$. In situation (3) $i$ prefers $\hat{\delta}$, while in situation (4) $i$ prefers $\delta$. Hence, $i$'s preference over discount factors $\delta, \hat{\delta}$ with $\delta < \delta_i < \hat{\delta}$ will depend on $i$'s belief of the relative frequency of facing the choice situations in (3) and (4). For our purposes it is sufficient to assume that every individual $i$ has a complete and transitive preference relation $\succsim_i$ over $(0, 1)$ such that for $\delta, \hat{\delta} \in (0, 1)$, (2) implies that $\delta \succsim_i \hat{\delta}$. From the above it then follows that $\succsim_i$ is single-peaked on $(0, 1)$ with a peak at $\delta_i$, i.e.

$$\hat{\delta} < \delta \le \delta_i \quad \text{or} \quad \delta_i \le \delta < \hat{\delta}$$
$$\Rightarrow \delta \succ_i \hat{\delta}. \tag{5}$$

Given that all individual preferences on the set of discount factors $(0, 1)$ are single-peaked, simple majority voting over the collective discount factor defines a transitive social preference relation on $(0, 1)$. Moreover, since $n$ is odd, at every profile $D = (\delta_1, \ldots, \delta_n)$ there exists a unique Condorcet winner $\delta_M$ which is the median discount factor in $D$. Finally, the voting rule that assigns to any profile $D$ the unique Condorcet winner at $D$ is coalitionally strategy-proof.

## 3 Outlook

The case of timed outcomes we have considered above is certainly special. With general consumption streams there can be choice situations, where an individual prefers a discount factor that is far away from her own discount factor over one that is close by, while in other situations she prefers the closer discount factor. The individual's preference over discount factors will then depend on her belief about facing choice situations where she has inconsistent preferences over the same pair of discount factors. Hence, in order to retain single-peakedness, one has to impose some restrictions on individuals' beliefs or on the set of available consumption streams.

We believe it is worthwhile following this route and exploring further the possibilities left open by the impossibility result in Jackson and Yariv (2015). To quote from Feldstein (1964), p. 367: "democratic theory does not require that each decision represent a consensus, but that government action as a whole be acceptable to the electorate."

# References

Arrow, K. J. (1951, 1963). *Social choice and individual values* (2nd ed.). New York: Wiley.

Feldstein, M. S. (1964). The social time preference discount rate in cost benefit analysis. *Economic Journal, 74*, 360–379.

Fishburn, P. C., & Rubinstein, A. (1982). Time preference. *International Economic Review, 23*, 677–694.

Frederick, S., Loewenstein, G., & O'Donoghue, T. (2002). Time discounting and time preference: A critical review. *Journal of Economic Literature, 40*, 351–401.

Gollier, C., & Zeckhauser, R. (2005). Aggregation of different time preferences. *Journal of Political Economy, 113*, 878–896.

Jackson, M. O., & Yariv, L. (2015). Collective dynamic choice: The necessity of time inconsistency. *American Economic Journal: Microeconomics, 7*, 150–178.

Marglin, S. A. (1963). The social rate of discount and the optimal rate of investment. *The Quarterly Journal of Economics, 77*, 95–111.

Zuber, S. (2011). The aggregation of preferences: Can we ignore the past? *Theory and Decision, 70*, 367–384.

# Economic Design for Effective Altruism

Dominik Peters

**Abstract** A growing movement of 'effective altruists' are trying to find the best ways of helping others as much as possible. The fields of mechanism design and social choice theory have the potential to provide tools allowing this community to make better decisions. In this article, we consider several avenues for such research. First, we discuss the problem of deciding who donates to what charity, and find that there are possible efficiency gains from donor coordination. We explore mechanisms for achieving more efficient donation allocations, and draw an analogy to participatory budgeting. Second, we discuss the problem of moral uncertainty, and propose open problems about the aggregation of moral theories. Finally, we consider the potential for economic design to open up opportunities for moral trade.

## 1 Effective Altruism

Effective Altruism is the project of using evidence and reason to figure out how to benefit others as much as possible, and taking action on that basis (MacAskill 2017). Thousands of people and many organisations now identify as Effective Altruists ("EAs"), and they have formed local groups in dozens of cities around the world and regularly meet at the Effective Altruism Global conferences. What do EAs do? Most of them donate a significant fraction of their annual income to charities that they view as particularly effective. Indeed, an early precursor to the EA movement was the Giving What We Can pledge to give at least 10% of one's income to effective charities; more than 3,000 people have now taken the pledge. EAs also invest significant effort into identifying which career path will best allow them to do the most good. But the defining and perhaps most challenging task of EAs is "cause prioritisation", that is, figuring out which actions (or jobs, or charities) actually do the most good. This usually involves designing cost-effectiveness models of various options, and gathering scientific research to inform those questions.

D. Peters (✉)
Department of Computer Science, University of Oxford, Oxford, UK
e-mail: dominik.peters@cs.ox.ac.uk

© Springer Nature Switzerland AG 2019                                                    381
J.-F. Laslier et al. (eds.), *The Future of Economic Design*, Studies in
Economic Design, https://doi.org/10.1007/978-3-030-18050-8_53

Effective Altruists tend to be most excited to tackle problems that have large scale (they severely affect many people), that are tractable (progress on solving the problems is possible), and that are neglected (marginal returns of working on the problems are high). Several broad fields of actions are commonly agreed to be promising on these criteria. For example, much attention is directed towards helping people in extreme poverty through interventions such as direct cash transfers or distributing malaria nets; towards farm animal welfare through lobbying corporations or funding research into meat replacements; and towards helping future generations through funding research into mitigating global catastrophic risks (such as biosecurity, runaway global warming, or risks from transformative AI).

As interest in Effective Altruism has grown since the inception of the movement in around 2011, there has been an increasing emphasis to optimise the impact of the whole *community*, and not just the impact of each individual separately. There are possible gains from cooperation, from compromise in case of disagreement, and from identifying individuals' comparative advantage.

In this article, I will outline several ways in which techniques from mechanism design and social choice theory could be used to develop tools that could help the Effective Altruism community make better decisions. For example, I will point out that there are some difficult coordination problems in deciding who should donate to what charities. Given the significant donation volume within the EA community, it is important to develop tools and conceptual insights that allow us to identify a good way of deciding where donations should go. I also outline several research questions in handling the aggregation of different moral views, which has applications in handling *moral uncertainty* and cases where a group of agents designs a systems making moral decisions (such as advanced AI system). In closing, I mention two other directions: facilitating *moral trade* and good *job allocations*.

The area of Economic Design has long studied ways in which we can improve society, make people better off, and implement outcomes that are socially optimal. Kidney exchange is a notable success story of this field, where matching theory and mechanism design have saved and improved hundreds of lives (Roth et al. 2004). Similarly, the Effective Altruism movement has the ambitious goal of improving the lives of as many people as possible. Helping EAs achieve these goals provides an exciting opportunity to apply our methods of economic design for social good.

## 2  Donation Allocation

Most effective altruists give a significant amount of money to charitable causes each year (Thomas 2017). At the time of writing, *Giving What We Can* has more than 3,000 members who have pledged to donate at least 10% of their annual income to effective charities; in 2015, they gave about $7 million (Centre for Effective Altruism 2016). There are also some larger organisations, notably the foundation *Good Ventures*, which since 2016 has granted more than $100 million per year to effective causes through the Open Philanthropy Project (GiveWell 2018).

What charities are 'effective'? Donors differ in their views on the value of donations to different charities, either due to value differences or due to different estimates of their effectiveness. This can give rise to gains from cooperation. For example, we can imagine two donors who both have $100 to donate. There are three charities under consideration, $A$, $B$, and $C$. Donor 1 thinks that charity $A$ does 10 utils of good per dollar donated while $C$ does no good; donor 2 thinks that charity $C$ does 10 utils of good per dollar while $A$ does no good. Both donors agree that charity $B$ does 9 utils of good per dollar. Situations of this kind can occur when charities $A$ and $C$ try to spread concern about certain issues in society, but only some donors think these issues are important. On the other hand, $B$ is a consensus choice.

In an uncoordinated outcome, it is likely that donor 1 will donate $100 to $A$ and 2 will donate $100 to $C$, so each thinks that 1000 utils of good have been done in total. However, had they cooperated, they could have sent $200 to $B$, and each would think that a total of 1800 utils of good had been done, a Pareto improvement. The two donors are playing, effectively, a prisoner's dilemma where each prefers to give to their pet cause. How can we achieve cooperation and coordination?

An intriguing possibility is a centralised mechanism, a *donation clearinghouse*, to which donors communicate their preferences over charitable organisations and their donation budget, and which then decides how these budgets are allocated to different charities. It is useful to think of the situation as a two-stage process. First, donors contribute their money to a *common* budget, and then we decide how to split this pie among causes based on donor's reports on how they would like the pie to be split. This makes our problem very similar to the problem of *participatory budgeting*, which is used by many local governments to allow their citizens to decide how to spend a part of public funds among projects (Cabannes 2004; Goel et al. 2016). The difference is in how the budget comes about: here, it is not exogenously given, but contributed by the donors.

What would a good mechanism for this setting look like? It needs to elicit donor's opinions about charities, which is a major problem in itself, given that the option space is very large. It then should select a split of the budget that is Pareto efficient, to avoid uncoordinated outcomes like in our example. Ideally, it should avoid incentives to strategically misreport one's preferences, though this will likely be impossible in the presence of other desiderata. And, importantly, it should have reasonable participation incentives: donors should want to contribute their money to the common pool rather than donate it directly and independently. (This need to incentivise pooling distinguishes our problem from standard participatory budgeting.) The design and implementation of such a mechanism has the potential to significantly improve the quality of donation decisions in society.

A good mechanism should also be able to accommodate donors with very different budget sizes (Muehlhauser 2017). For example, within the EA community, a significant fraction of donations currently comes from Good Ventures via the Open Philanthropy Project, accounting for about $100 million per year. This contrasts with donations by individuals, who typically donate less than $10,000 per year each. The mechanism should handle these differences gracefully, and make adequate use of the preference information of all donors.

The problem of donor coordination was first discussed in light of a different phenomenon in the charity world. Charities have a limited *room for more funding* (RFMF): any given organisation can only hire a limited number of staff and can only scale up a limited amount without diluting its cost-effectiveness, at least in the short run. In other words, donations to a given organisation have decreasing marginal returns. For example, the charity evaluator GiveWell (2016) estimates that its top recommended charity, the Against Malaria Foundation (AMF), could only productively use between $78 million and $191 million of funding for the year 2017.

Karnofsky (2014) noticed that this gives rise to a coordination problem: Following GiveWell's analysis, many donors think that AMF is a top giving opportunity, and collectively, these donors would be willing to donate more to AMF than its room for more funding. Not all of these donors should give to AMF, since not all the money could productively be used. Thus, some donors should give to their second-most preferred charities. However, donors may disagree about which cause is the best cause after AMF. In particular, they may prefer *other donors* to fill AMF's RFMF, while they donate to *their* next-best option. During the 'giving season' around Christmas time, this leads to donors waiting until the last minute to make their donation decision, only giving to a cause if it seems like other donors have not yet filled a relevant funding gap. This leads to inefficiencies such as wasteful fundraising expenses (Todd 2016). A donation clearinghouse could and should avoid such incentives, by taking limited RFMF into account, for example by allowing donors to specify decreasing marginal returns to additional donations to a charity.

There is an alternative approach to achieving more efficient donation allocations, which makes use of *matching offers*. In our introductory example, the uncoordinated outcome could be avoided if a new donor 3 enters the picture and offers to match every dollar donated to the compromise charity $B$. That is, donor 3 will donate the same number of dollars to $B$ that were donated by donors 1 and 2 together, perhaps up to some limit. Any unused money in the matching pot is burned. This changes the calculation for donors 1 and 2; now it is more attractive for them to donate to $B$ than to their favourite charity $A$ or $C$, achieving a more efficient result. A paper by Conitzer and Sandholm (2011) follows this approach, viewing donor coordination as a problem of *negotiation*. In their setting, donors can propose (potentially complicated) matching offers, which they can use to incentivise other agents to donate to certain charities. Conitzer and Sandholm (2011) study the computational complexity of clearing these markets, finding mostly hardness results, though they present integer linear programming formulations that should be efficient in practice. They also begin a study of the mechanism design problem, obtaining some impossibility results. Christiano (2016) proposes another mechanism using matching offers to incentivise donations to compromise charities.

# 3 Moral Uncertainty

To most effectively help others, one must be able to decide which of two actions better fulfils this goal. But this question is difficult to answer, because such comparisons involve many trade-offs. If some action improves the lives of many beings but violates some rights, is this worth it? If some action improves the lives of future generations but hurts those currently alive, is this worth it? How much do non-human animals count? Is it more important to prevent suffering, or more important to increase the number o f happy beings?

Different moral theories give different answers to these questions. Even after much reflection, many people find that they are uncertain about the correct answers. They have *moral uncertainty*. This uncertainty may be reflected by the decision-maker having credences (subjective probability judgements) about various moral theories being "correct".[1] For example, someone might spread 60% probability mass among various utilitarian theories, and allocate the rest among deontological (rights-based) theories and virtue ethics. Given these credences, the decision-maker asks the question: What is the morally right action for me to take?

If we formalise different theories as rankings of a set of feasible actions in order of "choice-worthiness", then the problem of selecting the right action is formally identical to the standard voting setting of choosing a winning alternative given a (fractional) profile of preference orderings (MacAskill 2016). Each theory becomes a voter, weighted by the credence the decision-maker has in that theory. Thus, we may evaluate the appropriateness of standard voting rules for this setting.[2] The standard tool for this is to employ *axiomatic analysis*, which identifies desirable properties of aggregation mechanisms and asks which rules satisfy them. MacAskill (2016) argues that, for the setting of moral uncertainty, the voting rule chosen should satisfy the *participation axiom* (Moulin 1988), which requires that if we increase our credence in a theory that recommends the currently recommended alternative, the recommendation does not change. He further argues that strategyproofness is not a concern, because theories are not strategic actors. Thus, he proposes to use *Borda's rule*, which is arguably the best of the known rules satisfying participation. However, there are clear disadvantages to Borda's rule, in particular its behaviour when there are clones,[3] and so the merits of different rules in this context needs further discussion.

---

[1]For a moral realist, this may refer to uncertainty over which moral theory is *true*. For others, it could for example refer to uncertainty about the moral view one would adopt after more reflection.

[2]Marcus Pivato pointed out to me that decision theory, rather than voting, is the usual tool to be applied to problems of this type. However, decision theory typically studies the case where the decision maker has *cardinal* utilities over actions in each state of the world (where states correspond to moral theories). In an informational setting with ordinal utilities, the theory of (homogeneous) voting rules is more appropriate.

[3]MacAskill (2016) develops a response to this criticism of Borda by enriching the description of the alternative space. Tarsney (2018) defends the use of the McKelvey uncovered set, rather than Borda, as an appropriate aggregation rule in this context.

Several other directions for analysing this setting seem promising. Many moral philosophers are *moral realists*, and thus believe in the existence of a ground truth of normative facts. This suggests the use of techniques from *epistemic social choice* (see the Chapter 16). Another direction is to study the aggregation of non-ordinal theories. Indeed, many moral theories are not adequately formalised as ordinal rankings. While plausibly, as MacAskill (2016) argues, *some* theories are 'merely ordinal', many theories encode their prescriptions in other forms: Most flavours of utilitarianism come with a natural cardinal structure; deontic theories only specify which actions are permissible, but give no further information; other theories may mainly be interested in forbidding certain actions, for example because they violate rights. Thus, for these settings, we need aggregation mechanisms that are able to combine inputs that come in fundamentally different formats.

Some voting rules may admit natural generalisations that allow this combination of different input formats. Range voting comes to mind: this rule would 'cardinalise' any ordinal input in some natural way, and sum the results. But such rules lack an axiomatic study, and it is not clear how exactly the cardinalisation should be performed.

Some people with uncertainty over moral theories may have all or most of their credences in moral theories that have a cardinal structure, such as utilitarian proposals. Such decision-makers need an aggregation rule for cardinal preferences. However, most of voting theory has considered the ordinal case only to avoid having to make inter-personal comparisons of utility, and pointing to concerns that it might be cognitively difficult for voters to figure out their own cardinal preferences, as well as concerns about avoiding strategic behaviour. The latter concerns are much less applicable in the setting of moral uncertainty. Since decision-makers could benefit from having cardinal aggregation procedures for this setting (and others), more formal analysis is needed.[4] Another way to view moral theories is via the trade-off ratios that they imply among various goods and bads; a recent literature studies ways of aggregating such trade-offs (Conitzer et al. 2015).

When voting, it is not uncommon to observe every possible preference ranking to be reported by some agent. In our case of moral theories, we should instead expect the input to be *structured*: most logically possible moral rankings are implausible or inconsistent and should receive (almost) zero credence. Maybe this structure can be used to evade impossibility results and design better aggregation rules, like is possible in the case of single-peaked preferences (Black 1948). For example, this approach may be relevant to the problem of aggregating different *population axiologies*, first considered by Greaves and Ord (2017). A population axiology ranks states of the world with different numbers of people and different welfare levels in order of goodness. Classical examples are *total* and *average utilitarianism*, which hold that a state is better than another iff the sum (resp. average) of the individual welfare levels is higher in the first state. There is a well-developed theory of properties that a single

---

[4]For some axiomatic results, see Pivato (2014) for a characterisation of range voting and 'formal utilitarianism' as the 'most-expressive' voting rules satisfying reinforcement, as well as some work on relative utilitarianism (Dhillon 1998; Dhillon and Mertens 1999; Börgers and Choo 2017).

axiology should satisfy (Blackorby et al. 2005), and these properties define domain restrictions for the problem of aggregating several axiologies. To analyse this setting, results from the literature on social choice on economic domains could be applicable (see Le Breton and Weymark 2010).

## 4 Further Problems

Toby Ord's (2015) influential article on *moral trade* contains several examples in which setting up markets and opportunities to trade can lead to morally better outcomes. He imagines a case where Alice is concerned about animal suffering and Bob is concerned about global poverty. He proposes a trade: Bob could become vegetarian in exchange for Alice's promise to donate some of her income to global poverty causes. Not many exchanges of this type occur today, and market designers may be able to set up institutions that allow people to access such moral gains.

Another setting in which economic design could help is in job allocation. Many EAs and others want to choose a career that maximises their altruistic impact. Often, people try to find their *comparative advantage* in choosing a certain job (Todd 2016). But it is difficult to know what one's comparative advantage is without having a lot of knowledge about other people's abilities. To help with this, it would be interesting to study matching mechanisms that have very low communication complexity: they only need to query few pairs of participants and jobs for their suitability, in order to determine a job assignment that is approximately optimal.

## 5 Conclusions

In this article, I have outlined several ways in which theorists in the area of economic design can support the Effective Altruism project. Progress on these problems may have high impact and improve many lives. In addition, the problems are technically and conceptually interesting, and may have applications in other areas. I am excited to see what we can do.

**Acknowledgements** This piece benefited from comments by Gabriel Carroll, Jean-Francois Laslier, Marcus Pivato, Ariel Procaccia, and Bill Zwicker.

## References

Black, D. (1948). On the rationale of group decision-making. *The Journal of Political Economy*, 56(1), 23–34.
Blackorby, C., Bossert, W., & Donaldson, D. J. (2005). *Population issues in social choice theory, welfare economics, and ethics*. Cambridge: Cambridge University Press.

Börgers, T., & Choo, Y.-M. (2017). A counterexample to Dhillon (1998). *Social Choice and Welfare, 48*(4), 837–843.

Cabannes, Y. (2004). Participatory budgeting: A significant contribution to participatory democracy. *Environment and Urbanization, 16*(1), 27–46.

Centre for Effective Altruism. (2016, December). *Fundraising prospectus—Winter 2017.* https://www.centreforeffectivealtruism.org/fundraising/.

Christiano, P. F. (2016, October). *Repledge++.* https://sideways-view.com/2016/10/31/repledge/.

Conitzer, V., & Sandholm, T. (2011). Expressive markets for donating to charities. *Artificial Intelligence, 175*(7–8), 1251–1271.

Conitzer, V., Brill, M., & Freeman, R. (2015). Crowdsourcing societal tradeoffs. In *Proceedings of the 14th International Conference on Autonomous Agents and Multiagent Systems (AAMAS)* (pp. 1213–1217).

Dhillon, A. (1998). Extended Pareto rules and relative utilitarianism. *Social Choice and Welfare, 15*(4), 521–542.

Dhillon, A., & Mertens, J.-F. (1999). Relative utilitarianism. *Econometrica, 67*(3), 471–498.

GiveWell. (2016, November). *Against malaria foundation—November 2016 version.* https://www.givewell.org/charities/against-malaria-foundation/november-2016-version.

GiveWell. (2018, March). *Our progress in 2017 and plans for 2018.* https://www.openphilanthropy.org/blog/our-progress-2017-and-plans-2018. (Written by Holden Karnofsky).

Goel, A., Krishnaswamy, A. K., & Sakshuwong, S. (2016). Budget aggregation via knapsack voting: Welfare-maximization and strategy-proofness. *Collective Intelligence,* 783–809.

Greaves, H., & Ord, T. (2017). Moral uncertainty about population ethics. *Journal of Ethics and Social Philosophy, 12*(2), 135–167.

Karnofsky, K. (2014, December). The value of coordination. *The GiveWell Blog.* https://blog.givewell.org/2014/12/02/donor-coordination-and-the-givers-dilemma/.

Le Breton, M., & Weymark, J. A. (2010). Arrovian social choice theory on economic domains. In K. J. Arrow, A. K. Sen, & K. Suzumura (Eds.), *Handbook of social choice and welfare* (Vol. 2, Chapter 17). Elsevier.

MacAskill, W. (2016). Normative uncertainty as a voting problem. *Mind, 125*(500), 967–1004.

MacAskill, W. (2017). Effective altruism: Introduction. *Essays in Philosophy, 18*(1), 1.

Moulin, H. (1988). Condorcet's principle implies the no show paradox. *Journal of Economic Theory, 45*(1), 53–64.

Muehlhauser, L. (2017, March). Technical and philosophical questions that might affect our grantmaking. *Open Philanthropy Project.* https://www.openphilanthropy.org/blog/technical-and-philosophical-questions-might-affect-our-grantmaking.

Ord, T. (2015). Moral trade. *Ethics, 126*(1), 118–138.

Pivato, M. (2014). Formal utilitarianism and range voting. *Mathematical Social Sciences, 67,* 50–56.

Roth, A. E., Sönmez, T., & Ünver, M. U. (2004). Kidney exchange. *The Quarterly Journal of Economics, 119*(2), 457–488.

Tarsney, C. (2018). Normative uncertainty and social choice. *Mind* (forthcoming).

Thomas, H. (2017, September). *EA survey 2017 series: Donation data.* http://effective-altruism.com/ea/1el/ea_survey_2017_series_donation_data/.

Todd, B. (2016, February). The value of coordination. *80,000 Hours.* https://80000hours.org/2016/02/the-value-of-coordination/.

# Is It Still Worth Exploring Social and Economic Networks?

Agnieszka Rusinowska

**Abstract** In this short note, I provide my personal thoughts on the future of network design. Since the network literature is extremely vast, is it useful and interesting enough to do more research on social and economic networks? And if there is a "future" of the economics of networks, where should the field go from here?

During the last couple of decades, research on social and economic networks has got a real explosion: thousands of papers, different approaches, and various phenomena studied from a network perspective. This is clearly not that surprising, since it would be difficult to avoid a network structure in the complete analysis of interactions. Various directions with a network dimension have been explored since the 1990s, which resulted in many important contributions concerning social learning, diffusion and contagion, games on networks, random networks, strategic network formation, network games and allocation rules, community networks, networked markets, among many others. Also the availability of various data on networks and the development of new computational methods and tools have strongly increased the use of an empirical approach to social and economic networks. Clearly, the network economics has become an established discipline.

The vast literature on social and economic networks, didn't it start suffering from some non-desirable network properties studied extensively in that literature, such as the congestion problems? And since so many key contributions have been presented, is it useful and interesting enough to do more research on social and economic networks? Is it still feasible to deliver crucial and important contributions of the network approach to better understand social and economic interactions? If my personal answer were "no", this would probably be the end of the present text, and therefore the readers can guess that I have a more optimistic view on the future of network design. Indeed, this is what is going to be presented in the following paragraphs.

First of all, let me make a general remark, which concerns any important field of research. What can be seen in many established disciplines is that despite numerous

A. Rusinowska (✉)
CNRS – Centre D'Economie de La Sorbonne, Paris School of Economics, University Paris 1, Paris, France
e-mail: agnieszka.rusinowska@univ-paris1.fr

© Springer Nature Switzerland AG 2019
J.-F. Laslier et al. (eds.), *The Future of Economic Design*, Studies in
Economic Design, https://doi.org/10.1007/978-3-030-18050-8_54

existing contributions, new research directions continue being developed, and many or even more and more scientists continue working on it, new PhD dissertations are defended and new important works continue being published. If an established discipline is broad enough, it has a chance to remain being alive and active. In my view, the economics of networks is such a discipline. Being by nature based on interactions and therefore appropriate to model all phenomena around us, it is not a field "without future".

Since there is a "future" of the economics of networks, where should (not) the field go from here? Obviously, many existing network frameworks can be still generalized to better meet real world interactions. In any of the even heavily "congested" topics related to social and economic networks, one can identify a number of important and worth investigating extensions of the existing models. However, I believe that sometimes this might not be the most desirable direction of future research based on the network analysis, in particular, if we lack a general objective to improve modeling of different phenomena and interactions. What seems to be more crucial in the future research on networks, is building new and useful bridges between the economics of networks and other scientific disciplines.

There exist several prominent areas of network applications that have not been sufficiently explored so far. Let me mention only some of them, as the aim of the present note is not to deliver the full list of topics worth being studied but rather to express some general thoughts. One of the areas where the network approach might be quite useful, in particular the issue of network externalities, is international trade. Among other prominent areas where networks should be straightforwardly applied, one can mention labor economics, development economics, political economy, macroeconomics, and not to forget about financial networks. Although the importance of the network analysis has already been recognized in these areas, there are numerous open questions and issues that can be successfully studied by the network approach.

Apart from identifying new areas and topics that are promising for the use of network theory, there are several other issues that should be mentioned, when thinking of the "future" of social and economic networks. One of them concerns the complexity of real life interactions. Obviously, many phenomena co-exist in a society, and therefore should be studied in a common framework. While not all extended models can be of particular use and interest, as already mentioned above, some extensions that combine different clearly co-existing phenomena might appear particularly important. For such more complex cases, the application of network theory should result in richer conclusions than when using the network approach to the separate analysis of each phenomenon.

Another concern related to the "future" of the economics of networks is related to the choice of appropriate tools and methods. Since the network theory has become very vast, the question of the use of appropriate network concepts, methods or algorithms for the analysis of a specific problem is of particular importance in theoretical as well as empirical studies. Although this sounds like an obvious fact, understanding of the appropriate choice of notions and tools when using the network approach can and should still be considerably improved.

Furthermore, an important issue for the "future" of network theory is related to the fact that networks are by nature interdisciplinary. A s i srêquently repeated in the literature, networks are studied in many different disciplines: economics, mathematics, statistics, physics, computer sciences, sociology, and finance, among others. Even if more and more attention is paid to interdisciplinary approaches, the literature on networks does not sufficiently take into account modeling and results from other disciplines. It is not surprising, since it becomes more and more difficult to follow all developments in social and economic networks, in particular, if they come from not our own scientific communities. However, it is not that unusual that the same concepts (possibly under different names) are introduced as "new concepts", and even the same results are presented as "new results" in different disciplines. One of the main challenges for the established and vast disciplines such as the economics of networks is indeed related to the fact that not enough bridges exist so far between different areas and researchers from different disciplines working on social and economic networks.

As a conclusion of this short note, let me repeat my optimistic message: yes, it is absolutely worth doing research on social and economic networks, despite the congestion feature of the discipline; yes, it is always worth teaching network theory and applications to new generations, and encouraging our students to explore social and economic networks.

# Part IX
# Law

# Political Hypotheses and Mathematical Conclusions

Paul H. Edelman

**Abstract** When modeling or analyzing democratic processes, mathematicians may find themselves in unfamiliar territory: politics. How we proceed mathematically may depend heavily on our conception of representative democracy and theory of government. I will give a number of illustrations to show how contestable political principles lead to differing mathematical analyses. Our mathematical conclusions are inherently governed by our political hypotheses.

## 1 Introduction

The "unreasonable effectiveness of mathematics" (Wigner 1960) is based at least in part on its ability to uncover the common elements of phenomena that appear in a multiplicity of contexts. But while it is easy to see if jettisoning context is important in modeling physical phenomena, it is a much more subtle issue when modeling social ones. In social situations context may play a more crucial role. So, for instance, while we could model judges selecting the best wine at a county fair the same way we model a polity choosing a president (Balinski and Laraki 2010) it seems to me to be inapt (Edelman 2012, p. 809).

In this essay I will discuss a more subtle contextual issue—Certain institutional choices that society has made are embedded in the context and these choices must be accounted for in any formal analysis. Such choices may be jurisprudential or political. That is, they may arise from legislative decisions or legal ones. Whatever their basis, these choices should be reflected in the structure of a model but are often ignored. That is, underlying legal or political commitments must be respected by the model even if they are not formally required.

I will give three examples of how institutional choices affect how a social science model should be structured. The examples all come from models of political representation. In the first I show how the US Supreme Court's view that the right to

P. H. Edelman (✉)
Department of Mathematics and the Law School, Vanderbilt University,
Nashville, TN 37203, USA
e-mail: paul.edelman@vanderbilt.edu

© Springer Nature Switzerland AG 2019
J.-F. Laslier et al. (eds.), *The Future of Economic Design*, Studies in
Economic Design, https://doi.org/10.1007/978-3-030-18050-8_55

vote is an individual right rather than a systemic one is reflected in how it measures representation and hence affects prescriptions of the most appropriate method for apportionment of representatives to the US House. The second example shows how the decision to treat the requirement of "one person, one vote" to apply to persons rather than voters is implicated in the analysis of weighted voting systems. In the third example I show how the question of whether a districting regime should seek accurate representation or promote competition has implications for how to best measure partisan gerrymander. I close with a call-to-arms to modelers to take these institutional choices seriously.

## 2   Individual or Systemic Right

Should we view the right to vote as an individual right, adhering to each participant, or as a systemic right guaranteeing a right to a democratic form of government? In the United States the Supreme Court has, with a few exceptions based on race, decided on the former interpretation. That is, when a voter claims a violation of voting rights he must show that he personally has been in some way impaired relative to some other voter. It is not enough to show a generalized harm to the working of democracy.

Such an abstract difference may seem remote from the concerns of modelers but it is relevant to a number of issues. When assessing whether the districts in a jurisdiction are sufficiently equinumerous to meet the "one person, one vote" (OPOV) requirement, what is important is the difference between the largest district and the smallest (essentially what the US Supreme Court calls *total deviation*, see (Edelman 2006, Sect. 2)), not the standard deviation of the districts overall. This follows because the claim of harm is that the person in the largest district is viewed as having a smaller amount of representation and hence is being short-changed with respect to the person in the smallest district. If voting was considered a systemic right, then a better measure of the harm might be standard deviation of the district sizes, rather than a pairwise comparison of the extremes.

As I have argued elsewhere (Edelman 2006), the same analysis extends to theories of apportionment. Most theorists, and the Supreme Court, claim that it is impossible to identify the appropriate apportionment method by way of choosing a function to optimize. For example, Balinski and Young, the bible of the subject, says, "The moral of this tale is that one cannot choose objective functions with impunity, despite current practices in applied mathematics. The choice of an objective is, by and large an *ad hoc* affair" (Balinski and Young 2001, p. 104). But of course, that is not true if the objective is endorsed by the courts as the appropriate measure as it has been for the total deviation measure. One can employ the total deviation measure to produce an apportionment which minimizes the gap between the most over-represented state and the most under-represented. While the resulting apportionment scheme has its perversities (Edelman 2006, Sect. V) it does more fully conform to the jurisprudence set out by the Court.

## 3  Equal Representation or Equal Voting Power

When districting a political entity we seek to make the districts "equal" but what exactly we seek to equalize is very unclear. There are two natural approaches to take. We might wish to equalize representation, i.e., we want the number of *persons* in the districts to be equal. This would produce districts in which every person has the same share of a representative. A different approach is to equalize voting power, i.e., we want the number of *voters* in the districts to be equal. In this case every voter has an equal opportunity to affect the outcome of an election.

These two goals are philosophically quite different. The choice between them implicates our view of how a representative democracy should function. If representatives are thought to be obligated to advance the interests of those who elect them then perhaps equalizing voters is the better choice. If representatives are expected to use their judgment as to what is in the best interests of their district, then perhaps equalizing the number of persons is a better choice. Which of these goals to pursue is clearly a political choice.

In the United States the distinction between these goals was ignored for close to 50 years. In the early OPOV cases the Supreme Court ruled that districts needed the same total population but in these same opinions it justified its decision by appealing to the importance of every voter having an equal influence over the election. The Court was able to avoid this conflict because, for the most part, the distribution of voters among the total population was sufficiently similar that equalizing population in the districts led to reasonably equal voting populations.[1]

After dodging the question for 50 years the Court was finally forced to confront it in the 2016 case *Evenwel v. Abbott* (2016). This case arose out of districting for the Texas state senate. Adhering to all previous precedent, the districts were drawn so that they all had equal numbers of persons within. However, there was a 40% disparity in the citizen voting age populations between the districts. The plaintiffs, who lived in a district with a large number of voters, claimed that this violated their right, under OPOV, to have equal influence on the election relative to those voters in districts with a smaller number of eligible voters.

Why did this problem come to head now? The influx of a large number of undocumented aliens to border states coupled with segregated housing patterns has led to ever larger disparities between total population and voting-eligible population. The facts of this disparity along with partisan interest in disfranchising certain immigrant groups led to the court case and the forcing of the Court's hand. Even when confronted directly with this problem, the Court still hesitated. It found that Texas was permitted to use total population as the basis of its redistricting but refused to opine on whether the use of voting-eligible population was prohibited or might be allowed as an alternative basis. No doubt this issue will arise in the future. As it turns out the Court need not choose between these measures, as I have shown that one can draw

---

[1] The one exception is *Burns v. Richardson* in which the Court allowed Hawaii to equalize resident voting-eligible populations. The Court was sure to note, however, that the resulting districting did not diverge very much from one that equalized total population.

districts that will simultaneously achieve equal total populations and equal voting populations (Edelman 2016).

The debate about the nature of representation has a broader impact than just in districting. Consider the question in the context of weighted voting. In upstate New York a number of county boards are run on the basis of weighted voting. Each town in the county has a single representative who casts a weighted ballot. Under the rule established in *Ianucci v. Board of Supervisors, Washington County, NY* (1967), to comply with OPOV the Banzhaf index of each representative should correspond to the percentage of the population of the county that is in that town. This decision has been roundly criticized because it seems to equate the Banzhaf power of the representative with the Banzhaf power of the citizens in the town. Most theorists criticize this decision, arguing that what is required is that each *citizen* have equal voting power which implies that the Banzhaf power of the representative should actually be proportional to the square root of the town's population, not to the population itself (Felsenthal and Machover 1998, p. 133).

But why should we assume that the real implication of OPOV is to equalize voting power? In *Evenwel* the Supreme Court contradicted that assessment, holding that equal representation was a perfectly suitable goal for districting. And under such a theory it is perfectly appropriate to require that a representative of a town have a Banzhaf power proportional to the town's population. As with districting, the requisite amount of power to be assigned in weighted voting is dependent on which goal of OPOV one chooses.

This also highlights an advantage of districting over weighted voting. As noted above, one can draw districts to simultaneously achieve both equality in the number of persons and in the number of voters. But in weighted voting one has to choose between equalizing representation and equalizing voting power.

## 4 Effective Representation or Competition

Finally, there is a large social choice literature on how to assess the fairness of districting schemes. A principle goal of this literature is to identify political gerrymanders, i.e., districts which are drawn to obtain unfair political advantage. Earlier work concentrated on identifying suspicious shapes of districts; more recent work has focused on the internal demographics of districts. Rather than discuss these developments, about which I am somewhat skeptical, I want to address the more fundamental question of what it is that we really want out of districting.

Consider the following situation: Suppose we have a rectangular state which is to be divided into two equinumerous voting districts. Suppose the left half of the state is uniformly Democratic and the right half is Republican, as illustrated in Fig. 1. How should we divide the state?

There are only two real choices. We can draw the division vertically to get two homogeneous political districts; the left half is a safe Democratic district and the right a safe Republican one. Or we can draw the division horizontally to result in

**Fig. 1**  A two district
example

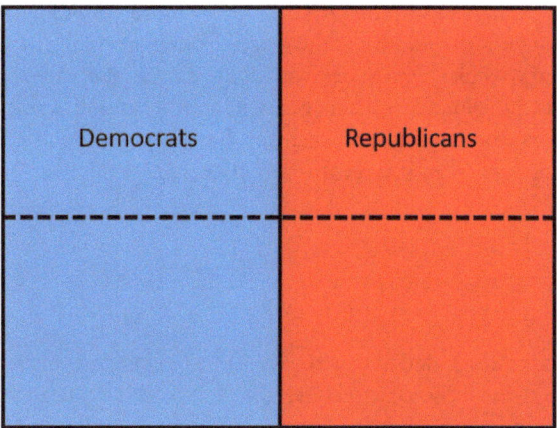

two identical districts each half Democratic and half Republican. Which of these
districtings is the right one?

If the goal of districting is to, as much as possible, have the views of the polity
be reflected by their representatives, then clearly the first is preferred. Everyone is
represented by their chosen party. But, as many have pointed out, such a districting is
not without its pitfalls. Safe districts can lead to corruption. They may also encourage
extremism among the representatives as the only competition they face is from their
own party. Traditional analysis would predict that the representatives will move to the
median position of their *party* rather than the median position of the entire population.

Many commentators would opt for the second districting as a way to encourage
competition among representatives and force the parties to compete for the median
voter. But half the time such a districting would result in one party winning both
seats even though they only represent half the population. It is an odd notion of
representational democracy that would opt for a districting that predictably fails to
represent half of the population.

I think that anyone in the business of detecting gerrymanders has to have an answer
to this puzzle. To see why, consider the measure of partisan gerrymander proposed
in *Gill v. Whitford* (2018), the partisan gerrymander case out of Wisconsin that
was heard by the Supreme Court. To support the claim of partisan gerrymander the
lower court employed the Efficiency Gap (EG) measure promoted by Eric McGhee
and Nicholas Stephanopolous (2015). The EG measure attempts to identify partisan
gerrymanders by computing the discrepancy in the number of wasted votes for each
candidate, i.e., votes that are unnecessary to determine the outcome. As it turns out,
this measure finds the most wasted votes in the situation of a close election, in which
all of the votes for the losing candidate (which would be close to half of the total)
are wasted but none of the votes for the winning candidate are wasted. In contrast,
EG finds a 75–25% vote to have an equal number of wasted votes: the 25% for the
losing candidate are wasted as are the 25% for the winning candidate in excess of
the necessary majority.

Thus the EG measure is naturally more suspicious of districtings that emphasize competition than it is of districtings that are safe. If EG is accepted as a measure of gerrymander then states will likely avoid competitive districts as much as possible since they could raise concerns about the validity of the scheme. This is fine if we want to encourage safe districts, but if the goal of districting is to create competitive elections, then it is a cause for concern.

# 5  Conclusion

I have illustrated three situations in which underlying institutional choices may affect modelers' choices in analyzing democratic institutions. Even a seemingly benign choice of the function used to measure disparity in OPOV implicates the political question of the nature of the right-to-vote. Whether a weighted voting scheme meets the requirements of OPOV depends on one's views on the nature of representation. The efficacy of certain objective measures of gerrymander may depend on one's political choice of a goal of representative democracy. If modelers wish their efforts to be useful, their institutional choices must be as transparent as their mathematical ones. This essay is a brief call to modelers to take on that challenge.

**Acknowledgements** The author thanks Ed Cheng, Marcus Pivato, Suzanna Sherry and Bill Zwicker for their helpful comments.

# References

Balinski, M., & Laraki, R. (2010). *Majority judgment: Measuring, ranking and electing.* Cambridge, MA: MIT Press.
Balinski, M., & Young, H. P. (2001). *Fair representation* (2nd ed.). Washington, DC: Brookings Institution Press.
Edelman, P. H. (2006). Getting the math right: Why California has too many seats in the house of representatives. *Vanderbilt Law Review, 59,* 297–346.
Edelman, P. H. (2012). Book review: Michel Balinski and Rida Laraki: Majority judgment: Measuring, ranking, and electing. *Public Choice, 151,* 807–810.
Edelman, P. H. (2016). Evenwel, voting power, and dual districting. *Journal of Legal Studies, 45,* 203–221.
Evenwel v. Abbott. (2016). 136 S. Ct. 1120.
Felsenthal, D. S., & Machover, M. (1998). *The measurement of voting power: Theory and practice, problems and paradoxes.* Cheltenham: Edward Elgar.
Gill v. Whitford. (2018). 585 US.
Ianucci v. Board of Supervisors of Washington County, NY. (1967). 282 NYS 2d 502.
Stephanopoulos, N. O., & McGhee, E. M. (2015). Partisan gerrymandering and the efficiency gap. *University of Chicago Law Review, 82,* 831–900.
Wigner, E. P. (1960). The unreasonable effectiveness of mathematics in the natural sciences. *Communications on Pure and Applied Mathematics, 13,* 1–14.

# Law and Economics and Economic Design

Nuno Garoupa

**Abstract**  My sense is that Economic Design has lost the influence it had in Law and Economics 30 years ago. The application of recent developments in Economic Design to Law and Economics are rare and fundamentally random. On the one hand, the relocation of the center of gravity of Law and Economics from economic departments to law schools has inevitably reduced the contact between Law and Economics and Economic Design. One the other hand, the general sense that Economic Design is too complex to offer appropriate and realistic answers to legal problems has inevitably made its way. Scholarship implications are discussed.

From the early days there has been a discussion about the nature of Law and Economics, particularly the extent to which Law and Economics and Economic Analysis of Law are the exact same thing. In fact, while the Journal of Law and Economics (started in 1958) tended to publish research developments in regulation, antitrust, taxation and other areas of law with an obvious economic content, the Journal of Legal Studies (launched by Judge Posner and others in 1972) developed its core business in the traditional areas of the common law (property, contract, torts, procedure) with an economic methodology (basically the application of microeconomic insights). Those who argued for a distinction between Law and Economics (as a discipline of Law) and Economic Analysis of Law (as an applied field of Economics) so many decades ago should by now be surprised with the developments. The field has flourished consistently in the last 50 years. The number of specialized journals and associations has expanded around the world. Particularly, while articles in the vein of the seminal work by Coase (1960) have moved to law reviews, the number of journals favoring formal models (International Review of Law and Economics; Journal of Law, Economics and Organization; American Review of Law and Economics; European Journal of Law and Economics; Review of Law and Economics),

I am grateful to Sofia Amaral-Garcia for suggestions that much improved the essay. The usual disclaimers apply.

N. Garoupa (✉)
George Mason University Scalia Law and Católica Global Law School, Arlington, VA, USA
e-mail: ngaroup@gmu.edu

© Springer Nature Switzerland AG 2019
J.-F. Laslier et al. (eds.), *The Future of Economic Design*, Studies in
Economic Design, https://doi.org/10.1007/978-3-030-18050-8_56

401

empirical work (Journal of Empirical Legal Studies) or both (Journal of Legal Analysis) has increased steadily since the 1980s. It is true that the success of Law and Economics in law schools and its influence in legal debates has not been echoed outside of the United States. In fact, this has led to an intense debate about the proper role of Law and Economics in legal academia around the world (Garoupa and Ulen 2007; Garoupa 2011). In short, it seems that while Law and Economics has taken over the Economic Analysis of Law in terms of scholarly focus, at the same time, the expansion of publication outlets has favored formal analysis.

In this short essay, in a different way, I want to take the view from the other side—the role of Law and Economics in the disciple of Economics and the use of economic modeling in Law and Economics. While the first generation of legal economists moved from economic departments to law schools in the 1970s and 1980s, with Economics as a major background and very limited training in Law, the following generations have very different profiles. In the United States, nowadays, Law and Economics is basically located in law schools (mainly in the elite law schools). Legal economists tend to have dual degrees (PhD Economics/JD) and many publish mainly in law reviews (which are not peer-refereed and are run by second-year law students; a weird system of publication that has been criticized by many scholars in Law and Economics, not least Judge Posner himself). Specialized journals have expanded vastly but mostly ignored by mainstream economists (with a few exceptions, they are usually not ranked highly in the discipline of Economics). In my own experience of 20 years, Law and Economics is now more dominated by (American) legal controversies and topics rather than recent developments in economic methodology. Non-American legal issues have very little bearing in Law and Economics (not surprisingly since the most prestigious journals are American based).

My sense is that Economic Design has lost the influence it had in Law and Economics 30 years ago. The pathbreaking contributions of legal economists such as Steve Shavell, Mitch Polinsky, Louis Kaplow, Lucian Bebchuk, Dan Rubinfeld, Lewis Kornhauser, Tom Ulen, Bob Cooter, Bill Landes, just to mention a few names that come immediately to mind, combined rigorous economic modeling while using the most sophisticated techniques available at the time. An intellectual giant like Judge Posner opted for both methods—verbal and discursive articles in law reviews (and books) and formal models in the specialized journals. The founding fathers of Law and Economics opened entire new economic applications to contracts, torts, property, procedure and litigation. Many were published in highly reputable economic journals such as the American Economic Review, Journal of Political Economy, Rand Journal of Economics, Journal of Public Economics and others. The publication strategy tended to encompass the technical version of the paper in a prestigious journal read by economists and a legal policy version of the paper in a law review. Things have changed in these three decades. Mainstream economic journals publish very few articles with a Law and Economics content. However, these few articles usually have no law review companion. The application of recent developments in Economic Design to Law and Economics are rare and fundamentally random. The

**Table 1** Publications by JLS and JLEO, 1996–2016, percentage of Theoretical Law and Economics out of total articles

|  | Journal of legal studies (%) | Journal of law, economics and organization (%) | Total (%) |
|---|---|---|---|
| 1996 | 44 | 60 | 53 |
| 2001 | 21 | 47 | 34 |
| 2006 | 29 | 48 | 39 |
| 2011 | 50 | 45 | 47 |
| 2016 | 43 | 48 | 46 |
| Total | 37 | 50 | 44 |

**Table 2** Publications by JLS and JLEO, 1996-2016, percentage of Empirical Law and Economics out of total articles

|  | Journal of legal studies (%) | Journal of law, economics and organization (%) | Total (%) |
|---|---|---|---|
| 1996 | 22 | 40 | 32 |
| 2001 | 63 | 53 | 58 |
| 2006 | 59 | 52 | 55 |
| 2011 | 50 | 50 | 50 |
| 2016 | 57 | 52 | 54 |
| Total | 50 | 50 | 50 |

Law and Economics conferences in the United States are populated more and more by law professors and less and less by economists (located in economic departments).

My perception is that the specialized journals in Law and Economics have also changed in terms of their priorities when it comes to publication record. Table 1 summarizes a few empirical observations about Theoretical Law and Economics. I have randomly selected five different years for two main journals, the Journal of Legal Studies (traditionally less friendly to complex Economic Design and more open to nontechnical articles, JLS hereafter) and the Journal of Law, Economics and Organization (considered very friendly to more technical work and highly regarded in economic departments, JLEO hereafter). While the trend in JLEO is for a clear decline in theory papers (from 60% in 1996 to less than 50% since 2001), the data shows an interesting pattern for JLS. There is a significant and sharp decrease initially (44% in 1996 to 21% in 2001), but the trend changes back in 2011 (to 50% in 2011 and 43% in 2016). Looking more closely at the data, on Table 2, summarizing the same percentages for Empirical Law and Economics, we understand the mechanism behind these changes—verbal or discursive articles have disappeared almost entirely from JLS while empirical articles have taken over in the last 20 years. Looking at both tables, we can say that technical articles (formal models and empirical analysis) accounted for 66% in 1996 and 100% after 2011 in JLS. They were always close to 100% in JLEO.

The perceived evolution of Theoretical Law and Economics and the loss of importance of Economic Design has been at the heart of significant concerns among legal
economists. A new Law and Economics Theory Conference is now organized annu-
ally to promote a gathering for scholars with emphasis on more technical and formal
articles.[1] However, its importance and role is by no means comparable to the annual
Conference in Empirical Legal Studies (not least, because the latter includes also
other social scientists while the former is mainly economists at top law schools).

In my view there are two different but mutually reinforcing explanations for
these trends, particularly for the decline of the role of Economic Design in Law and
Economics. On the one hand, the relocation of the center of gravity of Law and
Economics from economic departments to law schools (with the consequent switch
in publication strategy from economic journals to law reviews) has inevitably reduced
the contact between Law and Economics and Economic Design. Cross-pollination
of advanced economic methodologies to Law and Economics has been severely
damaged in my view (other forms of cross-pollination have taken over with the most
obvious example being Behavioral Law and Economics). One the other hand, the
general sense that Economic Design is too complex to offer appropriate and realistic
answers to legal problems has inevitably made its way. While verbal and discursive
work has found a natural home in law reviews, the specialized journals have gone
empirical to a large extent as indicated by Table 2. Moreover, my sense is that a
great part of the theory papers summarized on Table 1 use established principles of
Economic Design and not the most recent advances (unlike the empirical articles
which blend well-established econometrics with new empirical methods).

One possibility is that Law and Economics does not need Economic Design. If so,
the trends I have described before are of no particular concern. Law and Economics
has developed into an important field of legal analysis and a relevant source of
legal policymaking. Its detachment from Economic Design just confirms the priority
for empirical analysis and basic economic reasoning as a matter of pragmatism.
Detachment from Economic Design is unimportant.

This is a view that I do not share. What made Law and Economics attractive
30 years ago was the rigorous mathematical method. In my understanding, it was
Economic Design that empowered Law and Economics in the legal community and
explains why the field succeeded in ways that competing Law and Social Sciences
did not (Garoupa and Ulen 2007). If the theory of efficient contractual breach, the
model of optimal tort liability or the economic analysis of litigation is nowadays so
influential, it is because it was developed using the advanced concepts of principal-
agent theory and game theory from the 1980s. If Law and Economics had made no use
of Economic Design back then, most probably there would not be a prestigious and
important field of Law and Economics nowadays. Therefore, the current detachment
of Law and Economics from Economic Design is, in my view, of serious concern.

---

[1]Only the 2014 and 2017 meetings seem to have public webpages, https://www.law.berkeley.
edu/law-and-economics-theory-conference-iv/ and https://economics.rice.edu/sites/g/files/bxs876/
f/Law%20and%20Economic%20Theory%20Conference%20Program_3.pdf [both last checked
May 20, 2019].

Many advances in Economic Design may seem too unrealistic or too complex to be relevant for an applied field in Economics. We might even want to consider or debate whether future developments of Economic Design should take into account the needs of applied fields (of which Law and Economics is surely a mere example). However, as a general trend in a long period of time, I think the detachment between Law and Economics and Economic Design is more of a problem to Law and Economics. In that respect, a few measures might be considered. A change of gravity from law schools back to economic departments seems highly unlikely in the next decades, but a more systematic approach to conferences and journals with a focus on theory should be welcome. More cooperative research work between Economic Design and Law and Economics should also be promoted. For example, articles using advanced economic theory applied to evidence issues or litigation should actually be informed by legal concerns. In fact, I have experienced in the last couple of years reading papers on evidence and litigation by pure economists who have very little grasp of current legal issues. My explanation is that in many mainstream journals, while referees and editors are extremely demanding with formal technicalities, they tend to be soft with the actual policy context and ignore the current legal discussions. In order to assess or overcome this issue, an easy question should be asked—what is the current or actual evidence or litigation problem that this particular application of Economic Design intends to address?

I am not optimistic that the current gap between Law and Economics and Economic Design is easy to be overcome. I am also aware that many legal scholars think that such gap is a good and healthy development precisely because they dislike the rationality assumption, formal models or the parsimonious analysis that Economic Design inevitably promotes. However, as emphasized before, I take the view that Economic Design is one of the reasons why Law and Economics was so successful in the initial years. Therefore, I think the future of Law and Economics very much relies on its ability to be infused by developments in Economic Design. I am worried that if such methodological relationship is severely broken, the future of Law and Economics is somehow problematic.

# References

Coase, R. (1960). The problem of the social cost. *Journal of Law and Economics, 3,* 1–44.
Garoupa, N. (2011). The law and economics of legal parochialism. *University of Illinois Law Review, 2011,* 1517–1529.
Garoupa, N., & Ulen, Thomas S. (2007). The market for legal innovation: law and economics in Europe and in the United States. *Alabama Law Review, 59,* 1555–1633.

# Economic Design for Democracy

**Hans Gersbach**

**Abstract** Standard mechanism design has barely addressed how public good provision can be organized democratically. The much broader issue how the set of rules, the set of rights, and the organization of the state can be determined in a democratic society with a mechanism design perspective is awaiting in-depth examination. Using democratic mechanisms is a starting point for a more comprehensive economic design approach to the functioning of democracy and opens up new forms of potentially welfare improving democratic rules such as flexible majority rules or sampling with two-stage voting.

## 1 Introduction

Liberal democracies involve three classes of rights—property rights, political rights, and civil rights (see e.g. Mukand and Rodrik 2015), and a suitable organization of the state. The state itself stands on founding principles such as the monopoly of coercion, the power to levy taxes, the validation of property rights and contracts between citizens, the separation of the legislative, judicial and executive powers, as well as the definition of the range of governmental activities in the provision of public goods and services.

There are at least three different approaches to the functioning of a democratic state: (i) constitutional economics, (ii) election and voting, and (iii) mechanism design. While the first two branches of research have a long tradition and have provided valuable insights, the third branch has only just started to grow, as developing democracy has never been a main concern of standard mechanism design theory.

Of course, there is a rich literature on the mechanisms developed to provide public goods efficiently. Clarke (1971) and Groves (1973), for instance, made first suggestions for public good provision. Subsequent work uncovered conflicts between the requirements of incentive compatibility, efficiency and voluntary participation

H. Gersbach (✉)
CER-ETH – Center of Economic Research at ETH Zurich and CEPR, Zürichbergstrasse 18, 8092 Zurich, Switzerland
e-mail: hgersbach@ethz.ch

© Springer Nature Switzerland AG 2019
J.-F. Laslier et al. (eds.), *The Future of Economic Design*, Studies in Economic Design, https://doi.org/10.1007/978-3-030-18050-8_57

that exist in the context of public good provision.[1] However, standard mechanism design[2] has barely addressed how such public good provision can be organized democratically, and the much broader issues how the set of rules, the set of rights, and the organization of the state can be decided in a democratic society with a mechanism design perspective are still awaiting in-depth examination.

First attempts were made to develop democratic mechanisms. Gersbach (2009) has introduced the notion of democratic mechanisms, which are mechanisms that have to satisfy the liberal democracy constraints. Contrary to a standard mechanism design problem, the typical citizen in a jurisdiction can only send binary messages, that is, say "Yes" or "No" or a series of "Yes/No"-messages. Moreover, the issues or proposals are not made by an outside designer but are put forward by citizens, who are themselves members of the society. Moreover, citizens have the same voting rights and equal right to make proposals, be it in parliament in a representative democracy or in the society itself if direct-democracy procedures such as initiatives and referenda are allowed. A further feature of a democratic mechanism is that the status quo can only be replaced by an alternative via some type of a majority vote which guarantees that the anonymity principle holds. In addition, once the majority has come to a decision, all citizens must respect this decision and bear its financial consequences—participation constraints do not matter, at least not in the short term. Finally, it is often required that citizens are treated equally regarding taxation and other matters.

As outlined in Gersbach (2009, 2016), using democratic mechanisms opens up new forms of potentially welfare improving democratic rules such as flexible or double majority rules or rotating agenda setting. However, these investigations are restricted to particular types of public good provision problems. Recently, Britz and Gersbach (2016) have studied democratic mechanisms with sampling and two-stage voting which reliably reveal and implement the Condorcet winner in the presence of deep uncertainty about the benefits of public good provision.

## 2  Difficulties

A comprehensive economic design approach to the functioning of the democratic state is missing. There are obvious difficulties in finding a suitable method. First, any such mechanism has to satisfy liberal democracy constraints and sustain liberal democracy principles. Such constraints involve equal voting rights, decision rules that satisfy the anonymity principle, and equal rights to make proposals, be it in parliament or in the society. In addition, only a series of "Yes/No"-messages is allowed in voting stages. Moreover, principles of liberal democracies such as separation of powers or equal treatment rules and basic rights may need to be sustained.

---

[1] See e.g. Hellwig (2003).
[2] See Börgers (2015) for a thorough treatment.

Second, one has to operate with two distinct normative criteria in mind: a welfare-based one and a democratic one. Ideally, one would like to transcend the strict irreducibility of these two criteria, e.g. by some aggregate welfare justification for democratic and procedural requirements. In particular, having aggregate welfare as the only ultimate social maximandum would be an extreme form of reducibility. Although some of the democratic requirements can be justified this way (see e.g. Gersbach 2004), we are far away from such a reducibility. Moreover, a complete reducibility is probably impossible and not desirable, as there might be some tradeoffs inherent to the two criteria. In such cases, core democratic requirements may be chosen lexicographically over aggregate welfare maximization.

Third, since the message space is restricted, the revelation principle cannot be used, and thus we cannot simplify the task of analyzing optimal mechanisms. It is true that the government can enforce participation within its own jurisdiction. Thus, participation constraints can be neglected to some extent, which simplifies the task somewhat.[3]

Fourth, there is no designer who finds socially optimal mechanisms for public good provision. Such mechanisms, as well as proposals for particular public goods and financing schemes within these mechanisms, have to come from agents with particular self-interest.

Fifth, different sets of rules interact to govern democratic society, which go significantly beyond developing mechanisms for public good provision or investigations in social choice theory that thus has studied one rule at a time. For instance, there are rules how proposal-makers are chosen and treated as well as rules that put restrictions on proposals, amendments to proposals—a proposal consisting, for instance, of a level of public good and a tax/subsidy package, or either granting or restricting rights. Furthermore, there are rules how the society or the parliament decides about proposals. Moreover, there are rules how powers should be separated and/or delegated to branches of the state, and how different branches are involved in particular governmental activities. Finally, there is a separate, but large set of rules how citizens can challenge the outcomes of collective decision-making and how such disputes are resolved in a given court system. Each of these classes of rules is itself a space that cannot be easily aligned with well-known spaces such as n-dimensional Euclidean space ($n \geq 1$) or functional spaces. Up to now, it has been impossible to characterize all conceivable rules in each of these classes.

Sixth, communication is central to democracy, but is hampered by the rational ignorance of voters and the use of ideological shortcuts and narratives that form beliefs. This communication process is further complicated by the expectation—or the hope—that deliberation, exchanging arguments, fact-based reasoning and searching for the best alternative, should be central to achieving good collective decisions. Such a communication process cannot be phrased in a standard Bayesian framework alone, so that it would be essential to find a good alternative way to model deliberation.

---

[3]Citizens can switch jurisdictions and participation constraints reappear when citizens compare how they fare under different sets of rules across jurisdictions.

Seventh and last, democratic rules should help citizens identify with a jurisdiction or nation, and should motivate them to be active participants in democratic processes. Ideally, the citizens should report or challenge deficiencies in public good provision or governmental activities in general. They should articulate their preferences and participate in deliberation processes.

# 3 Which Approach?

In view of this complexity, it is difficult to imagine a comprehensive economic design approach to democracy that could be formulated successfully and would deliver significant results. As an alternative approach, one could—and should—try to modularize the grand design problem and to tackle all sub-problems sequentially. Still, the best way to modularize this issue and to perform a sequential economic design for democracy remains unclear. Moreover, departing from the Bayesian approach and modeling communication and reasoning with other tools seems a great challenge for economic design, as a well-working and widely applicable model of deliberation is missing.[4]

While such an alternative approach is itself a great challenge, recent research suggests that an economic design of democracy may emerge slowly, with justifications of existing rules and invention of entirely new rules for democracy that jointly improve democracy point by point. This might still improve the citizens' welfare and respect democratic requirements. Over time, an encompassing structure might emerge, and we might finally be able to build the bridge to a comprehensive economic design of democracy.

Overall, the economic design approach to democracy is still in its infancy, and to develop it is an ambitious undertaking. Yet, it could prove a most promising avenue for economic science—we should be on our way.

**Acknowledgements** I would like to thank Volker Britz, Gabrielle Demange, Klaus Nehring, and Oriol Tejada for valuable suggestions.

# References

Börgers, T. (2015). *An introduction to the theory of mechanism design*. New York, NY: Oxford University Press.
Britz, V., & Gersbach, H. (2016). *Information aggregation in democratic mechanisms*. CESifo Working Paper Series No. 5815.
Clarke, E. (1971). Multipart pricing of public goods. *Public Choice, 11*(1), 17–33.
Gersbach, H. (2004). Why one-person-one-vote? *Social Choice and Welfare, 23*(3), 449–464.

---

[4]See List (2017) for an innovative survey of possible directions to analyze deliberations beyond the Bayesian approach.

Gersbach, H. (2009). Democratic mechanisms. *Journal of the European Economic Association, 7*(6), 1436–1469.

Gersbach, H. (2016). Flexible majority rules in Democracyville: A guided tour. *Mathematical Social Sciences, 85,* 37–43.

Groves, T. (1973). Incentives in teams. *Econometrica, 14,* 617–631.

Hellwig, M. (2003). Public-good provision with many participants. *Review of Economic Studies, 70,* 589–614.

List, C. (2017). Democratic deliberation and social choice: A review. In: A. Bächtiger, J. Dryzek, J. Mansbridge, & M. Warren (Eds.), *Oxford handbook of deliberative democracy.* Oxford: Oxford University Press.

Mukand, S., & Rodrik, D. (2015). *The political economy of liberal democracy.* National Bureau of Economic Research Working Paper No. w21540.

# Collective Standards

**Alan D. Miller**

**Abstract** I describe the problem of collective standards used in negligence, contract, defamation, obscenity, and international law. I suggest how economic theorists might contribute to this problem.

Many legal rules depend on concepts that are commonly invoked yet little understood. I suggest that the tools of normative economics can be utilized to illuminate these concepts and to give practical guidance on how these legal rules can be applied. Here I describe one such concept, collective standards, used to judge behavior in a wide range of legal settings. I offer it as an "open problem" to researchers interested in law and economic theory.

I begin by describing five areas of law in which collective standards are used. It is not a matter of unanimous agreement that all of these legal concepts represent collective standards. However, all have been discussed in these terms in the caselaw and the academic literature.

The first standard, and perhaps the most famous, is the "reasonable person" used to ascertain negligence in the common law of torts (see Miller and Perry 2012). The standard is used, for example, to determine the liability of a driver who caused a crash: the driver must pay only if she failed to exercise the care of a reasonable person under the circumstances. There is very little agreement as to how a reasonable person behaves.

The second standard is that used to determine "good faith" in the law of contracts (see Miller and Perry 2013a). The dominant theories of good faith tie the term to collective standards of behavior. But like the reasonable person, this term is nebulous and legal scholars have struggled to define it.

The third standard comes from the common law tort of defamation (see Miller and Perry 2013b). The central question in defamation cases is whether a statement

I would like to thank Henrik Lando, Guido Maretto, and Vera Shikhelman for reading earlier versions of this work.

A. D. Miller (✉)
Faculty of Law and Department of Economics, University of Haifa,
31905 Mount Carmel, Haifa, Israel
e-mail: admiller@econ.haifa.ac.il
URL: http://econ.haifa.ac.il/~admiller/

© Springer Nature Switzerland AG 2019
J.-F. Laslier et al. (eds.), *The Future of Economic Design*, Studies in
Economic Design, https://doi.org/10.1007/978-3-030-18050-8_58

was of the type that tended to harm the plaintiff's reputation. The question of actual harm is irrelevant in many cases; it is an ancillary question in others. Lawyers have not developed a clear test to determine whether a statement tends to have this effect.

The fourth standard comes from First Amendment jurisprudence. American courts have held that pornographic materials do not qualify as protected speech if they are obscene under contemporary community standards (see Miller 2013). Courts have spent sixty years avoiding the question of how to define "community standards." This vague standard separates protected speech from criminal activity.

The fifth standard comes form international law. Some legal questions are decided according to "international custom, as evidence of a general practice accepted as law."[1] According to the theory, custom is determined from state practice, but it is not clear how to determine whether a custom exists.

In all five of these cases, there is a significant line of thought connecting the legal standards to collective. The relevant "collective" differs from standard to standard. For the reasonable person, it can be "the great mass of mankind" in cases of ordinary accidents and the group of doctors or specialists in cases of medical malpractice. For good faith, the collective is often a community of merchants. In defamation cases it may be either the general community or a "substantial and respectable minority." In obscenity cases the collective can be either a city, a county, or a state, or the collective can be left undefined. In international custom, the collective is the community of nations.

There are three major approaches to understanding collective standards: to deny that a problem exists, to assume the problem away, and to claim that collective standards do not exist.

The first approach, to deny the existence of the problem, is implicit in much of the caselaw and the literature. According to this view, there is a definition of collective standards that is understood by all with common sense. While people may disagree as to their content of these collective standards, this is a factual question that can be decided by a judge or a jury in the same way that we decide other factual matters, such as the speed of a car or the number of bullets that were fired. Economists have a word for this approach: "handwaving."

The second approach assumes the problem away by substituting a normative standard for that of the collective. The most well known approach in the context of negligence is that embodied by the famous "Hand Rule," which assumes, in essence, that the reasonable person is a utilitarian.[2] This is not because of any evidence (or even belief) that any individuals actually behave this way, but out of the belief that it would be better if they did.

The third approach neither denies the existence of a problem nor assumes it away. Instead, this approach acknowledges the existence of a problem but then claims that it can not be solved. The clearest exposition of this approach comes from theoretical economics in the form of impossibility theorems (Rubinstein 1983; Miller 2013), which provide clear arguments as to why collective standards cannot be defined.

---

[1] Statute of the International Court of Justice, Article 38(1)(b).
[2] See *United States v. Carroll Towing Co.* 159 F.2d. 169 (2nd. Cir. 1947).

These approaches have some merit. The first is the most difficult to justify. Ascertaining a collective standard is not similar to ascertaining facts like the weather; it involves the resolution of both a positive question (e.g. what people believe or how they act) and a normative question (how to determine the collective standard on the basis of this information). An argument in favor of the "handwaving" approach is that decisions need to be made. It is possible that telling judges and juries t o decide according to a collective standard yields "good" results, according to some external benchmark. However, this argument is not supported by any theory or evidence. It is also harder to justify in cases of obscenity and international custom, where it is not obvious what this external benchmark would be. (For example, one normative justification for the prohibition of obscenity is that it offends collective standards.)

The second approach can be justified by reference to the specific normative philosophy that underlies the chosen standard. It can also be justified by the argument that judges or legislators who developed the idea had specific normative standards in mind. Research in law and economics, for example, argues that common law judges made decisions that were wealth-maximizing (see Parisi 2004). This approach is especially difficult to justify, however, if the theory is explicitly a collective standard, such as in obscenity.

The third approach is easy to justify. It is meaningful to explore why no good theory of collective standards might exist.

However, courts continue to use the language of collective standards. An explanation of what courts are or could be doing would still be as valuable as ever. Such an explanation need not be entirely general; it may take into account features specific to defamation, for example, and not be applicable to the reasonable person. I suggest a few possibilities here.

First, the existence of impossibility theorems does not mean that possibilities do not exist. Potentially one could formalize these standards in a way different from Rubinstein (1983) and Miller (2013). This could lead to a new justification of collective standards or a new reason to reject them; either would be valuable.

Second, one could provide a more restricted solution, explaining how courts should make decisions in light of the impossibility results. Just as we still try to find ways to evaluate welfare despite Arrow's theorem, we might try to find imperfect but still useful definitions of collective standards. Perhaps one way to do this would be to develop a definition of collective standards that relies more on normative justifications, but which does not completely replace the collective standard with a single normative standard.

# References

Miller, A. D. (2013). Community standards. *Journal of Economic Theory, 148*, 2696–2705.
Miller, A. D., & Perry, R. (2012). The reasonable person. *New York University Law Review, 87*, 323–392.
Miller, A. D., & Perry, R. (2013a). Good faith revisited. *Iowa Law Review, 98*, 689–745.

Miller, A. D., & Perry, R. (2013b). A group's a group, no matter how small: An economic analysis of defamation. *Washington and Lee Law Review*, *70*, 2269–2336.

Parisi, F. (2004). *The efficiency of the common law hypothesis* (pp. 519–522). Boston: Springer.

Rubinstein, A. (1983). The reasonable man - A social choice approach. *Theory and Decision*, *15*, 151–159.

# The Present and Future of Judgement Aggregation Theory. A Law and Economics Perspective

Philippe Mongin

**Abstract** This chapter briefly reviews the present state of judgment aggregation theory and tentatively suggests a future direction for that theory. In the review, we start by emphasizing the difference between the *doctrinal paradox* and the *discursive dilemma*, two idealized examples which classically serve to motivate the theory, and then proceed to reconstruct it as a brand of *logical* theory, unlike in some other interpretations, using a single impossibility theorem as a key to its technical development. In the prospective part, having mentioned existing applications to social choice theory and computer science, which we do not discuss here, we consider a potential application to *law and economics*. This would be based on a deeper exploration of the doctrinal paradox and its relevance to the functioning of collegiate courts. On this topic, legal theorists have provided empirical observations and theoretical hints that judgment aggregation theorists would be in a position to clarify and further elaborate. As a general message, the chapter means to suggest that the future of judgment aggregation theory lies with its applications rather than its internal theoretical development.

## 1 A New Brand of Aggregation Theory

It is a commonplace idea that collegial institutions generally make better decisions than those in which a single individual is in charge. This optimistic view, which can be traced back to French Enlightment social theorists like Rousseau and Condorcet, permeates today's western judiciary organization, which normally entrusts collegial courts with the competence to rule on the more complex cases; think of constitutional courts like the U.S. Supreme Court with its nine judges. However, the following, by now classic example from legal theory challenges the orthodoxy. A plaintiff has

The author gratefully acknowledges comments by Franz Dietrich, Samuel Ferey, Umberto Grandi and Jean-François Laslier.

P. Mongin (✉)
CNRS & HEC Paris, 1 rue de la Libération, 78350 Jouy-en-Josas, France
e-mail: mongin@greg-hec.com

© Springer Nature Switzerland AG 2019
J.-F. Laslier et al. (eds.), *The Future of Economic Design*, Studies in
Economic Design, https://doi.org/10.1007/978-3-030-18050-8_59

417

brought a civil suit against a defendant, alleging a breach of contract between them. The court is composed of three judges $A$, $B$ and $C$, who will determine whether or not the defendant must pay damages to the plaintiff ($d$ or $\neg d$). The case brings up two issues, i.e., whether the contract was valid or not ($v$ or $\neg v$), and whether the defendant was or was not in breach of it ($b$ and $\neg b$). Contract law stipulates that the defendant must pay damages if, and only if, the contract was valid and the defendant was in breach of it. Suppose that the judges have the following views of the two issues and accordingly of the case:

| $A$ | $v$ | $\neg b$ | $\neg d$ |
|-----|-----|----------|----------|
| $B$ | $\neg v$ | $b$ | $\neg d$ |
| $C$ | $v$ | $b$ | $d$ |

In order to rule on the case, the court can either directly collect the judges' recommendations on it, or collect the judges' views of the issues and then solve the case by applying contract law to these data. If the court uses majority voting, the former, *case-based* method delivers $\neg d$, whereas the latter, *issue-based* method returns first $v$ and $b$, and then $d$. This elegant example is due to legal theorists Kornhauser and Sager (1993, with a preliminary sketch in 1986). They describe as a *doctrinal paradox* any similar occurrence in which the two methods give conflicting answers. What makes the discrepancy paradoxical is that each method is commendable on some ground, i.e., the former respects the judges' final views, while the latter provides the court with a rationale, so one would wish them always to be compatible. The legal literature has not come up with a clear-cut solution, although the issue-based method often attracts more sympathy. This persisting difficulty casts doubt on the belief that collegial courts would be wiser than individual ones. Clearly, with a single judge, the two methods coincide unproblematically.

An entire body of work, now referred to as *judgment aggregation theory,* has grown out of Kornhauser and Sager's doctrinal paradox. As an intermediate step, their problem was rephrased by political philosopher Pettit (2001), who wanted to make it both more widely applicable and more analytically tractable. What he calls the *discursive dilemma* is, first of all, the generalized version of the doctrinal paradox in which a group, whatever it is, can base its decision on either the *conclusion-based* or the *premiss-based* method, whatever the substance of conclusions and premisses. Any time the propositions of interest to the group can logically be divided into a set of premisses and a conclusion, the representative of the group can put to the vote either the conclusion directly or the premisses themselves, while performing the required inference in the latter case. What the court example illustrates can apply equally well to a political assembly, an expert committee, and many deliberating groups; as one of the promoters of the concept of deliberative democracy, Pettit would speculatively add political society as a whole. Second, and more technically, the discursive dilemma shifts the stress away from the conflict of two methods of decision to *the logical contradiction within the total set of propositions that the group accepts*. In the previous example, with $d \longleftrightarrow v \wedge b$ representing contract law, the contradictory set is

$$\{v, b, d \longleftrightarrow v \wedge b, \neg d\}.$$

This shift has far-reaching consequences since all propositions are now being treated alike; indeed, the very distinction between premises and conclusion vanishes. If one is concerned with developing a general theory, this move has clear advantages. Given a set of propositions, it may be tricky t o dvide it appropriately, and thus to say what the two methods exactly are. It is definitely simpler, and it may be sufficient, to pay attention to whole sets of propositions accepted by either the individuals or the group, more briefly *judgment sets*, and inquire how these relate to each other. The question raised by the discursive dilemma is why and when individual judgment sets that are consistent deliver a collective judgment set that is inconsistent. This is already the problem of judgment aggregation.

In a further step, List and Pettit (2002) introduced an aggregation mapping $F$, which takes profiles of individual judgment sets $(A_1, ..., A_n)$ to collective judgment sets $A$, and subject $F$ to formal conditions which they demonstrate are logically incompatible. Both the proposed formalism and impossibility conclusion are in the vein of social choice theory, but they are directed at the discursive dilemma, which the latter theory cannot handle in its own terms, since there is no question here of preferences and related social-choice-theoretic concepts. At this stage, judgment aggregation theory exists in full, having defined its object of study—the $F$ mapping, or *collective judgment function*—as well as its method of analysis—it consists in defining axiomatic conditions on $F$ and investigating when these conditions result into impossibilities, or to the contrary, support well-behaved rules, such as simple majority voting, the only rule considered in the above example.

List and Pettit's impossibility theorem was shortly superseded by others of grow-ing sophistication, due to Pauly and van Hees (2006), Dietrich (2006), Dietrich and List (2007a), Mongin (2008), Nehring and Puppe (2008, 2010), Dokow and Holz-man (2009, 2010a), Dokow and Holzman (2010b), Dietrich and Mongin (2010a). This lengthy, but still incomplete list, should be complemented by two papers that contributed differently to the progress of the field. Renewing the investigation of social choice rules, Nehring and Puppe (2002) characterized, in the sense of nec-essary and sufficient conditions, those preference domains for which relevant sets of axioms—strategy-proofness among them—entail dictatorship. Judgment aggre-gation theory has taken up this concern for characterizing impossibility domains. It does so by defining the *agenda* of propositions on which the individuals and the collective make judgments, and then by asking which properties of the agenda turn any $F$ satisfying certain axioms into a dictatorship (or some other unpalatable rules like oligarchy). On a very different score, Dietrich (2007) showed that the logical formalism of judgment aggregation theory could be deployed without making refer-ence to any specific logical calculus. Only a few elementary properties of the formal language and of the logic need assuming for the theorems to carry through. The so-called *general logic* states these requisites (see Dietrich and Mongin 2010a, for an up-to-date version). The first papers were unnecessarily restrictive in relying on the elementary propositional calculus alone.

Implicit in the previous comments is the view that judgment aggregation theory is *a brand of logical theory*. This view fits the work of many judgment aggregation theorists very well, but not so well the work of Nehring and Puppe (whose contributions often use the framework of social choice theory) or the work of Dokow and Holzman (who have devised a combinatorial framework of their own, drawing inspiration from Rubinstein and Fishburn 1986). The more thorough review by Mongin (2012) elaborates on this difference in perspectives while defending the use of logic (for an earlier French version, see Dietrich and Mongin 2010b). This review also points out the historical origin of judgment aggregation theory, arguably already understood as a logical theory, in Guilbaud (1952) and Condorcet himself, whose rediscovery in the mid-20th century was largely due to Guilbaud.

Consistently with this orientation towards logic, Sect. 2 introduces a syntactical framework for the $F$ function and states in this framework the axiomatic conditions on $F$ that have attracted most attention. The crucial issue of agendas arises in Sect. 3, which characterizes those which, for given conditions on $F$, turn this function into a dictatorship (oligarchy will not be discussed here). Essentially, there is a single theorem, but in three forms, each of which is instructive in itself. This theorem is the central achievement of judgment aggregation theory by common consent, so one can take it to represent, at least in very broad outlook, the *present* of that theory. Section 4 tentatively argues about its *future*, a challenging and unavoidably questionable exercise. It seems as if the years of high theorizing are over by now, and in the recent years, judgment aggregation theory has indeed moved forward mostly on the front of applications. In our view, this applied work delineates the near future of the theory, and we accordingly restrict attention to it in the prospective part of this chapter.

The first applications of judgment aggregation theory were to social choice theory. Having been a heuristic source for the framework and conditions on $F$, this field was also a convenient source of corollaries to the main theorems. It is sufficient to replace the abstract propositional symbols $p, q, r \ldots$ by formulas such as $xRy$, $zPw$, $uIv$, with the usual preference interpretations, to give rise to a tentative application. Whether a proper social choice corollary follows will depend in particular on whether the agenda conditions stated in the judgment aggregation theorems hold for the new set of formulas, given the logical connections between these formulas (reflecting transitivity, completeness, acyclicity, and the like). Arrow's theorem can be recovered in this way, along with variants that either already existed or are new to social choice theory. The second and actually larger group of applications has emerged at the crossroads of formal epistemology and theoretical computer science. Among others, it is concerned with relaxing those judgment aggregation conditions which are responsible for dictatorship. This has led to the flourishing "belief merging literature", which often adopts a syntactical form and often pays attention to computational issues, as befit its computer science connections. For this fairly large body of work, we refer to Grossi and Piggozzi (2014) and Pigozzi (2016). The present, purposefully short discussion deals with a third group of applications, concerning law and economics, which have not yet been explored. Given that judgment aggregation theory originates in a law and economics example, it is but natural to inquire what it can contribute to this field in return. Section 4 is entirely devoted to these applications, which will be seen to be mostly prospective, rather than already existing.

## 2   A Logical Framework for Judgment Aggregation Theory

By definition, a *language* $\mathcal{L}$ for judgment aggregation theory is any set of formulas $\varphi, \psi, \chi, \ldots$ that is constructed from a set of logical symbols $\mathcal{S}$ containing $\neg$, the Boolean negation symbol, and that is closed for this symbol (i.e., if $\varphi \in \mathcal{L}$, then $\neg\varphi \in \mathcal{L}$). Normally, $\mathcal{S}$ contains symbols for other Boolean connectives (conjunction $\wedge$, disjunction $\vee$, implication $\rightarrow$, bi-implication $\longleftrightarrow$) and in principle, it may contain modal operators (such as $\square$ interpreted as "it is obligatory that", $B_i$ interpreted as "individual $i$ believes that", or $\Longrightarrow$ interpreted as a non-Boolean conditional). Then, these added symbols will have to satisfy the relevant closure properties. A *logic* for judgment aggregation theory is any set of axioms and rules that regulates the inference relation $\vdash$ on $\mathcal{L}$ and associated technical notions—logical truth and contradiction, consistent and inconsistent sets—while satisfying the general logic. Informally, the latter stipulates that $\vdash$ be monotonic and compact, and that any consistent set of formulas can be extended to a complete consistent set ($S \subset \mathcal{L}$ is *complete* if, for all $\varphi \in \mathcal{L}$, either $\varphi \in S$ or $\neg\varphi \in S$). Monotonicity means that inductive logics are excluded, and compactness (which is needed only in specific proofs) that some deductive logics are. The extendability requisite is standard for deductive logics.

Among the many calculi that enter this framework, propositional examples stand out, but they can be richer than the elementary propositional calculus with Boolean connectors alone. Adding modal operators, one gets the many calculi of deontic, epistemic and conditional logics. Although this may not be so obvious, first-order calculi with quantifiers are also permitted.

In $\mathcal{L}$, a subset $X$ is fixed to represent the propositions that are in question for the group; this is the *agenda*, which is the focus of attention in judgment aggregation theory. In all generality, $X$ needs only to be non-empty with at least one contingent formula (not a logical truth or contradiction), and to be closed for negation. The court example, in the discursive dilemma version, involves the agenda

$$\overline{X} = \{v, b, d, d \leftrightarrow v \wedge b, \neg v, \neg b, \neg d, \neg(d \leftrightarrow v \wedge b)\}.$$

The theory represents judgments by subsets $B \subset X$ that are initially unrestricted. These *judgment sets* will be denoted by $A_i, A_i', \ldots$ for the individuals $i = 1, \ldots n$, and $A, A', \ldots$ for the collective. A formula $\varphi$ represents a proposition, in the sense of a semantic object endowed with a truth value, and it also serves to capture the associated judgment, in the sense of a cognitive operation, thanks to the informal translation:

*(R) individual i (the collective) makes the judgment $\varphi$ iff $\varphi$ belongs to the judgment set $A_i$ (resp. A).*

Judgment sets can be subjected to various logical properties. Here we consider only two sets of judgment sets, i.e., the unrestricted set $2^X$ and the set $D$ of *consistent* and *complete* judgment sets (with completeness being relative to $X$, not $\mathcal{L}$). The theory has also investigated the intermediate case of consistent, but not necessarily complete judgment sets. Thus far, it has not been able to relax consistency (see Dietrich and List 2008).

The last formal concept is the *collective judgment function F*, which associates a collective judgment set to each profile of judgment sets for the $n$ individuals:

$$A = F(A_1, \ldots, A_n).$$

We restrict attention to $F : D^n \to 2^X$ and $F : D^n \to D$. Taking $D$ to be the range means that the collective sets obey the same stringent logical constraints as the individual ones. This framework is sufficient to capture simple majority voting, as in the court example, and various generalizations. Formally, define *formula-wise majority voting* as $F_{maj} : D^n \to 2^X$ such that, for every profile $(A_1, \ldots, A_n) \in D^n$,

$$F_{maj}(A_1, \ldots, A_n) = \{\varphi \in X : |\{i : \varphi \in A_i\}| \geq q\},$$
$$\text{with } q = \tfrac{n+1}{2} \text{ if } n \text{ is odd and } q = \tfrac{n}{2} + 1 \text{ if } n \text{ is even.}$$

Here, the range is not $D$ because there can be unbroken ties, and so incomplete collective judgment sets, when $n$ is even. More strikingly, for many agendas, the range is not $D$ also when $n$ is odd, because there are inconsistent collective judgment sets, as the court example illustrates. By varying the value of $q$ between 1 and $n$ in the definition, one gets specific quota rules $F_{maj}^q$. One would expect inconsistency to occur with low $q$, and incompleteness with large $q$. Nehring and Puppe (2002, 2008) and Dietrich and List (2007b) investigate these rules in detail.

Having defined and exemplified $F$ functions, we introduce some axiomatic properties they may satisfy.

**Systematicity**. For all formulas $\varphi, \psi \in X$ and all profiles $(A_1, \ldots, A_n), (A'_1, \ldots, A'_n)$, if $\varphi \in A_i \Leftrightarrow \psi \in A'_i$ for every $i = 1, \ldots, n$, then

$$\varphi \in F(A_1, \ldots, A_n) \Leftrightarrow \psi \in F(A'_1, \ldots, A'_n).$$

**Independence**. For every formula $\varphi \in X$ and all profiles $(A_1, \ldots, A_n), (A'_1, \ldots, A'_n)$, if $\varphi \in A_i \Leftrightarrow \varphi \in A'_i$ for every $i = 1, \ldots, n$, then

$$\varphi \in F(A_1, \ldots, A_n) \Leftrightarrow \varphi \in F(A'_1, \ldots, A'_n).$$

**Monotonicity**. For every formula $\varphi \in X$ and all profiles $(A_1, \ldots, A_n), (A'_1, \ldots, A'_n)$, if $\varphi \in A_i \Rightarrow \varphi \in A'_i$ for every $i = 1, \ldots, n$, with $\varphi \notin A_j$ and $\varphi \in A'_j$ for at least one $j$, then

$$\varphi \in F(A_1, \dots, A_n) \Rightarrow \varphi \in F(A'_1, \dots, A'_n).$$

**Unanimity preservation**. For every formula $\varphi \in X$ and every profile $(A_1, \dots, A_n)$, if $\varphi \in A_i$ for every $i = 1, \dots, n$, then $\varphi \in F(A_1, \dots, A_n)$.

By definition, $F$ is a *dictatorship* if there is a $j$ such that, for every profile $(A_1, \dots, A_n)$,

$$F(A_1, \dots, A_n) = A_j.$$

There can only be one such $j$, to be called the *dictator*. The last property is

**Non-dictatorship**. $F$ is not a dictatorship.

It is routine to check that $F_{maj}$ satisfies all the list. *Unanimity preservation* and *Non-dictatorship* are self-explanatory. *Systematicity* means that the group, when faced with a profile of individual judgment sets, gives the same answer concerning a formula $\varphi$ as it would give concerning a possibly different formula $\psi$, when faced with a possibly different profile, supposing that the individual judgments concerning $\varphi$ in the first profile are the same as those concerning $\psi$ in the second profile. *Independence* amounts to restricting this requirement to $\varphi = \psi$. Thus, it eliminates one claim made by Systematicity—i.e., that the identity of the formula does not matter—while preserving another—i.e., that the collective judgment of $\varphi$ depends only on individual judgments of $\varphi$. *Monotonicity* requires that, when a collective result favours a subgroup's judgment, this still holds if more individuals join the subgroup.

Systematicity, Independence and Monotonicity are clearly reminiscent of neutrality, independence of irrelevant alternatives and positive responsiveness in social choice theory. Each can be defended by normative arguments that have already surfaced there, i.e., systematicity by the fact that many voting rules satisfy it, independence by its non-manipulability property, and monotonicity by its democratic responsiveness implication. The argument for Unanimity preservation follows the existing one for the Pareto principle, while Non-dictatorship is regarded as unpalatable, exactly as in social choice theory.

Systematicity was the condition underlying List and Pettit's (2002) impossibility theorem, but henceforth, the focus of attention shifted to Independence. The problem that has gradually raised to the fore is to characterize—i.e., find necessary and sufficient conditions for—the agendas $X$ such that no $F : D^n \to D$ satisfies Non-dictatorship, Independence, and Unanimity preservation. There is a variation of this problem with Monotonicity as a further condition, and another variation that involves drawing a distinction between premises and conclusions within the agenda $X$. The next section sketches the respective answers.

# 3  The Present of the Theory: A Characterization Theorem in Three Forms

The promised answers depend on further technical notions. First, a set of formulas $S \subset \mathcal{L}$ is called *minimally inconsistent* if it is inconsistent and all its proper subsets are consistent. With the elementary propositional calculus used for the court example, this is the case for

$$\{v, b, d \leftrightarrow v \wedge b, \neg d\},$$

but not for

$$\{\neg v, \neg b, d \leftrightarrow v \wedge b, d\}.$$

Second, for $\varphi, \psi \in X$, it is said that $\varphi$ *conditionally entails* $\psi$—denoted by $\varphi \vdash^* \psi$—if $\varphi \neq \neg\psi$ and there is some minimally inconsistent $Y \subset X$ with $\varphi, \neg\psi \in Y$. This is equivalent to requiring that $\{\varphi\} \cup Y' \vdash \psi$ hold for some minimal auxiliary set of premises $Y'$ that is contradictory neither with $\varphi$, nor with $\neg\psi$.

Now, an agenda $X$ is said to be *path-connected* (or *totally blocked*) if, for every pair of formulas $\varphi, \psi \in X$, there are formulas $\varphi_1, \ldots, \varphi_k \in X$ such that

$$\varphi = \varphi_1 \vdash^* \varphi_2 \vdash^* \ldots \vdash^* \varphi_k = \psi.$$

Loosely speaking, agendas with this property have many, possibly roundabout logical connections. Finite agendas can be represented by directed graphs: the formulas $\varphi, \psi$ are the nodes and there is an arrow pointing from $\varphi$ to $\psi$ for each conditional entailment $\varphi \vdash^* \psi$. It easy to check that the court agenda $\overline{X}$ is path-connected.

Now, we are in a position to state a first answer to the characterization problem (Dokow and Holzman 2010a; Nehring and Puppe 2010; this result originates in Nehring and Puppe 2002). From now we assume that $n \geq 2$.

**Theorem (first form).** If $X$ is path-connected, then no $F : D^n \to D$ satisfies Non-dictatorship, Unanimity preservation, Monotonicity and Independence. The agenda condition is also necessary for this conclusion.

The sufficiency part of this result generalizes the court example in the following sense. The profile of this example shows that, given the agenda $\overline{X}$, $F_{maj}$ does not have range $D$. This can be derived abstractly from the sufficiency part by noting that $F_{maj}$ satisfies the four axiomatic condition and $\overline{X}$ is path-connected, so that the range condition must fail. The necessity part delivers the significant information that, even keeping $F_{maj}$, the discursive dilemma cannot occur if $\overline{X}$ is not path-connected.

As it turns out, Monotonicity can be dropped from the list of axioms if the agenda is required to satisfy a further condition. Let us say that $X$ is *even-number negatable* if there is a minimally inconsistent set of formulas $Y \subseteq X$ and there are distinct $\varphi, \psi \in Y$ such that $Y_{\neg\{\varphi,\psi\}}$ is consistent, where the set $Y_{\neg\{\varphi,\psi\}}$ is obtained from $Y$ by replacing $\varphi, \psi$ by $\neg\varphi, \neg\psi$ and keeping the other formulas unchanged. This seemingly opaque condition is in fact not very demanding, and for instance $\overline{X}$ satisfies it. To see that, take

$$Y = \{v, b, d, \neg(d \leftrightarrow v \wedge b)\} \text{ and } \varphi = v, \psi = b.$$

The next result was proved by Dokow and Holzman (2010a). In showing how Arrow's theorem, in one version, can be recovered by judgment aggregation theory, Dietrich and List (2007a) indirectly proved the sufficiency part.

**Theorem (second form).** If $X$ is path-connected and even-number negatable, then no $F : D^n \to D$ satisfies Non-dictatorship, Unanimity preservation, and Independence. If $n \geq 3$, the agenda conditions are also necessary for this conclusion.

This is often referred to as the "canonical theorem" of judgment aggregation. However, a further step of generalization is available. Unlike the work reviewed so far, it is motivated by the court example in the doctrinal paradox version, as against its discursive dilemma restatement. It is specially devised to clarify the premiss-based method, which is often proposed as a solution to this paradox. Formally, we define the *set of premisses* to be a subset $P \subseteq X$, requiring only that it be non-empty and closed for negation, and we reconsider the framework to account for the difference between $P$ and its complement $X \setminus P$. Adapting the existing conditions, we define

**Independence on premisses**: same statement as for Independence, but holding only for every $p \in P$.

**Non-dictatorship for premisses**: there is no $j \in \{1, \ldots, n\}$ such that $F(A_1, \ldots, A_n) \cap P = A_j \cap P$ for every $(A_1, \ldots, A_n) \in D^n$.

Similarly revising the agenda conditions, we say that $X$ is *path-connected in $P$* if the definition of path-connectedness holds for formulas $\varphi, \psi, \varphi_1, \ldots, \varphi_k$ that are all in $P$, and say that $X$ is *even-number negatable in $P$* if the definition of even-number negatable holds with the negatable pair $\varphi, \psi$ being in $P$. These new conditions can be illustrated in terms of the agenda $\overline{X}$ after fixing a natural interpretation for the set of premisses:

$$\overline{P} = \{v, b, d \leftrightarrow v \wedge b, \neg v, \neg b, \neg(d \leftrightarrow v \wedge b)\}$$

As is easy to check, $\overline{X}$ is both path-connected and even-number negatable in $\overline{P}$.

We can now state the last result (Dietrich and Mongin 2010a).

**Theorem (third form).** If $X$ is path-connected and even-number negatable in $P$, there is no $F : D^n \to D$ that satisfies Non-dictatorship on premisses, Independence on premisses and Unanimity preservation. If $n \geq 3$, the agenda conditions are also necessary for this conclusion.

This third form of the theorem is more assertive than the second one (take $P = X$, a permitted case). Another relevant specification occurs by taking $X \setminus P$ to be a set of *conclusions* deducible from $P$ according to the underlying logic. This specification does not make the results any more precise, but it helps connect the theory with those applications—like the doctrinal paradox—in which a logical distinction holds between premisses and conclusions; more on this line below. Having explored our sample theorem in various ways, we now move to the evaluative and prospective part of this chapter.

# 4  Law and Economics and Some Future Directions for the Theory

Since judgment aggregation theory was motivated by the doctrinal paradox, a law and economics example, it is only natural to investigate its contribution, both current and potential, to that field. As we will show, the current contribution is modest, but various hints suggest a potential. Starting from Kornhauser and Sager's groundbreaking discussion, legal theorists have inquired about the prevalence of the doctrinal paradox in the actual working of U.S. collegiate courts, and usually concluded that the paradox, or related forms of conflict between the issue- and case-based methods, occurred rarely. Kornhauser and Sager (1993) found examples in the U.S. Supreme Court record, but striking as they are, they are small in number and even debatable as such (see Post and Salop 1992; Nash 2003, for a critical review). A major defect of these empirical investigations, they do not distinguish sufficiently between the low prevalence of the paradox and the fact that it is difficult to recognize. U.S. collegiate courts do not always state in their final judgments what the judges' individual positions are, and this even holds for the Supreme Court, which is more profuse than any other collegiate court when it comes to making individual positions known.

In view of these uncertainties, an obviously useful preliminary would be to investigate the combinatorics of the doctrinal paradox or related forms of conflict. On the court agenda defined above, granting that the formula $d \longleftrightarrow v \wedge b$ is unanimously agreed, the combinatorics reduces to the simple task of counting how many profiles of judgments on $b, d, v$ give rise to a conflict between the two methods when majority voting applies. It is easy to see that Kornhauser and Sager have exactly recognized which set of profiles precipitates the conflict and this set has a relatively low probability ratio (our computation based on standard independence and uniformity assumptions leads to a ratio of 3/32). Less trivially, one could let the court agendas vary, and for instance, count how many agendas are path-connected among all those which can be formulated in the propositional language based on the three propositional variables $b, d, v$. More abstract combinatorial investigations could be performed in connection with the theoretical results of last section, and they might deliver useful information for the legal theorists, over and beyond what can be said on the doctrinal paradox itself.

Whatever its mathematical and empirical prevalence, the doctrinal paradox serves the didactic purpose of pointing out two plausible adjudication methods that collective courts can employ, and the law and economics literature has actually spent more time on comparing these methods in general than on the paradox itself. While Kornhauser and Sager (1993) and Kornhauser (2008) express context-sensitive preferences between the two methods, Rogers (1996) supports the case-based method unreservedly, and Post and Salop (1992) support the issue-based method unreservedly. The arguments displayed in this controversy are far from being tight. They tend to entangle formal questions, e.g., which method is more likely to deliver a well-defined solution or more likely to resist strategic manipulation, with substantial legal questions, e.g., whether one method is more likely to protect the defendant in a criminal trial. They also tend to entangle static properties, such as the properties just said, with dynamic properties, such as the ability to define law by precedents, over and beyond the resolution of a particular case. Some writers go as far as to attribute the doctrinal paradox to one of the two methods, whereas the other is obviously co-responsible for it. All in all, despite some analytic efforts like Nash's (2003), much remains to be done in order to produce a decent list of normative considerations and weigh them against each other to decide between the two methods. Some formalism would help bring order, but legal theorists do not use any or provide sketches that fall below the mark.

They can be excused by the fact that judgment aggregation theorists have not sufficiently addressed their concerns. We have emphasized from the start that the doctrinal paradox was not the discursive dilemma, and that these theorists have taken the latter, not the former, to be the starting point of their formal investigations. The tendency in their field is to play down the significance of the distinction; this shows in some blurred accounts that motivate the theorems, and in such strange expressions as "the discursive paradox". But legal theorists have little use for the discursive dilemma and its elaboration; what they are really concerned with is the doctrinal paradox and the accompanying comparison of methods to adjudicate legal claims. This explains Dietrich and Mongin's (2010a) unconventional move of distinguishing between *premisses* and *conclusions* in their theorem statement. With this logical distinction, they make a step towards formalizing the legal distinction between *issues* and *case*. Clearly, the latter presupposes the former, although it goes beyond it semantically. This limited step makes it possible already to draw some lessons from judgment aggregation theory for the legal theorists (see also Mongin 2008).

One lesson concerns the highly generalized form of the doctrinal paradox this theorem delivers. It does not mention simple majority voting, which is the only decision rule considered by legal theorists, but abstract conditions that other rules can also obey; in this way, the theorem makes the paradox more acute than the legal theorists have conceived of it. Among these abstract conditions, Unanimity preservation stands apart in the theorem, because it is the only one that applies both to premisses and conclusions; for example, Independence is here restricted to premisses. Heuristically, the source of the impossibility lies in the fact that, when a profile exhibits unanimous endorsement of some conclusion, the collective concurs; this connects the impossibility with the classic problem of "spurious unanimity" in the case of collective preference under uncertainty (Mongin 1997). The implication for the legal theorists is that judges cannot at the same time apply majority voting on the issues and record their unanimous agreement on the case when this occurs. In other words, the issue-based method already clashes with a very weak form of the case-based method. Since such a limited addition may be difficult to avoid in practice, one may think that the issue-based method is not as safe as it has often been suggested.

Another lesson of interest, which is this time shared by judgment aggregation theory as a whole, is that the agenda plays an essential part in the origin of impossibilities. As the "canonical theorem" and its variants exactly characterize the troublesome agendas, they offer both an explanation of the impossibilities (through the sufficient conditions) and a way of avoiding them (through the necessary conditions). A quick implication for legal theorists is that courts could do away with some occurrences of the impossibilities by restructuring their agenda. Experimenting with toy examples suggests that slight changes might be sufficient. However, when they come to practical solutions, legal theorists consider *ex post* revisions of individual opinions rather than this attractive *ex ante* solution (see the examples of Supreme Court justices reversing their opinions in Kornhauser and Sager's 1993; Post and Salop 1992). To pursue this point appropriately, careful attention must be payed to the logical status of the *legal doctrine* (contract law in the court example of the doctrinal paradox). Does it belong to the agenda? If it does, should it also be subjected to a vote if the issue-based method is adopted? And what is the most suitable logic to capture its role in the judges' reasoning? More specifically, is elementary propositional logic sufficient, as in the usual restatement of the court example, or should one rather resort to some more sophisticated logic, say deontic or conditional logic? These are open questions, with some hints of answers in Mongin (2012) following technical advances by Dietrich (2010). Interestingly, some legal theorists—e.g., Landa and Lax (2009)—appear to have foreshadowed some of these apparently abstruse questions. Here is another subarea in which law and economics and judgment aggregation theory could fruitfully join forces.

# References

Dietrich, F. (2006). Judgment aggregation: (Im)Possibility theorems. *Journal of Economic Theory*, *126*, 286–298.

Dietrich, F. (2007). A generalized model of judgment aggregation. *Social Choice and Welfare*, *28*, 529–565.

Dietrich, F. (2010). The possibility of judgment aggregation on agendas with subjunctive implications. *Journal of Economic Theory*, *145*, 603–638.

Dietrich, F., & List, C. (2007a). Arrow's theorem in judgment aggregation. *Social Choice and Welfare*, *29*, 19–33.

Dietrich, F., & List, C. (2007b). Judgment aggregation by quota rules: Majority voting generalized. *Journal of Theoretical Politics*, *19*, 391–424.

Dietrich, F., & List, C. (2008). Judgment aggregation without full rationality. *Social Choice and Welfare*, *31*, 15–39.

Dietrich, F., & Mongin, P. (2010a). The premiss-based approach to judgment aggregation. *Journal of Economic Theory*, *145*, 562–582.

Dietrich, F., & Mongin, P. (2010b). Un bilan interprétatif de la théorie de l'agrégation logique. *Revue d'économie politique*, *120*, 929–972.

Dokow, E., & Holzman, R. (2009). Aggregation of binary evaluations for truth-functional agendas. *Social Choice and Welfare*, *32*, 221–241.

Dokow, E., & Holzman, R. (2010a). Aggregation of binary evaluations. *Journal of Economic Theory*, *145*, 495–511.

Dokow, E., & Holzman, R. (2010b). Aggregation of binary evaluations with abstentions. *Journal of Economic Theory*, *145*, 544–561.

Grossi, D., & Piggozzi, G. (2014). *Judgment aggregation: A primer*. San Rafael, CA: Morgan & Claypool.

Guilbaud, G. T. (1952). Les théories de l'intérêt général et le problème logique de l'agrégation. *Économie appliquée*, *5*, 501–584. Reprinted in *Revue économique*, *63*(2012), 659–720.

Kornhauser, L. A. (2008). Aggregate rationality in adjudication and legislation. *Politics, Philosophy, Economics*, *7*, 5–27.

Kornhauser, L. A., & Sager, L. G. (1986). Unpacking the court. *Yale Law Journal*, *96*, 82–117.

Kornhauser, L. A., & Sager, L. G. (1993). The one and the many: Adjudication in collegial courts. *California Law Review*, *81*, 1–59.

Landa, D., & Lax, J. R. (2009). Legal doctrine on collegial courts. *Journal of Politics*, *71*, 946–963.

List, C., & Pettit, P. (2002). Aggregating sets of judgments: An impossibility result. *Economics and Philosophy*, *18*, 89–110.

Mongin, P. (1997). Spurious Unanimity and the Pareto Principle. Thema Working Paper, Université of Cergy-Pontoise (published in *Economics and Philosophy*, *32*(2016), 511–532).

Mongin, P. (2008). Factoring out the impossibility of logical aggregation. *Journal of Economic Theory*, *141*, 100–113.

Mongin, P. (2012). The doctrinal paradox, the discursive dilemma, and logical aggregation theory. *Theory and Decision*, *73*, 315–345.

Nash, J. R. (2003). A context-sensitive voting protocol paradigm for multimembers courts. *Stanford Law Review*, *56*, 75–159.

Nehring, K., & Puppe, C. (2002). *Strategy-proof social choice on single-peaked domains: Possibility, impossibility and the space between*. WP of the Department of Economics: University of California, Davies.

Nehring, K., & Puppe, C. (2008). Consistent judgement aggregation: The truth-functional case. *Social Choice and Welfare*, *31*, 41–57.

Nehring, K., & Puppe, C. (2010). Abstract Arrowian aggregation. *Journal of Economic Theory*, *145*, 467–494.

Pauly, M., & van Hees, M. (2006). Logical constraints on judgment aggregation. *Journal of Philosophical Logic*, *35*, 569–585.

Pettit, P. (2001). Deliberative democracy and the discursive dilemma. *Philosophical Issues*, *11*, 268–299.

Pigozzi, G. (2016). Belief merging and judgment aggregation, In E. N. Zalta (Ed.), *The Stanford encyclopedia of philosophy* (Winter 2016 Edition). https://plato.stanford.edu/archives/win2016/entries/belief-merging/.

Post, D., & Salop, S. C. (1992). Rowing against the tidewater. A theory of voting by Multijudge Panels. *Georgetown Law Journal*, *80*, 743–774.

Rogers, J. M. (1996). "Issue-Voting" by multi-member appellate courts—A response to some radical proposals. *Vanderbilt Law Review*, *49*, 997–1044.

Rubinstein, A., & Fishburn, P. J. (1986). Algebraic aggregation theory. *Journal of Economic Theory*, *38*, 63–77.

# Thoughts on Social Design

## Walter Trockel and Claus-Jochen Haake

**Abstract** One of the fundamental problems in applications of methods and results from mechanism design and implementation theory is the effective enforcement of theoretically established equilibria by which social choice rules are implemented. Hurwicz (The American Economic Review 98(3):577–585, 2008) and Myerson (Review of Economic Design, 13(1–2):59, 2009) introduce different concepts of formalizing enforcement of institutional rules via the introduction of legal and illegal games. In this note the relation of their concepts with that of a social system defined in Debreu (Proceedings of the National Academy of Sciences, 38(10):886–893, 1952) is analyzed and its potential of being instrumental for modelling institution design is discussed. The existence proof for such a system, also known as generalized game or abstract economy had been the basis for the existence proof of a competitive equilibrium of an economy.

## 1 Introduction

One of the fundamental problems in applications of methods and results from Mechanism Design and Implementation Theory is the effective enforcement of theoretically established equilibria by which social choice rules are implemented.

A short reflection on what is going wrong on our planet in terms of environmental, political and human disasters will result into the insight to the necessity of intelligent social design of enforceable cooperation based structures.

One of the pillars of such "social design" that had been termed *genuine implementation* in several articles by the late Leo Hurwicz is the possibility of enforcement.

In his "Fundamental theory of institutions: a lecture in honor of Leo Hurwicz", that was presented at the North American Meeting of the Econometric Society on

W. Trockel (✉)
Bielefeld University, Bielefeld, Germany
e-mail: walter.trockel@uni-bielefeld.de

C.-J. Haake
Paderborn University, Paderborn, Germany
e-mail: cjhaake@wiwi.upb.de

© Springer Nature Switzerland AG 2019
J.-F. Laslier et al. (eds.), *The Future of Economic Design*, Studies in Economic Design, https://doi.org/10.1007/978-3-030-18050-8_60

June 22, 2006 and published in *Review of Economic Design* (2009, vol. 13, pp. 1–2), Roger Myerson starts the abstract by writing:

> We follow Hurwicz in considering fundamental questions about social institutions.

In section 6 of this article Myerson discusses mechanisms formalized as game forms and the way Hurwicz (2008) proposed to approach the enforcement problem via distinguishing legal games from true games.

In our present note we discuss Hurwicz's concept of enforcement and Myerson's proposed modification and compare them with the concept of a social system that had been introduced by Debreu (1952) also known as *abstract economy* or *generalized game*.

## 2  Enforcement

Section 8 of Hurwicz (2008) begins with the passage:

> A need for enforcement implies the possibility of behavior that violates the rules of the game. The point is that if there were no possibility of violation, then you wouldn't need enforcement. We refer to such strategies as illegal, as distinct from legal, of course.

Later we find:

> To say that the legal game rules are being successfully enforced means that the outcomes of the true game ensure that illegal strategies are less attractive than legal strategies. A strong formulation of successful enforcement might require that, for every player, every illegal strategy is dominated by (that is, is less attractive than) some legal strategy. A 'weak' domination would require only that a player at least be no worse off by staying within the law. In fact, however, if everyone else is acting illegally, a normally law-abiding player may not find it advantageous to remain law-abiding. It seems, therefore, more reasonable to adopt a somewhat weaker concept of successful enforcement.

Referring to this passage, Myerson (2009, pp. 67, 68) writes[1]:

> Hurwicz (1998) explains that, if our legal game $G = (N, (C_i)_{i \in N}, (U_i)_{i \in N})$ is embedded in some true game $H$, the structural relationship must be that $H = (N, (D_i)_{i \in N}, (U_i)_{i \in N})$ has a larger strategy spaces
>
> $$D_i \supset C_i \quad \forall i \in N$$
>
> and has utility functions that extend those of the legal game $G$ to the larger domain $D = \times_{j \in N} D_j$. Hurwicz (1998) then suggests that a strong formulation of successful enforcement could require that, for each player $i$, each illegal strategy outside $C_i$ should be dominated by some legal strategy in $C_i$, so that a player's best responses always take him into the legal game, even if others deviate.
>
> Hurwicz (2008) remarks, however, that a normally law-abiding player might not want to remain law-abiding when others are acting illegally, and so a weaker concept of enforcement may be appropriate. Thus, I would suggest that the definition of institutional enforcement should be weakened, to say that $G$ is enforceable in $H$ when

---

[1] An earlier version of Hurwicz's Nobel prize lecture (Hurwicz 2008) appeared in 1998 as working paper at the University of Minnesota.

$\forall i \in N, \forall c_{-i} \in \times_{j \in N-i} C_j, \forall d_i \in D_i \setminus C_i, \exists c_i \in C_i$  such that  $U_i(c_{-i}, c_i) > U_i(c_{-i}, d_i),$

so that each players optimal actions are in his legal strategy set when all other actions are expected to be in their legal sets. That is, G is enforceable when its strategy sets form a curb set in H, as defined by Basu and Weibull (1991). (Curb sets are closed under rational behavior.)

Strictly speaking, the relation between Myerson's definition of enforceability and the curb sets of Basu and Weibull (1991) is more subtle. In contrast to their definition that is based on mixed strategy profiles interpreted as players' beliefs about which pure strategies in compact strategy sets of the other players will be played and that thereby makes use of expected utilities there are no "expectations" in the less specific framework or in the definitions used by Hurwicz (2008) and Myerson (2009). The sentence "so that each player's optimal actions are in his legal strategy set when all others' actions are expected to be in their legal sets" used by Myerson adequately describes the situation in the model of Basu and Weibull (1991), but it is meaningless in his context and should be modified as follows in order to fit the pure strategy framework.

The argument in favor of a weaker definition of enforceability used by Hurwicz (2008) and followed by Myerson by his suggestion of the modified definition above does not completely convince us.

A designer's attitude that accepts deviation from law-abiding behavior of an individual provided the others act illegally, would not be consistent with the central goal of a mechanism to regulate activities in a society by restricting the sets of acceptable actions via unattractive payoffs. Such a weak enforcement fails, for instance, in cases of coordination of the players on a more advantageous illegal equilibrium. Only the trust of everybody in punishment of each single individual that deviates from legal behavior creates confidence in the effectiveness of the mechanism.

A weakness that we feel to be common to either of the above definitions of enforcement is the fact that it is totally determined on the mechanism side rather than also depending on the individuals' preferences as reflecting the respective society's values, customs and norms.

We shall deal with this aspect in the next section.

# 3   Social systems

Debreu (1952) introduced the notion of a social system in the first paragraph of his seminal article "A Social Equilibrium Existence Theorem" and used this notion again on page 397 of his Nobel Memorial Lecture on "Economic Theory in the Mathematical Mode" (Debreu 1984). The fact that in this purely mathematical analysis, references to Economics and Game Theory are made only in four footnotes may have been causal for the now prevailing habit in the related literature of denoting a social system as a generalized game or an abstract economy (e.g., Shafer and Sonnenschein 1975; Tian 1990; Tian and Zhou 1992). In particular in the Operations Research lit-

erature the problem of existence of equilibria in social systems has become popular
under the name *generalized Nash equilibrium problem (GNEP)*. Facchinei and Kanzow (2010) write:

> As we already mentioned, many researchers from different fields worked on the GNEP, and
> this explains why this problem has a number of different names in the literature including
> pseudo-game, social equilibrium problem, equilibrium programming, coupled constraint
> equilibrium problem, and abstract economy. We will stick to the term generalized Nash
> equilibrium problem that seems the favorite one by OR researchers in recent years.

They do not cite Debreu (1952) in their references but at least refer to Arrow and
Debreu (1954), where the Equilibrium Existence Theorem of Debreu (1952) was
first applied in Economics.

In sociology, where generalized games recently have gained some popularity, their
origins in Debreu (1952) seem to have remained widely unnoticed (cf. Burns and
Roszkowska 2005).

In this note we want to stress the relevance that Debreu's modeling of a social
system may have in an agenda of social and institutional design as propagated in
the work of the late Leo Hurwicz. Accordingly, we will introduce Debreu's social
system adding at the same time a way of interpreting it under the aspects of genuine
implementation and enforcement as discussed in our first two sections.

In Debreu (1952) a social system $S$ is defined as

$$S = (N, (D_i)_{i \in N}, (U_i)_{i \in N}, (\beta_i)_{i \in N})$$

The correspondence $\beta_i$ from $D_{-i} := \times_{j \in N-i} D_j$ to $D_i$ associates with each action
profile $d_{-i} \in D_{-i}$ the set of all those actions of individual $i$ in $D_i$ that are unanimously
declared legal by the individuals $j \in N - i$ via the joint choice of the action profile
$d_{-i}$. The correspondence $\beta$ from $D$ to $D$ is defined by $\beta(d) := \times_{i \in N} \beta_i(d_{-i})$.

This definition does not exclude the possibility that the mechanism declares additional actions or action profiles illegal. An action $d_i$ is illegal per se, if it is not
contained in $\beta_i(d_{-i})$ for any $d_{-i} \in D_{-i}$. We will not include this possibility, however, in our present considerations.

Debreu distinguished between *actions* and *choices* of agents $i \in N$ from the sets
$D_i$ and $\beta_i(d_{-i})$, respectively. We interpret here his actions as corresponding to strategies in Hurwicz's true game and choices as strategies in legal action profiles: An
action profile is legal if it is a choice profile, i.e. if each agent's action in the profile
is confirmed by the actions of all others in the profile.

Notice that in contrast to Hurwicz we do not have a legal game whose extension
is the true game! Our set of legal action profiles need not have a product structure.
So legality of an action is always contingent on all other agent's actions.

To illustrate this point, consider the scenario, in which a committee has to take
an ethically difficult decision. Each member can take the actions to "join" or "stay
away" from the committee meeting. The decision can only be taken, if half of the
members choose "join". As the decision has to be taken, illegal action profiles are

only those, in which more than one half of the individuals choose to stay away. Phrased differently, a single individual choosing "stay away" need not mean that the action profile itself is illegal, although everyone choosing to stay away should certainly be called an illegal action profile.

Any product of subsets of the players' action sets, however, that builds a set of legal action profiles would define a legal subgame game of the true game.

The similarity to curb sets is obvious but neither is such a legal game necessarily a curb set nor vice versa. The difference lies in the fact that legality in our context has nothing to do with rationality. The latter will enter when we start considering equilibria of legal games, an intention behind Hurwicz's idea of *genuine implementation*.

A detailed study of these relations could contribute a solid edifice of enforcement and social mechanism design.

**Acknowledgements** This work was partially supported by the German Research Foundation (DFG) within the Collaborative Research Centre "On-The-Fly Computing" (SFB 901).

# References

Arrow, K. J., & Debreu, G. (1954). Existence of an equilibrium for a competitive economy. *Econometrica: Journal of the Econometric Society*, 265–290.

Basu, K., & Weibull, J. W. (1991). Strategy subsets closed under rational behavior. *Economics Letters*, *36*(2), 141–146.

Burns, T. R., & Roszkowska, E. (2005). Generalized game theory: Assumptions, principles, and elaborations grounded in social theory. *Studies in Logic, Grammar and Rhetoric*, *8*(21), 7–40.

Debreu, G. (1952). A social equilibrium existence theorem. *Proceedings of the National Academy of Sciences*, *38*(10), 886–893.

Debreu, G. (1984). Economic theory in the mathematical mode. *The Scandinavian Journal of Economics*, *86*(4), 393–410.

Facchinei, F., & Kanzow, C. (2010). Generalized nash equilibrium problems. *Annals of Operations Research*, *175*(1), 177–211.

Hurwicz, L. (1998). But who will guard the guardians. University of Minnesota Working Paper, Revised for *Nobel Lecture in American Economic Review*, *98*(3), 577–585, (2008).

Hurwicz, L. (2008). But who will guard the guardians? *The American Economic Review*, *98*(3), 577–585.

Myerson, R. B. (2009). Fundamental theory of institutions: A lecture in honor of leo hurwicz. *Review of Economic Design*, *13*(1–2), 59.

Shafer, W., & Sonnenschein, H. (1975). Equilibrium in abstract economies without ordered preferences. *Journal of Mathematical Economics*, *2*(3), 345–348.

Tian, G. (1990). Equilibrium in abstract economies with a non-compact infinite dimensional strategy space, an infinite number of agents and without ordered preferences. *Economics Letters*, *33*(3), 203–206.

Tian, G., & Zhou, J. (1992). The maximum theorem and the existence of nash equilibrium of (generalized) games without lower semicontinuities. *Journal of Mathematical Analysis and Applications*, *166*(2), 351–364.

# Part X
# Matching and Markets

# The Data from Market Design Solutions for School Choice

Atila Abdulkadiroglu

**Abstract** Centralized school admissions systems produce data with quasi-experimental variation in student assignment at unprecedent levels. This note emphasizes consequences of such data on evidence-based policy design, and how it can lead to data driven market design and econometric market design for public school choice. In summary, data generated by centralized admissions systems can be used for (i) demand analysis, (ii) market assessment and data driven redesign (iii) enrolment and school portfolio planning, (iv) program evaluation, and (v) econometric market design for school choice.

Parental choice over public schools has become a major policy tool for education reform around the world. Increasingly, urban school districts have been moving away from traditional residence-based public-school assignment of pupils. The roots of such chance in the US trace back to inequality along income and racial lines that residence-based assignment tends to create through housing markets and public-school finance regimes. Recent concerns involve, among others, inadequacy of one-size-fits-all models of schooling for student populations with varying needs, and lack of competitive market forces on public schools in residence-based assignment. Consequently, school districts have been adopting alternative curricula and pedagogical approaches, as well as alternative school management models, such as charter schools. As districts offer more options for parents and students, meeting the needs of highly heterogeneous student populations via residence-based assignment becomes infeasible. Choice, and therefore, choice-based student assignment become an integral part of enrollment planning. Developments in matching theory and market design have led many urban school districts to rethink about how children are assigned to public schools. Starting with Boston Public Schools and New York City High Schools (Abdulkadiroglu and Sonmez 2003; Abdulkadiroglu et al. 2005a, b, 2009), matching theory has been influential in the design of centralized admissions processes for fair and efficient assignment of students to schools with straightforward incentives for families.

A. Abdulkadiroglu (✉)
Duke University, Durham, USA
e-mail: aa88@duke.edu

© Springer Nature Switzerland AG 2019                                          439
J.-F. Laslier et al. (eds.), *The Future of Economic Design*, Studies in
Economic Design, https://doi.org/10.1007/978-3-030-18050-8_61

This note focuses on data such centralized admissions systems produce. The note emphasizes consequences of micro level data on evidence-based policy design, and how it can make data driven market design and econometric market design a reality. In summary, data generated by centralized admissions systems can be used for (i) demand analysis, (ii) market assessment and data driven redesign (iii) enrolment and school portfolio planning, (iv) program evaluation, and (v) econometric market design for school choice.

(i) Parental preferences over schooling are central to any education debate in democratic societies. When schools are assigned based on residential location of families, parents may choose schools to the extent they are able to choose where to live. Consequently, such residence-based assignment systems weave information on parental preferences, family income and residential choices into data. Even when families can apply to schools outside of their neighborhood, ad hoc assignment mechanisms are usually susceptible to misreporting of preferences by parents. For instance, prior to 2003, high schools in New York City could see where applicants rank them in them choice lists. Certain schools used to announce that they would consider only applicants that rank them as their first choice, forcing families to rank such schools as first choice or not rank them at all (Abdulkadiroglu et al. 2005a). Similarly, an immediate acceptance algorithm that was in use in Boston until 2005 assigns students to schools in order of where applicants rank schools in their choice lists, then in order of assignment priority that schools grant to applicants. As a result, families could improve their likelihood of assignment to a school by ranking it higher in the application form (Abdulkadiroglu and Sonmez 2003). Parents are known to have adopted such preference manipulation schemes especially in affluent neighborhoods of Boston (Abdulkadiroglu et al. 2005b).

Data from such systems does not necessarily reflect true preferences of parents. Focusing on estimation challenges associated with such data, a growing empirical literature studies parental demand for schools utilizing micro data generated by such centralized admissions (Hastings et al. 2009; He 2012; Ajayi 2013; Agarwal and Somaini 2014; Calsamiglia et al. 2014; Hwang 2014).

In contrast, strategy-proof mechanisms, such as deferred acceptance and top trading cycles, provide families with straightforward incentives, making truthful reporting of preferences a dominant strategy for parents. Versions of the deferred acceptance algorithm have been adopted by several urban districts for their admissions. Certain implementation choices, such as limits on the number of choices an applicant can list in her application, force families to strategically determine the set of schools to list in their application forms (Haeringer and Klijn 2009; Calsamiglia et al. 2010). However, theoretically, strategic behavior does not extend to ranking of these schools in application. Therefore, data from strategy-proof mechanisms allow for identification assumptions that are easier to justify and implement in the study of parental preferences (e.g. see Burgess et al. 2015; Abdulkadiroglu, Agarwal and Pathak 2017).

Simple descriptive statistics of student choices in data generated by strategy-proof mechanisms reveal certain regularities. For example, high school students in New York City prefer schools closer to their home, as well as quality schools measured by student achievement levels, percentage of students attending a four-year college, fraction of teachers classified by experience level. Despite such common patterns, preferences tend to vary across subpopulations. For instance, Hastings et al. (2009) find that higher-SES families are more likely than lower-SES parents to choose higher-performing schools based on stated reports under Charlotte's school choice plan. This contrast to Jacob and Lefgren's (2007) finding that parents in low-income and minority schools value a teacher's ability to raise student achievement more than parents in high-income and nonminority schools. Motivated by such differences, Abdulkadiroglu, Agarwal and Pathak (2017) investigate how preferences change with baseline ability and neighborhood income in New York City. They find that high-achieving students tend to rank schools with higher Math achievement relative to low achievers, students from low-income neighborhoods tend to put less weight on Math achievement than do students from high-income neighborhoods, despite that both groups prefer higher achieving schools higher. These differences suggest the importance of potential preference heterogeneity across subpopulations.

(ii) The theoretical literature naturally remains silent on certain policy questions. Examples include market assessment, such as access to quality schools by various subpopulations and how access changes under various admissions systems, the impact on parental choice of factors, such as curriculum, peer composition, distance to school. Discrete choice models can be used to address such critical questions. For instance, Boston Public Schools' recent plan has altered choice sets for applicants in nontrivial ways. Pathak and Shi (2014) studied counterfactual policy simulations to assess several alternative plans for the Boston Public Schools. Such practice relies heavily on out-of-sample performance of discrete choice models, a topic of debate in the empirical IO literature. In a follow-up analysis, Pathak and Shi (2017) assess the performance of their ex ante counterfactual predictions by comparing them to the actual choices made under the admissions system. They show that the performance of structural choice models in their data rely on accurate forecasts about auxiliary input variables, and once conditioned on the characteristics of the actual applicants, the structural choice models predict outcomes and choice patterns better than ad hoc alternatives.

Even when theoretical conclusion is clear on an issue, its extent usually remains an empirical issue. For example, Gale and Shapley's celebrated deferred acceptance algorithm (Gale and Shapley 1962) optimizes student welfare subject to a stability constraint, therefore it is not Pareto efficient from the point of student welfare. Abdulkadiroglu, Agarwal and Pathak (2017) quantify aspects of such design trade-offs identified in the theoretical literature on school choice. They find that the magnitude of student welfare gains from any potential algorithmic improvement, such as moving to an ex post Pareto efficient assignment, is relatively very small in comparison to the effect of simply assigning students via a deferred acceptance algorithm in a coordinated manner.

(iii) Student populations have been shrinking in some American cities due to economic downturn, and in some OECD countries due to low birth rates (Epple et al. 2017). As a result, districts in such areas are forced to reduce capacity by closing schools. Similarly, certain areas in the US are facing higher concentration of low-income families. Pressure from such changing enrollment patterns necessitates closure of exiting schools or opening of new schools. Impact of school closures or openings on other schools cannot be easily predicted by simple descriptive statistics. Such market assessment requires understanding of substitution patterns in parental preferences for schooling, which can be predicted effectively via discrete choice models. For example, high-SES families prefer to travel further away for higher average peer achievement (e.g. Abdulkadiroglu, Agarwal and Pathak 2017). Understanding how families from different socioeconomic backgrounds substitute school characteristics in their preferences is critical in designing schools with optimal set of characteristics in serving certain subpopulations more effectively, such as location, extended day programs, extra-curricular activities.

(iv) Data generated by centralized admissions embeds lottery-based and regression discontinuity style quasi-experimental variation in student assignment to schools. This opens door for credible program evaluation. Examples include studies of schools in Boston (Abdulkadiroglu et al. 2011), Charlotte-Mecklenburg (Hastings et al. 2009; Deming 2011; Deming et al. 2014), New York (Bloom and Unterman 2014; Abdulkadiroglu et al. 2013), and the Louisiana Scholarship Program (Abdulkadiroglu et al. 2018; Mills and Wolf 2017). However, these studies fail to exploit the full randomization embedded in centralized assignment.

Recent developments in research design explore full extraction of such variation in centralized assignment (Abdulkadiroglu, Angrist, Narita and Pathak 2017a, b). Eliciting such randomization is complicated by the fact that assignment to schools is a complicated function of lotteries as well as student preferences, school priorities and preferences, and nonrandom tie breakers such as entrance exam scores and the choice of assignment algorithm. In a simple world without non-random tiebreakers, conditional on preferences and priorities, assignments are independent of potential outcomes. Theoretically this yields a straightforward research design. Yet, this strategy eliminates critical amount of data by creating too many cells with too few observations with no variation in treatment status. Alternatively, Abdulkadiroglu, Angrist, Narita and Pathak (2017a) propose conditioning on assignment probabilities, namely propensity score, conditional on preferences and priorities. As shown by Rosenbaum and Rubin (1983), propensity score is the coarsest balancing score that eliminates selection bias arising from the association between conditioning variables and potential outcomes. Multi-dimensional nature of the matching problem due to student preferences and school priorities prohibits analytical solutions of the propensity score. Abdulkadiroglu, Angrist, Narita and Pathak (2017a, b) offer asymptotically valid large market approximations to the deferred acceptance propensity scores with lottery-based randomization as well as regression discontinuity style randomization in the presence of non-random tiebreakers. These developments open the door for credible evaluation of schools, programs and aspects of curricula. In turn, they can be used for internal program evaluation and system diagnosis, such as value-added models for teacher and school evaluation, and offered as part of school report cards.

(v) Current market design practice in school choice focuses on trade-offs among efficiency, stability, incentives and institutional constraints such as balanced student demographics in school enrollment lists. Theoretical and empirical resolution of such trade-offs have been critical in market design decisions. Assignment algorithms, such as deferred acceptance and top trading cycles, as well as how ties among equal priority students are broken also impact the degree of quasi-experimental randomization generated in the data. Future market design practice may also incorporate such econometric concerns into designs aiming wider or more focused quasi-experimental variation in data for critical program evaluation, potentially leading to an econometric market design for school choice.

Developments in market design have led to rigorously designed assignment solutions for public schools in the US and around the world. The data produced by these systems is offering unprecedented opportunities for evidence-based policy design, data driven market design and econometric market design.

# References

Abdulkadiroglu, A., Agarwal, N., & Pathak, P. A. (2017). The welfare effects of coordinated assignment: Evidence from the New York City high school match. *American Economic Review, 107*(12), 3635–3689.

Abdukadiroglu, A., Angrist, J. D., Dynarski, S. M., Kane, T. J., & Pathak, P. A. (2011). Accountability and flexibility in public schools: Evidence from Boston's charters and pilots. *Quarterly Journal of Economics, 126*(2), 699–748.

Abdulkadiroglu, A., Angrist, J. D., Narita, Y., & Pathak, P. A. (2017a). Research design meets market design: Using centralized assignment for impact evaluation. *Econometrica, 85*(5), 1373–1432.

Abdulkadiroglu, A., Angrist, J. D., Narita, Y., & Pathak, P. A. (2017b). *Impact Evaluation in Matching Markets with General Tie-Breaking.* NBER Working Paper, 24172.

Abdulkadiroglu, A., Hu, W., & Pathak, P. (2013). *Small high schools and student achievement: Lottery-based evidence from New York City.* NBER Working Paper 19576.

Abdulkadiroglu, A., Pathak, P. A., & Roth, A. E. (2009). Strategyproofness versus Efficiency in Matching with Indifferences: Redesigning the NYC High School Match. *American Economic Review, 99*(5), 1954–1978.

Abdulkadiroglu, A., Pathak, P. A., & Roth, A. E. (2005a, May). The New York City High School Match. *American Economic Review, Papers and Proceedings, 95*, 364–367.

Abdulkadiroglu, A., Pathak, P. A., Roth, A. E., & Sonmez, T. (2005b). The Boston Public School Match. *American Economic Review, Papers and Proceedings, 95*, 368–371.

Abdulkadiroğlu, A., Pathak, P. A., & Walters, C. R. (2018). Free to choose: Can school choice reduce student achievement? *American Economic Journal: Applied Economics, 10*(1), 175–206.

Abdulkadiroglu, A., & Sonmez, T. (2003). School choice: A mechanism design approach. *American Economic Review, 93*, 729–747.

Agarwal, N., & Somaini, P. (2014). *Demand analysis using strategic reports: An application to a school choice mechanism.* National Bureau of Economic Research Working Paper 20775.

Ajayi, K. F. (2013). *School choice and educational mobility: Lessons from secondary school applications in Ghana.* Working Paper.

Bloom, H., & Unterman, R. (2014). Can small high schools of choice improve educational prospects for disadvantaged students? *Journal of Policy Analysis and Management, 33*(2), 290–319.

Burgess, S., Greaves, E., Vignoles, A., & Wilson, D. (2015). What parents want: School preferences and school choice. *Economic Journal, 125*(587), 1262–1289.

Calsamiglia, C., Fu, C., & Güell, M. (2014). *Structural estimation of a model of school choices: The Boston mechanism vs. its alternatives.* Unpublished.

Calsamiglia, C., Haeringer, G., & Klijn, F. (2010). Constrained school choice: An experimental study. *American Economic Review, 100*(4), 1860–1874.

Deming, D. (2011). Better schools, less crime? *Quarterly Journal of Economics, 126*(4), 2063–2115 [1374, 1410, 1413].

Deming, D., Hastings, J., Kane, T., & Staiger, D. (2014). School choice, school quality and post-secondary attainment. *American Economic Review, 104*(3), 991–1013.

Gale, D., & Shapley, L. S. (1962). College admissions and the stability of marriage. *American Mathematical Monthly, 69,* 9–15.

Haeringer, G., & Klijn, F. (2009). Constrained school choice. *Journal of Economic Theory, 144*(5), 1921–1947.

Hastings, J. S., Kane, T. J., & Staiger, D. O. (2009). *Heterogeneous preferences and the efficacy of public school choice*. Unpublished.

He, Y. (2012). *Gaming the Boston mechanism in Beijing*. TSE Working Paper 12-345.

Hwang, S. (2014). *A robust redesign of high school match*. Unpublished.

Jacob, B. A., & Lefgren, L. (2007). What do parents value in education? An empirical investigation of parents' revealed preferences for teachers. *Quarterly Journal of Economics, 122*(4), 1603–1637.

Mills, J. N., & Wolf, P. J. (2017). Vouchers in the Bayou: The effects of the Louisiana Scholarship Program on student achievement after 2 years. *Educational Evaluation and Policy Analysis, 39,* 464–484.

Pathak, P. A., & Shi, P. (2014). *Demand modeling, forecasting, and counterfactuals, Part I, January 2015*. Original January 2014, NBER Working Paper 19859.

Pathak, P. A., & Shi, P. (2017). *How Well Do Structural Demand Models Work? Counterfactual Predictions in School Choice*. NBER Working Paper, 24017.

Rosenbaum, P. R., & Rubin, D. B. (1983). The central role of the propensity score in observational studies for causal effects. *Biometrika, 70*(1), 41–55.

# Refugee Matching as a Market Design Application

Tommy Andersson

**Abstract** This note contains a few brief remarks on the similarities and differences between some standard market design applications (e.g., kidney exchange and school choice) and the refugee assignment problem. The main conclusion is that the refugee assignment problem is more complex in some dimensions than many of the standard market design applications. Consequently, classical mechanisms cannot be used to solve the problem and more research is needed to understand, e.g., how to model preferences, how to define relevant axioms, and how to specify multidimensional constraints.

## 1 Introduction and Statement of the Problem

The point of departure in this chapter is the ongoing European migrant crisis and other similar crises around the world, and, more precisely, the problem of assigning asylum seekers to different geographical areas. According to the UNHCR, the number of forcibly displaced people worldwide reached 65.3 million at the end of 2015 (the highest level ever recorded). Around 1.2 million of these people applied for asylum in European Union member states (more than a doubling compared to 2014). In an attempt to more fairly distribute asylum seekers between the European Union member states, the European Commission has formulated several European relocation schemes. These schemes specify a quota for each member state, based on criteria that reflect the capacity of the member states to absorb and integrate asylum seekers (see Grech 2017 for a systematic and critical analysis), but does not contain any details about which asylum seekers that should be relocated to what member state.

The above type of resettlement schemes share some fundamental properties with a standard two-sided matching market with capacity constraints (Jones and Teytelboym 2017b). More precisely, such matching market can be modelled with the European Union member states on one side of the market and asylum seekers on the other side of the market where the capacities of the member states are determined by

T. Andersson (✉)
Department of Economics, Lund University, Box 7082, 220 07 Lund, Sweden
e-mail: tommy.andersson@nek.lu.se

© Springer Nature Switzerland AG 2019                                                445
J.-F. Laslier et al. (eds.), *The Future of Economic Design*, Studies in
Economic Design, https://doi.org/10.1007/978-3-030-18050-8_62

the quotas in the resettlement scheme. Note, however, that the refugee assignment problem is not limited to a matching problem on supranational level as the basic structure of the problem is more or less identical independently of if it is specified for the supranational or the national level. For this reason, the remaining part of this chapter will, in general terms, discuss the refugee assignment problem within a given geographical region keeping in mind that local regulations may constrain the problem (in similarity with, e.g., the school choice problem where local regulations may impact the design of the matching mechanism).

To model refugee matching as a market design application, it is, however, not enough to know only the sets of agents (i.e., the geographical areas and the asylum seekers) and the capacities of the geographical areas. Additional information is needed, e.g., information related to how asylum seekers arrive to the geographical areas (static versus dynamic arrivals), how to describe asylum seekers in terms of their characteristics, and how to model local constrains and agent preferences. In addition, to identify and evaluate matchings, matching mechanisms and axioms that are tailored for the refugee assignment problem are needed. These issues are briefly discussed in the remaining part of the chapter under separate headings.

## 2  Static Versus Dynamic Refugee Assignment

A first distinction that separates any given refugee assignment problem into one of two possible directions is related to exactly *how* the asylum seekers enter the geographical area.

First, and as described in the above, asylum seekers may be resettled to a geographical area based on a redistribution key. In this case, the local authorities do not only know that they will be assigned a given number of asylum seekers during a pre-defined time period (e.g., the next year), they also know the characteristics of the asylum seekers resettled in their geographical area. Because this information is known and since the problem then lacks dynamics, this type of refugee assignment problem can be modelled as a version of a two-sided matching market with capacity constraints in similarity with the school choice problem (Abdulkadiroğlu and Sönmez 2003).[1] This type of refugee resettlement problem has recently been analyzed by Bansak et al. (2018), Delacretaz et al. (2016), Jones and Teytelboym (2017a, b), and van Basshuysen (2017).

Second, asylum seekers may arrive directly to a geographical area without being part of a resettlement program. For the European case, the Dublin Regulation then dictates that the member state in which an asylum seeker enters first is obliged to render asylum. This introduces dynamics to the problem since there is no way to predict exactly how many asylum seekers that will arrive directly to a specific geographical area, when they will arrive, and what characteristics they have. Hence, the problem to assign asylum seekers to a locality within the geographical area must

---

[1]Complications related preferences and constraints will be discussed in more detail below.

be solved upon each arrival, i.e., before knowing the identity and characteristics of all asylum seekers. Such dynamic refugee assignment system has recently been analyzed by Andersson et al. (2018).

Hence, refugee assignment problems can be though of as either static or a dynamic problem. The dynamic matching literature (not necessarily related to refugee matching) is, in general, far less developed than its static counterpart even if it can be argued that some classical market design applications have dynamic elements. For example, the kidney exchange problem is dynamic in its nature since patient-donor pairs enter the kidney exchange pool in a sequence and because there is no exact agreement on what the optimal size of the patient-donor pool should be before running the matching algorithm (Ünver 2009).

# 3  Multidimensional Characteristics and Constraints

In standard matching models (e.g., school choice and kidney exchange), agents and capacities are typically one-dimensional. In, for example, the standard school choice problem, students are usually completely described by their preferences over schools and schools are typically completely described by their priorities over students and their capacities (priorities may, however, be described by different priority classes to, e.g., reflect walking distance, sibling priority, or controlled school choice). As observed by Delacretaz et al. (2016) and Jones and Teytelboym (2017a), the basic structure in the refugee assignment problem is more complicated. For example, asylum seekers may be described by characteristics in several dimensions including, e.g., nationality, spoken languages, family constellation, need for health care, etc. It is not clear what the relevant characteristics are and how to identify them. For this reason, researchers interested in the refugee assignment problem must approach relevant authorities to seek input in this matter. Note also that it is not unlikely that relevant characteristics differ between geographical areas, e.g., depending on how the health care and education systems are organized.

The multidimensional description of asylum seekers also imposes additional constrains on local authorities. More specifically, a family of asylum seekers need not only accommodation in the geographical region where they are placed, but also a certain number of units of different public services related to, e.g., education and health care. In other words, asylum seekers are not only described by multiple characteristics, these characteristics also introduce explicit multidimensional constraints in local geographical areas. Hence, depending on how asylum seekers are described (in terms of characteristics), the computational as well as the modelling complexity of the matching problem varies. In particular, the description may introduce complementarities which is something that seldom is considered in the matching literature (a famous exception is the matching with couples problem, present in the National Resident Matching Program) and something which generally is difficult to deal with.

## 4 Agent Preferences

In many market design applications, agents can provide a ranking over the relevant objects. In, for example, the school choice problem, parents have access to information about the schools in their locality and can, based on this information, form preferences over schools. In the refugee assignment problem, however, it may be difficult for local authorities to form preferences over asylum seekers and vice versa even if there is an agreement about how to describe asylum seekers and geographical areas in terms of their characteristics. One reason for this is that there are thousands of asylum seekers (dozens of local geographical areas) and complete information about the asylum seekers (the geographical areas) may be unavailable and even if such information is available, it is not clear how to process it. It is therefore not unlikely that preferences in this type of application needs to be estimated or induced in some way.

Several different approaches may be taken. First, given the assumption that one side of the market can form preferences over agents on the other side of the market, preferences can be induced based on the concept of mutual acceptability. Such preferences has recently been theoretically studied by Haeringer and Iehlé (2017) and in a refugee assignment context by Andersson and Ehlers (2016). Second, preferences can be estimated based on econometric techniques. Here, one can, e.g., imagine that successful integration is the ultimate goal of both the local authorities and the asylum seekers. Then some type on integration score (like the ones published for Germany in The Economist 2016) for each type of asylum seeker in each geographical area may be estimated and preferences over geographical areas and asylum seekers may then be formed based on likelihood of successful integration (Andersson et al. 2018; Bansak et al. 2018). A third alternative is to induce preferences in a non-parametric way based only on characteristics of asylum seekers and geographical areas.

There are, of course, other alternatives for estimating or inducing preferences than the above mentioned, but the important conclusion is that it is unlikely that it is easy for the local authorities and asylum seekers to form preferences on their own. It is, therefore, not unreasonable that preferences must be estimated or induced in some way and this is typically not the case in the matching literature. Here, it should be mentioned that it can be argued that preferences, in fact, are induced in some existing market design applications. In the standard kidney exchange problem, for example, it can be argued that preferences are induced in a dichotomous fashion based on reported medical data such as tissue type antibodies and blood group.

## 5 Matching Mechanisms and Axioms

If local authorities are bounded by multidimensional constraints, the matching problem becomes a complex combinatorial problem and it may even contain complementarities. This may disqualify the use of some of the classical matching mechanisms, like the Deferred Acceptance Algorithm and the Top Trading Cycles Mechanism (see Delacretaz et al. 2016 for a detailed discussion). This may also call for the use of, e.g., combinatorial optimization techniques (Delacretaz et al. 2016) or graph optimization techniques (Andersson and Ehlers 2016).

Because the refugee assignment problem differs from classical market design applications, standard axioms for analyzing matching markets such as stability and strategy-proofness may have to be modified. For example, because complementarities may be present, the existence of a stable matching (as classically defined) is not guaranteed. Hence, some standard axioms must be defined and investigated from the perspective of the refugee assignment problem (for such analysis of the stability axiom in a refugee matching context, see Aziz et al. 2017) and new axioms need to be tailored for the refugee assignment context. Because the refugee assignment problem may also have a dynamic component, dynamic versions of existing axioms may need to be defined. To achieve this, researchers in the field may not only consult the existing market design literature but also literature related to, e.g., social choice, fairness and equity.

## 6   Summary and Conclusions

Even if this short chapter not has given any explicit advice on how to tackle the refugee assignment problem as a market design application, it has related the problem to some classical market design applications (such as school choice and kidney exchange) and pointed out a few differences and additional complications. Even if the discussed problem is on the highest political agenda in the European Union, there are not that many papers written on the subject in the market design literature. In fact, apart from the references mentioned in this paper, I am not aware of any other matching paper that investigates the refugee assignment problem.

It is clear that more research is needed to identify mechanisms that can be implemented to solve some of the problems that local authorities face in a refugee assignment context. The ultimate objective must be to provide tools that can be utilized to solve this important allocation problem exactly as the market design community has provided implementable mechanisms for school choice, kidney exchange and many other applications. Another direction to approach a solution to the problem is to regard the refugee assignment problem as a procurement problem (this was proposed to me by Lawrence Ausubel) or by approaching it as a system of tradable quotas like, e.g., emissions control. The latter approach is taken by Moraga and Rapoport (2014) for the refugee matching context and they demonstrated that tradable quotas can be designed based on matching techniques to solve some specific refugee resettlement problems.

# References

Abdulkadiroğlu, A., & Sönmez, T. (2003). School choice - A mechanism design approach. *American Economic Review, 93*, 729–747.

Andersson, T., & Ehlers, L. (2016). *Assigning refugees to landlords in Sweden: Efficient stable maximum matchings.* Lund University Department of Economics Working Paper 2016:18.

Andersson, T., Ehlers, L., & Martinello, A. (2018). *Dynamic refugee matching.* Lund University Department of Economics Working Paper 2018:7.

Aziz, H., Chen, J., Gaspers, S., & Sun Z. (2017). *Stability and Pareto optimality in refugee allocation matchings.* Working Paper.

Bansak, K., Ferwerda, J., Hainmueller, J., Dillon, A., Hangartner, D., & Lawrence, D. (2018). Improving refugee integration through data-driven algorithmic assignment. *Science, 359*, 325–329.

Delacretaz, D., Kominers, S., & Teytelboym, A. (2016). *Refugee resettlement.* Working Paper.

Economist (2016). Refugees in Germany may be seeking asylum in the wrong places. https://www.economist.com/blogs/graphicdetail/2016/04/daily-chart-8. Accessed April 25, 2016.

Grech, P. (2017). Undesired properties of the European Commission's refugee distribution key. *European Union Politics, 28*, 212–238.

Haeringer, G., & Iehlé, V. (2017). *Two-sided matching with (almost) one-sided preferences.* Working Paper.

Jones, W., & Teytelboym, A. (2017a). The local refugee match: Aligning refugees' preferences with the capacities and priorities of localities. *Journal of Refugee Studies.*

Jones, W., & Teytelboym, A. (2017b). The refugee match: A system that respects refugees' preferences and the priorities of states. *Refugee Survey Quarterly, 36*, 84–109.

Moraga, J., & Rapoport, H. (2014). Tradable immigration quotas. *Journal of Public Economics, 115*, 94–108.

Ünver, U. M. (2009). Dynamic kidney exchange. *The Review of Economic Studies, 77*, 372–414.

van Basshuysen, P. (2017). Towards a fair distribution mechanism for asylum. *Games 8*, article 41.

# Engineering Design of Matching Markets

## Péter Biró

**Abstract** Two-sided matching markets under preferences can be described by the stable matching model of Gale and Shapley, proposed for college admissions. Their deferred-acceptance algorithm always produces a student-optimal stable matching in linear time, and it is strategy-proof for the students. Mainly due to these desirable properties, this algorithm has been widely used in many applications across the world, including school choice, college admission, and resident allocation. Yet, having one extra custom feature can turn the corresponding problem intractable in a computational sense, the existence of a fair solution may not be guaranteed any more, and the strategic issues cannot be avoided either. One well-known example for such a market design challenge was the introduction of joint applications by couples in the US resident allocation program. In the 1990s Roth and Peranson managed to successfully resolve the case by taking an engineering approach, and constructing a Gale-Shapley based heuristic algorithm. In this writing we elaborate on this engineering concept by describing further examples for real-life design challenges in matching markets, the related computational and strategic issues, and the possible solution techniques from optimization and game theoretical points of view.

## 1 Introduction

Centrally coordinated matching markets have been established around the world in the last 65 years, first to allocate residents to hospitals in the US (Roth 1984), then for college admissions (Biró et al. 2010), school choice (Abdulkadiroglu and Sönmez 2003), course allocation (Budish and Kessler 2015), organ exchange (Roth et al. 2005), just to mention a few examples. For matching markets, where the agents on both sides have preferences, the stable matching model by Gale and Shapley (Gale and Shapley 1962) was the first scientific approach, published 56 years ago. In the context of college admissions, they showed that a so-called stable matching always exists, and a student-optimal solution can be found by the natural *deferred-*

P. Biró (✉)
Institute of Economics, Hungarian Academy of Sciences, Budapest, Hungary
e-mail: peter.biro@krtk.mta.hu

© Springer Nature Switzerland AG 2019                                              451
J.-F. Laslier et al. (eds.), *The Future of Economic Design*, Studies in
Economic Design, https://doi.org/10.1007/978-3-030-18050-8_63

*acceptance* (DA) algorithm. The concept of stability—which essentially means that the rejection of an application may only happen when the college filled its quota with better applicants—turned out to be the most important fairness criterion in two-sided matching markets (Roth 1991). Later, it was proved that the Gale-Shapley algorithm can be implemented in linear time (see e.g. Gusfield and Irving 1989), so it runs very fast in centrally coordinated programs, and it is also strategy-proof (Roth 1982), meaning that no student can get a better assignment by submitting false preferences. These desirable properties made this method a great success in many applications (Biró 2017; Roth 2008), and the paper of Gale and Shapley served as the foundation-stone of the interdisciplinary research on matchings under preferences carried out by economists/game theorists (Roth 2015; Roth and Sotomayor 1990) and computer scientists/mathematicians (Gusfield and Irving 1989; Manlove 2013). This work has also been recognized with the 2012 Nobel Memorial Award in Economic Sciences given to Roth and Shapley "for the theory of stable allocations and the practice of market design".

However, adding one custom feature to a centralized application process may force us to confront significant theoretical and practical challenges. As an example we describe briefly the issue of couples in resident allocation programs. In the US resident allocation program, called NRMP, the number of married couples became significant in the 1970s and most of them decided not to participate in the central scheme, since they were afraid of being assigned to places far from each other. So, they were looking for pairs of positions through private agreements with hospitals, which started to undermine the efficiency of the program.

The research community has recognized the importance of this issue, and started to investigate it. Game theorists showed that the existence of a stable matching is not guaranteed if couples may apply for pairs of positions (Roth 1984), and computer scientists showed that the problem of finding a stable solution is NP-hard (Ronn 1990), so it is unlikely that any fast algorithm would always find a stable matching, even if one exists. Yet, it was inevitable to let the couples join the central scheme, so after some unsuccessful attempts to handle the couples feature, the coordinating agency asked Al Roth to help in the redesign of the matching mechanism. Roth and Peranson took an engineering approach (Roth and Peranson 1999), they constructed a sophisticated heuristic based on the DA algorithm and conducted simulations, where they found that their algorithm was able to find stable solutions for the real data. This method was subsequently used in many similar applications in the US and Canada (Roth 2008).

The above experiment lead to a new research line in this field, rightly summarized by Roth (2002) with the title "The economist as engineer". The essence of this approach is that the market designer should not only study the theoretical properties and the tractability (more precisely the worse-case intractability) of a problem, but we should also consider heuristic mechanisms and test their performance in real markets. The most natural way to compare the performance of the heuristics is to run them on real and randomly generated instances, as typically done in applied computer science and engineering studies. However, when strategic agents are involved in the mechanism, the designer should also care about the potential manipulations, that might ruin the fairness or optimality of the solution.

Finally, if a mechanism seems to work well in practice then it is still an interesting task to understand the theoretical reasons behind its good behavior. For instance, in the case of couples further simulations have proved that stable matching almost always exists and can be found with sophisticated heuristics (Biró et al. 2011), while it was also shown that in large markets even a simple DA based algorithm is very likely to reach a stable solution (Kojima et al. 2013). In what follows we will elaborate on the engineering approach with regard to optimization, strategic behavior and social acceptability.

## 2   Optimization

If a problem is NP-hard then no polynomial time algorithm exists to provide the answer, unless $P = NP$, which is believed to be highly unlikely. Yet, there are many theoretical and practical approaches to tackle these seemingly hard problems.

If the problem is an optimization problem then one can always ask how close can we get with a polynomial time algorithm? For instance, coloring the edges of a graph with the minimum number of colors in such a way that no two adjacent edges share the same color is an NP-hard problem. Yet, by the theorem of Vizing we know that the answer is either the maximum degree in the graph or one more, and based on this proof in was subsequently proved that we can also construct an edge-coloring in polynomial time where the number of colors used is the latter number. So an almost optimal solution can be found efficiently here.

One important example in the literature on matching under preferences with approximation results is the problem of finding a maximum size weakly stable matching when ties are present. This problem has been present in the Scottish resident allocation scheme, where hospitals could express indifferences among doctors, and a pair was considered blocking when both parties would strictly improve by deviating. It is very easy to see that weakly stable matchings can have different sizes; just take two doctors applying to hospital A as their first choices with the same score, and only one of these doctors applying to hospital B as her second choice. The problem was proved to be NP-hard in Manlove et al. (2002), and the best approximation ratio was gradually improved until (McDermid 2009) reached the best ratio at the time of writing: 3/2. In the meantime, is was also shown that the problem is not approximable within a factor better than 33/29, unless $P = NP$ (Yanagisawa 2007). Regarding the engineering side, heuristics were developed and used for the Scottish application (Irving and Manlove 2009), that later were extended with the use of integer programming (IP) technique (Kwanashie and Manlove 2013). The interested reader is recommended to consult Manlove's book (Manlove 2013) about this problem.

Parameterized complexity is another important theoretical approach when studying NP-hard problems. The question is whether we can find some parameter, e.g., the number of couples in a resident allocation program, such that there exists a so-called

FPT algorithm, whose running time is exponential only in this parameter. This turned out to be impossible for the couples problem (unless P = NP), as Marx and Schlotter proved that this problem is W[1]-hard (Marx and Schlotter 2011). However, when the hospitals use a common master list to rank the doctors, as in Scotland, the problem becomes FPT solvable (Biró et al. 2011).

When a problem is NP-hard, but the practical instances are relatively small and sparse then we may consider the usage of exponential time exact algorithms. This is what we did when we designed a matching algorithm for the UK kidney exchange program (Biró et al. 2009), which was later replaced with a more robust IP method (Manlove and O'Malley 2012).

We have already seen some examples of heuristics used to find maximum size weakly stable matchings or solving the resident allocation problem with couples. A heuristic algorithm is also used in the Hungarian college admission scheme. Here we have at least four special features, namely, paired applications for teachers' programs (which is similar in nature to the couples' problem), the presence of lower quotas for the programs, and common quotas for sets of programs, and finally the usage of equal treatment policy for the students with the same score. Each of the latter first three features makes the problem NP-hard (Biró et al. 2010), with only the last issue being tractable (Biró and Kiselgof 2015), but even that can cause strategic issues. Currently a complex algorithm is used that provides cutoff scores and implies a matching (i.e., everyone is admitted to the best program on her list, among those for which she achieved the cutoffs). Recently we have investigated the usage of IP technique for these special features (Ágoston et al. 2016).

Getting back to the case of couples, heuristic solutions have been developed, motivated by the American (Roth 2002) and Scottish applications (Biró et al. 2011), and finally IP techniques were also tested with encouraging results (Biró et al. 2014).

To summarize, the NP-hardness of a problem does not necessarily mean that it is impossible to find an optimal or a reasonably good solution in practice. So, besides the theoretical research on the approximation and parameterized complexity of these problems I am expecting a more intensive research by the optimization community with an engineering approach, conducting simulations with heuristics and developing IP solutions for two-sided matching problems under preferences.

## 3   Strategic Issues

Strategy-proofness is a core desideratum in mechanism design also when considering matching markets. However, here again the practical aspect and the average case behavior of the agents seem to be much more important than the theoretical properties of the mechanism in the success of the applications.

First, we point out that the strategy-proofness of a mechanism does not imply that the agents will not try to manipulate it in practice. One reason is that non-manipulability is sometimes *not obvious* for the agents. There are many papers documenting that manipulations are attempted for the DA both in the laboratory and in the field, see e.g., Chen and Sönmez (2006) and Rees-Jones (2018), respectively. It is also well-known in the auction literature that bidders play differently in the English and the second-price sealed-bid auctions, even though these mechanisms should produce the same results in theory in a strategy-proof way. A novel attempt for understanding this phenomenon is by Li (2017), who defined *obvious strategy-proofness*, and recently it was shown that no mechanism that implements a stable matching is obviously strategyproof for any side of the market (Ashlagi and Gonczarowski 2016).

Secondly, a theoretically manipulable mechanism may still provide truth-telling in practice. For instance, the college-proposing deferred-acceptance mechanism is manipulable by some students whenever the core fails (as it usually does) to be a singleton in a market. However, when the market is large, the proportion of students having an opportunity to manipulate successfully tends to zero, as seen in the field e.g., in Roth and Peranson (1999), and also in the theory for various large market models, e.g., in Kojima and Pathak (2009). Thus, some mechanisms that are manipulable in small markets may behave as a strategy-proof mechanism in large markets. Azevedo and Budish have proposed a new theoretical concept to analyze this property of matching mechanisms, that they called *strategy-proofness in the large* (Azevedo and Budish 2013).

Related to the above concept, truth-telling may also be explained by the Bayesian approach. Even if many students have some chance to successfully manipulate a mechanism, they will not do it if the risk of getting a worse allocation is high. Until 2007 the Hungarian college admission scheme used the college-oriented DA, and many other countries still do (http://www.matching-in-practice.eu), but I would never advise a student to remove her last application from her list to create some chances to get a better match, since the risk of getting no seat at all is higher and the loss would be much more severe than the benefit of an unlikely improvement. The problem with the Bayesian approach is that it is very challenging to analyze, but personally I would find this concept the most reasonable to use in the study of practical applications.

For some special features, there may not exist any strategy-proof mechanism. E.g., in the resident allocation problem with couples no mechanism that returns a stable matching (if one exists) can be strategy-proof (Biró and Klijn 2013). But in practice, a stable solution almost always exists (Roth and Peranson 1999), which was proved theoretically for large markets under some assumptions (Kojima and Pathak 2009), and the agents are unlikely to be able to successfully manipulate the heuristic used. The same is true for the Hungarian college admission scheme, where the mechanism is vulnerable to manipulation by many reasons (see e.g. Biró and Kiselgof 2015), but since the solution is essentially stable by the usage of cutoff scores (as we argue next), no student has a reasonable chance to successfully manipulate it.

The cutoff scores can play a crucial role not only to transparently prove stability (fairness) of the solution, but also to make the mechanism strategy-proof in practice. This is because the students believe that they cannot affect the cutoffs by submitting false preferences (which is a quite reasonable belief in large markets, even if there are some special features), so truth-telling becomes an obvious best reply, when cutoffs are considered fixed. Furthermore, when the cutoffs of the programs are relatively stable over the years, as they are in Hungary, they convey information about the prestige of the programs and the applicants can also better estimate their chances of getting admission. Therefore, I would advocate for the usage of cutoff scores in priority based systems, no matter how complex these admission systems are.

Finally, manipulability is just one property of the mechanisms, in some cases the outcomes by a manipulable mechanism may be better in average than by an alternative strategy-proof mechanism. We can mention the example of the VCG mechanism, which is strategy-proof in a very general model that covers most of the practical markets with auctions, yet, in many operating applications the auctions are based on alternative, manipulable mechanisms. A much more controversial case is the *immediate acceptance* (IA) mechanism (or 'Boston') for school choice when compared with the DA. The difference between the two mechanisms is that the acceptances in the IA are final, so when a school becomes full during the process then they will not accept further applications. The manipulability of IA is obvious, IA is also manipulable in large markets, and most parents do indeed strategize. It is believed that the affluent parents with more information can better navigate the system, so this was the main reason why the mechanism was replaced in Boston (Abdulkadiroglu et al. 2005b), and was banned in England (Pathak and Sönmez 2013). However, there are simple examples and reasonable theories showing that the IA may significantly improve the overall expected welfare of the participants (Abdulkadiroglu et al. 2011), which can also be seen in laboratory experiments (Abdulkadiroglu et al. 2017).

Thus, restricting ourselves to use theoretically strategy-proof mechanisms, whenever possible, seems to be unreasonable. On the one hand, agents may attempt to manipulate mechanisms that are provably strategyproof, and on the other hand some theoretically manipulable mechanisms are not manipulated in practice. Moreover, even if some mechanisms are manipulated, their outcomes may be better than what a strategy-proof alternative would produce. So, I would expect to see many more scientific works on documenting the performance and manipulability of different mechanisms in practice, in the lab and in the field, together with the corresponding theoretical explanations, e.g., by the Bayesian approach.

## 4   Fairness, Efficiency and Social Welfare

Finally, let me get back to the original question: what kind of solutions are considered to be socially acceptable/desirable in matching markets, and how can we obtain these with transparent mechanisms? Take the classical example of school choice. Here, the school seats can be considered as objects that we want to allocate to the students by taking into account their preferences and other objective factors. The objective factors, such as the distances to the schools, whether the children have already siblings somewhere, and maybe the ethnic and socio-economic background of the children, can be used in different ways.

One reasonable solution is to create priorities at the schools based on the above factors and compute a fair solution, where no student has justified envy over another student. A student-optimal envy-free allocation can be simply obtained by the DA, as organized in many US cities, starting in New York (Abdulkadiroglu et al. 2005a) and Boston (Abdulkadiroglu et al. 2005b).

A potential criticism against the above classical approach is that the solution may not be Pareto-efficient. Just take the following simple example: we have a student having high priority in the nearby school A, and another student having high priority at her close-to-home school B, but they both would be happier to swap because of personal reasons. However, if there is a third student applying to school A with a priority in between the two students then even though she can never get admission to A, she will block the exchange in an envy-free solution. Forbidding Pareto-improving exchanges is even more controversial when the creation of a blocking pair is due to a lottery, as in New York.

An alternative solution to DA, that guarantees efficiency in a strategy-proof way, is TTC (Abdulkadiroglu and Sönmez 2003). This was used in New Orleans (although only for one year) (Abdulkadiroglu et al. 2017), but this mechanism is not compatible with envy-freeness. However, there are many new ideas in the recent school choice literature recommending solutions in between these two concepts, see e.g. Morrill (2015).

Finally, we may also reconsider the whole question by forgetting about priorities and efficiency, but focusing on the overall social welfare of the solution. Here we may also consider the interest of larger groups of agents (including also agents that are seemingly outside of the market), such as the taxpayers of a country or city represented by the government of the local authority, respectively. A very simple measure of the solution can be the total travel distance, but there are also more complex measures, such as diversity considerations. One interesting approach is to use the priority-based method with carefully constructed choice-sets, or 'menus', offered to the parents. This was proposed by Ashlagi and Shi (2015) and was later implemented in Boston, where the City wanted to reduce the busing costs. We did a similar investigation in an Estonian kindergarten allocation program, where we simulated different priority-based policies under the DA and the TTC and compared their performance with respect to total travel distance and fair access (Veski et al. 2017). Another solution concept that we investigated in a very recent paper (Biró, and Gudmondsson 2018) is the constrained welfare-maximizing solution. Here we represented the social welfare as the sum of edge-weights over the school-applicant pairs in the allocation, and we restricted the solution to be Pareto-efficient in order to take the preferences of the parents/children into account. When the ethnical and socio-economic composition of the children is also important to consider in the allocation then the problem becomes much more complicated. In this case some form of controlled school choice through quotas or other affirmative actions may be used (Echenique and Yenmez 2015). Complex distributional requirements and goals can also be modeled and solved with integer programming techniques, as we demonstrated in a recent application at Corvinus University of Budapest, where students were allocated to business projects under various distributional constraints in a fair way, taking into account the preferences of the students and the rankings by the companies (Ágoston et al. 2018).

## 5   Final Notes

The ultimate goal of the engineering approach described in this writing is to understand which social and economic goals can be implemented in practice with sophisticated mechanism design. The practical design of matching mechanisms and the underlying theoretical investigations require the interaction of multiple disciplines. The starting steps were made by economists/game theorists, led by Roth (although note that both Roth and Shapley had PhDs in Operations Research), but in the last decades the computer science community also became very active in the field of matching markets. This research line will definitely continue to rise in the coming years under the umbrella of the emerging topics of *Computational Social Choice* and *Algorithmic Game Theory*. We have established the Matching in Practice research network in 2010 devoted to study the practical applications with regular workshops and an informative website (http://www.matching-in-practice.eu) where we collect the descriptions of centrally coordinated matching programs in Europe.

**Acknowledgements**  The author acknowledges the support of the Hungarian Academy of Sciences under its Momentum Programme (LP2016-3/2018) and Cooperation of Excellences Grant (KEP-6/2018), and the Hungarian Scientific Research Fund, OTKA, Grant No.\ K128611.

## References

Abdulkadiroglu, A., & Sönmez, T. (2003). School choice: A mechanism design approach. *American Economic Review, 93*(3), 729–747.

Abdulkadiroglu, A., Pathak, P. A., & Roth, A. E. (2005a). The New York city high school match. *American Economic Review, Papers and Proceedings, 95*(2), 364–367.

Abdulkadiroglu, A., Pathak, P. A., & Roth, A. E. (2005b). Boston public school match. *American Economic Review, Papers and Proceedings, 95*(2), 368–371.

Abdulkadiroglu, A., Che, Y. K., & Yasuda, Y. (2011). Resolving conflicting preferences in school choice: The "Boston mechanism" reconsidered. *American Economic Review, 101*(1), 399–410.

Abdulkadiroglu, A., Che, Y. K., Pathak, P. A., Roth, A. E., & Tercieux, O. (2017). *Minimizing Justified Envy in School Choice: The Design of New Orleans' OneApp* (No. w23265). National Bureau of Economic Research.

Ágoston, K. Cs., Biró, P., & McBride, I. (2016). Integer programming methods for special college admissions problems. *Journal of Combinatorial Optimization, 32*(4), 1371–1399.

Ágoston, K Cs, Biró, P., & Szántó, R. (2018). Stable project allocation under distributional constraints. *Operations Research Perspectives, 5*, 59–68.

Ashlagi, I., & Gonczarowski, Y. A. (2016). Stable matching mechanisms are not obviously strategy-proof. arXiv:1511.00452.

Ashlagi, I., & Shi, P. (2015). Optimal allocation without money: An engineering approach. *Management Science, 62*(4), 1078–1097.

Azevedo, E. M., & Budish, E. (2013). Strategy-proofness in the large. *The Review of Economic Studies, 86*(1), 81–116.

Biró, P. (2017). Applications of matching models under preferences. In U. Endriss (Ed.), *Trends in Computational Social Choice*, Chapter 18 (pp. 345–373). AI Access.

Biró, P., & Gudmondsson, J. (2018). Complexity of finding pareto-efficient allocations of highest welfare. *Working paper*.

Biró, P., & Kiselgof, S. (2015). College admissions with stable score limits. *Central European Journal of Operations Research, 23*(4), 727–741.

Biró, P., & Klijn, F. (2013). Matching with couples: A multidisciplinary survey. *International Game Theory Review, 15*(02), 1340008.

Biró, P., Manlove, D. F., & Rizzi, R. (2009). Maximum weight cycle packing in directed graphs, with application to the kidney exchange programs. *Discrete Mathematics, Algorithms and Applications, 1*(4), 499–517.

Biró, P., Fleiner, T., Irving, R. W., & Manlove, D. F. (2010). The College admissions problem with lower and common quotas. *Theoretical Computer Science, 411,* 3136–3153.

Biró, P., Irving, R. W., & Schlotter, I. (2011). Stable matching with couples: An empirical study. *ACM Journal of Experimental Algorithmics, 16,* 1–2.

Biró, P., McBride, I., & Manlove, D. F. (2014). The hospitals/residents problem with couples: Complexity and integer programming models. *In Proceedings of SEA 2014: The 13th International Symposium on Experimental Algorithms, 8504* (pp. 10–21). LNCS, Springer.

Budish, E., & Kessler, J. (2015). *Experiments as a Bridge from Market Design Theory to Market Design Practice: Changing the Course Allocation Mechanism at Wharton*. Working paper.

Chen, Y., & Sönmez, T. (2006). School choice: An experimental study. *Journal of Economic Theory, 127*(1), 202–231.

Echenique, F., & Yenmez, M. B. (2015). How to control controlled school choice. *American Economic Review, 105*(8), 2679–2694.

Gale, D., & Shapley, L. S. (1962). College admissions and the stability of marriage. *American Mathematical Monthly, 69,* 9–15.

Gusfield, D., & Irving, R. W. (1989). *The Stable Marriage Problem: Structure and Algorithms.* MIT press.

Irving, R. W., & Manlove, D. F. (2009). Finding large stable matchings. *ACM Journal of Experimental Algorithmics, 14,* Section 1, Article 2, 30.

Kojima, F., & Pathak, P. A. (2009). Incentives and stability in large two-sided matching markets. *American Economic Review, 99*(3), 608–627.

Kojima, F., Pathak, P. A., & Roth, A. E. (2013). Matching with couples: Stability and incentives in large markets. *The Quarterly Journal of Economics, 128*(4), 1585–1632.

Kwanashie, A., & Manlove, D. F. (2014). An integer programming approach to the hospitals/residents problem with ties. In *Operations Research Proceedings 2013* (pp. 263–269). Springer.

Li, S. (2017). Obviously Strategy-proof mechanisms. *American Economic Review, 107*(11), 3257–3287.

Manlove, D. F. (2013). *Algorithms of Matching Under Preferences.* World Scientific.

Manlove, D. F., & O'Malley, G. (2012). Paired and altruistic kidney donation in the UK: Algorithms and experimentation. *In proceeding of SEA 2012, 7276,* (pp. 271–282). LNCS.

Manlove, D. F., Irving, R. W., Iwama, K., Miyazaki, S., & Morita, Y. (2002). Hard variants of stable marriage. *Theoretical Computer Science, 276*(1–2), 261–279.

Marx, D., & Schlotter, I. (2011). Stable assignment with couples: parameterized complexity and local search. *Discrete Optimization, 8,* 25–40.

Mcdermid, E. (2009). A 3/2 approximation algorithm for general stable marriage. In *Proceedings of ICALP '09: The 36th International Colloquium on Automata, Languages and Programming, 5555* (pp. 689–700). LNCS, (Springer).

Morrill, T. (2015). Making just school assignments. *Games and Economic Behavior, 92,* 18–27.

Pathak, P. A., & Sönmez, T. (2013). School admissions reform in Chicago and England: Comparing mechanisms by their vulnerability to manipulation. *American Economic Review, 103*(1), 80–106.

Rees-Jones, A. (2018). Suboptimal behavior in strategy-proof mechanisms: Evidence from the residency match. *Games and Economic Behavior, 108*, 317–330.

Ronn, E. (1990). NP-complete stable matching problems. *Journal of Algorithms, 11*, 285–304.

Roth, A. E. (1982). The Economics of matching: Stability and incentives. *Mathematics of Operations Research, 7*, 617–628.

Roth, A. E. (1984). The Evolution of the labor market for medical interns and residents: A case study in game theory. *Journal of Political Economy, 92*, 991–1016.

Roth, A. E. (1991). A Natural experiment in the organization of entry level labor markets: Regional markets for new physicians and surgeons in the U.K. *American Economic Review, 81*, 415–440.

Roth, A. E. (2002). The economist as engineer: Game theory, experimentation, and computation as tools for design economics. *Econometrica, 70*(4), 1341–1378.

Roth, A. E. (2008). Deferred acceptance algorithms: History, theory, practice, and open questions. *International Journal of Game Theory, 36*, 537–569.

Roth, A. E. (2015). *Who Gets What and Why: The New Economics of Matchmaking and Market Design*. Eamon Dolan.

Roth, A. E., & Peranson, E. (1999). The redesign of the matching market for American physicians: Some engineering aspects of economic design. *American Economic Review, 89*(4), 748–780.

Roth, A. E., & Sotomayor, M. O. (1990). *Two-Sided Matching: A Study in Game Theoretic Modelling and Analysis*. Cambridge University Press.

Roth, A. E., Sönmez, T., & Utku Ünver, M. (2005). A kidney exchange clearinghouse in New England. *American Economic Review, 95*(2), 376–380.

Veski, A., Biró, P., Pöder, K., & Lauri, T. (2017). Efficiency and fair access in kindergarten allocation policy design. *Journal of Mechanism and Institutional Design, 2*(1), 57–104.

Website of the Matching in Practice network. http://www.matching-in-practice.eu.

Yanagisawa, H. (2007). *Approximation algorithms for stable marriage problems*. Ph.D. thesis, Kyoto University, School of Informatics.

# Matthews–Moore Single- and Double-Crossing

Craig Brett and John A. Weymark

**Abstract** This article provides an introduction to the Matthews–Moore single- and double-crossing properties for screening problems with one-dimensional types. The relationship of Matthews–Moore single–crossing to the Mirrlees single-crossing property is discussed.

The single-crossing property of preferences introduced by Mirrlees (1971) is extensively used in screening models with one-dimensional type spaces. However, its use is limited to applications in which contracts are two dimensional. It is not widely known that a methodology due to Matthews and Moore (1987) provides a fruitful way of analyzing screening problems with a one-dimensional type space that does not require contracts to be two dimensional. The Matthews–Moore methodology employs either a single-crossing property distinct from that of Mirrlees or a related double-crossing property. Since they were introduced, the Matthews–Moore single- and double-crossing properties have rarely been employed. Matthews–Moore single-crossing is used by Bohn and Stuart (2013) and by van Egteren (1996) to study majority voting over income tax schedules and the regulation of a public utility, respectively.[1] Brett (1998) uses Matthews–Moore double-crossing to investigate when workfare should supplement income taxation. Here, we provide an introduction to Matthews–Moore single- and double-crossing in the hope that we will thereby encourage the use of these tools in future analyses of screening problems with multidimensional contracts.

---

[1]Bohn and Stuart do not cite Matthews and Moore (1987), which suggests that they independently recognized the significance of having single-crossing contract utility curves.

---

C. Brett
Department of Economics, Mount Allison University, 144 Main Street, Sackville,
New Brunswick E4L 1A7, Canada
e-mail: cbrett@mta.ca

J. A. Weymark (✉)
Department of Economics, Vanderbilt University, VU Station B #35189,
2301 Vanderbilt Place, Nashville, TN 37235-1819, USA
e-mail: john.weymark@vanderbilt.edu

© Springer Nature Switzerland AG 2019                                           461
J.-F. Laslier et al. (eds.), *The Future of Economic Design*, Studies in
Economic Design, https://doi.org/10.1007/978-3-030-18050-8_64

In a screening problem, individuals have private information about some characteristics of themselves that is of value to, but not known by, the principal. Individuals who share the same private characteristics are said to have the same *type*. Each type chooses its most preferred contract from the set of contracts on offer which, because of the asymmetric information, is the same for every type. This is the *incentive constraint*. An *allocation* consists of a contract for each type. A principal maximizes an objective function subject to the incentive constraint and one or more additional constraints. For example, in the Mussa and Rosen (1978) monopoly nonlinear pricing problem, a firm maximizes profit by choosing a schedule that specifies the payment as a function of the quality of the good chosen subject to the incentive constraint and a *participation constraint* that requires each type to obtain some common minimum utility level. In the Mirrlees (1971) optimal nonlinear income tax problem, a government chooses an income tax schedule to maximize a utilitarian social welfare function subject to the incentive constraint and a *materials balance constraint* that requires the total amount consumed of the single good not to exceed the amount produced.

The principal's optimization problem is complex in part because the incentive constraint is itself a set of optimization problems, one for each type. This complexity can be somewhat mitigated by noting that when types choose from a common schedule of options, each type weakly prefers what it chooses to what any other type chooses. A set of contracts with this property is said to satisfy the *self-selection* constraints. As a consequence, instead of having the principal offer a set of contracts from which each type chooses its most preferred contract, the principal can equivalently directly specify its own most preferred allocation subject to the self-selection constraints and any other constraints that might apply without any explicit optimization on the part of the types. Unfortunately, because of the nature of the self-selection constraints, the set of feasible allocations in this problem is non-convex. Hence, knowing that an allocation is locally optimal does not guarantee that it is globally optimal, as would be the case with a convex optimization problem. Moreover, it may be difficult to determine which of the self-selection constraints bind at a solution to the principal's optimization problem. Identifying the pattern of binding incentive constraints is important for characterizing the direction of distortions. This is a further source of complexity.

One of the virtues of the Mirrlees and Matthews–Moore single-crossing properties is that a local approach is nevertheless possible in spite of the non-convexity. When there are a finite number of types, the *local approach* proceeds by analyzing a *relaxed problem* in which only the adjacent self-selection constraints are considered. This approach is valid if all of the self-selection constraints are satisfied whenever the adjacent ones hold. These two single-crossing properties also facilitate the identification of the pattern of binding self-selection constraints. The validity of the local approach can be established using only the self-selection constraints without analyzing the principal's full optimization problem. However, to identify which self-selection constraints bind, it is not sufficient to only consider these constraints. In order to make these ideas precise, we now proceed more formally. Our formal discussion draws extensively on Matthews and Moore (1987).

There are a finite number of types of individuals, $i = 1, \ldots, n$. The $i$th type's private information is characterized by a scalar $\theta^i$, with the types ordered so that $\theta^1 < \theta^2 < \cdots < \theta^n$. A *contract* is a vector $\mathbf{c} = (a_1, \ldots, a_m, b) \in \mathbb{R}_+^{m+1}$. We use $k$ to index the $m$ attributes. A *menu of contracts* is a set $\mathscr{C} = \{\mathbf{c}^1, \ldots, \mathbf{c}^n\}$, where the $i$th of these contracts is the one designed for type $i$. The *utility function* $U : \mathbb{R}_+^{m+1} \times [\theta^1, \theta^n]$ specifies the utility $U(\mathbf{c}, \theta)$ that an individual of type $\theta$ obtains with the contract $\mathbf{c}$. It is assumed that $U$ is continuously differentiable in $\mathbf{c}$ with $U_b < 0$. Whether $U$ is monotone in any of its other arguments depends on the application. The first $m$ components of a contract are the *attributes* and the last component is the *outlay*. For example, in the Mussa and Rosen (1978) monopoly pricing problem, the single attribute is the quality of a good and the outlay is the payment. In the Mirrlees (1971) income tax problem, the single attribute is after-tax consumption and the outlay is pre-tax income.

The menu $\mathscr{C}$ satisfies the *self-selection constraints* if

$$U(\mathbf{c}^i, \theta^i) \geq U(\mathbf{c}^j, \theta^i) \text{ for all } j \neq i$$

and it satisfies the *adjacent self-selection constraints* if

$$U(\mathbf{c}^i, \theta^i) \geq U(\mathbf{c}^j, \theta^i) \text{ for all } i \text{ and for } j \in \{i - 1, i + 1\}.$$

In the first case, a type weakly prefers the contract designed for it to the contract designed for any other type, whereas in the second, a type is only required to weakly prefer its own contract to those of the adjacent types.

The utility function $U$ satisfies the MRS-*ordering property* if for all $\mathbf{c}$,

$$-\frac{U_k(\mathbf{c}, \theta)}{U_b(\mathbf{c}, \theta)} \text{ is (i) increasing in } \theta \text{ for all } k \text{ or (ii) decreasing in } \theta \text{ for all } k.$$

In other words, the marginal rates of substitution between an attribute and the outlay are ordered by type in the same way for each attribute. When $m = 1$ (in which case contracts are two dimensional), the utility function $U$ satisfies the *Mirrlees single-crossing property* if

an indifference curve of any type intersects an indifference curve of any other type at most once.

This is simply the MRS-ordering property for $m = 1$. Mirrlees single-crossing is illustrated in Fig. 1 for the case in which the MRS is decreasing in type.

The *contract utility curve* for the contract $\mathbf{c}$ is the graph of the function $U(\mathbf{c}, \cdot)$ on the domain $[\theta^1, \theta^n]$. It shows what the utility is of each possible type in the interval $[\theta^1, \theta^n]$ with the contract $\mathbf{c}$. Associated with the menu of contracts $\mathscr{C} = \{\mathbf{c}^1, \ldots, \mathbf{c}^n\}$ is the *menu of contract utility curves* $\mathscr{U}^{\mathscr{C}} = \{U(\mathbf{c}^1, \cdot), \ldots, U(\mathbf{c}^n, \cdot)\}$. A menu of contracts $\mathscr{C}$ satisfies the *Matthews–Moore single-crossing property* if

no two distinct contract utility curves in $\mathscr{U}^{\mathscr{C}}$ (i) intersect more than once or (ii) are tangent to each other at any point in $(\theta^1, \theta^n)$.

Matthews–Moore single-crossing is illustrated in Fig. 2.

**Fig. 1** Mirrlees
single-crossing

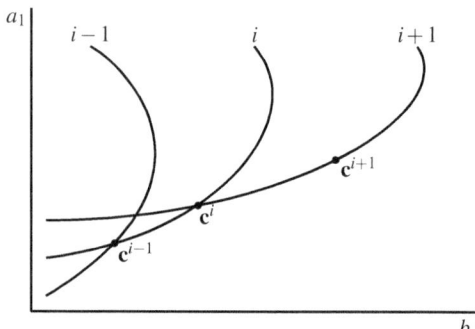

**Fig. 2** Matthews–Moore
single-crossing

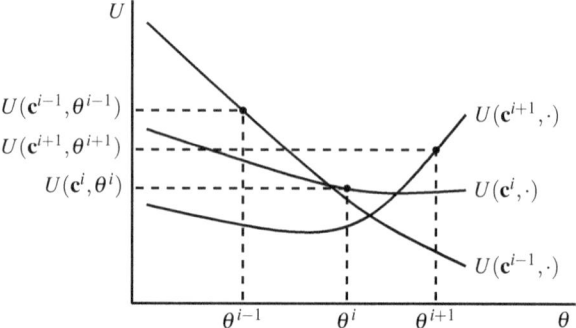

Because utility levels and types are both scalars, it is possible to satisfy the
Matthews–Moore single-crossing property no matter how many attributes there are
in a contract. In contrast, it is not possible for indifference contours to cross only
once when there is more than one attribute, so it is not possible to extend the Mirrlees
single-crossing property to higher dimensions.

The menu $\mathscr{C}$ is *weakly attribute ordered* if

$$\text{for each pair } \{i, j\}, \text{ (i) } a_k^i \leq a_k^j \text{ for all } k \text{ or (ii) } a_k^i \geq a_k^j \text{ for all } k$$

and it is *attribute ordered* if

$$\text{(i) for all } i < j, a_k^i \leq a_k^j \text{ for all } k \text{ or (ii) for all } i < j, a_k^i \geq a_k^j \text{ for all } k.$$

In the first case, for any pair of types, one of them has weakly more of every attribute
than the other. In the second, the attributes are either all nondecreasing or all nonin-
creasing in type. Clearly, a menu is weakly attribute ordered if it is attribute ordered.

Matthews and Moore (1987) show that weak attribute ordering provides a link between the MRS-ordering property and their single-crossing property.

**Theorem 1** *If the utility function U satisfies the* MRS-*ordering property and the menu of contracts $\mathscr{C}$ is weakly attribute ordered, then $\mathscr{C}$ satisfies the Matthews–Moore single-crossing property.*

When $m = 1$, a menu of contracts is necessarily weakly attribute ordered.[2] Thus, the Mirrlees single-crossing property implies the Matthews–Moore single-crossing property and, hence, the latter is less restrictive.[3]

The following theorem due to Matthews and Moore (1987) shows that the local approach is justified if their single-crossing property holds.

**Theorem 2** *If the menu of contracts $\mathscr{C}$ satisfies the adjacent self-selection constraints and the Matthews–Moore single-crossing property, then $\mathscr{C}$ satisfies the self-selection constraints.*

By Theorem 1, the analogous result holds for the Mirrlees single-crossing property when $m = 1$. Figure 2 can be used to illustrate Theorem 2. For any type $\theta$, satisfaction of the self-selection constraints requires that the contract utility curve for this type's contract must lie weakly above the curves for the other contracts at $\theta$. This is the case for the three types in the figure. If the adjacent self-constraints are satisfied but, say, type $\theta^{i-1}$ prefers $\mathbf{c}^{i+1}$ to its own contract, then the contract utility curve for $\mathbf{c}^{i+1}$ would have to intersect the contract utility curve for $\mathbf{c}^i$ at least twice for this to be possible, violating the Matthews–Moore single-crossing property.

Even when the local approach cannot be employed, it may nevertheless be possible to characterize the solution to a screening problem by solving a relaxed problem in which some of the self-selection constraints are not considered, particularly if it can be determined which of them are binding. For example, when $m = 1$, if the Mirrlees single-crossing property is satisfied and the menu of contracts is attribute ordered, all of the self-selection constraints are satisfied if all of the adjacent downward (or adjacent upward) self-section constraints are binding (see Fig. 1).

Matthews and Moore (1987) consider an extension of the Mussa and Rosen (1978) monopoly pricing problem in which a warranty level is an additional attribute. In their analysis, they introduce the following property. A menu of contracts $\mathscr{C}$ satisfies the *Matthews–Moore double-crossing property* if

---

[2]A menu of contracts is attribute ordered if the Mirrlees single-crossing property and the adjacent self-selection constraints are satisfied. However, if, as in Fig. 1, the utility function is not monotonic in this attribute, the outlays need not be monotone in type.

[3]When a set of alternatives and preference types are both one dimensional, a profile of preferences satisfies the *preference single-crossing property* if the direction of preference between any two alternatives reverses only once as the type increases. In a screening problem with $m = 1$, if the two contract components are monotone in type, the menu of contracts is effectively one dimensional, in which case the preference single-crossing property is satisfied if and only if the Mirrlees single-crossing property is as well. See Gans and Smart (1996).

no two distinct contract utility curves in $\mathscr{U}^{\mathscr{C}}$ intersect more than twice.

In their model, this double-crossing property of contract utility curves enables Matthews and Moore to characterize the optimal solution without considering the upward self-selection constraints.[4] In order to show that the solution to this relaxed problem solves the unrelaxed problem, they first show that all of the adjacent downward self-section constraints bind.

The MRS-ordering property is a natural assumption in many screening problems with multi-dimensional contracts and a one-dimensional type space. When it holds, it is only necessary to determine if the menu of contracts is weakly attribute ordered in order to conclude that the Matthews–Moore single-crossing property is satisfied and, hence, that the local approach is valid. It is more difficult to identify situations in which the Matthews–Moore double-crossing property is satisfied, but as in the models of Matthews and Moore (1987) and Brett (1998), reasonable restrictions may imply that it is.

An obvious limitation of the single- and double-crossing properties considered here is that they assume that the type space is one dimensional. With a multi-dimensional type space, it is possible for the projections of the indifference curves or the utility contract curves for each dimension of the type space to satisfy one of these properties. For the case of the Mirrlees single-crossing property, Armstrong and Rochet (1999) and Brett (2007) illustrate the usefulness of this observation for two-dimensional type spaces.

# References

Armstrong, M., & Rochet, J. C. (1999). Multi-dimensional screening: A user's guide. *European Economic Review, 43*, 959–979.

Bohn, H., & Stuart, C. (2013). Revenue extraction by median voters, unpublished manuscript, Department of Economics, University of California, Santa Barbara.

Brett, C. (1998). Who should be on workfare? The use of work requirements as part of an optimal tax mix. *Oxford Economic Papers, 50*, 607–622.

Brett, C. (2007). Optimal nonlinear taxes for families. *International Tax and Public Finance, 14*, 225–261.

Gans, J. S., & Smart, M. (1996). Majority voting with single-crossing preferences. *Journal of Public Economics, 59*, 219–237.

Matthews, S., & Moore, J. (1987). Monopoly provision of quality and warranties: An exploration in the theory of multi-dimensional screening. *Econometrica, 55*, 441–467.

Mirrlees, J. A. (1971). An exploration in the theory of optimum income taxation. *Review of Economic Studies, 38*, 175–208.

Moore, J. (1984). Global incentive constraints in auction design. *Econometrica, 52*, 1523–1535.

---

[4]In an auction model in which a contract consists of the probability of a buyer winning an object, the payment if he wins, and the payment if he loses, Moore (1984) shows that it is only necessary to consider the downward self-selection constraints. He does not employ contract utility curves or the Matthews–Moore double-crossing property to do so. We are grateful to Steve Matthews for drawing our attention to Moore's article.

Mussa, M., & Rosen, S. (1978). Monopoly and product quality. *Journal of Economic Theory, 18,* 301–317.
van Egteren, H. (1996). Regulating an externality-generating public utility: A multi-dimensional screening approach. *European Economic Review, 40,* 1773–1797.

# Mechanism Design Approach to School Choice: One Versus Many

Battal Doğan

**Abstract** A vast majority of the school choice literature focuses on designing mechanisms to simultaneously assign students to *many* schools, and employs a "make it up as you go along" approach when it comes to each school's admissions policy. An alternative approach is to focus on the admissions policy for *one* school. This is especially relevant for effectively communicating policy objectives such as achieving a diverse student body or implementing affirmative action. I argue that the latter approach is relatively under-examined and deserves more attention in the future.

Mechanism Design has been especially successful in allocating students into schools. In a seminal study, Abdulkadiroğlu and Sönmez (2003) formulate desirable properties for assignment mechanisms, and in light of this desiderata show that some school districts use *deficient* mechanisms and propose *better* alternatives. Since then, economists have been studying assignment mechanisms for many school districts around the world, formulating desirable properties motivated by various policy objectives, and proposing alternative mechanisms.

This design process has two essential parts, one of which has been well-studied and understood, while the other has been relatively underestimated. The well-studied and understood part is the formulation of desirable properties for assignment mechanisms and figuring out which properties can be satisfied simultaneously and by which mechanisms. Although design objectives may differ from district to district, the reforms have been centered around two objectives: (i) Optimizing student welfare while achieving fairness and (ii) making it safe for the students to report their preferences truthfully. The Student Optimal Stable Mechanism (SOSM), proposed by Abdulkadiroğlu and Sönmez (2003), achieves the two goals and is now in use in many school districts around the world. Economists have been very successful in not only proposing alternative mechanisms, but also in convincing the public and the policy makers about the *desirability* of the formulated properties. It is important to note

I gratefully acknowledge financial support from the Swiss National Science Foundation (SNSF).

B. Doğan (✉)
Department of Economics, University of Bristol, Bristol, UK
e-mail: battal.dogan@bristol.ac.uk

© Springer Nature Switzerland AG 2019
J.-F. Laslier et al. (eds.), *The Future of Economic Design*, Studies in Economic Design, https://doi.org/10.1007/978-3-030-18050-8_65

469

here that the formulation and successful communication of the desirable properties is very crucial to induce reforms in practice, since these desirable properties of the mechanisms pave the way for the economists and the policy makers to convince the public for the necessity of a change. A very striking example is the case of England. In 2007, the *nationwide* School Admissions Code *prohibited* authorities from using the older mechanism which did not make it safe for the students to report their preferences truthfully. The justification for the reform given by Department for Education and Skills was that the earlier mechanism "made the system unnecessarily complex to parents", pointing to a specific desirable property violated by the old mechanism.

The other, relatively underestimated, part of the design process is endowing each school with a choice rule (or an admission rule). Although student preferences are elicited from the students, endowing each school with a choice rule is an essential part of the design process. In the earlier school choice literature, the main focus has been on assignment problems where each school is already endowed with a priority ordering over students, and the choice rule of a school is to simply admit highest-priority students up to the capacity. Although there are indeed school districts where there are no policy objectives that would preclude us from endowing each school with such a simple choice rule and directly proceeding with the other part of the design process, there are many school districts with additional concerns, which calls for a non-trivial design of an appropriate choice rule for each school. An important example is achieving a diverse student body or affirmative action. Many school districts are concerned with maintaining a diverse student body at each school while assigning students to schools when each student belongs to one of multiple types (based on factors such as gender, socioeconomic status, or ethnicity). In this case, what is a *good* choice rule that reconciles diversity objectives with other objectives such as admitting students with higher test scores? The answer to this question is not evident.

To this end, most of the literature so far has employed a "make it up as you go along" approach by keeping the focus on the design of a mechanism to assign students to *many* schools and, as a *detail* of the assignment mechanism, endowed schools with choice rules that guarantee assignments with certain properties. An alternative approach, which has first been employed by Echenique and Yenmez (2015), is to separately focus on the choice problem of *one* school, formulate desirable properties of choice rules for *one* school, figure out which properties can be satisfied simultaneously and by which choice rules, and then study implications of using these choice rules in assigning students to *many* schools.

The second approach has an important advantage over the first approach. In the first approach, the objectives are solely stated as properties of the assignment mechanism, and it may be difficult to understand according to which principals each school admits students. The second approach provides a clear foundation for each school's choice behaviour in the assignment process. This is a very important advantage because for some objectives, such as achieving diversity, it is either *easier* or more *natural* to communicate policies towards achieving such objectives through properties of choice rules for *one* school, rather than properties of mechanisms assigning students to *many* schools. As an example, consider the Boston school district where diversity is an objective, based on neighbourhood boundaries. The policy of the Boston school district is the following: at each school in Boston, half of the seats at each school are made open to all applicants, while the other half are reserved seats that prioritize applicants from the local neighborhood, and reserved seats are filled ahead of open seats. Note that the Boston school district describes its diversity policy to the public by referring to the choice rule of one school, rather than referring to the assignment mechanism which is the SOSM where each school is endowed with a choice rule as described above. To further justify why the current policy is a good policy to achieve diversity, a natural way to proceed is to discover desirable properties underlying the choice rules of the schools in Boston, which we address in Doğan et al. (2017).

It is also important to understand how properties of choice rules for one school relate to the properties of an assignment mechanism that uses these choice rules. In fact, there are properties of choice rules for one school such that the natural counterparts of the properties may not be satisfied by an assignment mechanism that uses choice rules satisfying the properties. For example, some school districts aim to favor the underrepresented minority students, and in order to achieve that objective, there is a majority-quota at each school and it is required that no school be assigned more majority students than its majority-quota. Suppose that there is a priority ordering over students based on their test scores. Define a majority-quota choice rule for a school as the choice rule that admits students according to priority until the majority-quota is reached and from then on only admits minority students according to priority until the capacity is reached. Such a choice rule satisfies the following property: if the majority-quota is lowered, a chosen minority student still remains to be chosen. Now, when assigning students to many schools, a natural counterpart of the property that we have defined above is the following: if the majority-quota at each school is lowered, no minority student is worse off. However, it is known that this property is not satisfied by the student optimal stable mechanism which endows each school with a majority-quota choice rule; even worse, when majority-quotas at all schools are lowered, it is possible that all the minority students are worse off (Kojima 2012). In other words, a choice rule which unambiguously favors minority students when there is one school may have perverse welfare consequences for minority students when endowed by schools in a school district that uses the student optimal stable mechanism.

I believe that the field should grow more in the direction of formulating desirable properties for one school, figuring out which properties can be satisfied simultaneously and by which choice rules, and understanding whether these properties translate into desirable properties of an assignment mechanism that endows schools with these choice rules. I believe that this will facilitate the communication of design objectives and pave the way for further school choice reforms.

# References

Abdulkadiroğlu, A., & Sönmez, T. (2003). School choice: A mechanism design approach. *American Economic Review, 93*, 729–747.

Doğan, B., Doğan, S., & Yildiz, K. (2017). *Lexicographic choice under variable capacity constraints* (Working Paper Available at SSRN: https://ssrn.com/abstract=2886494).

Echenique, F., & Yenmez, M. B. (2015). How to control controlled school choice. *American Economic Review, 105*, 2679–2694.

Kojima, F. (2012). School choice: Impossibilities for affirmative action. *Games and Economic Behavior, 75*, 685–693.

# Equity and Diversity in College Admissions

Eun Jeong Heo

**Abstract** Equity and diversity issues are becoming more important in higher education and colleges are adopting various policies to address them. I am going to focus on college admissions and discuss two policies that are most commonly used in practice: affirmative action and need/merit-based financial aid. Here are what we have come up with so far and what can be done in the future.

## 1 Introduction

In college admissions, special considerations are taken into account to address concerns on equity and diversity. These policies have drawn economists' attention in theory and in empirical studies. But whether they indeed benefit the purported beneficiaries is still being debated. Apart from ethical and legal issues, I am going to make a brief survey on several theoretical and empirical findings and discuss what questions remain.

## 2 Affirmative Action

Affirmative action is often known as "positive discrimination" that provides a favorable consideration to historically excluded or socioeconomically disadvantaged groups—e.g., racial minorities, native American, women, or low-income families—in college admissions or employment.

**Market Design Approach**

In market design literature, affirmative action policy is accommodated to the mechanisms that match students to colleges. There are three forms by and large:

E. J. Heo (✉)
Department of Economics, Vanderbilt University, VU Station B, 2301 Vanderbilt Place,
Box #351819, Nashville, TN 37235, USA
e-mail: eun.jeong.heo@vanderbilt.edu

© Springer Nature Switzerland AG 2019
J.-F. Laslier et al. (eds.), *The Future of Economic Design*, Studies in
Economic Design, https://doi.org/10.1007/978-3-030-18050-8_66

1. Majority quota: A hard upper bound for non-minority students is set and each college cannot admit non-minority students above this bound. The remaining seats are reserved for minority students and should be disposed if not taken by minority students.
2. Minority reserve: A block of seats are reserved for minority students as in the majority quota, but when minority applicants do not take all of these seats, the remaining seats can be assigned to majority students.
3. Promoting priorities of minority students: when colleges rank students based on their exam score, GPA, and other factors, they assign extra points to minority students to promote these students' rankings.

Economists have been studying whether each policy benefits all (or at least some) minority students and whether the policy stands along with other economic objectives such as efficiency, "stability", and incentives. The answer is negative.

Kojima (2012) shows that the majority quota can make *all* minority students worse off and Hafalir et al. (2013) find that this adverse effect can be quite common in some circumstances. When the majority quota is implemented, some majority students cannot get into colleges that they could have made and therefore should search for other colleges. The competition at other colleges increases as a result and this can make minority students who want to get into these colleges worse off. Promoting priorities of minority students also results in a similar adverse effect.

Hafalir et al. (2013) instead work with minority reserves and show that there is a significant efficiency gain under the minority reserve in comparison to the majority quota when we consider stable mechanisms. However, Dogan (2016) shows that even minority reserve policy conflicts with other economic objectives.

Matching mechanisms with affirmative action are still being proposed and debated. But no consensus has been reached and its welfare effect remains indefinite. More recently, there are attempts to connect these theoretical findings to experiments and real world data (Dur et al. 2016; Klijn et al. 2016).

**Contest Model**

College admissions can be viewed as a "contest" problem where students make efforts to win the limited seats available at colleges. This approach is quite interesting because we can talk about the eventual goal of affirmative action policy: to help minority/disadvantaged students achieve competency by exerting more effort.

There are only a handful of papers analyzing affirmative action in this context. Among others, Fu (2006) formulates a model in which minority and non-minority groups have differential returns on higher education. When affirmative action is implemented, more minority students pursue college education, expecting a higher probability of winning the prize. Both minority and non-minority students face increased competition as a result and choose to work harder. This result invalidates the concern that affirmative action lowers the academic quality of higher education institutions.[1] Despite this positive effect, Fu (2006) also shows that non-minority stu-

---

[1] A similar "positive cross-group interaction" from affirmative action is also found in labor tournament in Fain (2009).

dents react to this change more aggressively and increase their effort even more than minority students do. After all, the existing test score gap between the two groups can be widened even further.

The conclusion of the wider gap is drawn from a stylized contest setting, but this is an important policy implication. I think this type of effort investment stage should be considered when we design college-student matching mechanisms, if we do care about what affirmative action brings to these groups in the end. Hafalir et al. (2017) make a good start on this line of research, but their focus is not to address equity issue, but to compare different systems of college admissions.

**Empirical Studies**

The effect of affirmative action on diversity or equal access also remains controversial in the empirical literature. Long (2004) suggests that *application* behaviors of minority and non-minority students have changed significantly when California and Texas eliminated their affirmative action policies in late 1990s. As the probabilities of acceptance change, minority students are less likely to apply for top-tier colleges, while non-minority students increase the number of applications to those colleges, widening the gap between the two groups. Antonovics and Backes (2013), however, claim that this change does not prevail at large, but rather limited to a couple of colleges among UC campuses in the case of California.

In terms of *enrollment* of minority students, similar observations are found. It is shown that, when affirmative action is eliminated, the number of minority students decreases significantly in simulations and estimations (Rothstein and Yoon 2008). Other studies however show that the effect of affirmative action is quite limited and it only affects admissions at highly selective universities (Kane 1998; Hinrichs 2012). The total enrollment of minorities may not change much, but there is a qualitative change in enrollment of minority groups: a significant fraction of minority students shifts from highly selective colleges to less selective colleges (Bodoh-Creed and Hickman 2017).

With heterogeneity within minority groups and with different measures of accounting the welfare effect, we cannot tell exactly what would result from implementing affirmative action. However, these empirical findings give us insights on what is happening in real world and would help theorists construct more plausible models.

# 3 Need/Merit-Based Financial Aid

Nowadays almost all universities are offering different types of financial aid so as to recruit more competent students as well as to address equity and diversity issues. The need-based aid is usually provided to socioeconomically disadvantaged students. The eligibility is entirely based on the financial specifics of their families and FAFSA (Free Application for Federal Student Aid application) is used to determine the grant eligibility in practice. The merit-based financial aid is, on the other hand, provided to students with outstanding academic or artistic/athletic talents. As the cost of tuition and fees rises, colleges have increased merit-based supports in 1990s. But this move brought up a serious concern that it takes away opportunities to get need-based support.

## Empirical Studies

There are many empirical studies on welfare effect of each financial aid program. Myong (2016) provides a quantitative analysis of how need-based aid affects students' effort and achievement in high schools. The need-based aid increases the competition among the students who are qualified enough to get into colleges. The overall effort level of these students therefore increases, irrespective of their socioeconomic backgrounds. The need-based aid also narrows the *effort* gap as well as *achievement* gap (in terms of the SAT score) between low-income and high-income groups.

There are studies on merit-based financial aid program as well. Kane (2003) find that the enrollment rate increases when the merit-based aid becomes available, but the beneficiaries are more likely to be students from middle or high income families, possibly enlarging the achievement gap between different groups.

Jackson (1990) points out another important effect of financial aid: the benefits from financial aid are distributed unevenly among minority populations. Financial aid is more likely to increase the enrollment of African American students, for instance, while the enrollment of Hispanics may actually decrease.What causes such disparity is yet to be answered in a formal analysis.

## Decentralized College Admission Model

Along with these empirical findings, Heo (2017) introduces financial aid programs to the decentralized college admission model (Avery and Levin 2010). Two groups of students—the advantaged and the disadvantaged—make effort to increase their entrance exam scores (or GPA) paying different effort costs. The applicants take an entrance exam and then apply for colleges. Colleges set admission cutoffs and financial aid is provided to eligible applicants.

This approach emphasizes economic incentive of applicants. Expecting the available financial aid and the admission cutoffs, students figure out whether it is worthwhile paying effort costs, given their academic abilities and own preferences. Heo (2017) analyzes such incentives under three possible schemes—no financial aid, need-based aid, and merit-based aid. She first identifies equilibria of this game and then compares admission cutoffs and students' optimal effort/enrollment choices. The standard welfare analysis follows, but in addition to that, this approach allows us to analyze colleges' optimal strategies in allocating their limited budget.

There are several interesting findings. Surprisingly, the merit-based aid lowers the cutoff o f highly selective (or highly-ranked) colleges, while it raises the cutoff of less selective (or lower-ranked) colleges. In contrast, the need-based aid increases the cutoff of *all* colleges. In terms of enrollment, the need-based aid strictly increases the aggregate enrollment of the disadvantaged group. However, the merit-based aid is not too bad in this respect—it does not lower the aggregate enrollment of the disadvantaged group.

## Market Design Approach

Financial aid is also studied in market design literature in the form of stipend/resource allocation. Abizada (2016) first proposes an admission model with budget-constrained colleges and looks for desirable mechanisms assigning seats and stipend to the applicants. Afacan (2017) and Hsu (2017) also stand on this line of work, but with different specifications of the problem. They do care about *fairness* of matching mechanisms, but this notion is completely independent of applicants' socioeconomic backgrounds: it only involves how students rank colleges and how colleges rank students. To the best of my knowledge, I rarely find welfare analyses across different groups in market design literature.

## 4 Conclusion

As shown above, colleges are taking actions to provide minority/disadvantaged students equal access to higher education. These policies are introduced with the best of intentions to level the playing field. But whether they indeed redress socioeconomic inequality should be carefully verified. Among a number of different approaches, I only presented a fraction very briefly. Many open questions still remain and more follow-ups are to come.

## References

Abizada, A. (2016). Stability and incentives for college admissions with budget constraints. *Theoretical Economics*, *11*, 735–756.

Afacan, M. O. (2017). *Graduate admission with financial support*. New York: Mimeo.

Antonovics, K., & Backes, B. (2013). Were minority students discouraged from applying to University of California campuses after the affirmative action ban? *Education*, *8*, 208–250.

Avery, C., & Levin, J. (2010). Early admissions at selective colleges. *American Economic Review*, *100*, 2125–2156.

Bodoh-Creed, A., & Hickman, B. R. (2017). *College assignment as a large contest*. New York: Mimeo.

Dogan, B. (2016). Responsive affirmative action in school choice. *Journal of Economic Theory*, *165*, 69–105.

Dur, U., Pathak, P. A., & Sonmez, T. (2016). *Explicit vs. statistical preferential treatment in affirmative action: Theory and evidence from Chicagos exam schools*. New York: Mimeo.

Fain, J. R. (2009). Affirmative action can increase effort. *Journal of Labor Research*, *30*, 168–175.

Fu, C. (2006). A theory of affirmative action in college admissions. *Economic Inquiry*, *44*, 420–428.

Hafalir, I. E., Yenmez, M. B., & Yildirim, M. A. (2013). Effective affirmative action in school choice. *Theoretical Economics*, *8*, 325–363.

Hafalir, I. E., Hakimov, R., Kubler, D., & Kurino, M. (2017). *College admissions with entrance exams: Centralized versus decentralized*. New York: Mimeo.

Heo, E. J. (2017). *Financial aid in college admissions: Need-based versus merit-based*. New York: Mimeo.

Hinrichs, P. (2012). The effects of affirmative action bans on college enrollment, educational attainment, and the demographic composition of universities. *Review of Economics and Statistics, 94*, 712–722.

Hsu, C. L. (2017). *Promoting diversity of talents: A market design approach.* New York: Mimeo.

Jackson, G. A. (1990). Financial aid, college entry, and affirmative action. *American Journal of Education, 98*, 523–550.

Kane, T. J. (1998). Racial and ethnic preferences in college admissions. *Ohio State Law Journal, 59*, 971–996.

Kane, T. J. (2003). *A quasi-experimental estimate of the impact of financial aid on college-going.* New York: Mimeo.

Klijn, F., Pais, J., & Vorsatz, M. (2016). Affirmative action through minority reserves: An experimental study on school choice. *Economics Letters, 139*, 72–75.

Kojima, F. (2012). School choice: Impossibilities for affirmative action. *Games and Economic Behavior, 75*, 685–693.

Long, B. (2004). How have college decisions changed over time? An application of the conditional logistic choice model. *Journal of Econometrics, 121*, 271–296.

Myong, S. (2016). *Need-based aid from selective universities and the achievement gap between rich and poor.* New York: Mimeo.

Rothstein, J., & Yoon, A. H. (2008). Affirmative action in law school admissions: What do racial preferences do? *University of Chicago Law Review, 75*, 649–714.

# New Directions of Study in Matching with Constraints

Fuhito Kojima

**Abstract** This piece surveys a recent body of studies in "matching with constraints." It discusses relationship with traditional matching theory and new applications, and concludes with some open questions.

I am very excited about matching theory and its applications to economic design. Much of recent advance in matching theory has been motivated by very practical applications, while the theory itself is intellectually stimulating. Simple theory of matching proved to be useful for not only formulating and analyzing markets, but also even guiding a better design of an economic mechanism. As economists began to study more and more practical problems through the lenses of matching theory, however, they have found more and more markets some of whose features violate assumptions made in the standard model(s). These "realistic features" pose challenges for researchers because they can not necessarily be analyzed adequately by the simple, standard theory. Classical problems of this kind include analysis of labor markets in which some workers are part of a couple who need a pair of jobs—a feature which violates the standard assumption that each worker is single and only demands one job. Although stability—the canonical solution concept in matching problems—is incompatible with the general existence in the presence of couples (Roth 1984), progress has been made in various directions, such as finding domain restrictions (Klaus and Klijn 2005), numerical and algorithmic analysis (McDermid and Manlove 2009), and probabilistic existence results (Kojima 2013; Ashlagi et al. 2014), among others.

One of the "realistic features" found in many matching markets are the existence of constraints. Among the most extensively studied are diversity (or affirmative action) constraints which require a certain balance of the student body in each school in terms of socioeconomic class (Abdulkadiroğlu and Sönmez 2003; Abdulkadiroğlu 2005; Kojima 2012; Hafalir et al. 2013; Ehlers et al. 2014; Fragiadakis and Troyan 2016). In a series of papers (Kamada and Kojima 2015, 2017, 2018a; Kojima et al. 2018; Goto et al. 2016), my coauthors and I studied a different kind of constraints from

F. Kojima (✉)
Stanford University, Stanford, CA, USA
e-mail: fuhitokojima1979@gmail.com

© Springer Nature Switzerland AG 2019
J.-F. Laslier et al. (eds.), *The Future of Economic Design*, Studies in
Economic Design, https://doi.org/10.1007/978-3-030-18050-8_67

those in the diversity constraint literature. In these papers, constraints are imposed on subsets of institutions as joint restrictions as opposed to constraints at each individual institution. A leading example is medical match in Japan, in which the government imposes "regional caps," a n upper bound on the number of doctors who can be placed in each region of the country. Constraints on subsets of institutions are present in many other contexts such as Chinese graduate admission, college admission in Europe, UK medical match, among others. It turned out that allocation rules tried in practice or in consideration suffer from lack of efficiency, stability, and/or strategy-proofness. Papers referenced above prove the existence of mechanisms that achieve versions of these desirable properties and improve on those used in practice.

I think that a lot remains to be done in matching problems with constraints given that there are still only few existing studies despite the prevalence of various constraints in practice. For instance, in an ongoing project, Kamada and Kojima (2018b) study matching under constraints imposed on individual institutions (as in school choice with diversity and unlike Japanese medical match). Departing from existing research, we allow for fully general constraints, which we define simply as an arbitrary family of subsets of applicants, which are interpreted as all sets of applicants feasible at the given institution. In that setting, we first observe that a stable matching typically does not exist. Given this limitation, we focus on feasible, individually rational, and fair matchings. We characterize such matchings by fixed points of a certain function. Building on this result, we characterize the class of constraints on individual schools under which there exists a student-optimal fair matching (SOFM), the matching that is the most preferred by every student among those satisfying the three desirable properties. The characterized class of constraints, called "general upper bounds," require that any subset of a feasible set of applicants is also feasible. Thus, under general upper bound, there is a clear "most desirable" matching to be selected despite the generality of the problem (recall that a stable matching does not necessarily exist in this class of problem). Thanks to the generality of the class of constraints allowed here, this result can be applied to a wide variety of problems such as the standard model of school choice with diversity concerns mentioned above. In addition, the setting also includes more recent problems such as refugee resettlement (Delacrétaz et al. 2016) where refugee families may be heterogeneous in their sizes and demand for various resources, as well as school choice with anti-bullying goal (Kasuya 2016), where the allocation should not place a bully and his or he victim to the same school.

We are especially interested in the allocation of daycare/nursery school slots for small kids. In many countries, daycare services are highly subsidized with regulated prices (and, in many cases, the service is provided by government), and the local government is in charge of the assignment of daycare slots, making the problem a prime target of application of matching theory. The problem is particularly important in contemporary Japan because there is large demand for daycare services, but the supply is very limited. One of the ways in which some local governments are trying to cope with this problem is to "transfer" a capacity pre-committed for one age to another if the demand for the former is small and the demand for the latter is large. This is a matching problem with general upper bounds, in which a daycare center's feasibility is defined over vectors of kids of different ages rather than having capacity constraints for each age separately (Okumura 2018). This constraint cannot be written as a simple capacity constraint because the necessary staff-child ratio is higher for younger kids than for older kids and space needed for each kid also depends on age, but this is still an example of our general upper bound constraint. To apply our analysis to daycare matching, we obtained administrative data from Yamagata City and Bunkyo City, rural and urban cities in Japan. First, we compare SOFMs under the more flexible (real) daycare constraint and the rigid (artificial) constraint where resources are committed to different ages in advance. We find that the effect of allowing flexibility in constraints is substantial in our data from both municipalities. Second, we compare the SOFMs with the real allocations in those municipalities. We find that, relative to the mechanisms that are used in reality, our proposed mechanism may result in a mild improvement in efficiency while eliminating justified envy completely.

To summarize, I think that constraints in matching theory offer a rich set of interesting research topics. Existing studies have analyzed specific types of constraints such as diversity constraints in school choice and regional caps in medical match. One of the exciting directions being considered is to advance a general theory that aims to not only encompass these important but specific examples but also give a unified understanding in a greater generality. And the general theory could also shed light on new kinds of practical applications such as flexible daycare slot assignment. I think there are many open questions and room for study in this subject.

# References

Abdulkadiroğlu, A. (2005). College admissions with affirmative action. *International Journal of Game Theory, 33*(4), 535–549.

Abdulkadiroğlu, A., & Sönmez, T. (2003). School choice: A mechanism design approach. *American Economic Review, 93*, 729–747.

Ashlagi, I., Braverman, M., & Hassidim, A. (2014). Stability in large matching markets with complementarities. *Operations Research, 62*, 713–732.

Delacrétaz, D., Kominers, S. D., & Teytelboym, A. (2016). Refugee resettlement, working paper.

Ehlers, L., Hafalir, I. E., Yenmez, M. B., & Yildirim, M. A. (2014). School choice with controlled choice constraints: Hard bounds versus soft bounds. *Journal of Economic Theory, 153*, 648–683.

Fragiadakis, D., & Troyan, P. (2017). Improving matching under hard distributional constraints. *Theoretical Economics, 12*, 863–908.

Goto, M., Kojima, F., Kurata, R., Tamura, A., & Yokoo, M. (2017). Designing matching mechanisms under general distributional constraints. *American Economic Journal: Microeconomics, 9*, 226–262.

Hafalir, I. E., Yenmez, M. B., & Yildirim, M. A. (2013). Effective affirmative action in school choice. *Theoretical Economics, 8*(2), 325–363.

Kamada, Y., & Kojima, F. (2015). Efficient matching under distributional constraints: Theory and applications. *American Economic Review, 105*(1), 67–99.

Kamada, Y., & Kojima, F. (2017). Stability concepts in matching with distributional constraints. *Journal of Economic theory, 168*, 107–142.

Kamada, Y., & Kojima, F. (2018a). Stability and strategy-proofness for matching with constraints: A necessary and sufficient condition. *Theoretical Economics, 13*, 761–793.

Kamada, Y., & Kojima, F. (2018b). Fair matching under constraints: Theory and applications, working paper.

Kasuya, Y. (2016). Anti-bullying school choice mechanism design, working paper.

Klaus, B., & Klijn, F. (2005). Stable matchings and preferences of couples. *Journal of Economic Theory, 121*, 75–106.

Kojima, F. (2012). School choice: Impossibilities for affirmative action. *Games and Economic Behavior, 75*(2), 685–693.

Kojima, F., Pathak, P. A., & Roth, A. E. (2013). Matching with couples: Stability and incentives in large markets. *Quarterly Journal of Economics, 128*, 1585–1632.

Kojima, F., Tamura, A., & Yokoo, M. (2018). Designing matching mechanisms under constraints: An approach from discrete convex analysis. *Journal of Economic Theory, 176*, 803–833.

McDermid, E. J., & Manlove, D. (2009). Keeping partners together: Algorithmic results for the hospitals/resident problem with couples. *Journal of Combinatorial Optimization, 19*, 279–303.

Okumura, Y. (2018). School choice with general constraints: A market design approach for nursery school waiting lists problem in Japan," forthcoming. *Japanese Economic Review*.

Roth, A. E. (1984). The evolution of the labor market for medical interns and residents: A case study in game theory. *Journal of Political Economy, 92*, 991–1016.

# Beyond Strategyproofness

**Thayer Morrill**

**Abstract** Currently a mechanism is either viewed as strategyproof or manipulable. But surely manipulability is not a binary concept. Can we move beyond strategyproofness to identify manipulations that will be hard to identify or risky to implement? Since strategyproofness is a constraint, this would allow us to increase the mechanisms we consider without sacrificing their performance in real world implementations.

School assignment has been one of the great success stories of market design. Since the groundbreaking research of Abdulkadiroglu and Sonmez, a large number of school-choice programs have been redesigned using mechanism design principles (for example, New York City, Boston, New Orleans, Denver, Washington DC, Wake County NC, England, Amsterdam, and many others). We are lucky that two strategyproof mechanisms exist: Deferred Acceptance and Top Trading Cycles. We are unlucky in that a fair assignment does not need to be Pareto efficient. Our poor luck extends further in that there is no strategyproof mechanism that Pareto improves a fair assignment.

In some sense, this parallels auction theory. There is a strategyproof auction: VCG. However, for multiple units and complementary preferences, the VCG outcome may not be in the core. If auction theory insisted on strategyproofness, then its theory would consist of two papers: Vickrey (1961) and Myerson (1981). After all, we can unambiguously conclude that the VCG auction is the best strategyproof auction. But of course, auction theory is much larger than these two papers, and in fact, VCG is rarely used in real-world, multi-unit auctions.

However, the theory of school assignment has not really extended past the orthodoxy of strategyproofness. Most of the discussion on incentives in school assignment has compared the Boston mechanism to Deferred Acceptance. A number of studies have concluded that students do in fact manipulate the Boston mechanism and do so at the expense of sincere students. The conclusion that is often reached from this evidence is that a good assignment mechanism should be strategyproof. I do not

T. Morrill (✉)
North Carolina State University, Raleigh, USA
e-mail: thayermorrill@gmail.com

© Springer Nature Switzerland AG 2019
J.-F. Laslier et al. (eds.), *The Future of Economic Design*, Studies in
Economic Design, https://doi.org/10.1007/978-3-030-18050-8_68

agree with this. Rather, what I draw from this exercise is that a good assignment mechanism should be hard to manipulate.

In theory, we often treat manipulability as a binary property. A mechanism is either strategyproof or it is not. But in practice manipulability is far more nuanced than this would suggest. Does a manipulation require a great deal of information about other preferences? Does manipulating your preferences expose you to very bad outcomes? Does a manipulation require coordination?

For example, consider the male-proposing Deferred Acceptance algorithm. As is well known, it is not strategyproof for the women. The opportunity for manipulation occurs when a woman is assigned to different men under two stable assignments. She manipulates the algorithm by declaring that her less preferred of these two men is unacceptable. This requires that she know a great deal of information (the set of stable assignments). Moreover, it is quite risky. If she was wrong about the set of stable assignments, then she risks being unassigned.

Compare this to the manipulations that occur under the Boston mechanism. The Boston mechanism is easy to manipulate. It is clear to parents (or at least some parents) that by ranking their neighborhood school first they are guaranteed to get it, and if they don't rank their neighborhood school first, they risk losing it. This requires no information about other student preferences except knowing whether or not their neighborhood school is generally popular. Moreover, there is no risk of them being unassigned when employing this strategy.

I believe there is a need to create a theory that describes this nuance. Suppose we could find a tractable way of categorizing how hard or how risky a mechanism is to manipulate. Then we could fully examine the trade-offs we face. We would like to make fair assignments more efficient. It is impossible to do this with a strategyproof mechanism, but can it be done with a mechanism that is "hard" to manipulate? It is estimated that 5% of the students assigned in New York City could be Pareto improved. This is a heavy cost to pay. My conjecture is that this cost is not necessary. I believe we can find an algorithm that Pareto improves the deferred acceptance assignment, achieves close to the full level of efficiency, and is hard to manipulate (in the sense of a deviation requiring precise information and being risky for the student). Perhaps Kesten has already done so with his Efficiency Adjusted Deferred Acceptance. But I think it is up to us to develop a theory that allows us to articulate why one manipulable mechanism is better than another.

# References

Myerson, R. B. (1981). Optimal auction design. *Mathematics of Operations Research, 6*(1), 58–73.
Vickrey, W. (1961). Counterspeculation, auctions, and competitive sealed tenders. *Journal of Finance, 16*, 8–37.

# Part XI
# New Technologies

# Economic Design of Things

**Burak Can**

**Abstract** Economics is a social science, so is economic design as a field. This short article discusses, in particular, the future of economic design, and of economic theory in general. By suggesting some examples, I hope to convince the readers that the recent technological advances in science and technology will not only be disruptive to the social machinery that surrounds us but also to the future of economic design as a field. Economic design, however, has the potential to add value to the society by offering an axiomatic framework to the design of the future through a social sciences perspective.

## Introduction

Economics as a scientific discipline is a study within the domain of social sciences. Therefore economic theory is about building theoretical foundations for social phenomena that we experience around us. Economic design as a theoretical subdiscipline is about analyzing, designing, implementing, or improving markets, systems, institutions, -broadly speaking any socio-economic, legal, financial platform- under which agents interact and engage in transactions that are financial or otherwise.

The rules of engagement in economic systems are based on our assumptions about agents, their payoff functions, and social norms that define what is desirable in a system. We consider agents to be utility maximizers and hence the more the cargo, the merrier. We assume that agents will be in a pursuit of happiness, whatever that concept might mean. This implies that agents would (if they could) try to manipulate

The author is grateful to Deniz Iren, Orhan Erdem, Gabriel Carroll, Umberto Grandi, Conchita D'Ambrosio, and Erkan Yonder for their comments on earlier drafts of this manuscript. Financial support from the Fonds National de la Recherche Luxembourg and the Netherlands Organisation for Scientific Research (NWO) under project no. 451-13-017 (VENI) is also gratefully acknowledged.

B. Can (✉)
INSIDE, University of Luxembourg, Luxembourg City, Luxembourg
e-mail: burakcan1@gmail.com

and *game* the system.[1] We deal with cases involving uncertainty, risk, asymmetric information. We also worry about efficiency and fairness in the design of our systems, along with other desirable norms in our society. We define mathematical axioms that relate to or reflect those noble notions.

The design of economic systems is often about axiomatic quests for finding mechanisms with normatively appealing features proposed by the economic designer such as finding efficient, non-manipulable, and fair mechanisms for a given problem with certain assumptions on agents' (humans') behavior and level of information which is available to the agents. The future of economic design, however, is not necessarily in the design of these systems under which humans interact. It is in the design of systems for phenomena that are not necessarily within the scope of social sciences. For instance:

1. The design of mechanisms under which *internet of things* (IoT), *artificial intelligence* (AI), and *robots* interact on platforms featuring properties desirable to these things. The design of collective decision mechanisms that can handle *Big Data* produced by these machines and that has Big Data as its main concern, e.g., computational complexity, aggregation via delegation, aggregation on networks, clustering algorithms etc.
2. The design of digital transaction/interaction methods with advances in digital technologies in mind, e.g., *Blockchain*, *Ethereum* and other *distributed ledger technologies* (DLT) and cryptocurrencies. The design of *digital platform economies*, under which the assumptions on information structures are different than traditional systems, and agents are typically on a multi-sided network, e.g., Uber, AirBnB, App Store, Social Media.

These disruptive technologies will eventually lead to drastic changes in the society, and in the economy, hence also in the way w e design the latter. I n what follows, I discuss the possible effects of these innovations on the study of economic design with some examples.

# 1 Disruptive Technologies and Their Design

## 1.1 Untraditional Agents: IoT, AI, Big Data and Robots

IoT broadly refers to devices that connect to the internet or a network, deliver data, take actions and typically comprise of MCUs (micro controller units), sensors and actuators. These devices might also incorporate AI and can be deployed en masse and engage in collective decision making scenarios. Types of these devices range from home appliances such as refrigerators, light bulbs, and stereos to urban infrastructures

---

[1] There is an undeniably growing literature on behavioral economics where the rationality assumption is seriously challenged, and there are other proposed concepts such as bounded rationality, altruism, reciprocity, k-level reasoning etc., which are beyond the scope of this article.

such as automated streetlights, and recycling sensors, or even military applications e.g., UAVs, drones. As of 2016, there are approximately 3,4 billion people who are connected to the internet, while the number of IoT devices the same year is estimated to be 6,4 billion. The latter is estimated at least to triple (or quadriple) by 2020.[2] Some of the reasons for this massive boost in the deployment of IoT devices are the advances in the microchip production amazingly consistent with (Gordon) Moore's law[3] that lead to decrease in costs. Hence the design of such immense networks of IoT devices, possibly communicating with one another, and making collective decisions on what type of actions to take is an essential scientific endeavor.

One implication of the increase in the availability of cheap IoT devices is the massive increase in the "volume", the "variety" and the "velocity" of available data. The term *Big Data* is used to refer to the type of *data sets that are so large or complex that traditional data processing application software is inadequate to deal with.*[4] Imagine an agricultural application where tens of thousands of sensors are deployed on a field to measure certain characteristics of the soil. One typical problem in these applications occur when sensors go out of calibration. It is therefore imperative for the user to decide which sensors are facing such a problem and clean the raw data by removing the input from these particular sensors.[5] Given the advances in AI, Machine Learning and the connectivity capabilities, it is possible to let the devices undertake this task. The sensors can find out which "fellow" sensors in the network are not working well, and then decide "collectively" to shut those down in order to produce a somewhat cleaner data. The devices can also aggregate their own data based on location, day, season etc., and produce already processed data so as to decrease the workload of the researcher doing his Ph.D. in agricultural sciences.

In parallel to the development in IoT devices and systems producing Big Data, there is an increase in the automation in production, and a shift from human employment to robotics in many traditional occupations. Acemoglu and Restrepo (2017) show negative effects of robots on employment and wages depending on the exposure to robotics in different zones. Frey and Osborne (2017) study the probability of job loss due to digitalization for 702 detailed occupations. Some jobs are already being taken over by robots and AI, or digital platforms. The issue is actually so critical that AliBaba founder and chairman Jack Ma warns of "decades of pain" referring to the job disruptions that would be created by automation and digitalization and calls for educational reform (Solon 2017). Perhaps, it is tempting to think that most technological changes favor the skilled labor, e.g., those with computer skills etc. Nevertheless, Acemoglu (2002) discusses that the technological advances in the 19th

---

[2]See http://www.internetlivestats.com/internet-users and http://www.gartner.com/newsroom/id/3165317.

[3](Gordon) Moore's law is the observation that the number of transistors in a dense integrated circuit doubles approximately every two years. The original paper in 1965 can be also be found at Moore (1998).

[4]https://en.wikipedia.org/wiki/Big_data.

[5]This particular problem is analogous to the Condorcet Jury Theorem, i.e., there is a true value of the characteristic to be measured, and the sensors estimate this value with a varying degree of precision (competence).

century favored the unskilled labor against the skilled artisans, due the interchange-
able parts technology. Today, we face a situation where the direction of the demand
for labor, in terms of skills, is not that clear.[6]

The analysis of all the aforementioned technological advancements from socio-
economic, legal, and ethical perspectives falls within the scope of economic design.
The design of coordination and communication between IoT devices, the design of
hierarchical frameworks for interaction among robots,[7] and the design of interaction
methods between humans and collaborative robots (a.k.a., cobots, or co-robots) are all
possible research questions for our field. Economic design can address these problems
not only from a human-agent perspective but also from a collective AI perspective, or
by introducing new delegation methods and layered aggregation procedures so as to
efficiently deal with the computational complexity issues that may come along with
the Big Data. For the latter, tools from clustering theory can be enhanced by axioms
from economic design literature on a scenario basis.[8] With its strong theoretical
foundation, economic design has a lot to offer to the society of future via revisiting
its postulates, axioms, and theorems to analyze, design, implement, and improve
systems for machines, and thereby for humans, perhaps starting from Asimov's
famous *three laws of robotics*.[9]

## 1.2 Untraditional Markets: DLT, and (Digital) Platform Economies

Markets are typically physical places where exchanges occur and trust is the most
crucial aspect for a market and its participants to thrive. Traditionally banks serve a
centralized medium for trust and binds the lender and the borrower with contracts.
Distributed Ledger Technologies (DLT) changes the very structure of this central-
ization. As the name suggests, instead of a central ledger, such as a bank, in DLT
systems, each transaction is written in decentralized ledgers that cannot be altered
without predefined consensus mechanisms. Among the DLTs, a.k.a. Blockchain, Bit-
Coin and Ethereum are the most well-known examples. The latter especially has a lot
of applications with practical solutions for democratic decision making procedures,
participatory and possibly direct democracy platforms. There are also existing financial applications based on the Ether, the cryptocurrency based on Ethereum, where
people build their own investment funds, and collectively decide (without delegating

---

[6]Consider for instance the digital platforms such as Uber or AirBnB which allows unskilled people
to offer commuting or hospitality services.

[7]For some interesting applications on coordination among drones while making installations such
as bridges, see Augugliaro et al. (2013), Augugliaro et al. (2014).

[8]See Kleinberg (2003) and Ben-David and Ackerman (2009) for a brief introduction on axiomatization of clustering algorithms.

[9]Isaac Asimov, a scholar in chemistry by profession, was one of the first to propose robotics ethics
in his famous science-fiction story series *I, Robot* (1950). For an interesting read on the new field
"machine ethics" see Anderson and Anderson (2011).

that decision power to the bankers) which start-ups they would like to support as a financial investment or digital Blockchain markets, where people can buy and sell, without any intermediaries hence avoiding unnecessary costs.

A relatively recent concept, (digital) platforms, is another important subject of interest for economic design. Platforms are essentially markets where different types of agents meet and conduct transactions. However platforms themselves are also products. To illustrate platforms, consider two simple -and rather nostalgic- examples of the concept that were actually not digital: VHS and Betamax video formats. Under these formats, movie producers and consumers were matched. The choice of platform for all agents in this multi-sided network, e.g., movie producers, consumers, movie rental shops etc., has always been an interesting game. Agents, with a sunk cost perhaps, could leave the network and go for the other platform (just like it did happen at the end of the Videotape format war and resulted in the end of Betamax). Today we have the so-called digital platforms. We have Uber, a digital platform where drivers meet passengers. We use Netflix, or Hulu (or other platforms) to watch our favorite shows and meet content producers. Even larger platforms where consumers and software developers meet are iOS AppStore and Google Play. Both of these digital platforms are massive and thick markets with multiple safe payment systems. The simplicity of smoothly buying-downloading-installing apps via these digital platforms is incomparable to that of buying software on a real store and installing them via CDs. Finally, without any real physical congestion like a store on a weekend would have, there is effectively little issue of congestion.

DLT systems, digital platforms, and cryptocurrencies potentially exhibit all the features that (Roth 2015) thinks are essential for a design to work successfully. Eventually economic design will be more and more about platform design and the design of other digital technologies, and markets with digital features. Therefore, the future of economic design lies in the ambition to be the "scientific platform" under which the designers and the implementers of digital platforms are matched.

## 2   Conclusion

As a final concluding thought, let us take a moment and ask ourselves: "why do we not use direct democracy?". One of the most accepted arguments against a Greek style direct democracy is that it is practically impossible to ask every individual their opinion on all issues in a society. However, given the possibilities with Ethereum and other digital platforms, and the advances in Machine Learning and AI, it is only a matter of time before each individual is able to choose their own AI representative and delegate collective decisions on every issue in their society to their personal AI representative, should he or she wishes so. Perhaps in a few decades, parliamentary systems may completely disappear and be taken over by these machine delegation systems, -a truly digital democratic platform where representatives are never fraud and strictly represent individuals' perspectives. Direct democracies under AI-representation might be much more desirable also because it can, in principle, protect democracies from populism that often threaten universal values, human rights, minorities. It is possible to offer a rich and diverse set of AI softwares for representing all colors of the political spectrum, on the condition that it does not violate universally agreed principles and does not conflict with, or challenge other pre-agreed values, such as constitutions (for machines). Such changes should not be taken as science fiction. It is already happening in hedge-fund investments where AI, Big Data and Machine Learning meet, and investors are clearly better of delegating some financial decisions to machines[10] rather than humans.

To sum up, the frontier of research on smart and autonomous systems has been mainly pushed by computer scientists. However, there is an urgent need in addressing the recent technological advances from a social sciences perspective, by using economic design and collaborating with psychologists, philosophers and of course computer scientists. Implementation of most technologies requires an ethical foundation. In case these technologies are endowed with AI and learning capacity, these ethical, and social considerations must be encoded in the AI[11] to create machines *in our own image*. Nationwide smart electricity grid implementations, for instance, require prioritization of services and certain locations, e.g., hospitals, water systems, and public buildings. What are social, ethical, political consequences of different smart implementations? How can we build smart but also social devices, networks, and smart services? How can we ensure innovation goes hand in hand with social inclusion? How can we use (digital) platforms to that end? These priorities should not depend only on concepts such as cost-efficiency but also on other values, such as fairness, neutrality, consistency etc., -the very axioms that economic design deals with.

The world is on the verge of yet another industrial revolution characterized by digitalization, automation and other technologies. Times are a changing, and the future has many possibilities for the mankind, some of which may be dreadful. Economic design, however, has the potential to incorporate human values in the design of every prospect that may be initiated by this *creative destruction*. A well designed prospect is always better than one unexpected and unforeseen, hence, the sooner the better that economic design starts to invest in the "economic design of things".

---

[10] AI and Machine Learning based hedge funds have been outperforming quants and other traditional hedge funds in a persistent way since 2010 (Eurekahedge 2017).

[11] For a concrete application of this idea, see MIT Media Lab's "Moral Machine", a web application on which answers to ethical dilemmas are crowdsourced via humans vis-à-vis Amazon's Mechanical Turk. This is one of the first steps to incorporate human values on controversial cases into machines (see Bonnefon et al. (2016).

# References

Acemoglu, D. (2002). Technical change, inequality, and the labor market. *Journal of Economic Literature, 40*(1), 7–72.

Acemoglu, D., & Restrepo, P. (2017). Robots and jobs: Evidence from US labor markets.

Anderson, M., & Anderson, S. L. (2011). *Machine ethics*. Cambridge: Cambridge University Press.

Asimov, I. (1950). *I, Robot*. New York: Gnome Press.

Augugliaro, F., Lupashin, S., Hamer, M., Male, C., Hehn, M., Mueller, M. W., et al. (2014). The flight assembled architecture installation: Cooperative construction with flying machines. *IEEE Control Systems, 34*(4), 46–64.

Augugliaro, F., Mirjan, A., Gramazio, F., Kohler, M., D'Andrea, R. (2013). Building tensile structures with flying machines. In *2013 IEEE/RSJ International Conference on Intelligent Robots and Systems (IROS)* (pp. 3487–3492). IEEE.

Ben-David, S., & Ackerman (2009). Measures of clustering quality: A working set of axioms for clustering. In *Advances in neural information processing systems* (pp. 121–128).

Bonnefon, J.-F., Shariff, A., & Rahwan, I. (2016). The social dilemma of autonomous vehicles. *Science, 352*(6293), 1573–1576.

Eurekahedge. (2017). Artificial intelligence: The new frontier for hedge funds. Technical report, Eurekahedge. http://www.eurekahedge.com/Research/News/1614/Artificial-Intelligence-AI-Hedge-Fund-Index-Strategy-Profile.

Frey, C. B., & Osborne, M. A. (2017). The future of employment: how susceptible are jobs to computerisation? *Technological Forecasting and Social Change, 114*, 254–280.

Kleinberg, J. M. (2003). An impossibility theorem for clustering. *Advances in neural information processing systems* (pp. 463–470).

Moore, G. E. (1998). Cramming more components onto integrated circuits. *Proceedings of the IEEE, 86*(1), 82–85.

Roth, A. E. (2015). *Who gets what—and why: The new economics of matchmaking and market design*. Boston: Houghton Mifflin Harcourt.

Solon, O. (2017). Alibaba founder Jack Ma: AI will cause people 'more pain than happiness'. https://www.theguardian.com/technology/2017/apr/24/alibaba-jack-ma-artificial-intelligence-more-pain-than-happiness.

# Machine Learning for Optimal Economic Design

Paul Dütting, Zhe Feng, Noah Golowich, Harikrishna Narasimhan,
David C. Parkes and Sai Srivatsa Ravindranath

**Abstract** This position paper anticipates ways in which the disruptive developments in machine learning over the past few years could be leveraged for a new generation of computational methods that automate the process of designing optimal economic mechanisms.

## 1 Introduction

Mechanism design is the problem of designing incentives to achieve an outcome that satisfies desired objectives in the presence of self-interested participants. Because the participants are assumed to act rationally, and play an equilibrium, it can also

We would like to thank Yang Cai, Vincent Conitzer, Constantinos Daskalakis, Scott Kominers, Alexander Rush, and participants at seminars at the Simons Institute for the Theory of Computing, Dagstuhl, London School of Economics, IJCAI'18, AAMAS'18, The Technion, the EC'18 WADE workshop, the German Economic Association annual meeting, Google, MIT/Harvard Theory Seminar, WWW'19 workshop, and HBS for their helpful feedback.

P. Dütting
Department of Mathematics, London School of Economics, Houghton Street,
London WC2A 2AE, UK
e-mail: p.d.duetting@lse.ac.uk

Z. Feng · N. Golowich · H. Narasimhan · D. C. Parkes (✉) · S. S. Ravindranath
Paulson School of Engineering and Applied Sciences, Harvard University,
33 Oxford Street, Cambridge, MA 02138, USA
e-mail: parkes@eecs.harvard.edu

Z. Feng
e-mail: zhe_feng@g.harvard.edu

N. Golowich
e-mail: ngolowich@college.harvard.edu

H. Narasimhan
e-mail: hnarasimhan@g.harvard.edu

S. S. Ravindranath
e-mail: saisr@g.harvard.edu

© Springer Nature Switzerland AG 2019
J.-F. Laslier et al. (eds.), *The Future of Economic Design*, Studies in
Economic Design, https://doi.org/10.1007/978-3-030-18050-8_70

be thought about as the problem of inverse game theory. A mechanism designer creates the rules of a game, by which an outcome will be selected based on messages sent by participants. Mechanism design has developed into a beautiful theory that has influenced thinking across a range of problems, including auctions and voting procedures, but despite more than 40 years of intense research, several fundamental questions remain open.

Revenue-optimal auction design is the prime example, both for having elegant theoretical results and also for seemingly simple and important-to-practice cases that remain unsolved. The central result is the characterization of revenue-optimal single-item auctions as *virtual value maximizers* (Myerson 1981). We know, for example, that second price auctions with a suitably chosen reserve price are optimal when selling to bidders with i.i.d. values, and how to prioritize one bidder over another in settings with bidder asymmetry. But Myerson's theory is as beautiful as it is rare. Indeed, the design of optimal auctions for multiple items is much more difficult, and has defied a thorough theoretical understanding.

The contours of the available analytical results bear witness to the severe analytical challenges in going beyond single-item auctions. Even the design of the optimal auction for selling two items to just a single buyer is not fully understood.[1] For a single additive buyer with values on items i.i.d. $U(0, 1)$, Manelli and Vincent (2006) handle three items, and Giannakopoulos and Koutsoupias (2014) up to six items. Yao (2017) provides the optimal design for any number of additive bidders and two items, buy only as long as item values can take on one of two possible values. Decades after Myerson's result, we do not have a precise description of optimal auctions with two or more bidders and more than two items.

A promising alternative is to use computers to solve problems of optimal economic design. The framework of *automated mechanism design* (Conitzer and Sandholm 2002, 2003) suggests to use algorithms for the design of optimal mechanisms. Early approaches required an explicit representation of all possible type profiles, which is exponential in the number of agents and does not scale (see also Albert et al. 2017). Others have proposed more restricted approaches, that search through a parametric family of mechanisms (Guo and Conitzer 2009, 2010; Sandholm and Likhodedov 2015; Narasimhan et al. 2016).

In recent years, efficient algorithms have been developed for the design of optimal, *Bayesian incentive compatible (BIC)* auctions in multi-bidder, multi-item settings (Cai and Daskalakis 2015; Alaei et al. 2012, 2013; Cai et al. 2012a, b, 2013a; Bhalgat et al. 2013; Cai and Huang 2013; Daskalakis et al. 2017). But while there is a characterization of optimal mechanisms as virtual-value maximizers (Cai et al.

---

[1]Results are known for additive i.i.d. $U(0, 1)$ values on items (Manelli and Vincent 2006), additive, independent and asymmetric distributions on item values (Daskalakis et al. 2017; Giannakopoulos and Koutsoupias 2015; Thirumulanathan et al. 2016), additive, i.i.d. exponentially distributed item values (Daskalakis et al. 2017) and extended to multiple items (Giannakopoulos 2015), additive, i.i.d. Pareto distributions on item values (Hart and Nisan 2012), unit-demand valuations with item values i.i.d. $U(c, c + 1)$, $c > 0$ (Pavlov 2011), and unit-demand, independent, uniform and asymmetric distributions on item values (Thirumulanathan et al. 2017).

2012a, 2013b), relatively little is known about the structure of optimal mechanisms; see Daskalakis (2015) for an overview.

Moreover, these algorithms leverage a reduced-form representation that makes them unsuitable for the design of *dominant-strategy incentive compatible* (DSIC) mechanisms, and similar progress has not been made for this setting. DSIC is of special interest because of the robustness it provides, relative to BIC. The recent literature has focused instead on understanding when simple mechanisms can approximate the performance of optimal designs.[2]

Where do we go from here? Thanks to the disruptive developments in machine learning, we believe that there is a powerful opportunity to use its tools for the design of optimal economic mechanisms. The essential idea is to repurpose the training problem from machine learning for the purpose of optimal design. In what follows, we will highlight some recent results that we have in support of this agenda. The question we ask is:

*Can machine learning be used to design optimal economic mechanisms, including optimal DSIC mechanisms, and without the need to leverage characterization results?*

The illustrative examples will be drawn from optimal auction design, including optimal design with private budget constraints, as well as a problem in social choice— the multi-facility location problem. We believe the framework is considerably more general, and will extend to address problems in matching and non-linear pricing, for example.

## 2   Adopting the Lens of Machine Learning

To understand the opportunity, we start with optimization-based formulations for each of the problems of mechanism design and machine learning.

A typical problem in mechanism design is to find a function from inputs (a type profile) to outputs (say an allocation and payments) that maximizes the expected value of an objective, defined for a distribution on inputs. Global constraints are also imposed, for example *incentive compatibility* (IC).[3] Illustrating this for the design of an *allocation rule g* and *payment rule p* (mapping reported types to an allocation and payments, respectively) of an auction, we would solve:

---

[2]Working in increasingly general settings, relevant results on DSIC auction design include Chawla et al. (2007, 2010), Alaei (2014), Kleinberg and Weinberg (2012), Hart and Nisan (2012), Li and Yao (2013), Babaioff et al. (2014), Yao (2015), Rubinstein and Weinberg (2015), Cai et al. (2016), Cai and Zhao (2017), Dütting et al. (2017). These mechanisms are simple, and reveal the structural ingredients that are important for the design of mechanisms with good revenue properties.

[3]IC means that no agent can benefit, in equilibrium, by misreporting its type, and can hold in a dominant-strategy equilibrium (DSIC) or a Bayes-Nash equilibrium (BIC). We will generally be interested in DSIC.

$$\max_{g,p} \mathbf{E}_{v \sim F_V} \mathcal{O}(v; g, p) \tag{1}$$

$$\text{s.t.} \quad (g, p) \in IC.$$

This maximizes the expected value of objective $\mathcal{O}(v; g, p)$, where $v=(v_1, \ldots, v_n)$ denotes the type profile for $n$ bidders, and $F_V$ the distribution from which type profiles are sampled. For revenue optimality, the objective would be $\mathcal{O}(v; g, p) = \sum_{i=1}^{n} p_i(v)$. Here, $IC$ denotes the set of IC rules. Following Myerson (1981), this kind of problem can be solved in simple cases through a characterization of allocation rules for which there exists a payment rule that provides IC, allowing the objective to be expressed in terms of the allocation rule alone, and then proceeding analytically. But this approach is very challenging to extend to general, multi-item problems.

A typical problem in machine learning is to find a function from inputs (a vector of features) to outputs (say an image label) that minimizes the expected value of an objective. A typical objective is to minimize the expected loss on input-output pairs sampled from some distribution, where the loss might be defined to be 0 if the predicted output is correct and 1 if it is incorrect. For parametric models, with function $f^w$ defined through parameters $w \in \mathbb{R}^d$ (for some $d \geq 1$), we would solve

$$\min_{w} \mathbf{E}_{(x,y) \sim F_{XY}} \mathcal{L}(x, y; f^w). \tag{2}$$

The objective is to minimize expected loss, for *loss function* $\mathcal{L}(x, y; f^w)$, where $(x, y)$ is an input-output pair sampled i.i.d. from some distribution $F_{XY}$. The input-output pair could be *feature vector* $x \in \mathbb{R}^k$ for $k \geq 1$, and *target value* $y \in \mathbb{R}$, respectively. Here, $f^w : \mathbb{R}^k \mapsto \mathbb{R}$ is a parameterized function (the target can also be categorical, in which case $f^w$ would map to a finite set). A typical approach to solve (2) is to use training data sampled from $F_{XY}$, together with an optimization method such as stochastic gradient descent to minimize the loss on the training data (perhaps along with regularization, to prefer simple solutions over complex solutions that might over-fit to the training data).

Comparing formulations (1) and (2), and considering the particular setting of revenue-optimal auction design, this suggests the following representation of a problem of optimal economic design as one of machine learning:

| | |
|---|---|
| Feature vector | $x \longrightarrow (v_1, \ldots, v_n)$ |
| Target value | not needed |
| Hypothesis | $f^w \longrightarrow (g^w, p^w)$ (parameterized allocation rule and payment rule) |
| Loss function | $\mathcal{L}(v; g^w, p^w) = -\sum_{i=1}^{n} p_i^w(v)$ |
| Constraints (new) | IC |

The loss function becomes the negated revenue, and thus minimizing expected loss is equivalent to maximizing expected revenue. There is no need for labeled training data: rather, the required training data is samples of type profiles, and the loss function is defined to directly capture the economic objective (e.g., negated revenue). For this reason, there is no object that corresponds to the target value.

The technical challenge, relative to standard training problems in machine learning, is to formulate the IC constraint. In some settings, IC can be directly achieved by constraining the set of functions (the hypothesis class). In other settings, we have found it useful to work with quantities that capture the degree of violation of the constraint. Fixing the bids of others, the *ex post regret* to a bidder is the maximum increase in the bidder's utility, considering all possible non-truthful bids. The *expected ex post regret* for bidder $i$, given mechanism parameters $w$, is defined as

$$rgt_i(w) = \mathbf{E}_{v \sim F_V}\left[\max_{v_i' \in V_i} u_i(v_i', v_{-i}; v_i, g^w, p^w) - u_i(v_i, v_{-i}; v_i, g^w, p^w)\right], \quad (3)$$

where $V_i$ is the valuation domain for bidder $i$, and $u_i(v_i', v_{-i}; v_i, g^w, p^w)$ is the utility (value minus price) to bidder $i$ with valuation $v_i$ when reporting $v_i'$, when others report $v_{-i} = (v_1, \ldots, v_{i-1}, v_{i+1}, \ldots, v_n)$, and with allocation and payment rule $g^w$ and $p^w$, respectively.

For a suitably expressive, parameterized set of functions $g^w$ and $p^w$, the problem of optimal auction design can be formulated as:

$$\min_w \mathbf{E}_{v \sim F_V} \mathcal{L}(v; g^w, p^w) \quad (4)$$

$$\text{s.t. } rgt_i(w) = 0, \quad \forall i \in N.$$

This allows for ex post regret only on measure zero events. We will additionally require *individual rationality*, a property that every agent has a weak incentive to participate in a mechanism. This can be ensured by restricting our search space to a class of parametrized mechanisms $(g^w, p^w)$ that charge no agent more than its expected utility for an allocation.

Let us suppose this can be made to work— that machine learning can be used in this way for optimal economic design. Before continuing, we will discuss some objections that could be raised about this research agenda:

*(1) "As theorists, we care about understanding the structure of optimal designs, we're not interested in black-box solutions."* In fact, we expect that a machine learning framework can provide a useful complement to theory, used for example to support or refute conjectures on the structure of optimal designs, or to identify parts of the theory landscape where current designs are far from optimal. Asking that learned designs are interpretable is also an interesting research agenda in its own right, and one that should find synergy with a growing attention to interpretability in machine learning (Doshi-Velez et al. 2015; Wang and Rudin 2015; Caruana et al. 2015; Ribeiro et al. 2016; Smilkov et al. 2016; Raghu et al. 2016; Andrew Slavin Ross 2017).

*(2) "Simplicity is important. Participants need to understand mechanisms."* While this is undoubtedly important in some settings, we believe that participants in many kinds of economic mechanisms will be increasingly automated (consider, for example the use of automated bidding for advertising and other problems in marketing, and automated trading in finance). Mechanisms populated by automated agents do not need to be simple in the same way as those intended for use by people. Rather, it

seems to use that robust game-theoretic properties such as DSIC are more important than descriptive simplicity, and especially if a mechanism is accompanied by a proof of its economic properties.[4]

(3) *"What if incentive compatibility is only approximately achieved? What good is this from an equilibrium perspective?"* We have some sympathy for this concern, in that when the expected, *ex post* regret of a learned mechanism is small but positive, some types may still have a large incentive to deviate. But this is only a first step. Going forward, we can think about other notions of approximate DSIC.[5] Moreover, this concern can be tempered by also imposing additional structural properties that are necessary for IC, thus tightening the approximation.[6]

(4) *"What if there are computation-theoretic or learning-theoretic barriers to optimal design?"* Any such barrier is intrinsic, and holds whether the design problem is left to human ingenuity or formulated in a way that is amenable to solution by an algorithm. Barriers, where they exist, will require the design of second-best mechanisms, that are optimal given not only incentive constraints but also these computational or learning-theoretic constraints.[7] As such, we see this not as an objection to using machine learning for optimal economic design, but as a broader objection to the agenda of optimal economic design.

(5) *"What if it is the rules of the optimal mechanism entail solving an intractable computational problem?"* We think this presents the most serious complaint, in that we already know of settings such as those of combinatorial auctions where the allocation rule requires solving an **NP**-hard optimization problem (Rothkopf et al. 1998). Still, because we may be interested in solving problems of a fixed size (in terms of the number of items and bidders), these kinds of complexity barriers do not immediately bite. Moreover, many complexity barriers are worst-case, and there is an increasing attention to using neural networks to solve problems of combinatorial optimization for distributions on inputs (Niepert et al. 2016; Vinyals et al. 2015; Orhan and Ma 2017), and progress there will also benefit the use of machine learning for automated economic design.

(6) *"What if the learned design is brittle, with its incentives or optimality properties not robust to a small change in the type distribution?"* On one hand, DSIC designs are intrinsically more robust than BIC designs in that incentive compatibility does not depend on the distribution. On the other hand, empirical observations

---

[4]See Parkes and Wellman (2015) for a discussion on the role of AI in the mediation of economic transactions.

[5]See Carroll (2013), Mennle and Seuken (2014), Lubin and Parkes (2012), Mennle and Seuken (2017) for some discussion of approximate notions of incentive compatibility.

[6]For example, we could also penalize failure of weak-monotonicity (Bikhchandani et al. 2006), or insist that the implied pricing-function is agent independent (with prices to an agent that are do not depend on its report, conditioned on an allocation).

[7]Daskalakis et al. (2014) give a complexity result for optimal mechanism design. There is also a recent literature on the sample complexity of auctions and mechanisms, including revenue-optimal auctions (Elkind 2007; Cole and Roughgarden 2014; Dughmi et al. 2014; Morgenstern and Roughgarden 2015, 2016; Huang et al. 2015; Devanur et al. 2016; Narasimhan and Parkes 2016; Gonczarowski and Nisan 2017; Cai and Daskalakis 2017).

about the use of highly non-linear models in other settings, such as those of *deep learning* (Goodfellow et al. 2016), suggest that robustness to small perturbations in the inputs can indeed be a concern (Szegedy et al. 2014; Fawzi et al. 2018; Moosavi-Dezfooli et al. 2016). The robustness of learned models is gaining attention within machine learning (Chen et al. 2017; Shalev-Shwartz and Wexler 2016; Goodfellow et al. 2015; Abadi et al. 2016), and progress there will also bring benefits here. At the same time, it will be important to conduct thorough studies of learned mechanisms to validate their robustness.

# 3   Deep Learning for Optimal Auction Design

We have initiated the study of multi-layer, feed-forward neural networks for the design of optimal auctions (Dütting et al. 2019).[8] These networks provide differentiable, non-linear function approximations to auction rules, and the training problem—the problem of optimal design—is solved through stochastic gradient descent.[9]

We focus here on describing a "fully agnostic" approach, which proceeds without the use of characterization results and, because of this, holds the most promise in discovering new economic designs.[10] The input layer of the REGRETNET architecture represents bids, and the network has two logically distinct components: the allocation network and the payment network (see Fig. 1). The networks consist of multiple "hidden layers" (denoted $h^{(r)}$ and $c^{(t)}$ in the figure) and an output layer. Each unit in a hidden layer and each unit in an output layer may be a non-linear activation function, applied to a weighted sum of outputs from the previous layer. These weights form the parameters o f he network.

The allocation rule $g$ is modeled with $R$ fully-connected hidden layers (we have used $R = 2$ and 100 units in each layer in our experiments), each with *tanh activations*, and a fully-connected output layer. For a given bid profile $b$, illustrated here as providing a number for each bidder for each of $m$ items, the network outputs a vector of allocation probabilities $z_{1j}(b), \ldots, z_{nj}(b)$, for each item $j \in [m]$, through a

---

[8]The use of machine learning for mechanism design was earlier pioneered by Dütting et al. (2015), who use support vector machines to design payment rules for a given allocation rule (which can be designed to be scalable). But their framework can fail to even closely approximate incentive compatibility then the rule is not implementable, and does not support design objectives that are stated on payments. Earlier, Procaccia et al. (2009) studied the learnability of specific classes of voting rules, but without considering incentives; see also Xia (2013), who suggests a learning framework that incorporates specific axiomatic properties.

[9]Deep learning, which refers typically to the use of multi-layer neural networks, has gained a great deal of attention in recent years. This is because of the existence of large data sets, the development of tool chains that make experimentation easy, optimized hardware to speed-up training (GPUs), as well as massive investment from the private sector. Whether a network is considered 'deep' or not is a matter of taste.

[10]We have also explored network architectures that leverage characterization results; Myerson (1981) and Rochet (1987) for optimal auction design, and Moulin (1980) for facility location problems.

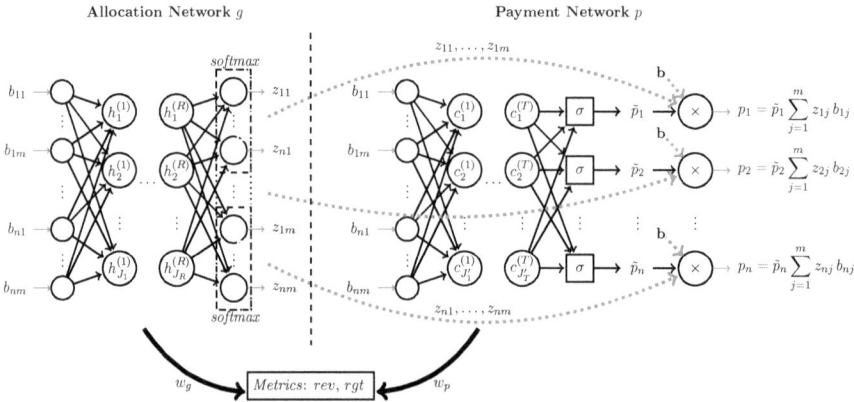

**Fig. 1** REGRETNET: The allocation network $g$ and payment network $p$ for a setting with multiple bidders $(1, \ldots, n)$ and multiple items $(1, \ldots, m)$ (Dütting et al. 2019). The *rev* and *rgt* are defined as a function of the parameters of the allocation and payment networks $w = (w_g, w_p)$

*softmax activation function*, with $\sum_{i=1}^{n} z_{ij}(b) \leq 1$ for each item $j \in [m]$.[11] Bundling of items is possible because the value on output units corresponding to allocating each of two different items to the same bidder can be correlated.

The payment rule is modeled using a feed-forward neural network with $T$ fully-connected hidden layers (we use $T = 2$ and 100 units in each layer in our experiments), each with tanh activations, and a fully-connected output layer. The output layer defines the payment for each bidder $i$ given a type profile. To ensure that the auction satisfies individual rationality (IR), i.e. does not charge a bidder more than its expected value for the allocation, the network first computes a fractional payment $\tilde{p}_i \in [0, 1]$ for each bidder $i$ using a sigmoid unit, and outputs a payment $p_i = \tilde{p}_i \sum_{j=1}^{m} z_{ij} b_{ij}$, where $z_{ij}$'s are the allocation probabilities output by the allocation network.[12]

Altogether, we can adopt $w \in \mathbb{R}^d$ to denote the vector of parameters, including parameters in both the payment and allocation parts of the network.

In practice, the loss and regret involved in formulating (4) are estimated from a sample of value profiles, $S = \{v^{(1)}, \ldots, v^{(L)}\}$, drawn i.i.d. from $F_V$. In place of expected loss, we adopt the *empirical loss*, defined as

---

[11] The sigmoid activation function is $\sigma(z) = 1/(1 + e^{-z})$. The softmax activation function for item $j$ is $softmax_j(s_{1j}, \ldots, s_{nj}, s_{n+1,j}) = e^{s_{ij}} / \sum_{k=1}^{n+1} e^{s_{kj}}$, where $s_{n+1,j}$ is a dummy input that corresponds to the item not being allocated to any bidder. In another variation, we handle unit-demand valuations of bidders by using an additional set of softmax activation functions, one per agent, and taking the minimum of these item-wise and agent-wise softmax components in defining the output layer.

[12] The output $p_i$ that corresponds to bidder $i$ is the amount the bidder should pay in expectation, for a particular bid profile. This can be converted into an equivalent lottery on payments, such that a bidder's payment is no greater than her value for any realized allocation (the property of *ex post* IR).

$$\widehat{\mathcal{L}}(g^w, p^w) = -\frac{1}{L} \sum_{\ell=1}^{L} \sum_{i=1}^{n} p_i^w(v^{(\ell)}). \tag{5}$$

To estimate the regret, we use additional samples of valuation profiles $S_\ell$ drawn i.i.d. from $F_V$ for each profile $v^{(\ell)}$ in $S$, and compute the maximum utility gain over these alternate profiles. The regret penalty is estimated as:

$$\widehat{rgt}_i(w) = \frac{1}{L} \sum_{\ell=1}^{L} \max_{v' \in S_\ell} (u_i(v_i', v_{-i}^{(\ell)}; v_i^{(\ell)}, g^w, p^w) - u_i(v^{(\ell)}; v_i^{(\ell)}, g^w, p^w)). \tag{6}$$

For large samples $S$ and $S_\ell$ (for each $\ell$), a mechanism with very low empirical regret, will, with high probability, have very low regret.[13]

The training problem becomes:

$$\min_w \widehat{\mathcal{L}}(g^w, p^w)$$
$$\text{s.t. } \widehat{rgt}_i(w) = 0, \quad \forall i \in N. \tag{7}$$

We can optimize (7) via the method of *augmented Lagrangian optimization*. This uses a sequence of unconstrained optimization problems, where the regret constraints are enforced through a weighted term in the objective. The solver works with the Lagrangian function, augmented with a quadratic penalty term for violating the constraints:

$$\mathcal{C}_\rho(w; \lambda_{rgt}) = \widehat{\mathcal{L}}(g^w, p^w) + \sum_{i \in N} \lambda_{rgt,i} \widehat{rgt}_i(w) + \frac{\rho}{2} \left( \sum_{i \in N} \widehat{rgt}_i(w)^2 \right), \tag{8}$$

where $\lambda_{rgt} \in \mathbb{R}^n$ is a vector of Lagrange multipliers, and $\rho > 0$ is a fixed parameter that controls the weight on the quadratic penalty. The solver operates across multiple iterations, and performs the following updates in each iteration $t$:

$$w^{t+1} \in \operatorname{argmin}_w \mathcal{C}_\rho(w; \lambda_{rgt}^t) \tag{9}$$
$$\lambda_{rgt,i}^{t+1} = \lambda_{rgt,i}^t + \rho \, \widehat{rgt}_i(w^{t+1}), \quad \forall i \in N, \tag{10}$$

where the inner optimization in (9) is approximated through multiple iterations of stochastic subgradient descent; in particular, the gradient is pushed through the loss function as well as the empirical measure of regret. The Lagrange multipliers are initialized to zero.[14]

---

[13] In more recent work (Dütting et al. 2019) we take an adversarial-style approach, using a gradient-based approach for estimating regret for a given profile. The gradient-based approach requires that the valuation space is continuous and the utility function is differentiable, but is more scalable and stable for larger settings.

[14] With a suitably large penalty parameter $\rho$, the method of augmented Lagrangian is guaranteed to converge to a (locally) optimal solution to the original problem (Wright and Nocedal 1999). In practice we find that even for small values of $\rho$ and enough iterations, the solver converges to auction designs that yield near-optimal revenue while closely satisfying the regret constraints.

## 3.1  Illustrative Results

Through this approach, almost optimal auctions with almost zero expected *ex post* regret can be obtained across a number of different economic environments.

For the results presented here, we set $\rho = 0.05$ and sample 5000 training and 5000 value profiles i.i.d from a known distribution. We use the *TensorFlow* deep learning library, solving the inner optimization in the augmented Lagrangian method using the ADAM solver (Kingma and Ba 2015) with learning rate 0.001 and mini-batch size 64. All the experiments are run on a cluster of NVIDIA GPU cores.

We first present results for the following two item, single-bidder settings, for which there exist theoretical results (this provides an optimal benchmark):

- 2 items, single additive bidder, with item values $x_1, x_2 \sim U[0, 1]$. See Fig. 2a. The optimal DSIC mechanism is due to Manelli and Vincent (2006).
- 2 items, single additive bidder, with item values $x_1 \sim U[4, 16]$ and $x_2 \sim U[4, 7]$. The optimal DSIC mechanism is due to Daskalakis et al. (2017).
- 2 items, single unit-demand bidder, with item values $x_1, x_2 \sim U[2, 3]$. The optimal DSIC mechanism is due to Pavlov (2011).
- 2 items, single unit-demand bidder, with item values $x_1, x_2 \sim U[0, 1]$. The optimal DSIC mechanism is due to Pavlov (2011).

Table 1 summarizes the revenue and regret for the learned mechanisms (all measured on data sampled from $F_V$ and distinct from training data, and with regret normalized to be stated per-agent). The revenue from the learned auctions is very close to the optimal designs from the theoretical literature. In two cases, the revenue from REGRETNET is slightly higher than optimal. This can be explained by the non-zero regret, which makes these auctions not quite DSIC when training was terminated.

Figure 2a–d provide a visualization of the allocation rules in the learned mechanisms, comparing them with the optimal rules. In each case, we plot the probability of allocating item 1 and item 2 to the bidder in the learned mechanism, as a function of the bidder's value on each item. The design of the optimal allocation rule

**Table 1** Revenue and regret for REGRETNET, comparing to the expected revenue of the optimal DSIC auction. We also state the normalized revenue, as a fraction of the revenue from the optimal auction (Dütting et al. 2019)

| Auction environment | Optimal | REGRETNET | |
|---|---|---|---|
| | rev | rev (norm) | Regret |
| 2 item, 1 additive bidder, $x_1, x_2 \sim U[0, 1]$ | 0.550 | 0.557 (101.3%) | <0.001 |
| 2 item, 1 additive bidder, $x_1 \sim U[4, 16]$, $x_2 \sim [4, 7]$ | 9.781 | 9.722 (99.4%) | <0.004 |
| 2 item, 1 unit-demand bidder, $x_1, x_2 \sim U[0, 1]$ | 0.384 | 0.386 (100.5%) | <0.001 |
| 2 item, 1 unit-demand bidder, uniform, $x_1, x_2 \sim U[2, 3]$ | 2.137 | 2.124 (99.4%) | <0.001 |

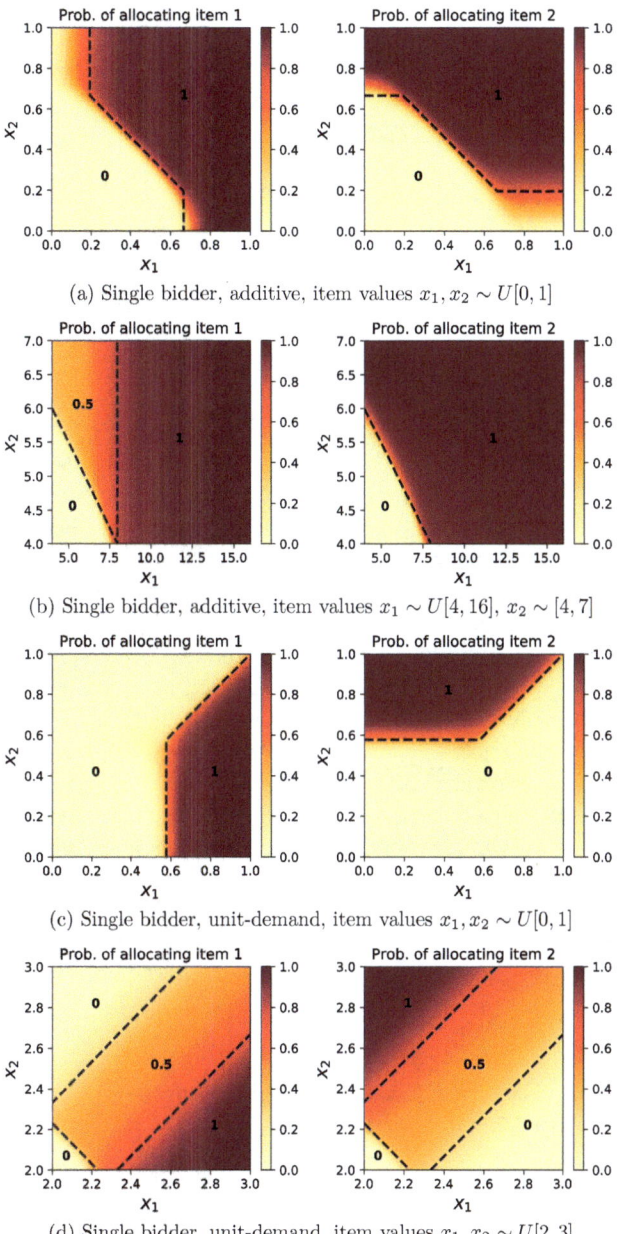

**Fig. 2** A comparison of the allocation rules learned by REGRETNET to those of optimal auctions (Dütting et al. 2019). These are all single bidder environments. We plot the probability of allocating item 1 and item 2, as a function of the bidder's value on each item. The design of the optimal allocation rule is superimposed, with different allocation regions separated by dashed lines (the number in a region gives the probability the item is allocated in the optimal solution)

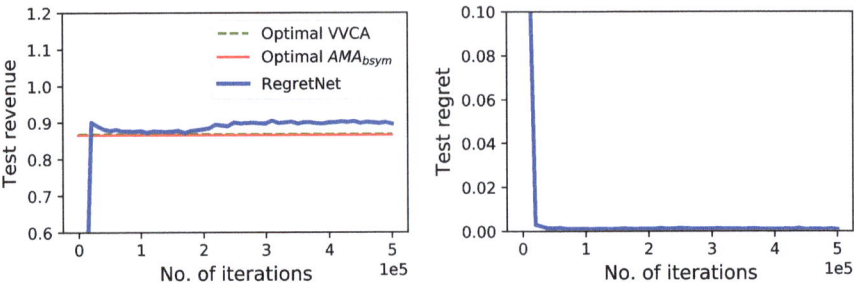

**Fig. 3** The test revenue and regret of REGRETNET as a function of solver iterations for the two items, two additive bidders setting, where item values are i.i.d. uniform on [0, 1]. (Dütting et al. 2019)

is superimposed, with different allocation regions separated by dashed lines (with the number in a region giving the probability the item is allocated in the optimal solution). We can confirm visually a very close correspondence between the result of machine learning and the optimal design.

We have also used this approach to design auctions for economic environments that are out of reach of theoretical analysis.

These include the single, additive bidder environment with ten items (there is no analytical solution with more than six items), as well as a setting with two items and two additive bidders, where item values are i.i.d. uniform on interval [0, 1]. Figure 3 illustrates the effect of additional training iterations on revenue and regret in this example, where revenue and regret are computed on hold out (test) data.[15] This progress of successively lower regret across training iterations is representative of the experimental results reported in Dütting et al. (2019). In this case, we learn an auction with effectively zero regret, and expected revenue of 0.895 (compared with two state-of-art results from the computational literature, namely 0.867 for the optimal *VVCA auction* and 0.864 for the optimal *symmetric AMA auction* (Sandholm and Likhodedov 2015).

## 4 Extensions of the Deep Learning Approach to Other Domains

To illustrate the generality of the framework, we briefly describe two different extensions.

First, we are exploring the use of machine learning for the design of optimal auctions in the presence of private *budget constraints* (Feng et al. 2018). This is a particularly difficult problem, as can be understood from the contours of the few

---

[15] A training iteration is one mini-batch gradient updated in the solver that we use for stochastic gradient descent.

theoretical results that are available. Even the optimal DSIC single-item, multiple bidder problem is not fully understood.[16]

The REGRETNET architecture can be extended to handle bidders with budget constraints, as well as to allow for BIC rather than DSIC where that is of interest. In regard to private budgets, these are handled by introducing an additional penalty (a *budget penalty*) to penalize payments above reported budget, and regret is modified to only consider deviations for which the payment is no greater than a bidder's true budget. In regard to BIC, a bidder's *interim* regret is estimated as the maximum, over a set of possible misreports, of the average increase in utility from deviation given a set of possible reports by others.

We obtain positive results for various auction environments, including settings for which there are no theoretical results for optimal design (Feng et al. 2018). An illustrative result is shown in Fig. 4. This is for auctioning a single item to two bidders, each with value uniform on set $\{1, 2, \ldots, 10\}$, where the first bidder is unconstrained and the second bidder has a budget of 4. In this case, we only consider the case of downward misreports of budget (conditional DSIC) and the optimal result is shown in Malakhov and Vohra (2008). Figure 4a shows the test revenue and *ex post* regret for the mechanism learned by REGRETNET, as a function of the number of solver iterations. The trained mechanism yields revenue very close to the optimal revenue, while yielding negligible regret. Figure 4b, c show that the learned allocation rule closely matches the optimal allocation rule.

Sticking within the context of auctions, we believe the REGRETNET architecture (or related approaches) will allow rapid experimentation in the following of directions:

- Correlated, private values; interdependent values.
- Comparing the revenue properties of DSIC and BIC auctions.
- Various kinds of budgeted settings, including budgets that depend on outcomes, and settings where there are correlations between budgets and values.
- Auctions that are robust against deviations by groups of bidders (i.e., properties such as *group strategy-proofness* and its relaxations).
- Auctions that are *envy-free*, so that one bidder does not envy the allocation of another.
- Auctions that satisfy *core* properties, so that no group of participants can do better by breaking away from the auction.
- Revenue-optimal, combinatorial auctions.

The REGRETNET architecture can also be used for problems of mechanism design without money. To illustrate this, we have results for the *K-facility location problem* (Golowich et al. 2018). This problem generalizes the single-facility, 1-dimensional location problem under single-peaked preferences (Moulin 1980) to

---

[16]Pai and Vohra (2014) design the optimal, single-item BIC auction, while Malakhov and Vohra (2008) provide the state-of-the-art result for the optimal, single-item DSIC auction (for two bidders, and with a weaker "conditional" form of DSIC). These results build on earlier results for more stylized settings (Che and Gale 1998, 2000; Maskin 2000; Laffont and Robert 1996). There are also a few approximation results for DSIC and BIC designs (Borgs et al. 2005; Bhattacharya et al. 2012; Chawla et al. 2011).

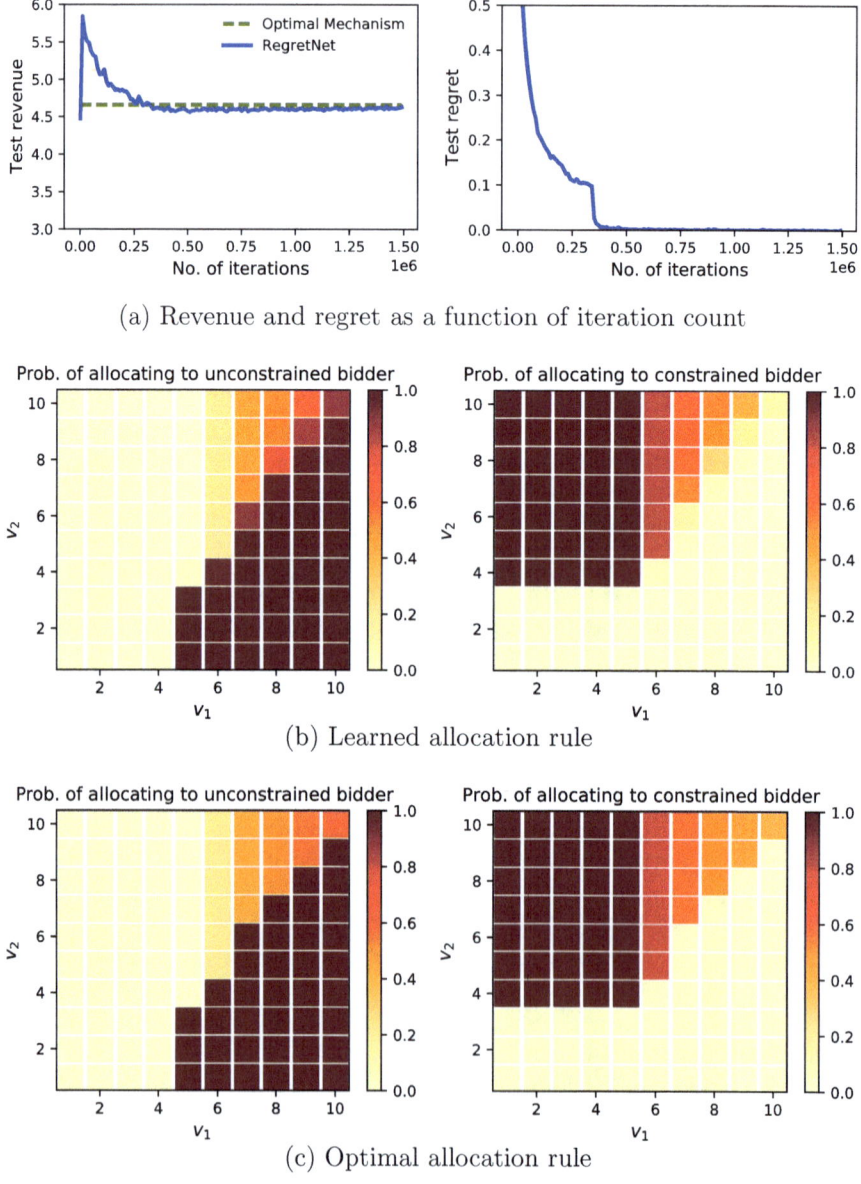

(a) Revenue and regret as a function of iteration count

(b) Learned allocation rule

(c) Optimal allocation rule

**Fig. 4** REGRETNET (extended to handle private budgets) for a single item, two bidder auction, where bidder values $v_1, v_2 \sim Unif\{1, 2, ..., 10\}$, bidder 1 is unconstrained, and bidder 2 has a budget of 4. (b) and (c) compare the trained allocation rule and the optimal allocation rule, illustrating the probability of assigning the item to each bidder for different values $(v_1, v_2)$. Based on based on Fig. 4 in Feng et al. (2018)

consider multiple facilities. An agent's utility depends on the distance between its peak and the closest facility. No general characterization result is available for DSIC mechanisms for facility location with $K \geq 2$ facilities.[17] Moreover, a series of negative results show the impossibility of achieving a good, worst-case approximation to the optimal social cost (Procaccia and Tennenholtz 2013; Fotakis and Tzamos 2014, 2016). Here, social cost is the total, negated distance from each agent to its closest facility.

We assume that each agent has a single-peaked utility function, i.i.d. drawn from some distribution, and look to minimize the expected social cost, i.e. the sum of the agents' costs. The architecture is modified to REGRETNET-NM, where the input layer is defined to include one unit for the position of each agent's peak and the output layer includes one unit for each facility, representing its location.[18] The training problem is solved through augmented Lagrangian.

We compare the social cost with that of the best percentile, dictatorial, and constant rules, as well as the optimal (non DSIC) rule.[19] We vary the number of facilities and number of agents. In each case, agent peaks are i.i.d. sampled, uniform on $[0, 1]$. In one variation, weighted social cost is also considered, where the designer may associate a different weight with each agent. Table 2 summarizes the results for the case of $K = 4, n = 5$, as well as $K = 3, n = 9$ for a weighted objective. The expected, per-agent *ex post* regret is low, and the performance of REGRETNET-NM is competitive with the best percentile rule for the unweighted case and better than the best percentile rule for the weighted case. Figure 5 illustrates the social choice rule that is learned in each of these two environments. This shows the histogram on percentiles of reports for each of the facilities.[20]

For $K = 4$ and $n = 5$, the concentration around $i/4$ for $i \in \{0, 1, 2, 3, 4\}$ in Fig. 5a indicates that the behavior of the learned rule is close to that of a percentile rule, almost always choosing the min and the max peaks, but making different choices about which reports to use for the other two facilities. For $K = 3$ and $n = 9$, and with a weighted objective that places a high weight on agents 1 and 2, we see in Fig. 5b that

---

[17] Ehlers and Gordon (2011) and Heo (2013) provide characterizations for the special case of $K = 2$ under additional assumptions. Heo (2013) assumes anonymity and an additional property, *users only*, which means that agents cannot influence the locations of facilities they will not use. Ehlers and Gordon (2011) assume that agents have *lexmax* preferences over facilities, and thus do not only care about the peak closest to them.

[18] For the single facility location problem, it is w.l.o.g. to consider mechanisms that operate on agent peaks (Border and Jordan 1983). This extends also to a more general "voting under constraints" setting (Barberà et al. 1997). For multiple facilities there are DSIC mechanisms that do not depend only on agent peaks. For example, one can consider the example of $K = 2$ facilities and $n = 2$, where one facility is placed at the peak of agent 1 and the other at some location that depends on the shape of agent 1's report. Still, we retain this simple representation in our current work.

[19] In a multi-facility, dictatorial rule, each facility is determined by a separate dictatorial rule. This is equivalent to having a serial-dictatorial rule for all $K$ facilities. In a multi-facility, percentile rule, each facility is determined by a separate percentile rule (Sui et al. 2013). A constant rule places each facility in the same location all the time.

[20] If $p_1, \ldots, p_n$ are the agent peaks in sorted order, then a facility at location $x$ has percentile 0 if $x \leq p_1$, has percentile 1 if $x \geq p_n$, and has percentile $\frac{i-1}{n-1} + \frac{x-p_i}{(n-1)(p_{i+1}-p_i)}$ if $p_i \leq x < p_{i+1}$.

**Table 2** (Weighted) social cost and regret for REGRETNET-NM, comparing to the best percentile, best dictatorial, best constant, and optimal (non-DSIC) rules. $K$ is the number of facilities, $n$ the number of agents (Golowich et al. 2018)

| Environment | Percentile | Dictatorial | Constant | Optimal | REGRETNET | |
|---|---|---|---|---|---|---|
| | Social cost | Social cost | Social cost | Social cost | Social cost | Regret |
| $K = 4, n = 5$ | **0.017** | 0.024 | 0.064 | 0.0083 | 0.018 | 0.0024 |
| $K = 3, n = 9$, weighted | 0.056 | 0.053 | 0.085 | 0.032 | **0.041** | 0.0005 |

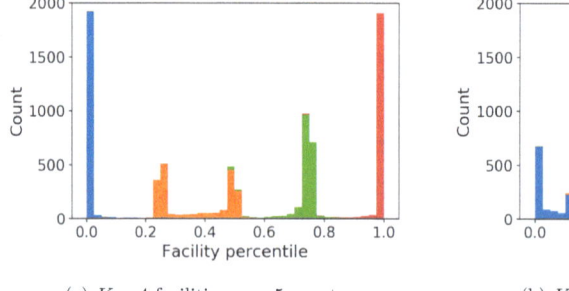 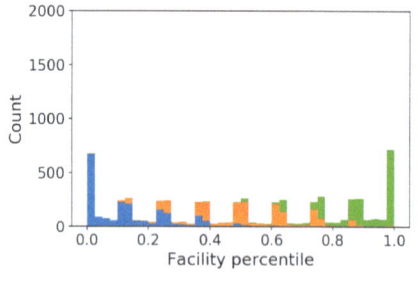

(a) $K = 4$ facilities, $n = 5$ agents                       (b) $K = 3$ facilities, $n = 9$ agents

**Fig. 5** Histograms of facility percentiles chosen by REGRETNET-NM. For each instance, the locations selected by the network are sorted and shown in different colors. Blue is the percentile of the left-most facility, then orange, then green, then red (for $K = 4$). Figure 5a is from Golowich et al. (2018)

the learned rule frequently places each facility at a location that corresponds to one of the reported peaks. In fact, we can confirm that the network learns to approximately treat agents 1 and 2 as dictators.

More generally, we anticipate that this framework can be successfully applied to the following variations on facility location, as well as more general problems of mechanism design without money:

- Facility location problems in multiple dimensions, as well as problems with multiple capacitated facilities (see Aziz et al. 2018 for a single facility version).
- Matching problems, for example optimizing expected welfare in one-sided and two-sided matching problems (including considerations of incentive alignment or stability).
- Voting problems, for example to maximize expected welfare subject to various axiomatic properties.

# 5 Looking Forward

We believe that computational methods, and machine learning in particular, can power a new generation of optimal economic design. We look to learn optimal designs from data that we generate that represents the distribution on agent types, and capture economic constraints such as incentive compatibility within the training problem, and an objective such as revenue via an appropriate loss function. There are many fundamental research questions to pursue, including questions that are responsive to the earlier discussion— robustness, interpretability, and understanding computation-theoretic and learning-theoretic barriers. The approach is not limited to auction design, but can extend to other problems of economic design such as those of optimal contract theory as well as to problems without money. Success will enable bespoke, optimal designs to be deployed in all corners of the rapidly emerging digital economy. Indeed, automated design looks like a necessary response to a future in which we should expect the increasing adoption of algorithmic and AI methods for economic decision making. Advances in the use of computational methods for design will also provide tools with which to validate and advance the theory of optimal economic design.

**Acknowledgements** This research is supported in part by NSF EAGER #124110.

# References

Abadi, M., Chu, A., Goodfellow, I. J., McMahan, H. B., Mironov, I., Talwar, K., et al. (2016). Deep learning with differential privacy. *Proceedings of the ACM Conference on Computer and Communications Security* (pp. 308–318).

Alaei, S., Fu, H., Haghpanah, N., & Hartline, J. D. (2013). The simple economics of approximately optimal auctions. *Proceedings of the 54th IEEE Symposium on Foundations of Computer Science* (pp. 628–637).

Alaei, S., Fu, H., Haghpanah, N., Hartline, J. D., & Malekian, A. (2012). Bayesian optimal auctions via multi- to single-agent reduction. *Proceedings of the 13th ACM Conference on Electronic Commerce* (p. 17).

Alaei, S. (2014). Bayesian combinatorial auctions: Expanding single buyer mechanisms to many buyers. *SIAM Journal on Computing, 43*, 930–972.

Albert, M., Conitzer, V., & Stone, P. (2017). Automated design of robust mechanisms. *Proceedings of the 31st AAAI Conference on Artificial Intelligence* (pp. 298–304).

Aziz, H., Chan, H., Lee, B. E., & Parkes, D. C. (2018). Mechanism design without money for common goods. arXiv:1806.00960.

Babaioff, M., Immorlica, N., Lucier, B., & Weinberg, S. M. (2014). A simple and approximately optimal mechanism for an additive buyer. *Proceedings of the 55th IEEE Symposium on Foundations of Computer Science* (pp. 21–30).

Barberà, S., Massó, J., & Neme, A. (1997). Voting under constraints. *Journal of Economic Theory, 76*, 298–321.

Bhalgat, A., Gollapudi, S., & Munagala, K. (2013). Optimal auctions via the multiplicative weight method. *Proceedings of the 14th ACM Conference on Economics and Computation* (pp. 73–90).

Bhattacharya, S., Goel, G., Gollapudi, S., & Munagala, K. (2012). Budget-constrained auctions with heterogeneous items. *Theory of Computing, 8*, 429–460.

Bikhchandani, S., Chatterji, S., Lavi, R., Mu'alem, A., Nisan, N., & Sen, A. (2006). Weak monotonicity characterizes deterministic dominant-strategy implementation. *Econometrica, 74*, 1109–1132.

Border, K. C., & Jordan, J. S. (1983). Straightforward elections, unanimity and phantom voters. *Review of Economic Studies, 50*, 153.

Borgs, C., Chayes, J., Immorlica, N., Mahdian, M., & Saberi, A. (2005). Multi-unit auctions with budget-constrained bidders. *Proceedings of the 6th ACM Conference on Electronic Commerce* (pp. 44–51).

Cai, Y., & Daskalakis, C. (2017). Learning Multi-Item Auctions with (or without) Samples. *Proceedings of the 58th IEEE Symposium on Foundations of Computer Science* (pp. 516–527).

Cai, Y., & Huang, Z. (2013). Simple and nearly optimal multi-item auctions. *Proceedings of the 24th ACM-SIAM Symposium on Discrete Algorithms* (pp. 564–577).

Cai, Y., & Zhao, M. (2017). Simple mechanisms for subadditive buyers via duality. *Proceedings of the 49th ACM Symposium on Theory of Computing* (pp. 170–183).

Cai, Y., Daskalakis, C., & Weinberg, M. S. (2012a). Optimal multi-dimensional mechanism design: Reducing revenue to welfare maximization. *Proceedings of the 53rd IEEE Symposium on Foundations of Computer Science* (pp. 130–139).

Cai, Y., Daskalakis, C., & Weinberg, S. M. (2012b). An algorithmic characterization of multi-dimensional mechanisms. *Proceedings of the 44th ACM Symposium on Theory of Computing* (pp. 459–478).

Cai, Y., Daskalakis, C., & Weinberg, S. M. (2013a). Reducing revenue to welfare maximization: approximation algorithms and other generalizations. *Proceedings of the 24th ACM-SIAM Symposium on Discrete Algorithms* (pp. 578–595).

Cai, Y., Daskalakis, C., & Weinberg, S. M. (2013b). Understanding incentives: Mechanism design becomes algorithm design. *Proceedings of the 54th IEEE Symposium on Foundations of Computer Science* (pp. 618–627).

Cai, Y., Devanur, N. R., & Weinberg, S. M. (2016). A duality based unified approach to bayesian mechanism design. *Proceedings of the 48th ACM Symposium on Theory of Computing* (pp. 926–939).

Cai, Y., & Daskalakis, C. (2015). Extreme value theorems for optimal multidimensional pricing. *Games and Economic Behavior, 92*, 266–305.

Carroll, G. (2013). *A Quantitative Approach to Incentives: Application to Voting Rules.* Technical report, Microsoft Research and Stanford University.

Caruana, R., Lou, Y., Gehrke, J., Koch, P., Sturm, M., & Elhadad, N. (2015). Intelligible models for healthcare: Predicting pneumonia risk and hospital 30-day readmission. *Proceedings of the 21sth ACM Conference on Knowledge Discovery and Data Mining* (pp. 1721–1730).

Chawla, S., Hartline, J.D., & Kleinberg, R. D. (2007). Algorithmic pricing via virtual valuations. *Proceedings of the 8th ACM Conference on Electronic Commerce* (pp. 243–251).

Chawla, S., Hartline, J. D. Malec, D. L., & Sivan, B. (2010). Multi-parameter mechanism design and sequential posted pricing. *Proceedings of the 42th ACM Symposium on Theory of Computing* (pp. 311–320).

Chawla, S., Malec, D. L., & Malekian, A. (2011). Bayesian mechanism design for budget-constrained agents. *Proceedings of the 12th ACM Conference on Electronic Commerce* (pp. 253–262).

Che, Y.-K., & Gale, I. (1998). Standard auctions with financially constrained bidders. *The Review of Economic Studies, 65*, 1–21.

Che, Y.-K., & Gale, I. (2000). The optimal mechanism for selling to a budget-constrained buyer. *Journal of Economic Theory, 92*, 198–233.

Chen, R., Lucier, B., Singer, Y., & Syrgkanis, V. (2017). Robust optimization for non-convex objectives. *Proceedings of the 21st Conference on Neural Information Processing Systems* (pp. 4708–4717).

Cole, R., & Roughgarden, T. (2014). The sample complexity of revenue maximization. *Proceedings of the 46th ACM Symposium on Theory of Computing* (pp. 243–252).

Conitzer, V., & Sandholm, T. (2002). Complexity of mechanism design. *Proceedings of the 18th Conference in Uncertainty in Artificial Intelligence* (pp. 103–110).

Conitzer, V., & Sandholm, T. (2003). Applications of automated mechanism design. *Proceedings of the 4th Bayesian Modelling Applications Workshop*.

Daskalakis, C., Deckelbaum, A., & Tzamos, C. (2014). The complexity of optimal mechanism design. *Proceedings of the 25th Annual ACM-SIAM Symposium on Discrete Algorithms* (pp. 1302–1318).

Daskalakis, C., Deckelbaum, A., & Tzamos, C. (2017). Strong duality for a multiple-good monopolist. *Econometrica* (pp. 735–767).

Daskalakis, C. (2015). Multi-item auctions defying intuition? *SIGecom Exchanges*, *14*, 41–75.

Devanur, N. R., Huang, Z., & Psomas, C. -A. (2016). The sample complexity of auctions with side information. *Proceedings of the 48th ACM Symposium on Theory of Computing* (pp. 426–439).

Doshi-Velez, F., Wallace, B. C. & Adams, R. (2015). Graph-sparse LDA: A topic model with structured sparsity. *Proceedings of the 29th AAAI Conference on Artificial Intelligence* (pp. 2575–2581).

Dughmi, S., Han, L., & Nisan, N. (2014). Sampling and representation complexity of revenue maximization. *Proceedings of the 10th Conference on Web and Internet Economics* (pp. 277–291).

Dütting, P., Feldman, M., Kesselheim, T., & Lucier, B. (2017). Prophet inequalities made easy: stochastic optimization by pricing non-stochastic inputs. *Proceedings of the 58th IEEE Symposium on Foundations of Computer Science* (pp. 540–551).

Dütting, P., Feng, Z., Narasimhan, H., Parkes, D. C., & Ravindranath, S. S. (2019). Optimal auctions through deep learning. *Proceedings of the 36th International Conference on Machine Learning* (first version, 2017)

Dütting, P., Fischer, F. A., Jirapinyo, P., Lai, J. K., Lubin, B., & Parkes, D. C. (2015). Payment rules through discriminant-based classifiers. *ACM Transactions on Economics and Computation*, *3*, 5:1–5:41.

Ehlers, L., & Gordon, S. (2011). *Strategy-Proof Provision of Two Public Goods: The LexMax Extension*. Technical report, Mimeo Université de Montréal.

Elkind, E. (2007). Designing and learning optimal finite support auctions. *Proceedings of the 18th ACM-SIAM Conference on Discrete Algorithms* (pp. 736–745).

Fawzi, A., Fawzi, O., & Frossard, P. (2018). Analysis of classifiers' robustness to adversarial perturbations. *Machine Learning*, *107*, 481–508.

Feng, Z., Narasimhan, H., & Parkes, D. C. (2018). Deep learning for revenue-optimal auctions with budgets. *Proceedings of the 17th Conference on Autonomous Agents and Multi-Agent Systems* (pp. 354–362).

Fotakis, D., & Tzamos, C. (2014). On the power of deterministic mechanisms for facility location games. *ACM Transactions on Economics and Computation*, *2*, 15:1–15:37.

Fotakis, D., & Tzamos, C. (2016). Strategyproof facility location for concave cost functions. *Algorithmica*, *76*, 143–167.

Giannakopoulos, Y. & Koutsoupias, E. (2014). Duality and optimality of auctions for uniform distributions. *Proceedings of the 15th ACM Conference on Economics and Computation* (pp. 259–276).

Giannakopoulos, Y. & Koutsoupias, E. (2015). Selling two goods optimally. *Proceedings of the 42nd International Colloquium on Automata, Languages and Programming* (pp. 650–662).

Giannakopoulos, Y. (2015). Bounding the optimal revenue of selling multiple goods. *Theoretical Computer Science*, *581*, 83–96.

Golowich, N., Narasimhan, H., & Parkes, D.C. (2018). Deep learning for multi-facility location mechanism design. *Proceedings of the 27th International Joint Conference on Artificial Intelligence* (pp. 261–267).

Gonczarowski, Y. A., & Nisan, N. (2017). Efficient empirical revenue maximization in single-parameter auction environments. *Proceedings of the 49th ACM Symposium on Theory of Computing* (pp. 856–868).

Goodfellow, I. J., Bengio, Y., & Courville, A. C. (2016). *Deep Learning, Adaptive Computation and Machine Learning*. MIT Press: Cambridge.

Goodfellow, I. J., Shlens, J., & Szegedy, C. (2015). Explaining and harnessing adversarial examples. *Proceedings of the 3rd International Conference on Learning Representations*.

Guo, M., & Conitzer, V. (2010). Computationally feasible automated mechanism design: general approach and case studies. *Proceedings of the 24th AAAI Conference on Artificial Intelligence*.

Guo, M., & Conitzer, V. (2009). Worst-case optimal redistribution of VCG payments in multi-unit auctions. *Games and Economic Behavior, 67*, 69–98.

Hart, S., & Nisan, N. (2012). Approximate revenue maximization with multiple items. *Proceedings of the 13th ACM Conference on Economics and Computation* (p. 656).

Heo, E. J. (2013). Strategy-proof rules for two public goods: double median rules. *Social Choice and Welfare, 41*, 895–922.

Huang, Z., Mansour, Y., & Roughgarden, T. (2015). Making the most of your samples. *Proceedings of the 16th ACM Conference on Economics and Computation* (pp. 45–60).

Kingma, D. P., & Ba, J. (2015). Adam: A method for stochastic optimization. *Proceedings of the 3rd International Conference on Learning Representations*.

Kleinberg, R., & Weinberg, S. M. (2012). Matroid prophet inequalities. *Proceedings of the 44th ACM Symposium on Theory of Computing* (pp. 123–136).

Laffont, J.-J., & Robert, J. (1996). Optimal auction with financially constrained buyers. *Economics Letters, 52*, 181–186.

Li, X., & Yao, A. C.-C. (2013). On revenue maximization for selling multiple independently distributed items. *Proceedings of the National Academy of Sciences, 110*, 11232–11237.

Lubin, B., & Parkes, D. C. (2012). Approximate strategyproofness. *Current Science, 103*, 1021–1032.

Malakhov, A., & Vohra, R. V. (2008). Optimal auctions for asymmetrically budget constrained bidders. *Review of Economic Design, 12*, 245.

Manelli, A., & Vincent, D. (2006). Bundling as an optimal selling mechanism for a multiple-good monopolist. *Journal of Economic Theory, 127*, 1–35.

Maskin, E. S. (2000). Auctions, development, and privatization: efficient auctions with liquidity-constrained buyers. *European Economic Review, 44*, 667–681.

Mennle, T., & Seuken, S. (2014). An axiomatic approach to characterizing and relaxing strategyproofness of one-sided matching mechanisms. *Proceedings of the 15th ACM Conference on Economics and Computation* (pp. 37–38).

Mennle, T., & Seuken, S. (2017). Partial strategyproofness: Relaxing strategyproofness for the random assignment problem. arXiv:1401.3675v4.

Moosavi-Dezfooli, S., Fawzi, A., & Frossard, P. (2016). DeepFool: A simple and accurate method to fool deep neural networks. *Proceedings of the IEEE Conference on Computer Vision and Pattern Recognition* (pp. 2574–2582).

Morgenstern, J., & Roughgarden, T. (2015). On the pseudo-dimension of nearly optimal auctions. *Proceedings of the 29th Conference on Neural Information Processing Systems*.

Morgenstern, J., & Roughgarden, T. (2016). Learning simple auctions. *Proceedings of the 29th Conference on Learning Theory* (pp. 1298–1318).

Moulin, H. (1980). On strategy-proofness and single-peakedness. *Public Choice, 35*, 55–74.

Myerson, R. (1981). Optimal auction design. *Mathematics of Operations Research, 6*, 58–73.

Narasimhan, H., & Parkes, D. C. (2016). A general statistical framework for designing strategy-proof assignment mechanisms. *Proceedings of the 32nd Conference on Uncertainty in Artificial Intelligence* (pp. 527–536).

Narasimhan, H., Agarwal, S., & Parkes, D. C. (2016). Automated mechanism design without money via machine learning. *Proceedings of the 25th International Joint Conference on Artificial Intelligence* (pp. 433–439).

Niepert, M., Ahmed, M., & Kutzkov, K. (2016). Learning convolutional neural networks for graphs. *Proceedings of The 33rd International Conference on Machine Learning* (Vol. 48, pp. 2014–2023).

Orhan, A. E., & Ma, W. J. (2017). Efficient probabilistic inference in generic neural networks trained with non-probabilistic feedback. *Nature Communications, 8*, 138.

Pai, M. M., & Vohra, R. (2014). Optimal auctions with financially constrained buyers. *Journal of Economic Theory, 150*, 383–425.

Parkes, D. C., & Wellman, M. P. (2015). Economic reasoning and artificial intelligence. *Science*, *349*, 267–272.

Pavlov, G. (2011). Optimal mechanism for selling two goods. *B.E Journal of Theoretical Economics*, *11*, 1–35.

Procaccia, A. D., & Tennenholtz, M. (2013). Approximate mechanism design without money. *ACM Transactions on Economics and Computation*, *1*, 1–26.

Procaccia, A., Zohar, A., Peleg, Y., & Rosenschein, J. (2009). The learnability of voting rules. *Artificial Intelligence*, *173*, 1133–1149.

Raghu, M., Poole, B., Kleinberg, J. M., Ganguli, S., & Sohl-Dickstein, J. (2016). Survey of expressivity in deep neural networks. arXiv:1611.08083.

Ribeiro, M. T., Singh, S., & Guestrin, C. (2016). "Why should i trust you?": explaining the predictions of any classifier. *Proceedings of the 22nd ACM Conferecne on Knowlegde Discovery and Data Mining* (pp. 1135–1144).

Rochet, J.-C. (1987). A necessary and sufficient condition for rationalizability in a quasilinear context. *Journal of Mathematical Economics*, *16*, 191–200.

Ross, A. S., Hughes, M. C., & Doshi-Velez, F. (2017). Right for the Right Reasons: Training Differentiable Models by Constraining their Explanations. *Proceedings of the 26th International Joint Conference on Artificial Intelligence* (pp. 2662–2670).

Rothkopf, M. H., Pekec, A., & Harstad, R. M. (1998). Computationally manageable combinational auctions. *Management Science*, *44*, 1131–1147.

Rubinstein, A., & Weinberg, S. M. (2015). Simple mechanisms for a subadditive buyer and applications to revenue monotonicity. *Proceedings of the 16th ACM Conference on Economics and Computation* (pp. 377–394).

Sandholm, T., & Likhodedov, A. (2015). Automated design of revenue-maximizing combinatorial auctions. *Operations Research*, *63*, 1000–1025.

Shalev-Shwartz, S., & Wexler, Y. (2016). Minimizing the maximal loss: how and why. *Proceedings of the 33rd International Conference on Machine learning* (pp. 793–801).

Smilkov, D., Thorat, N., Nicholson, C., Reif, E., Viégas, F. B., & Wattenberg, M. (2016). Embedding projector: interactive visualization and interpretation of embeddings. arXiv:1611.05469.

Sui, X., Boutilier, C., & Sandholm, T. (2013). Analysis and optimization of multi-dimensional percentile mechanisms. *Proceedings of the 23rd International Joint Conference on Artificial Intelligence* (pp. 367–374).

Szegedy, C., Zaremba, W., Sutskever, I., Bruna, J., Erhan, D., Goodfellow, I. J., & Fergus, R. (2014). Intriguing properties of neural networks. *Proceedings of the 2nd International Conference on Learning Representations*.

Thirumulanathan, D., Sundaresan, R., & Narahari, Y. (2016). Optimal mechanism for selling two items to a single buyer having uniformly distributed valuations. *Proceedings of the 12th International Conference on Web and Internet Economics* (pp. 174–187).

Thirumulanathan, D., Sundaresan, R., & Narahari, Y. (2017): On optimal mechanisms in the two-item single-buyer unit-demand setting. arXiv:1705.01821.

Vinyals, O., Fortunato, M., & Jaitly, N. (2015). Pointer networks. *Proceedings of the 29th Conference on Neural Information Processing Systems* (pp. 2692–2700).

Wang, F., & Rudin, C. (2015). Falling rule lists. *Proceedings of the 18th Conference on Artificial Intelligence and Statistics*.

Wright, S., & Nocedal, J. (1999). Numerical optimization. *Springer Science*, *35*, 67–68.

Xia, L. (2013). Designing social choice mechanisms using machine learning. *Proceedings of the 13th Conference on Autonomous Agents and Multi-Agent Systems* (pp. 471–474).

Yao, A. C.-C. (2015). An *n*-to-1 bidder reduction for multi-item auctions and its applications. *Proceedings of the 26th ACM-SIAM Symposium on Discrete Algorithms* (pp. 92–109).

Yao, A. C.-C. (2017). Dominant-strategy versus bayesian multi-item auctions: maximum revenue determination and comparison. *Proceedings of the 18th ACM Conference on Economics and Computation* (pp. 3–20).

# The Design of Everyday Markets

Nicole Immorlica

*The problem with the designs of most engineers is that they are too logical. We have to accept human behavior the way it is, not the way we would wish it to be.*

–Don Norman, *The Design of Everyday Things*

**Abstract** Economic design is based on a deep and instrumental theory of optimal strategic behavior. This approach, while highly valuable in complex economic environments involving sophisticated agents with clear objectives, sometimes results in design that is unintuitive to humans. This article promotes the study of everyday markets which are easy for humans to navigate. Such an agenda requires new interdisciplinary interactions and metrics of success.

The past century of economic design developed powerful tools and techniques to optimize complex markets. Cloud computing, with a combination of carefully constructed on-demand and reservation pricing schemes, delivers top-of-the-line computer technologies to millions of companies (Babaioff et al. 2017). Revenue from online advertising fuels the Internet; the underlying sale mechanisms are inspired by the theory of optimal auction design (Ostrovsky and Schwarz 2019). Spectrum auctions leverage combinatorial auction theory and SAT solvers to allocate a valuable scarce resource to those companies that are best positioned to make use of it (Leyton-Brown et al. 2017; Newman et al. 2017). National kidney exchanges save hundreds of lives nationwide using top-trading cycles and integer programming (Anderson et al. 2015; Roth et al. 2004). These markets and the tools that power them have transformed industries, especially industries involving sophisticated agents, who themselves are often advised by economists and algorithmic game theorists and computerized agents.

But these are not the markets ordinary people encounter in their daily lives. They can not be, for they are too complicated. People have a limited understanding of the strategic choices in such markets, the impacts of their actions, and even their own

N. Immorlica (✉)
Microsoft Research, 1 Memorial Drive, Cambridge, MA, USA
e-mail: nicimm@gmail.com

© Springer Nature Switzerland AG 2019
J.-F. Laslier et al. (eds.), *The Future of Economic Design*, Studies in
Economic Design, https://doi.org/10.1007/978-3-030-18050-8_71

preferences over the outcomes (Kahneman 2011). Until recently, this was of little importance. The markets people encountered in their daily lives were large, with comparatively few choices. High school sweethearts on a movie date selected from around a dozen movies screened at the local theater. A college student moving into a dormitory in need of hangers bought one of the two-to-three brands sold at the local drug store. A new employee selected a health plan from the limited choices offered by her employer. And the strategic consequences of those choices was easy to grasp. Parents moved to neighborhoods, confident of the schools their children would attend. Singles met and wooed each other at social events through well-established social norms. Young professionals found jobs based on recommendations, applications, and interviews. There was little opportunity for incomprehensible mistakes in this world, and the mistakes people made were unlikely to be significant in aggregate.

In today's world, aided by the increased presence of increasingly sophisticated technologies, choice is ubiquitous. People stream movies through sites like Netflix, buy hangers online through sites like Amazon, and join exchanges to shop for health care. Most of these markets have hundreds or thousands of choices. And, as economic design increasingly permeates everyday life, people encounter more-and-more choices with significant strategic considerations. They must submit preference orderings over schools in city-wide school choice programs, stream movies through sites like Netflix, create profiles and search for love in apps, and find work through computerized matching services like the National Residency Matching Program, Uber, or Mechanical Turk.

Such richly customizable lives can be tailored to individual preferences and offer opportunities for greatly increased satisfaction. But the number of choices and lack of clarity in the consequence of each choice is overwhelming. This leads people to make mistakes in their choices, or sometimes to make no choices at all. In retirement savings plans, for every 10-option increase in 401(k) investment funds, participation in the plans decreases by 2% (Iyengar et al. 2004). In strategyproof school choice programs as diverse as the Israeli Psychology Master's Match and the Hungarian college admissions process, significant fractions of participants appear to occasionally misrepresent their preferences (Hassidim et al. 2017; Shorrer and Sovago 2017). As there are more choices, these mistakes can lead to greater dissatisfaction, and as there are more agents, the compounded impact of these mistakes can be dire.

Economic design in such a world can not be optimal if it is not navigable by humans. Navigable design requires rules that are *interpretable by humans*, paired with trusted technologies that *delegate strategic choices to machines* which they are better-equipped to make. Economists and computer scientists have proposed several definitions of interpretable[1] mechanisms. Historically, the implicit definition included mechanisms that are deterministic, strategyproof, and/or solvable by iterated deletion of dominated strategies. Recently, more nuanced definitions have emerged. Researchers distinguish flavors of strategyproofness (Li 2017), define degrees of

---

[1]Some prefer the word simple, but I personally find this misleading. Simple, in opposition to complex, is a machine-centric word in computer science. It suggests the implementation is concise, which is quite different from saying that it is easy for humans to interpret.

complexity in menu pricing (Hart and Nisan 2013), and measure the ease with which auction formats can be learned (Balcan et al. 2005; Elkind 2007; Morgenstern and Roughgarden 2015). Others, in the field of behavioral economics, have elected to forgo definitions of simplicity and instead model sub-optimal strategic choices mathematically.

This research forces us to re-evaluate our design techniques, and rightfully so. A broadly accepted conclusion of this research is that incentive compatible mechanisms, even strategyproof ones, are not necessarily interpretable. Significantly, this means the revelation principle is not without loss. Non-revelation mechanism design, although apparently strategically more complicated, could potentially out-perform optimal incentive compatible mechanisms in practice. For example, people often overbid in second price auctions, but rarely do so in first-price auctions where the implication of a bid is clear (Kagel and Levin 1993). While it is possible to take an incentive compatible mechanism and apply the revelation principle in reverse to obtain one with desired non-incentive compatible, e.g., pay-as-bid, semantics, mechanisms obtained in this manner are often quite perverse. This suggests the need for the explicit design of non-revelation mechanisms (Hartline and Taggart 2016), accompanied by a quantitative statement of the tradeoff between optimality and interpretability. Results along these lines dominate the combinatorial auction literature of late, where researchers have proved, for example, that the equilibria of simultaneous single-price auctions approximate the optimal welfare in a variety of settings (Bhawalkar and Roughgarden 2011; Christodoulou et al. 2008; de Keijzer et al. 2013; Feldman et al. 2013, 2016; Hassidim et al. 2011; Markakis and Telelis 2015).

These promising research directions provide valuable guidelines for good economic design. However, they are clearly neither necessary nor sufficient for interpretability. Games like chess, second-price auctions, and famous problems like Fermat's Last Theorem each satisfy some or all proposed definitions of interpretability, but they are hard for humans. Conversely, playing a new video game, participating in first price auctions, walking down a crowded street, and navigating a new social setting all fail some of the proposed definitions of interpretability and yet are games that people play well relative to the aforementioned purportedly simple ones.

The flaw underlying standard economic intuition is the assumption that humans choose actions based on (perhaps limited) strategic reasoning, or that they learn (usually slowly) from repeated play. In fact, people can play even unfamiliar games quite well based on expectations, analogy, and social cues (Norman 2013). To incorporate these elements into economic design, we must move beyond our crutch of mathematical formality and our comfortable collaborations with our standard set of economists and computer scientists. Psychology, sociology, and, perhaps most relevant, design and human-computer interaction are all rich fields with substantial research and methodology developed around creating improved interfaces for human navigation. Economic designers should leverage these fields to create interpretable mechanisms.

There are several examples in which economic design, paired with human design, created highly successful mechanisms. In participatory budgeting, citizens collectively vote on city budget proposals. Stanford's Crowdsourced Democracy Team implemented a participatory budgeting platform based on knapsack voting and $k$-approval voting (Goel et al. 2016) which is now being used by cities nationwide. In the context of market survey research, the company Collective Decision Engines, co-founded by economists and designers, developed an app to elicit opinions based on quadratic voting (Quarfoot et al. 2017; Glen Weyl 2012). In the context of fair division, Spliddit uses cake-cutting algorithms to propose how to split rent among roommates, among other tasks (Goldman and Procaccia 2015; Procaccia and Wang 2014). In the context of polling and predictions, Predictwise uses sophisticated prediction market algorithms to predict the outcome of presidential elections, among other things, with alarming accuracy (Rothschild 2015). While only some of these mechanisms are explicit collaborations between economists, computer scientists, and other social sciences, all benefit greatly from careful interface design and implementation. The resulting combined product—mathematical machines plus humanistic design—is much more powerful than either element alone. Such holistic work is disappointingly infrequent and undervalued within economics and computer science.

Even interpretable mechanism design is of little use if people are uncertain of their preferences over potential outcomes or paralyzed by the number of options. In the long-term, economic design must come to rely on the *automated economic self*. Machine learning technologies, living on our cell phones in our pockets, or maybe even implanted in our brains, will learn our preferences over time by sensing our biological reactions to the choices we make. Does our heartbeat quicken when we see the new Star Wars, or do our eyelids droop? Does our temperature rise when we enter our child's school, or do we smile? If this seem far-fetched, we already live in such a world to a limited degree. Music streaming services like Spotify choose our playlists for us based on limited feedback from our reactions to the songs played. Movie streaming services like Netflix curate movie selections for us based on the similarity of our preferences with our peers. Advertisers sense our location through our cell phones to offer nearby deals.

Such scenarios may seem dystopian today. People desire transparency and agency. The price of an apple in a supermarket is transparent, and a person has the agency to choose whether to purchase it or pass it by. Both these properties could be lost in a world where an automated economic self uses fancy sensors and machine learning technologies choose our fruit (or our movies or our retirement plans) for us. But, if the technology is transparent, people may be able to accept it. Indeed, there is a nascent movement in the machine learning community aiming to do precisely that (Doshi-Velez and Kim 2017; Lipton 2016). And then, by delegating the multitude of mundane choices in our daily lives to machines, we can free ourselves to contemplate, innovate, and act in more significant spaces, providing us with more meaningful agency than we have today.

The future of economic design requires improved models of behavior, mathematical ones of the sort currently being developed by economists and computer scientists, but also intuitive humanistic ones developed in the other social sciences. Combining these models, we must design interpretable mechanisms that smartly delegate mundane choices to machines and important ones to humans. The resulting designs must be vetted, both at a mathematical level and at a social level. This research necessitates new interdisciplinary interactions and a broadening of accepted methodologies. Only through such a shift in our field's values can we achieve the full potential of economic design.

**Acknowledgements** I would like to thank Nima Haghpanah, Jason Hartline, Jenn Wortman Vaughn, Glen Weyl and James Wright for many lovely conversations that inspired ideas in this piece. The main thesis of this work echoes that of Chen et al. (2016), which similarly emphasizes the importance of interdisciplinary collaboration and an extended set of tools and techniques for the field of social computing.

# References

Anderson, R., Ashlagi, I., Gamarnik, D., Rees, M., Roth, A., Snmez, T., et al. (2015). Kidney exchange and the alliance for paired donation: Operations research changes the way kidneys are transplanted. *Interfaces, 45*(1), 26–42.

Babaioff, M., Mansour, Y., Nisan, N., Noti, G., Curino, C., Ganapathy, N. et al. (2017). Era: A framework for economic resource allocation for the cloud. In *International Conference on World Wide Web*, WWW (pp. 635–642).

Balcan, M.-F., Blum, A., Hartline, J. D. & Mansour, Y. (2005). Mechanism design via machine learning. In *IEEE Symposium on Foundations of Computer Science* (pp. 605–614).

Bhawalkar, K., & Roughgarden, T. (2011). Welfare guarantees for combinatorial auctions with item bidding. In *ACM-SIAM Symposium on Discrete Algorithms* (pp. 700–709).

Chen, Y., Ghosh, A., Kearns, M., Roughgarden, T., & Vaughan, J. W. (2016). Mathematical foundations for social computing. *Communications of the ACM, 59*(12), 102–108.

Christodoulou, G., Kovács, A., & Schapira, M. (2008). Bayesian combinatorial auctions. In *International Colloquium on Automata, Languages and Programming* (pp. 820–832).

de Keijzer, B., Markakis, E., Schäfer, G., & Telelis, O. (2013). Inefficiency of standard multi-unit auctions. In *European Symposium on Algorithms* (pp. 385–396).

Doshi-Velez, F., & Kim, B. (2017). Towards a rigorous science of interpretable machine learning, working paper.

Elkind, E. (2007). Designing and learning optimal finite support auctions. In *ACM-SIAM symposium on Discrete algorithms* (pp. 736–745).

Feldman, M., Fu, H., Gravin, N., & Lucier, B. (2013). Simultaneous auctions are (almost) efficient. In *ACM Symposium on Theory of Computing* (pp. 201–210).

Feldman, M., Immorlica, N., Lucier, B., Roughgarden, T., & Syrgkanis, V. (2016). The price of anarchy in large games. *ACM symposium on Theory of Computing* (pp. 963–976). Providence: ACM.

Goel, A., Krishnaswamy, A. K., Sakshuwong, S., & Aitamurto, T. (2016). Knapsack voting: Voting mechanisms for participatory budgeting, working paper.

Goldman, J., & Procaccia, A. D. (2015). Spliddit: Unleashing fair division algorithms. *ACM SIGecom Exchanges, 13*(2), 41–46.

Hart, S., & Nisan, N. (2013). The menu-size complexity of auctions. In *ACM Conference on Economics and Computation*.

Hartline, J., & Taggart, S. (2016). Non-revelation mechanism design, working paper.

Hassidim, A., Kaplan, H., Mansour, Y., & Nisan, N. (2011). Non-price equilibria in markets of discrete goods. In *ACM Conference on Electronic Commerce* (pp. 295–296).

Hassidim, A., Marciano, D., Romm, A., & Shorrer, R. (2017). The mechanism is truthful, why aren't you? *American Economic Review: Papers & Proceedings*.

Iyengar, S., Jiang, W., & Huberman, G. (2004). How much choice is too much: Determinants of individual contributions in 401k retirement plans. In O. Mitchell & S. Utkus (Eds.), *Pension design and structure: new lessons from behavioral finance* (pp. 83–97). Oxford: Oxford University Press.

Kagel, J. H., & Levin, D. (1993). Independent private value auctions: Bidder behaviour in first-, second-and third-price auctions with varying numbers of bidders. *The Economic Journal, 103*(419), 868–879.

Kahneman, D. (2011). *Thinking, fast and slow*. London: Macmillan.

Leyton-Brown, K., Milgrom, P., & Segal, I. (2017). Economics and computer science of a radio spectrum reallocation. *Proceedings of the National Academy of Sciences, 114*(28).

Li, S. (2017). Obviously strategy-proof mechanisms. *American Economic Review.*

Lipton, Z. C. (2016). The mythos of model interpretability, working paper.

Markakis, E., & Telelis, O. (2015). Uniform price auctions: Equilibria and efficiency. *Theory of Computing Systems, 57*(3), 549–575.

Morgenstern, J. H., & Roughgarden, T. (2015). On the pseudo-dimension of nearly optimal auctions. In *Advances in Neural Information Processing Systems* (pp. 136–144).

Newman, N., Frechette, A., & Leyton-Brown, K. (2017). Deep optimization for spectrum repacking. *Communications of the ACM.*

Norman, D. (2013). *The design of everyday things: Revised and expanded edition*. New York: Basic Books (AZ).

Ostrovsky, M., & Schwarz, M. Reserve prices in internet advertising auctions: A field experiment. *Journal of Politcal Economy* revise and resubmit.

Procaccia, A. D., & Wang, J. (2014). Fair enough: Guaranteeing approximate maximin shares. In *Proceedings of the fifteenth ACM conference on Economics and computation* (pp. 675–692). Providence: ACM.

Quarfoot, D., von Kohorn, D., Slavin, K., Sutherland, R., Goldstein, D., & Konar, E. (2017). Quadratic voting in the wild: Real people, real votes. *Public Choice, 172*(1–2), 283–303.

Roth, A., Snmez, T., & nver. U., (2004). Kidney exchange. *Quarterly Journal of Economics, 119*(2), 457–488.

Rothschild, D. (2015). Combining forecasts for elections: Accurate, relevant, and timely. *International Journal of Forecasting, 31*(3), 952–964.

Shorrer, R., & Sovago, S. (2017). Obvious mistakes in a strategically simple college-admissions environment, working paper.

Weyl, E. G. (2012). Quadratic vote buying, working paper.

# The Role of Theory in an Age of Design and Big Data

Matthew O. Jackson

**Abstract** I discuss economists' views of what they do, and the particular roles of models and theory. The focus is on theory's role in the design of social and economic systems—past, present, and future.

## 1 Economists Searching for an Identity

To assess what the future holds for economic theory and economic design, it makes sense to first think about what economists actually do and where economic theory and design fit in. There is a remarkable variety of views on these questions, as well as some natural historical progression.[1]

- **Artists and Ethicists**

  John Neville Keynes (John Maynard Keynes' father, an economist and philosopher) is known for distinguishing positive economics—'what is'—from normative economics—'what should be' (Keynes 1904). In making this distinction, he separated the positive role of economists who seek to uncover 'laws', from the normative role of economists who serve as 'ethicists' and describe ideals, as well as 'artists' who 'formulate precepts'. Keynes the Elder's distinction between positive and normative economics remains fundamental to all economists who study the design of institutions to achieve some ideal. Moreover, the normative role pro-

---

[1] This list is an expanded and different take from one offered by Beatrice Cherrier in her Blog: "From physicists to engineers to meds to plumbers: Esther Duflo rediscovering the lost art of economics @ASSA2017". Posted on January 7, 2017. But you will see that the inspiration for this list comes directly from hers. See also Backhouse and Cherrier (2017) for more discussion of the changes in the study of economics over time.

External faculty member at the Santa Fe Institute, and a fellow of CIFAR. I gratefully acknowledge financial support under NSF grant SES-1629446. I thank Salvador Barbera, Gabe Carroll, Ashley Piggins, and Ariel Rubinstein for helpful comments.

M. O. Jackson (✉)
Department of Economics, Stanford University, Stanford, CA 94305-6072, USA
e-mail: jacksonm@stanford.edu

© Springer Nature Switzerland AG 2019                                523
J.-F. Laslier et al. (eds.), *The Future of Economic Design*, Studies in
Economic Design, https://doi.org/10.1007/978-3-030-18050-8_72

vides the early seeds of economic design. Nonetheless, his perspective emerged from an era of academic debate and reasoning with some distance from practical design, and stops far short of the modern economic design paradigm.

- **Scientists**

Paul Samuelson is recognized as a key figure in transforming economics from the realm of essays to that of a 'science'.[2] It is not just that he laid more rigorous and careful mathematical formulations and foundations than had existed before. It is really that Samuelson had a different perspective on the endeavor: to uncover cause and effect via use of the scientific method, which requires careful specifications and precise reasoning given the complexity of the systems in play. The rigorous modeling that Samuelson espoused remains central to modern day economic design.

- **Fablists**

There has been push back on the view of economics as a science, both from those outside who doubt the value of some of its formality as well from those within who see models in a different light. An important perspective on a more basic role of economic models has been articulated by Ariel Rubinstein who states (Rubinstein 2005), "For me, economics is a collection of ideas and conventions which economists accept and use to reason with. Namely, it is a culture." He views economic models as 'fables' (Rubinstein 2012): "hovering between fantasy and reality ... modeling is essential because it is the only method we have of clarifying concepts, evaluating assumptions, verifying conclusions and acquiring insights that will serve us when we return from the model to real life."

- **Engineers**

A more practically-minded view of economic modeling emerges clearly in Al Roth's (2002) view of the economist as an engineer. Roth views economic design's relationship to economics as being analogous to engineering's relationship to physics. The microeconomist's role as engineer had been growing along with the mechanism design literature. Questions raised by Leo Hurwicz (1972, 1973)

---

[2]Samuelson's views on nuances of the economics as a science, and cautions concerning the view, can be glimpsed in this excerpt from an interview of Samuelson by Paul Solman of the PBS News Hour, published December 24, 2009 ("Samuelson on Whether Economics Is a Science").

"SAMUELSON: Economics is not an exact science, it's a combination of an art and elements of science. And that's almost the first and last lesson to be learned about economics: that in my judgment, we are not converging toward exactitude, but we're improving our data bases and our ways of reasoning about them.

SOLMAN: I asked [philosopher] Nelson Goodman this question once, and Nelson Goodman said economics is as much of a science as physics. I said, well, how could that be? He said: Physics can explain how a leaf falls from a tree and everything that happens to it, but it can't tell you where the leaf's going to land. Economics is the same.

SAMUELSON: I think that it's more important for an economist to be wise and sophisticated in scientific method than it is for a physicist because with controlled laboratory experiments possible, they practically guide you, you couldn't go astray. Whereas in economics, by dogma and misunderstanding, you can go very sadly astray."

There are many others who have thought about and discussed economics as a science and the role of models and economic theory, from John Stuart Mill to Milton Friedman (see, for instance Hausman 1989, as well as Davidson 2012).

in the late 1960s into the early 1970s about decentralized economic systems laid a foundation. Studies of mechanisms, implementation, auction design, matching, and agency theory—by a who's who list of economic theorists from the 1970s, 80s, and 90s—advanced the understanding in a variety of directions.[3]

However, what ultimately set the design arena of economic theory apart was a push to put knowledge directly into practice. This was partly serendipitous in terms of facing applications in which theory applied fairly cleanly, such as matching markets and auctions. Still, without a relentless drive to overcome institutional hurdles in actually implementing new market designs, a number of seminal success stories would never have occurred.

- **Plumbers**

Esther Duflo's perspective (Duflo 2017) takes Al Roth's one step further. She views economists not only as doing practical design, but actually being the people who make sure that things work in the field, just as a plumber would; tinkering as unforseen issues emerge.[4] This reflects the perspective of a 'new wave' of development economists who are facing urgent and practical challenges in instituting a long list of programs to improve people's lives.

It is natural that economists' practical ambitions have grown along with available tools and data. Economists have gone from isolated individual debaters exploring simple logical deductions, to teams tackling specific and practical problems that involve varied expertise. It is easy to see this in publication patterns. In 1983 only 46% of articles in top journals were co-authored, while by 2011 almost 80% were, with more than 30% involving three, four, or more authors (Hamermesh 2013).

Based on my own experience, I would expect this trend to grow as we increasingly deal with problems in which it is necessary to gather data, build models, analyze the data, and test new policy designs. For instance, a project with Abhijit Banerjee, Arun Chandrasekhar, and Esther Duflo (Banerjee et al. 2013), in which we studied the diffusion of microfinance in a series of Indian villages, required: theorizing about and modeling information spread, gathering data about people's network structures, and then simulating and fitting the models to the data. A few decades ago gathering network data from 75 villages would have been beyond our means,[5] and there were no computers that could have fit models as rich as ours to the data. What we learned in the first part of the study also led us to design new techniques for identifying central individuals, which we were able to subsequently field-test in another 213 villages. Identifying how microfinance diffused and how t o better diffuse it could

---

[3]I provide surveys of the mechanism design, implementation, and matching and auction design literatures in Jackson (2001, 2013, 2014). For a recent discussion of market design, see Kominers et al. (2017).

[4]Ashly Piggns pointed me to a predecessor of this view, due to John Maynard Keynes (the younger): "If economists could manage to get themselves thought of as humble, competent people on a level with dentists, that would be splendid." "The Future", Chap. 5, *Essays in Persuasion* (1931).

[5]The first network data collection of such a scale was the Add Health data set in the mid 1990s, but involved enormous grants and was a unique project. Now such data gathering is feasible for many teams of researchers at manageable cost levels—as tablets and other technological advances make gathering and entering data easier.

not have been done without the coupled theory, data, and statistical analysis; nor without modern technology and computing capabilities.

As technological advances continue to expand our reach to more complex systems, we need to maintain the humility of Rubinstein's fablists, but still be willing to roll up our sleeves and get our hands dirty when it is needed, just as Roth and Duflo suggest. If economists are not involved in the design of policies and institutions, they will be designed without us, and often by someone with less knowledge of the forces at play, lacking the tools to develop and test prototypes, and without awareness of potential pitfalls.

## 2   Why Theory Is Needed in an Era of Design and Big Data

With the growing availability of large and detailed data sets, and the improvement in the computing technology and methodology to mine and analyze those data, is economic theory doomed to extinction? One can find those who claim that the data will do *all* of the speaking, and that theory will become obsolete. If one looks at trends in publications, one might even believe such claims. According to Hamermesh (2013), more than 57% of papers published in a set of top economics journals involved economic theory in 1983. By 2011 that number had dropped to 19%. Such a three-fold drop is dramatic!

Much of that decline may be natural as there are fads and fashions in science just as in our broader lives. Increased availability of data has led to a growth in empirical analyses. But the absurdity of theory's extinction becomes clear to anyone who gives the question any serious thought and who understands why scientists work with models. Theory will continue to play a vital role, especially in the realm of economic design.

The more intricate the design problems become, the more essential modeling becomes to make sense of all the moving parts. With more dimensions come more design decisions. One cannot simply pull a few designs out of a hat and test them. How would one hope to design a combinatorial auction by trial and error, without any guide (at least from economic fables)? Theory suggests which designs to test, and models provide the first steps in testing a design. An economic model provides the prototype and wind tunnel for the development of any economic design. A new economic system or institution may face nuances in the field that take it beyond wind tunnel testing, and will require Duflo's plumbing and tinkering, but without those first steps the problem becomes hopeless. One would never design a large airliner without carefully modeling its aeronautic properties, and testing it thoroughly via simulations and test flights of prototypes, before loading it with passengers. Why should designing a market for health insurance be any different? Models have the virtue of offering us insight in to what should we expect in scenarios that have never been tried before. Perhaps most importantly, they also allow us to analyze welfare and evaluate different systems' performances.

Given that making a case for theory requires requires explaining why models can be essential, it is worth discussing at least one example in more than just passing. In the microfinance study mentioned above, one thing that we wanted to understand was whether getting people to take advantage of loans: (i) simply required having them be aware of microfinance availability, or (ii) whether their decisions to participate were also being influenced by the decisions of their friends—possibly due to imitation, inference, or other complementarities. Knowing which of these is at play becomes essential if one wants to design a policy to improve participation. These alternative hypotheses both result in correlations between people's participation—the first because awareness flows through the network and so friends' awareness is correlated, and the second because people react to their friends' behaviors. Thus, simply testing for correlated behavior does not distinguish between these hypotheses. Modeling information flow and awareness coupled with potential peer influence allowed us to figure out how the patterns of correlation would differ depending on how important each of these factors were—and how these would map out in the village networks. The basic intuition is that it only takes one friend to make someone aware, while complementarities depend on more people's decisions; but exactly how to measure and test for such differences requires some sitting down and modeling. Moreover, in our case, the network-conditional correlations that we needed to test statistically required simulating the model and backing out what was seen in the data. Without a model there would have been no hope.[6]

# 3    Looking Backward and Forward

Economic theory experienced two golden eras in the twentieth century. The first was fueled by the development of general equilibrium theory which became the backbone of many models, allowing economists to see how changes in one market would affect others. This enabled dramatic advances in the modeling of everything from dynamic macroeconomics, to international trade, and public and private finance. The second era came with an explosion in game theory and its ability to model interaction in settings with asymmetric information. This fueled advances in industrial organization, political economy, health economics, mechanism design, labor economics, and many other fields.

Even though it is now very clear why these major theoretical advances were so useful, and drove much of the development of modern economics, it is always hard to see what the next major wave in any endeavor will be before it occurs. If one went back to the 1950s and polled economists as to what the next major advances would

---

[6]One can argue that an alternative is to simply go out and design new experiments to test every hypothesis. However, it can become prohibitively expensive to design an experiment to test every alternative hypothesis, while existing data can yield many insights and help point out which experiments are worth running. Moreover, modeling can be a useful tool in the design of many experiments, especially as hypotheses are becoming increasing subtle and nuanced. Ultimately, some combination of theory, empirics, and experimentation, is the answer.

be, it is unlikely that many would foresee how powerful a tool general equilibrium theory would become. Similarly, when game theory was beginning to blossom in the late 1970s, it was happening out of the mainstream and would have made few top-ten lists of major advances for the next decades. So, prognosticating what the next major advances in economic theory will be is not easy. Nonetheless it is always fun to try.

There are two important ways in which the neoclassical economic paradigm is being loosened and where new tools are emerging.

One is in 'behavioral economics', which opens up the behavior of individuals to a variety of models that move beyond the fully-rational and computationally-unchallenged benchmark agents that economists have used to build models for some decades. While economists are acutely aware that full rationality is far from an accurate depiction, it still offers a useful benchmark and in many instances is a good first order approximation to how experienced agents act. Nonetheless, there are settings in which rational modeling gets things substantially wrong, and building models that provide better predictions as to how humans behave is making waves not just in the lab, but also in explaining many things, from consumer behavior to systematic anomalies in finance.[7]

Another development comes from modeling the patterns in which people interact. Most economic interactions take place in decentralized networks of relationships—from the spread of information and formation of opinions among individuals to the trade of goods and services between firms. In many instances the network patterns of interactions, and the externalities and constraints that they impose, are fundamental drivers of behavior. This is a 'social' parallel to the 'psychological' underpinnings of behavioral economics. Network models have become increasingly important in many areas of economics, including understanding information flows and frictions in development economics, persistent inequality in job markets, political polarization, contagions in financial markets, and international trade and conflict.[8] Simple network models can become complex quickly, but with increased computing power, they are increasingly easy to incorporate into our thinking. Given how strong network patterns are in most human interactions, there is strong promise for further advances in economic understanding when accounting for the externalities that networks embody and their impact on people's information and opportunities.

Both behavioral and network economics face lingering challenges in widespread acceptance, as people naturally approach new tools with understandable caution, but both areas are gelling in terms of making clear the value of their expanding tool boxes. Moreover, the more plumbing that is done, the clearer it becomes how

---

[7]Here, I am lumping several things together, including: models of bounded rationality with some limits or constraints on computation or memory (e.g., see the discussion in Simon 1982; Rubinstein 1998; Mullainathan and Thaler 2000), models that build upon some observed behavioral bias or psychological primitive (e.g., see the discussion in Camerer and Loewenstein 2004; Fudenberg 2006; DellaVigna 2009), and modeling individual preferences that depend on comparisons to others and their welfare (e.g., see the discussion in Cooper and Kagel 2016). Although these are quite distinct in motivation and methodology, they all are expand modeling beyond the classical paradigm that was based on a specific and unlimited definition of rationality.

[8]See Jackson (2019) for an overview.

essential it is to have good models of how people learn, how they are affected by their surroundings, how they choose their interactions, and how they make decisions—the heart of behavioral economics and network economics. Thus, there is a healthy feedback between economic design and economic theory.[9, 10]

Beyond these new theoretical advances, computing capabilities that are enabling analyses of models and data that were once unthinkable. There are undoubtedly new tools that will emerge to drive future modeling to new heights.

In addition to new tools and technological advances, it is also important to consider the new design questions that are emerging. Economic and social interactions are taking place via a number of new online platforms. At this point many platforms are ad hoc. Why do they take the forms that they do? How can platforms be improved? Should we be concerned about the enormous economies of scope and scale that they embody, as they often lead to significant concentrations of market power? What do economists have to say about privacy and how to design it into platforms? ...

As a last thought on the future, I offer a caution that points to another important role for social and economic theory. With increased availability of data and computational power, we will no doubt be inundated with miners digging for surprising and unexpected patterns. With enough data and time, many intriguing relationships will appear, many of which will turn out to be entirely spurious. It is important to approach data with some direction and hypotheses in hand, so as not to become distracted by the many shiny and yet ephemeral facts that may be unearthed along the way. There is ample room for exploratory analyses, provided that we are properly aware and cautious in interpreting what is found as we unleash armies of hungry researchers on data concerning all of human behavior. Models and theory of human interaction will help guide us as we explore the existing world, tinker with it, and endeavor to design its future. We must not get derailed.

## References

Backhouse, R. E., & Cherrier, B. (2017). The age of the applied economist: The transformation of economics since the 1970s. *History of Political Economy, 49*, 1–33.

Banerjee, A., Chandrasekhar, A., Duflo, E., & Jackson, M. O. (2013). Diffusion of microfinance. *Science, 341* (26 July 2013). https://doi.org/10.1126/science.1236498.

---

[9]There is one area of theory which has long been ripe for expansive use in design and plumbing: social choice theory. Although one might think of Arrow's Theorem as a fable, it quite clearly provides testable implications on imperfections in collective decision making. Given how much research has followed on voting systems and mechanism design more generally, it is actually surprising that more social choice theorists have not been involved in the design of voting systems and legislatures around the world. But it seems inevitable.

[10]There are other areas of theory that are being pushed by plumbing that I have not discussed here. For instance, designing institutions and mechanisms that are robust and operate in a variety of environments and circumstances. See for instance, the discussion in other contributions to this volume, e.g., Carroll (2019).

Banerjee, A., Chandrasekhar, A., Duflo, E., & Jackson, M. O. (2015). Using Gossips to Spread Information: Theory and Evidence from Two Randomized Controlled Trials, forthcoming: the Review of Economic Studies.

Camerer, C. F., & Loewenstein, G. (2004). Behavioral economics: Past, present, future. In C. F. Camerer, G. Loewenstein, & M. Rabin (Eds.), *Advances in behavioral economics*. Princeton: Princeton University Press.

Carroll, G. (2019). Design for weakly structured environments. *The future of economic design*.

Cooper, D. J., & Kagel, J. H. (2016). Other-regarding preferences. *The handbook of experimental economics, Volume 2: The handbook of experimental economics* (p. 217).

Davidson, P. (2012). Is economics a science? Should economics be rigorous?" *Real-World Economics Review, 59*, 58–66.

DellaVigna, S. (2009). Psychology and economics: Evidence from the field. *Journal of Economic Literature, 47*, 315–372.

Duflo, E. (2017). The Economist as Plumber. Technical report. National Bureau of Economic Research.

Fudenberg, D. (2006). Advancing beyond advances in behavioral economics. *Journal of Economic Literature, 44*, 2.

Hamermesh, D. S. (2013). Six decades of top economics publishing: Who and how? *Journal of Economic Literature, 51*, 162–172.

Hausman, D. M. (1989). Economic methodology in a nutshell. *The Journal of Economic Perspectives, 3*, 115–127.

Hurwicz, L. (1972). On informationally decentralized systems. *Decision and Organization*, 297–336.

Hurwicz, L. (1973). The design of mechanisms for resource allocation. *The American Economic Review, 63*, 1–30.

Jackson, M. O. (2001). A crash course in implementation theory. *Social Choice and Welfare, 18*, 655–708.

Jackson, M. O. (2013). Matching, auctions, and market design, Parts appear in economic engineering and the design of matching markets: the contributions of Alvin E. Roth. *The Scandinavian Journal of Economics 115*(3), 619–639. https://doi.org/10.2139/ssrn.2263502.

Jackson, M. O. (2014). Mechanism theory. In U. Derigs (Ed.), appeared in *The encyclopedia of life support systems*. Oxford, UK: EOLSS Publishers (2003). https://ssrn.com/abstract=2542983.

Jackson, M. O. (2019). *The human network*. New York: Pantheon Books.

Keynes, J. N. (1904). *The scope and method of political economy*. Macmillan.

Kominers, S. D., Teytelboym, A., & Crawford, V. P. (2017). An invitation to market design. *Oxford Review of Economic Policy, 33*, 541–571.

Mullainathan, S., & Thaler, R. H. (2000). Behavioral Economics. Technical report. National Bureau of Economic Research.

Roth, A. E. (2002). The economist as engineer: Game theory, experimentation, and computation as tools for design economics. *Econometrica, 70*, 1341–1378.

Rubinstein, A. (1998). *Modeling bounded rationality*. Cambridge: MIT Press.

Rubinstein, A. (2005). Discussion of 'Behavioral Economics'. In R. Blundell, W. Newey, & T. Persson (Eds.), *Advances in economics and econometrics, theory and applications: Ninth world congress of the econometric society* (pp. 246–254).

Rubinstein, A. (2012). *Economic fables*. Open Book Publishers.

Simon, H. A. (1982). *Models of bounded rationality: Empirically grounded economic reason* (Vol. 3). Cambridge: MIT Press.

# Social Choice Theory and Data Science: The Beginning of a Beautiful Friendship

**Ugur Ozdemir**

**Abstract** In this note, I advocate the development of a symbiotic relationship between the theoretical world of economic design and that of quantitative data analysis. I illustrate both directions of this symbiosis by looking at how the axiomatic approach of social choice theory can provide foundations for the quantitative methodological choices and, how data science tools can help testing assumptions of formal models. For the former, I focus on the two important stages of any empirical research design: measurement and methods selection. For the latter, I assert that machine learning algorithms can be employed to test assumptions regarding to the nature of individual preferences. The spatial model of electoral competition serves as a workhorse example throughout the paper.

Suppose a researcher would like to use the spatial model of voting in order to shed some light on the nature of electoral competition in a given country. She is interested in comparing the game theoretical estimates of party positions on the ideological space with the empirical estimates and, in understanding how important issue voting is in determining voters' choices. For this sort of an analysis, she needs to have the following: (i) policy space on which the electoral competition on taking place, (ii) an estimate of the empirical distribution of the voter ideal points (iii) estimates of party positions on that space.

For the first two, the researcher will use data from a representative survey experiment. The idea is to use the issue opinion questions in this survey to determine the latent ideological dimensions in the polity and respondents' scores on these dimensions. There are many different feature extraction and dimensionality reduction techniques available for this purpose. Looking at the earlier literature, the researcher decides that she should use either principal component analysis or common factor analysis. She runs both algorithms and observes that the results differ substantively. This does not come as a surprise once she looks at a bit more closely to these techniques because even if both principal component analysis and factor analysis aim to reduce the dimensionality of a set of data, the approaches taken to do so are very

U. Ozdemir (✉)
University of Edinburgh, Edinburgh, Scotland
e-mail: ugur.ozdemir@ed.ac.uk

© Springer Nature Switzerland AG 2019
J.-F. Laslier et al. (eds.), *The Future of Economic Design*, Studies in
Economic Design, https://doi.org/10.1007/978-3-030-18050-8_73

much different (Jolliffe 2010). They assume completely different models relating the underlying variables to latent dimensions. So which method is the most appropriate? Can there be theoretical reasons to choose one method over the other? Which axioms does each of these methods satisfy? This is, after all, an aggregation problem, and this is the point where, I argue, social choice theory can help with.

The existing research on the axiomatisation of the data analysis methods is quite limited and focuses entirely on the clustering algorithms. One of the reasons for this is that this strand of work is dominated by computer scientists and data scientists and, has not received attention from social scientists yet. Very much like social choice theory, this literature started with an impossibility result. Kleinberg (2002) proposed three axioms[1] and showed that there can be no clustering method which satisfies all of these at the same time.

Ben-David and Ackerman (2009) criticise Kleinberg's result by arguing that impossibility is not an inherent feature of clustering, but rather it is an artefact of the specific formalism. In contrast to Kleinberg's setup, they propose to focus on the clustering quality measures as the object to be axiomatised rather than clustering methods and provide several clustering quality measures all satisfying the proposed axioms. Another "positive" result for clustering is by Zadeh and Ben-David (2009) who relax one of Kleinberg's axioms and show that there exists a unique clustering method which satisfy their axioms.

As mentioned above, the existing literature on the theoretical foundations of statistical methods is entirely on clustering procedures and there is no work on other data analysis methods frequently used by social scientists, such as feature extraction our researcher used, or classification (clustering with known, predefined categories, i.e, *supervised* clustering.). Social choice theory can contribute to providing guidelines to help empirical social scientist make better and more conscious methodological choices. This is particularly important for empirical social sciences where theoretical justification is expected to be more important than better predictions. In practice however, researchers make this choice in quite an ad-hoc manner with reasons such as, "freely available code","ease of use", "it has worked for another paper" etc. With the increasing "super large N" datasets, this becomes even a bigger problem, because if you have a sufficiently large dataset, you can reach almost any conclusion with a "smart" choice of methods.

Let's now go back to the our spatial model example. Since our researcher now has an empirical distribution of the voters, the next step is to get the estimates for party positions on the same policy space. One standard way of doing this is to use expert surveys. In these surveys, a group of experts is asked to locate the policy positions of political parties on different issues. So she decides to use the most commonly

---

[1] Here are Kleinberfg's axioms stated rather informally:

- Clustering function is not sensitive to changes in the units of distance measurement.
- Any desired clustering structure should be attainable by some distance measure.
- If we reduce distances within the clusters and enlarge distances between the clusters then the clustering structure should not change.

.

used one in the literature (Bakker et al. 2015). This survey includes the scores from individual experts. But what should be the method of aggregation? Simply taking the average across all experts as it is done by almost everyone? That seems to be a bit problematic because, looking at the distributions of the scores at the expert level, she observes a considerable level of disagreement among experts. Moreover, there is significant variation in this disagreement both at the party and issue dimension level. Note that statistical inference problem in expert surveys differs quite substantially from that in public opinion surveys because we are not aggregating information over a randomly selected sample (Benoit et al. 2006). That means we cannot use the standard "confidence intervals around the mean" to deal with uncertainties.

There are in fact some statistics used in the psychometrics to assess levels of agreement and reliability among experts such as intraclass correlation (James et al. 1984). But again, none of these measures are theoretically justified. So, we have another aggregation problem which can certainly benefit from a theoretical foundation, in particular, from the axiomatic approach that social choice has been founded on. Expert surveys are just one example of composite indices used in social sciences. There many other similar measures social scientists use such as measures of democracy, inequality, poverty, power and environmental pollution responsibility. Just like the axiomatisation of methods, axiomatisation of measurement has not received much attention from social scientists as well. An interesting recent exception is Patty and Penn (2015) who does argue that formal theory, social choice theory in particular, is the heart of measurement and present an axiomatic analysis of network centrality measures to demonstrate the idea.

In order to illustrate the opposite direction of the symbiotic relationship, i.e., how data science can help us test our theoretical assumptions, let's go back to our example once again. Our researcher is now ready to run a multinomial logistic regression on voter choice,[2] the distance between the voters and the parties being the independent variable of interest together with some sociodemographic variables as controls. She is however, a skeptical type and enjoys questioning her theoretical assumptions as much as the methodological ones. One thing catches her eye is the Euclidean preferences assumption she made. This assumption is operationalised through the choice of the metric used to measure the distances between the political parties and the individuals. By employing the Euclidean metric, she realizes that she is assuming circular indifference curves. Is this really a reasonable assumption? How can we test this. The brand new tools of machine learning can help us test this assumption and help us choose "the best distance metric" (Yang 2006). We can use supervised metric learning algorithms on a subsample of our dataset (training data) to see which metric best explains the underlying behaviour, and use that metric to run the regression analysis instead of simply assuming the Euclidean metric.

---

[2]Why logistic regression? Note that, similar to our investigation regarding to principal component analysis versus factor analysis above, we might want to question this methodological choice as well. After all, logistic regression is nothing but a classification algorithm. There are other methods such as random forests or support vector machines suitable for the same task. So, the axiomatic study of quantitative methods can go as far as to include regression analysis.

**Conclusion**

We have transitioned into a world of complex, multidimensional data which changed the lives of empirical social scientists dramatically. This has increased the need for new analytical and statistical techniques and, aggregated measures to reduce the inherent complexity and discover the patterns buried in the data. In fact, an entirely new research field-that of data science was born. This world is mainly dominated by computer scientists and statisticians, but it is transforming from a set of mysterious techniques to an essential toolkit for a much broader social science community.

In this note, I pointed into a direction for a rather surprising collaboration between social choice theory and data science. I argued that social choice theory can help bridging the gap between empirical social scientists and statisticians in order to strengthen the theoretical basis of the social scientific enquiry through development of an axiomatic understructure. This is increasingly becoming important as ad-hoc choices of methods and measures with no theoretical justification can transform empirical social sciences into data mining in the age of big data. I further argued that brand-new data science algorithms can provide opportunities to test the assumptions of theoretical models.

I see this as the beginning of a beautiful friendship between two seemingly unrelated disciplines and looking forward to being a part of this exciting journey.

# References

Bakker, R., De Vries, C., Edwards, E., Hooghe, L., Jolly, S., Marks, G., et al. (2015). Measuring party positions in europe: The chapel hill expert survey trend file, 1999–2010. *Party Politics*, *21*(1), 143–152.

Ben-David, S., & Ackerman, M. (2009). Measures of clustering quality: A working set of axioms for clustering. *Advances in Neural Information Processing Systems*, 121–128.

Benoit, K., Laver, M., et al. (2006). *Party policy in modern democracies*. Routledge.

James, L. R., Demaree, R. G., & Wolf, G. (1984). Estimating within-group interrater reliability with and without response bias. *Journal of Applied Psychology*, *69*(1), 85.

Jolliffe, I. T. (2010). *Principal component analysis*. Springer.

Kleinberg, J. (2002). An impossibility theorem for clustering. *NIPS*, *15*, 463–470.

Patty, J. W. & Penn, E. M. (2015). Analyzing big data: Social choice and measurement. *PS: Political Science & Politics*, 48(01):95–101.

Yang, L., R. J. (2006). Distance metric learning: A comprehensive survey. Technical report, Michigan State University.

Zadeh, R. B. and Ben-David, S. (2009). A uniqueness theorem for clustering. In *Proceedings of the twenty-fifth conference on uncertainty in artificial intelligence*, pp. 639–646. AUAI Press.

# Technological Change and Market Design

Marek Pycia

**Abstract** Technological innovations lead to new market designs and new designs catalyze the emergence of new technologies. Building on examples drawn from recent advances in medical, electricity, car, computing, and data collection technologies, this note discusses the relationship between technological change and market design with an emphasis on new questions for market design theory.

Recent developments in market design have been largely driven by changes in technology, and new market designs catalyzed the emergence of new technologies as well as changes in legal and ethical environments. Consider, for instance, kidney exchange. The design of exchange (pairwise donation) markets was made possible by developments in transplantation technology, and legal and ethical resolutions that deemed kidney exchange not to violate the proscription against exchanging organs for valuable consideration (Roth et al. 2004, 2005, 2007). Various important refinements and innovations then followed: organs are now being shipped via commercial flights (Butt et al. 2009) and altruistic kidney donations led to the introduction of donation chains (Roth et al. 2006; Rees et al. 2009). Kidney chains made possible for hospitals to issue vouchers for future kidney donations (Veale et al. 2017). The vouchers create the possibility of further innovations in designing the market and its underlying contractual structure (e.g. Sönmez et al. 2018). Other transplant markets are also developing in response to emerging technological possibilities, c.f. Bergstrom et al. (2009) work on bone marrow transplants, and Ergin et al. (2017) on lung transplants.

The future will bring more instances of such a feedback loop between market design and technology developments and new technological developments will pose new market design questions. For instance, in California, new sources of electricity are already changing electricity markets. The supply of electricity from new sources

U. Zürich. For their comments, I would like to thank Ernst Fehr, Todd Hare, Anna Myjak-Pycia, Nick Netzer, Christian Ruff, Manuela Steinauer, and Utku Unver.

M. Pycia (✉)
University of Zürich, Blümlisalpstrasse 10, 8006 Zürich, Switzerland
e-mail: marek.pycia@econ.uzh.ch

such as solar and wind is much harder to forecast than the relatively predictable traditional electricity sources such as oil, coal, and nuclear. The electricity from these new sources flows into the grid at many entry points, rather than few big plants. Price patterns change: while in the past electricity was cheapest at night, this is no longer true in California where the price of electricity is lowest, and occasionally negative, mid-day because of the inflow of solar-source electricity to the grid. In addition, new demand for electricity comes from users of hybrid and electric cars. Without new market design, the current grid will not be able to balance supply and demand (Davis 2014). Regulators and entrepreneurs are aware of this looming crisis, and are developing solutions such as balancing the supply and demand through the development of electricity storage, e.g. by enabling car batteries to trade with the grid (Davis 2016).

There is a clear opening for economists to offer new market design solutions to the grid crisis. Even if the car-battery solution works in the near future, new designs might be needed when driverless cars become standard because the car-battery solution relies on the cars being connected to and trading with the grid most of the time. This assumption might fail in the presence of driverless cars and new market designs that their presence enables: the driverless cars that are in motion almost all the time will not be able to play substantial storage role unless new technology developments allow them to trade with the grid while there are in motion.

A related set of market design opportunities is being brought on by the development of technologies that allow sophisticated congestion charges for road use. Vickrey (1969) initiated the economic analysis of congestion charges (see also e.g. Arnott et al. 1993), and technology might soon allow traffic optimization through more sophisticated price schemes than were possible in Vickrey's times.

The changes in electricity and transportation markets are just two examples of how technological changes destabilize some markets and make other markets possible, creating opportunities for market design. These new problems will be overlaid on classical challenges: despite some progress (e.g. Ausubel et al. 2014; Anderson et al. 2013; Pycia and Woodward 2016), we still do not understand equilibria in pay-as-bid auctions despite their important role in markets for electricity and other commodities.

Technological changes also generate new foundational questions for market design theory. Increases in computing power mean that in future markets—like in the electricity trading and transportation examples above—market participants might be able to perform more complex strategizing than seems reasonable to expect of them today. Increased strategic sophistication might both upend the rational for merely approximately strategy-proof mechanisms as well as weaken the case for fully strategy-proof or otherwise simple mechanism. In particular, the increase in market participants' ability to perform complex strategic calculations might enable market designs that lead to higher aggregate welfare than currently used simple designs. While a general analysis of such problems remains open, the first inroads were made in some special markets; see, for instance, the literature on possible welfare gains in school choice market that could be obtained by moving away from strategy-proof mechanisms (e.g. Bogomolnaia and Moulin 2001; Miralles 2008; Abdulkadiroglu et al. 2011; Featherstone and Niederle 2016; Troyan 2012; Pycia 2011; Ashlagi and Shi

2015; He et al. 2018; Abdulkadiroglu et al. 2017). The first inroads are also made into the analysis of sophisticated multiplayer games in which participants can offer contracts that depend on other's contracts (e.g. Peters and Severinov 1997; Peck 1997; Epstein and Peters 1999; Peters 2001; Martimort and Stole 2002; Calzolari and Pavan 2006; Pavan and Calzolari 2009; Peters and Szentes 2012). Technological and legal innovations might make such complex contracting arrangements feasible, and in some environments such complex mechanisms might be able to achieve substantial welfare gains. I conjecture that this might be the case particularly in settings in which externalities play substantial role.

The general impact of new technologies goes beyond the increased computational power. The current technology already makes it possible to collect unprecedented amounts of data about people, e.g. based on their online and offline behavior (see e.g. Taylor 2004). Leaving aside the question how to draw conclusions from rich, multi-dimensional data, the data abundance opens the possibility that many buyers might have little price-relevant information that would not be available to the sellers.

The technology also creates new types of data and new possibilities for mechanism design. In markets in which buyers (or other market participants) are physically present, the past behavior data might some day be supplemented by contemporaneous measurement of neurological and other bodily processes (e.g. eye movements). Recent extensive research in neuroeconomics demonstrated that such data correlate with buyers' valuations; for surveys see Rangel and Hare (2010), Ruff and Fehr (2014). This correlation enables the use of new market mechanisms that rely on neurodata. For instance, the mechanism designer might be able to costlessly elicit a buyer's valuation by offering him or her fair bets on the neurodata that would be subsequently recorded: because the neurodata is hard for the buyer to control and correlated with the buyer's valuation, such bets might incentivize the buyer to reveal the valuation. Krajbich et al. (2017) make this theoretical point and test it experimentally. While such bets resemble and are inspired by Cremer and McLean (1988), here the bets do not rely on any assumptions on the behavior of agents other than the buyer. Furthermore, whenever the correlation between the neurodata and the valuation is strong, the offered bets might have realistic magnitude.

These new mechanisms are bound to lead to refinements in the legal and ethical environments in which they will be used, and the diminished role of individuals' private information will increase the importance of resale markets and regulation. In these tasks, the new technologies, and the neurodata in particular, might turn out to be useful in allowing the society to refine our normative welfare concepts. Indeed, we already see the first explorations of welfare concepts that are motivated by the availability of neurodata and other non-choice data (see e.g. Salant and Rubinstein 2008; Bernheim and Rangel 2009; Fehr and Rangel 2011).

# References

Abdulkadiroglu, A., Che, Y.-K., & Yasuda, Y. (2011). Resolving conflicting preferences in school choice: The Boston mechanism reconsidered. *American Economic Review, 101*, 1–14.

Abdulkadiroglu, A., Agarwal, N., & Pathak, P. (2017). The welfare effects of coordinated assignment: Evidence from the New York City high school match. *American Economic Review, 107*, 3635–3689.

Anderson, E. J., Holmberg, P., & Philpott, A. B. (2013). Mixed strategies in discriminatory divisible-good auctions. *RAND Journal of Economics, 44*, 1–32.

Arnott, R., de Palma, A., & Lindsey, R. (1993). A structural model of peak-period congestion: A traffic bottleneck with elastic demand. *The American Economic Review, 83*, 161–179.

Ashlagi, I., & Shi, P. (2015). Improving community cohesion in school choice via correlated-lottery implementation. *Operations Research, 62*, 1247–1264.

Ausubel, L. M., Cramton, P., Pycia, M., Rostek, M., & Weretka, M. (2014). Demand reduction and inefficiency in multi-unit auctions. *The Review of Economic Studies, 81*, 1366–1400.

Bergstrom, T. C., Garratt, R. J., & Sheehan-Connor, D. (2009). One chance in a million: Altruism and the bone marrow registry. *American Economic Review, 99*, 1309–1334.

Bernheim, B. D., & Rangel, A. (2009). Beyond revealed preference: Choice-theoretic foundations for behavioral welfare economics. *The Quarterly Journal of Economics, 124*, 51–104.

Bogomolnaia, A., & Moulin, H. (2001). A new solution to the random assignment problem. *Journal of Economic Theory, 100*, 295–328.

Butt, F. K., Gritsch, H. A., Schulam, P., Danovitch, G. M., Wilkinson, A., Pizzo, J. D., et al. (2009). Asynchronous, out-of-sequence, transcontinental chain kidney transplantation: A novel concept. *American Journal of Transplantation, 9*, 2180–2185.

Calzolari, G., & Pavan, A. (2006). On the optimality of privacy in sequential contracting. *Journal of Economic theory, 130*, 168–204.

Cremer, J., & McLean, R. P. (1988). Full extraction of the surplus in Bayesian and dominant strategy auctions. *Econometrica: Journal of the Econometric Society*, 1247–1257.

Davis, S. (2014). *Presentation - Kn-Grid - Lead commissioner workshop on electric and natural gas vehicles in California*. Technical report TN 73170, California Energy Commission.

Davis, S. (2016). *Presentation - Kn-Grid - California energy commission vehicle-grid integration workshop SB 350 transportation electrification (Publicly Owned Utilities)*. Technical report 16-TRAN-01, California Energy Commission.

Epstein, L. G., & Peters, M. (1999). A revelation principle for competing mechanisms. *Journal of Economic Theory, 88*, 119–160.

Ergin, H., Sonmez, T., & Unver, M. U. (2017). Dual-donor organ exchange. ECMA forthcoming.

Featherstone, C. R., & Niederle, M. (2016). Boston versus deferred acceptance in an interim setting: An experimental investigation. *Games and Economic Behavior, 100*, 353–375.

Fehr, E., & Rangel, A. (2011). Neuroeconomic foundations of economic choice - recent advances. *The Journal of Economic Perspectives, 25*, 3–30.

He, Y., Miralles, A., Pycia, M., & Yan, J. (2018). A pseudo-market approach to allocation with priorities. *American Economic Journal: Microeconomics, 10*, 272–314.

Krajbich, I., Camerer, C., & Rangel, A. (2017). Exploring the scope of neurometrically informed mechanism design. *Games and Economic Behavior, 101*, 49–62.

Martimort, D., & Stole, L. (2002). The revelation and delegation principles in common agency games. *Econometrica, 70*, 1659–1673.

Miralles, A. (2008). School choice: The case for the Boston mechanism, Boston University. Working paper.

Pavan, A., & Calzolari, G. (2009). Sequential contracting with multiple principals. *Journal of Economic Theory, 144*, 503–531.

Peck, J. (1997). A note on competing mechanisms and the revelation principle.

Peters, M. (2001). Common agency and the revelation principle. *Econometrica, 69*, 1349–1372.

Peters, M., & Severinov, S. (1997). Competition among sellers who offer auctions instead of prices. *Journal of Economic Theory, 75,* 141–179.

Peters, M., & Szentes, B. (2012). Definable and contractible contracts. *Econometrica, 80,* 363–411.

Pycia, M. (2011). The cost of ordinality, Working paper.

Pycia, M., & Woodward, K. (2016). Pay-as-bid: Selling divisible goods.

Rangel, A., & Hare, T. (2010). Neural computations associated with goal-directed choice. *Current Opinion in Neurobiology, 20,* 262–270. Cognitive neuroscience.

Rees, M. A., Kopke, J. E., Pelletier, R. P., Segev, D. L., Rutter, M. E., Fabrega, A. J., et al. (2009). A non-simultaneous extended altruistic donor chain. *The New England Journal of Medicine, 360,* 1096–1101.

Roth, A. E., Sonmez, T., & Unver, M. U. (2004). Kidney exchange. *Quarterly Journal of Economics, 119,* 457–488.

Roth, A. E., Sonmez, T., & Unver, M. U. (2005). Pairwise kidney exchange. *Journal of Economic Theory, 125,* 151–188.

Roth, A. E., Sonmez, T., Unver, U., Delmonico, F. L., & Saidman, S. L. (2006). Utilizing list exchange and non-directed donation through chain paired kidney donations. *American Journal of Transplantation, 6,* 2694–2705.

Roth, A. E., Sonmez, T., & Unver, M. U. (2007). Efficient kidney exchange: Coincidence of wants in markets with compatibility-based preferences. *American Economic Review, 97,* 828–851.

Ruff, C. C., & Fehr, E. (2014). The neurobiology of rewards and values in social decision making. *Nature Reviews Neuroscience, 15,* 549–562.

Salant, Y., & Rubinstein, A. (2008). (A, f): Choice with frames. *The Review of Economic Studies, 75,* 1287–1296.

Sönmez, T., Ünver, M. U., & Yenmez, M. B. (2018). Incentivized kidney exchange. Working paper.

Taylor, C. (2004). Consumer privacy and the market for customer information. *RAND Journal of Economics, 35,* 631–650.

Troyan, P. (2012). Comparing school choice mechanisms by interim and ex-ante welfare. *Games and Economic Behavior, 75,* 936–947.

Veale, J. L., Capron, A. M., Nassiri, N., Danovitch, G., Gritsch, H. A., AmyWaterman, J. D., et al. (2017). Vouchers for future kidney transplants to overcome "Chronological Incompatibility" between living donors and recipients. *Transplantation, 101,* 2115–2119.

Vickrey, W. S. (1969). Congestion theory and transport investment. *The American Economic Review, 59.*